International Conference on Extraction, Transport, Storage and Processing of Hydrocarbons and Minerals (ETSaP 2019)

IOP Conference Series: Materials Science and Engineering
Volume 663

Tyumen, Russia
19 - 20 August 2019

ISBN: 978-1-7138-0918-0
ISSN: 1757-8981

Printed from e-media with permission by:

Curran Associates, Inc.
57 Morehouse Lane
Red Hook, NY 12571

Some format issues inherent in the e-media version may also appear in this print version.

Institute of Physics
Dirac House, Temple Back
Bristol BS1 6BE UK

Phone: 44 1 17 929 7481
Fax: 44 1 17 920 0979

techtracking@iop.org

Additional copies of this publication are available from:

Curran Associates, Inc.
57 Morehouse Lane
Red Hook, NY 12571 USA
Phone: 845-758-0400
Fax: 845-758-2633
Email: curran@proceedings.com
Web: www.proceedings.com

TABLE OF CONTENTS

IOP Conf. Series: Materials Science and Engineering **663** (2019) 011001 doi:10.1088/1757-899X/663/1/011001

Preface

The International Conference on Extraction, Transport, Storage and Processing of Hydrocarbons and Minerals (ETSaP 2019) successfully took place in Tyumen, Russia on 19–20 August 2019. ETSaP 2019 was held through the cooperative efforts of Tyumen industrial University, North-Caucasian Institute of Mining and Metallurgy (State Technological University) (Vladikavkaz), and Nizhnevartovsk branch of Tyumen industrial University.

This year, the conference has covered a wide range of subject matters, and the geography has expanded greatly. The participants presented the results of their research on minerals mining and hydrocarbons extraction, as well as on transportation, storage and processing of minerals and hydrocarbons. Moreover, issues of infrastructure development in oil-producing and mining regions as well as responsible resource extraction and environment-friendly industry were discussed. Two days of the conference became the time of intense learning for all the participants and surely gave them the impetus to further research.

We are pleased to present to our readers the latest research on extraction and processing of hydrocarbons and minerals. After the careful reviewing, 69 papers were selected for publication that present the most valuable and significant research results.

We would like to thank all the participants of ETSaP 2019 as well as the organizing committee for their efforts in holding the conference. We also express out deep gratitude to the reviewers and editors for their invaluable contribution to publication of the ETSaP 2019 proceedings.

IOP Conf. Series: Materials Science and Engineering **663** (2019) 011002 doi:10.1088/1757-899X/663/1/011002

Peer review statement

All papers published in this volume of *IOP Conference Series: Materials Science and Engineering* have been peer reviewed through processes administered by the proceedings Editors. Reviews were conducted by expert referees to the professional and scientific standards expected of a proceedings journal published by IOP Publishing.

IOP Conf. Series: Materials Science and Engineering **663** (2019) 012001 doi:10.1088/1757-899X/663/1/012001

Methods for calculating thermal fields using modern software products

V V Golik[1,2], Yu D Zemenkov [1], V Ya Subbotin[1] and S G Belsky [1]

[1] Industrial University of Tyumen, 38 Volodarskogo str., Tyumen, 625000, Russia
[2] darkpaint@mail.ru

Abstract. The paper presents methods of mathematical modelling of buried multilayer oil pipelines laid in areas of Western Siberia and the Arctic zone of the Russian Federation. Modelling was carried out on the basis of data from several sections of oil pipelines under various burying conditions. In the work, we also used a wide range of different software programs for modelling in order to identify a software product suitable for the goals and objectives of this work.

1. Introduction

Currently, the development of the north is the basis of the "Energy Strategy of Russia for the period until 2030" and the order of the Government of the Russian Federation of November 13, 2009 No. 1715-r, therefore, ensuring the reliability and safety of pipeline transport facilities is a priority in the design, construction and the operation of the oil transportation system. Modern software products allow the engineer-designer to predict thermal processes at the stage of the project development, and to prevent further behaviour of the pipe-soil system and emergency situations during operation. [1]

The selection of reliable and economical thermally insulated pipeline structures when designing linear sections is carried out considering operating conditions, types of laying and conditions on site.

Thermal design of such structures often takes a lot of time due to many changing conditions along the length of the pipeline and looks rather cumbersome.

Mathematical modelling of these processes on modern computers allows you to set many changing parameters over time, and to predict the behaviour of nearby engineering structures over a long period of time.

2. Objects and methods of research

The first stage of modelling is to search for a specific object for modelling and processing field data which include the main characteristics of the studied area. In this work, we took a section of the pipeline laid in difficult geocryological conditions of the Yarudeyskoye field. This section is indicative of the large geological diversity of frozen rocks and many cases of displacement of pipelines from the design position (Figure 1). [2]

Figure 1. Oil pipeline deformation

There are many programs that allow you to create a two-dimensional model of the area under consideration, among which Autodesk products such as AutoCAD, and built-in CAD packages (computer-aided design and drawing tools) have proven themselves. [3]

Two-dimensional modelling has the following advantages:

- does not require large computing power;
- simplicity of model building.

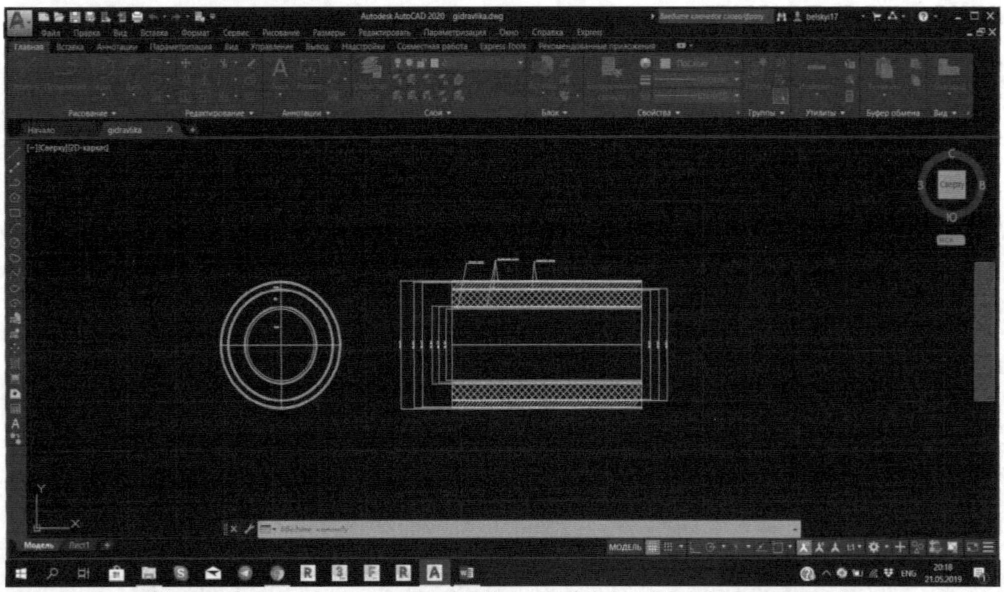

Figure 2. A multi-layer oil pipeline model developed in AutoCAD

Another option is to create a three-dimensional model of the area under consideration, giving a visual representation of nearby engineering objects and structures that are part of the general system. The most widely known CAD programs are Inventor, Fusion 360, 3ds Max (Figures 3,4,5). [4]

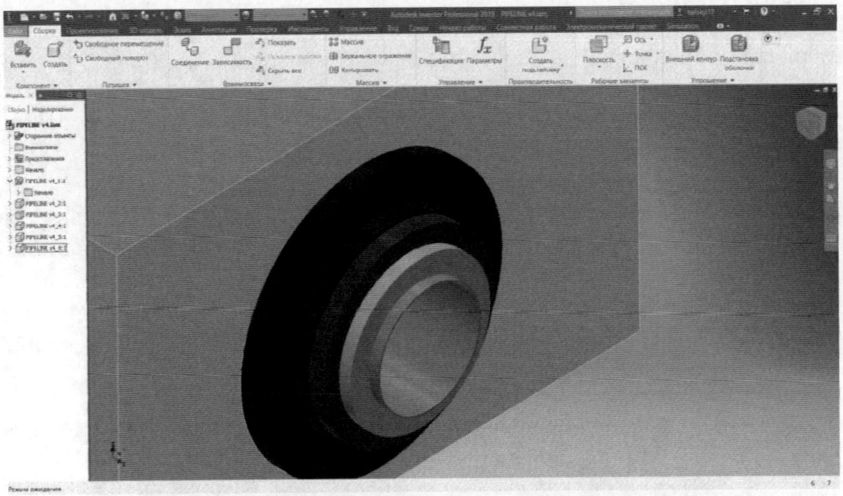

Figure 3. A model developed in Autodesk Inventor

Figure 4. A model developed in Fusion 360

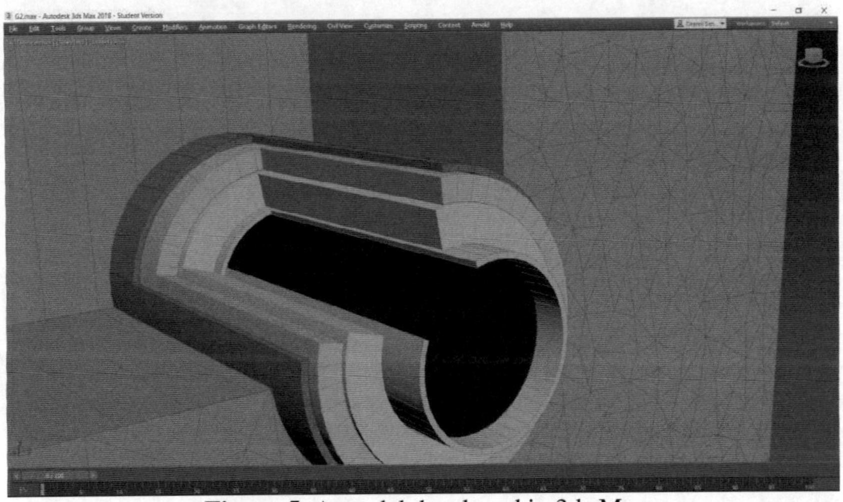

Figure 5. A model developed in 3ds Max

IOP Conf. Series: Materials Science and Engineering **663** (2019) 012001 doi:10.1088/1757-899X/663/1/012001

The second stage is mathematical modelling of heat transfer processes, if we know most of the required parameters. Currently, the following programs provide modelling of thermal processes as close to real conditions as possible: a universal software system for finite element modelling - ANSYS, a system for modelling fluid flows, gases and heat transfer processes - Autodesk CFD and Frost 3D Universal - a software package for modelling thermal processes in soil. [5,6]

At the modelling stage, in the ANSYS system and the Transient Thermal package for a preloaded model, triangles are created and set up for more correct calculation of heat fluxes. The boundary conditions are also determined, and the parameters of insulating materials and soil are set. For soil, the parameters of the phase transition from the frozen state to the thawed state and vice versa are set. After completing the calculation in this software package, you can display not only a graphic image of the temperature halo, but also graphs and various dependencies (Figures 6,7).

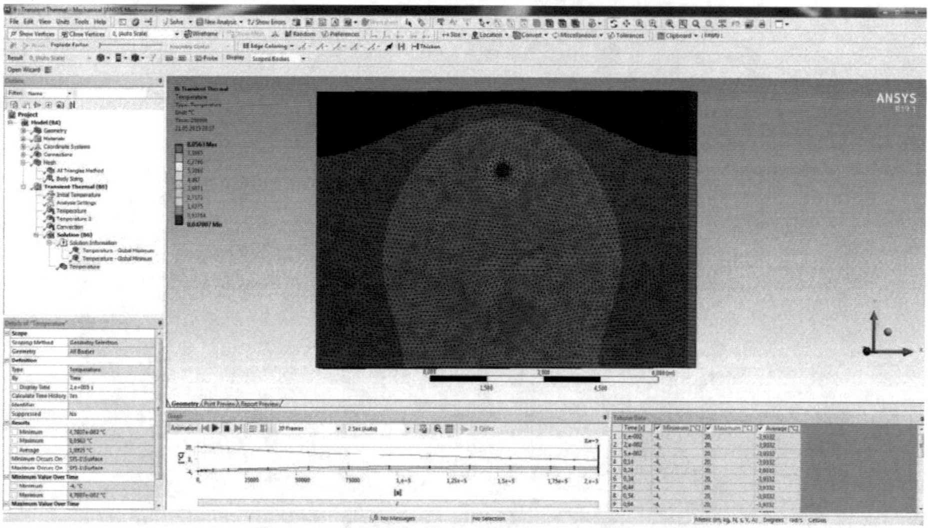

Figure 6. Modelling results in ANSYS

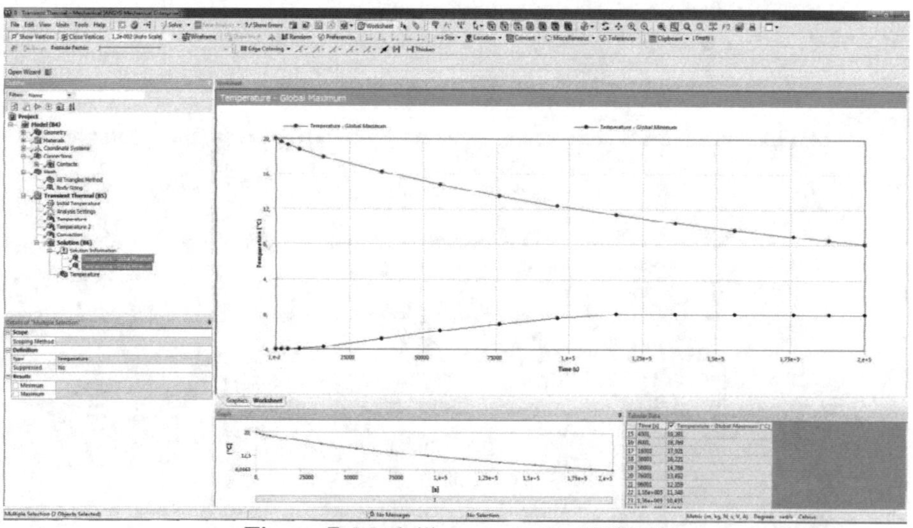

Figure 7. Modelling results in ANSYS

Modelling of thermal processes in Autodesk CFD differs markedly from the previous one in fewer functions and less accurate results.

When creating a 3D body in the first stages, you can specify unnecessary small elements and objects that the program will not take into account in the calculations, which positively affects the processing speed, but negatively affects the accuracy of the final result. [7,8]

Next, it is necessary to refine the primary model, to exclude microvoids and structural defects from the calculation in order to avoid errors in the calculations, each insulation layer here is taken as a continuous medium. After finishing work with geometry, each design layer is given a specific parameter of density and temperature. These parameters are set only from predefined materials, which is a significant drawback (Figure 8).

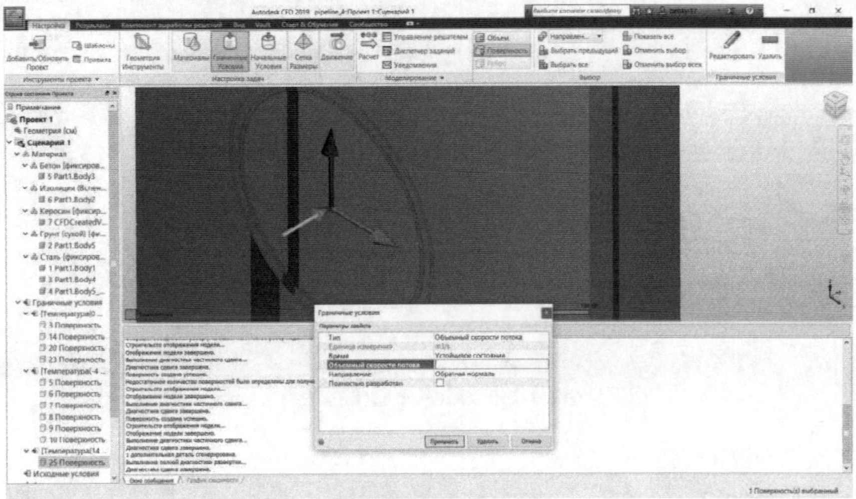

Figure 8. Determination of the model parameters.

After determining the material of each of the elements of the model, constructing a cloud of points and a grid for calculation, the grid is calculated automatically, only the size can be adjusted (Figure 9).

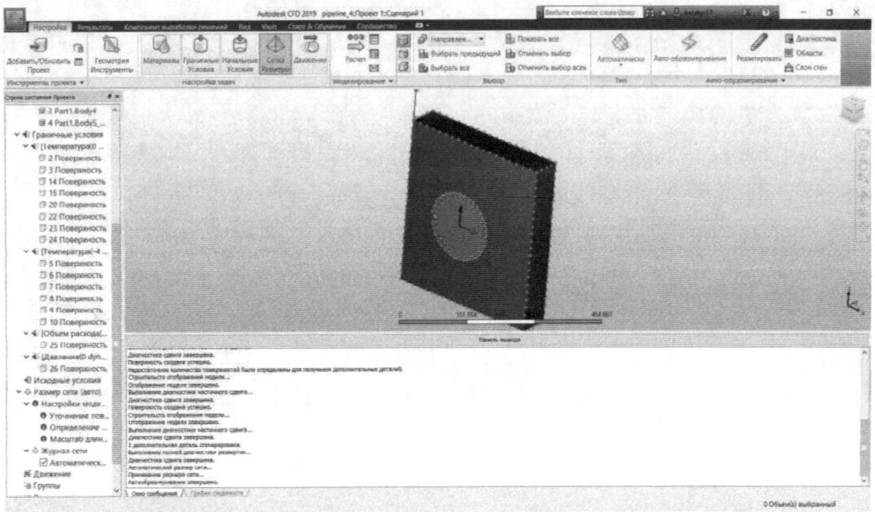

Figure 9. Creating a grid for calculation in Autodesk CFD

At the last stage, the simulation of the movement of heat fluxes and data output are performed. Calculations are performed for a certain number of iterations, and it does not allow calculating the thermal effect of the pipeline in the long term (Figure 10).

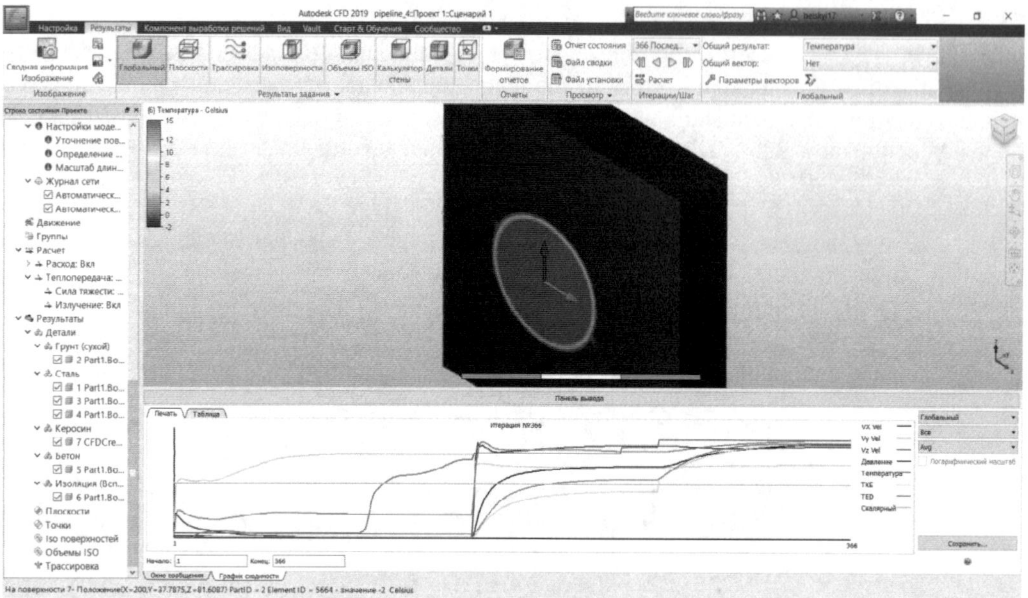

Figure 10. Modelling results in CFD

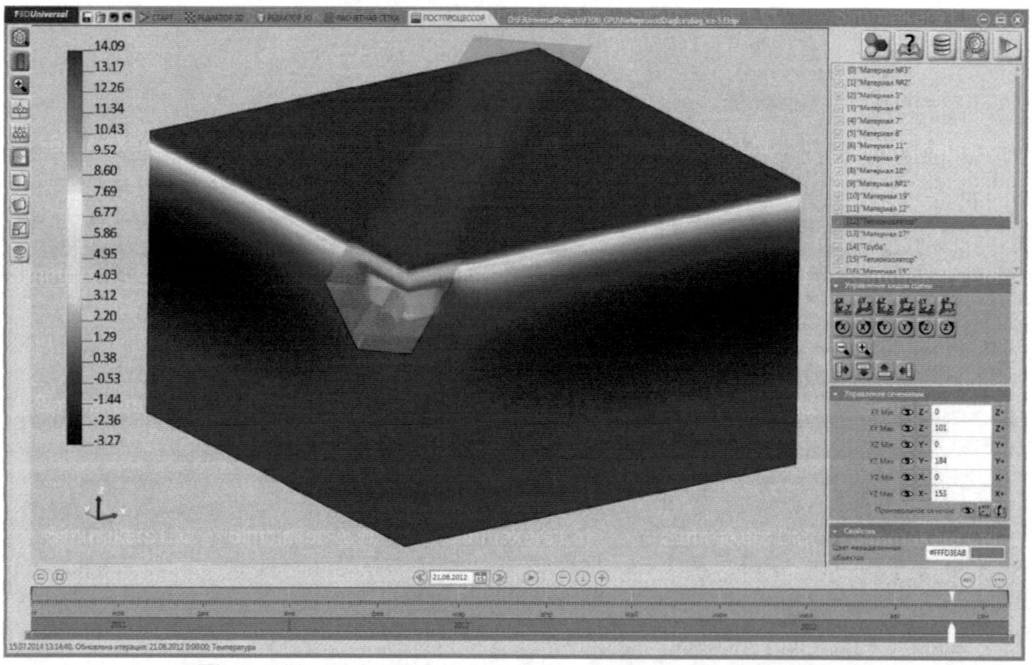

Figure 11. An example of modelling in Frost 3D Universal

3. Conclusion

Mathematical modelling, as a research method, allows you to get an idea of the physicochemical processes taking place in the places of the experiment without directly interfering with the transport of the oil product, or evaluate the degree of reliability of structures at the design and construction stages. The mathematical model, in contrast to the actual experiment, has several advantages:

- it saves the material resources required for the formulation and conduct of a physical experiment;
- performance assessment of systems with long technological cycles (weeks, months, years) occurs in a much shorter time.

References

[1] *The energy strategy of Russia for the period until 2030* [Electronic resource] URL: https://minenergo.gov.ru/node/1026 (Date of reference: 20.05.2019)

[2] Mathematical modeling of the interaction of land pipelines with the environment Zemenkov Yu D, Moiseev B V, Ilyukhin K N and Nalobin N V *News of higher educational institutions. Oil and gas* 2014 No 2 pp 51-56

[3] Porkhaev G V and Aleksandrov Yu A 1971 *The study guide for heat engineering calculations of sanitary networks laid in permafrost soils* Stroyizdat 73

[4] Modelling of freezing-thawing boundary fluctuations in soils and external building structures Aksenov B G, Koryakina S V, Moiseev B V, Shapoval A F and Stepanov O A 2012 *Journal "Modern problems of science and education"* **2** (URL: http://science-education.ru/ru/article/view?id=6088 (Date of reference: 20.05.2019))

[5] Golik V V, Moiseev B V, Zemenkov Y D and Kabes E N 2018 Assessment of geo-cryological conditions in the design and operation of pipelines in the Arctic zone of the Russian Federation *IOP Conference Series: Materials Science and Engineering*

[6] Moiseev B V, Zemenkov Y D, Nalobin N and Dudin S Thermal calculations of underground oil pipelines 2017 *MATEC Web of Conferences*

[7] Zemenkova M Y, Shastunova U, Shabarov A, Kislitsyn A and Shuvaev A 2018 Physical and mathematical modeling of process of frozen ground thawing under hot tank IOP *Conference Series: Materials Science and Engineering*

[8] Stepanov O, Moiseev B, Chekardovskiy M, Aksenov B and Shapoval A Physical and mathematical conditions of non-stationary thermal conditions of the underground air channels 2017 *International Journal of Applied Engineering Research*

[9] Calculation of the thermal interaction of the oil pipeline and permafrost in the frost 3d universal program URL:*http://frost3d.ru/raschyot-truboprovoda-mnogoletnemyorzlaya-poroda/10-15/*

[10] Fedorova O B and Chizhevskaya E L 2014 Adapting intellectual property evaluation methods to the region brand evaluation *Middle-East Journal of Scientific Research* DOI: 10.5829/idosi.mejsr.2014.19.1.12477

[11] Balikaeva M B, Chizhevskaya E L, Grevtseva Ya G, Kotlyarov I O and Volkova M A 2018 Innovative technologies as a means of the development of future engineers' professional mobility abroad *IOP Conference Series: Materials Science and Engineering* doi:10.1088/1757-899X/441/1/012007

IOP Conf. Series: Materials Science and Engineering **663** (2019) 012002 doi:10.1088/1757-899X/663/1/012002

Measurement of phase flow in water, oil and gas media using infrared radiation

Yu D Zemenkov[1], E V Zaitsev[2,3] and A B Shabarov[2,4]

[1] Industrial University of Tyumen, 38 Volodarskogo str., Tyumen, 625000, Russia
[2] University of Tyumen, Tyumen, 6 Volodarskogo str., Tyumen, 625000, Russia
[3] welin151992@mail.ru
[4] a.b.shabarov@utmn.ru

Abstract. This paper analyzes the problem of multiphase flow metering in the oil and gas industry. The main types of multiphase flow meters, their advantages and disadvantages are considered. A schematic diagram of a multiphase flow meter is proposed, which includes a mixture homogenizer, a Coriolis flow meter, an IR fraction meter and is based on the concept of intelligent measuring systems. An experimental pouring stand has been developed for conducting studies of the absorption of near-infrared radiation in a gas-liquid mixture stream. Studies have been carried out on the absorption of near-infrared radiation in a multiphase flow for a volumetric gas content of 5 and 10% in the range of liquid moisture content from 10 to 90% in increments of 20%. Based on the results of the study, a generalized correlation dependence is obtained. A computational and experimental method is proposed for determining the volumetric ratio of phases in a flow using the parameters of absorption of near-infrared radiation, which can be applied to various ranges of gas content.

1. Introduction

The task of measuring the flow rate of phases of well products is one of the urgent problems of oil and gas production. There are two main types of multiphase flow meters (MFMs): separating and non-separating. At the initial stage of accounting for well production, separating MFMs, the so-called automated group metering units (AGMU) [1-3], were used, since separating the multiphase flow into liquid and gas and measuring the flow rates of these fluids separately is the easiest way to account for produced hydrocarbons from the point of view of the principle of action. However, separating MFMs have such disadvantages as large dimensions, high metal consumption, a significant number of moving parts, and significant maintenance costs. Due to this and scientific and technological development, since the 90s of the last century, non-separating MFMs have been actively developed, based on various physical principles of action, which make it possible to measure the flow of oil, gas and water in a gas-liquid mixture flow (GLM) without their separation. Currently, compact non-separating MFMs are produced [4-7], which have satisfactory metrological characteristics. However, these units are quite complex in their principle of operation and have a high cost, and some of them use unsafe radioactive radiation sources. [8]

In existing non-separating measuring systems, in most cases, a Venturi tube is used, which allows determining the flow rate of the entire GLM flow at a known mixture density from the measured pressure drop with the required accuracy, which was discussed in [9-12]. The need for preliminary data on the density of the mixture to determine the flow rate is the main disadvantage of the Venturi tube. In this

IOP Conf. Series: Materials Science and Engineering **663** (2019) 012002 doi:10.1088/1757-899X/663/1/012002

regard, the use of Coriolis flow meters, which make it possible to carry out independent measurements of mass flow and mixture density, seems more promising for this purpose. The vast majority of existing commercial Coriolis flow meters are designed for measuring single-phase media. The difficulties in measuring multiphase media with such devices are due to the bubble effect, resonator effects, attenuation, etc. [13]

However, the development of a Coriolis flow meter for measuring the flow rate of a multiphase mixture is known [14,15]. These developments allow us to consider the development of flow metering as a promising area - improving the accuracy of measuring the parameters of a multiphase mixture using Coriolis flow meters.

In [16], the concept of intelligent measuring systems was proposed, in accordance with which the method of physical and mathematical modelling is used in conjunction with measurements of the necessary indirect parameters. The intelligent system, which includes measuring blocks of signals from sensors, digital processing, multiphase hydrodynamics ratios, and specially developed algorithm and program code, allows you to determine the desired parameters of a multiphase mixture (phase composition and phase flow). This paper proposes and substantiates the concept of a multiphase flow meter that implements this approach, including a Coriolis flow meter and a method for determining the phase flow in a stream using the parameters of absorption of near-infrared (NIR) radiation.

The basics of NIR spectroscopy, and in particular the absorption of infrared radiation, are currently widely used to determine the moisture content of various substances, including liquids. For example, the use of NIR spectroscopy for determining moisture content in a biodiesel emulsion is described in [17]. And in [18], the NIR spectroscopy method was used to determine the proportion of water in crude oil in the moisture content range from 0 to 5%. Currently, there is a flow meter for oil and oil products - Red Eye Multiphase [19,20], based on the principle of absorption NIR spectroscopy. This moisture meter has high accuracy but provided that the gas content in the stream should not exceed 10%, and the structure of the oil and gas stream should be homogeneous since water cut measurements are carried out at only one point.

2. Schematic diagram of a multiphase flow meter

The schematic diagram of the proposed non-separating flow meter is presented in Figure 1. A homogenizer (mixer) is installed to create a uniformly mixed flow. The mixer also serves to equalize the speed of the phases in the flow. The design of the homogenizer should ensure high-quality homogenization of the flow on the one hand and a low pressure drop on the other. After the mixer, a representative sample is taken, which passes through the NIR fraction meter and returns to the general flow. The NIR fraction meter measures the optical density at two wavelengths of NIR radiation over the entire cross-section of a representative sample using the patented technology. [21]

Next, a Coriolis mass meter is installed, designed to measure the total mass flow rate and density of the mixture. To control the thermobaric parameters, pressure and temperature meters are installed.

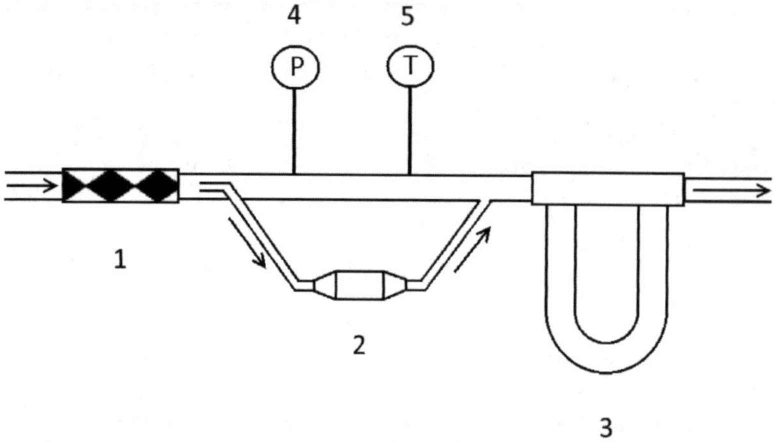

Figure 1. Diagram of a multiphase non-separating flow meter with a Coriolis flow meter and an IR fraction meter. 1 - homogenizer; 2 - IR fraction meter; 3 - Coriolis mass meter; 4 - pressure meter; 5 - temperature meter.

This system provides measurements of the following parameters: mixture mass flow G_{mix} and density ρ_{mix}, temperature T and pressure P of the mixture, optical densities of the medium D^{λ_1} at wavelength λ_1 and D^{λ_2} at wavelength λ_2.

3. Study of the absorption of NIR radiation in a gas-liquid mixture flow

Experimental installation (pouring stand) and experimental technique. The installation diagram is shown in Figure 2. The principle of operation of the stand is as follows. By means of a pump 3, a gas-liquid flow with a gas content of $\alpha_g \leq 10\%$ is circulated in a closed circuit having a certain volume *V*. To homogenize the flow, a mixer 2 is used, which is a pipe section in which spiral-shaped flow bodies are installed. After the mixer, the flow passes through a measuring IR cell, where the optical density of the mixture is measured at wavelengths of 880 and 1660 nm. The thickness of the translucent layer is 2.5 mm and can be adjusted. IR LEDs are used as a source of NIR radiation, IR photodiodes are used as detectors.

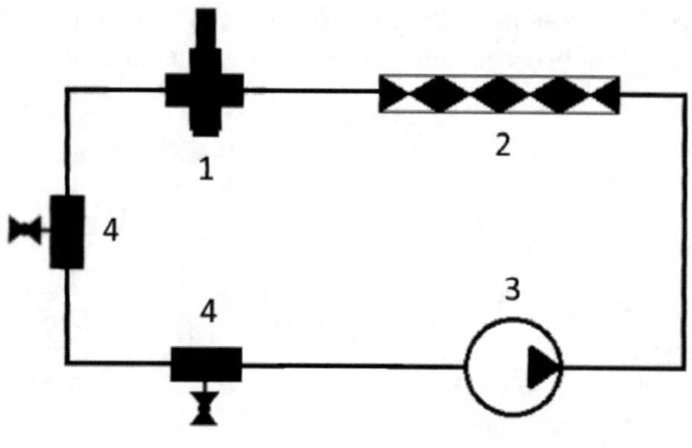

Figure 2. Schematic diagram of the pouring stand. 1 – measuring IR cell;
2 - homogenizer 3 – fluid pump; 4 – drain tap

The components in the gas-liquid flow were water, exxsol D100 (oil simulator) and air. The required ratio of these components in the flow was set by installing the corresponding volumes of oil simulator and water on the stand. The required volumes were set through the formula for the connection of mass and density using high-precision scales. The optical density measurements were based on measurements of the zero signal (when filling the measuring IR cell with air) and the signal when the NIR radiation passed through the studied flow. It should be noted that the installation in question is a prototype installation where a mixture of oil, water and associated petroleum gas can be studied.

Experiment results. The experiments were carried out for a volumetric gas content of 5 and 10% in the range of the liquid moisture content from 10 to 90% in increments of 20%. The results are shown in Figure 3, which shows graphs of the dependence of the difference in optical density $\Delta D = D^{\lambda_1} - D^{\lambda_2}$ on the liquid moisture content α_{wl}.

a)

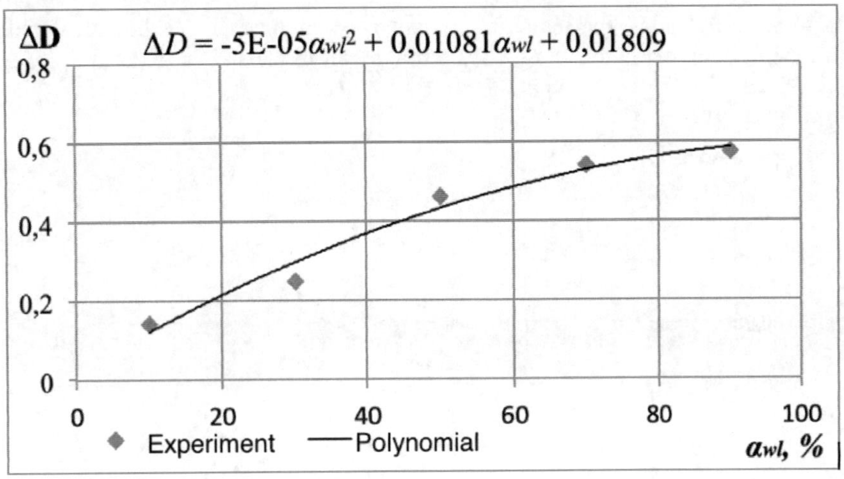

b)

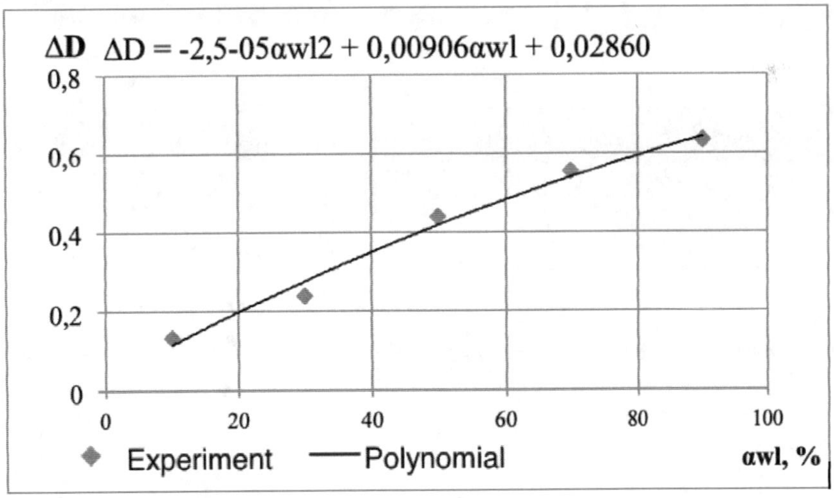

Figure 3. Dependence of the difference in optical density ΔD on the liquid moisture content α_w:

a) $\alpha_g = 0{,}05$; б) $\alpha_g = 0{,}1$.

The obtained experimental data are approximated by second-order polynomials:

for $\alpha_g = 0{,}05$ $\quad \Delta D = -0{,}00005\,\alpha_{wl}^2 + 0{,}01081\,\alpha_{wl} + 0{,}01809$ (1)

for $\alpha_g = 0{,}1$ $\quad \Delta D = -0{,}000025\,\alpha_{wl}^2 + 0{,}00906\,\alpha_{wl} + 0{,}02860$ (2)

Expressing from here the quantity α_{wl}, we obtain:

for $\alpha_g = 0{,}05$ $\quad \alpha_{wl} = 108{,}06 - \sqrt{12038{,}8 - 20000\,\Delta D}$ (3)

for $\alpha_g = 0{,}1$ $\quad \alpha_{wl} = 181{,}28 - \sqrt{34006{,}2 - 40000\,\Delta D}$ (4)

According to these formulas, the moisture content was calculated, and the calculated values were compared with the true ones (Figure 4). Quality estimates of models (1) and (2) are presented in Table 1.

a)

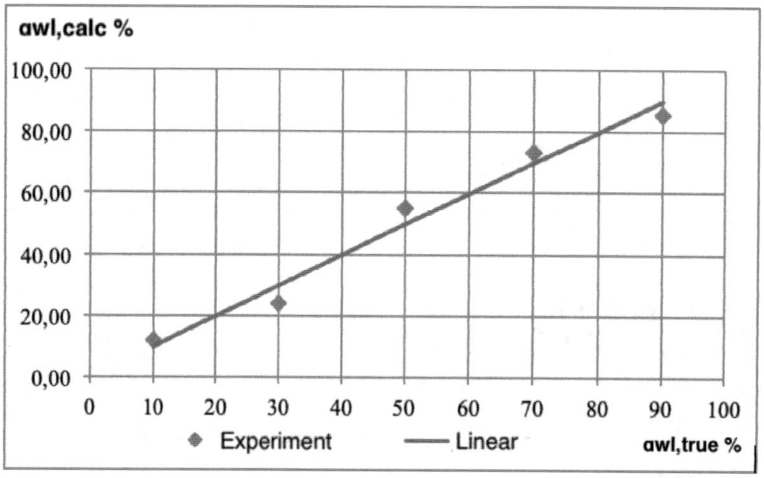

b)

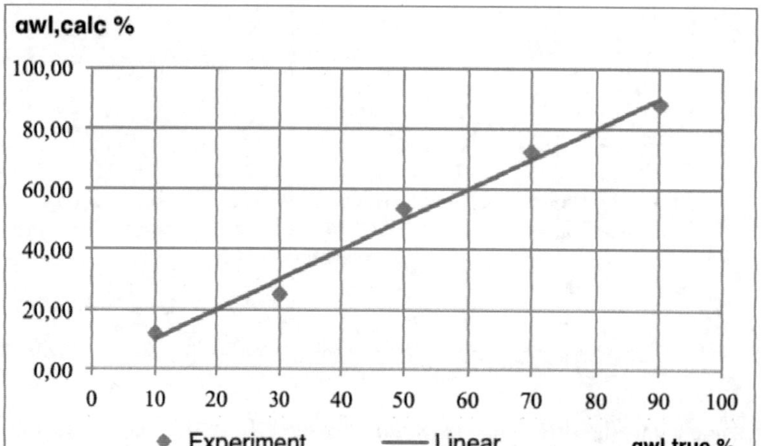

Figure 4. Comparison of the calculated water cut values with true values: a) $\alpha_g = 0{,}05$; b) $\alpha_g = 0{,}1$.

From formulas (3), (4), we can obtain a generalized dependence for the gas content range from 5 to 10%:

$$\alpha_{wl} = 34{,}84 + 1464{,}4\alpha_g - \sqrt{439348\,\alpha_g - 9928{,}6 - 400000\,\alpha_g \Delta D} \tag{5}$$

The given error in determining the liquid moisture content by the generalized expression (5) was 8%.

Table 1. Model quality assessment parameters

Model	Correlation index R	Determination index R^2	Mean approximation error ε_{mean}, %
Model 1 ($\alpha_g = 0.05$)	0.987	0.974	8.65
Model 2 ($\alpha_g = 0.1$)	0.993	0.986	7.32

4. Computational and experimental method

Thus, a computational and experimental method has been developed for determining the phase composition and phase flow, justified for the gas content range $\alpha_{g1} \leq \alpha_g \leq \alpha_{g2}$, where in this case α_{g1} = 0.05, α_{g2} = 0.1. The proposed method can be applied for any range of gas content in the flow of water, oil and gas mixture. This method, together with the measurements provided by the proposed multiphase flow meter scheme, and data from laboratory studies (true oil and water densities, component composition of reservoir oil) is reduced to the following calculations.

The volume fraction of phases is determined from the system of equations:

– mixture density

$$\rho_{mix} = \rho_w^{(0)}\alpha_w + \rho_o^{(0)}\alpha_o + \rho_g^{(0)}\alpha_g \tag{6}$$

where $\rho_i^{(0)} = \frac{m_i}{V_i}$ - true density of the i-th component, m_i - mass of the i-th component occupying the volume V_i. The true density of the gas is determined by the expression:

$$\rho_g^{(0)} = \frac{P}{Z(P,T)RT}, \tag{7}$$

where the gas compressibility factor $Z(P,T)$ can be determined from the equations of state of para-liquid equilibrium [22].

– a condition of equality of the sum of volume fractions of phases to one:

$$\alpha_w + \alpha_o + \alpha_g = 1 \tag{8}$$

– a volume fraction of water by liquid, determined by the NIR method according to the formula (5):

$$\alpha_{wl} = a(\alpha_g) - \sqrt{b(\alpha_g) - c(\alpha_g)\Delta D}, \tag{9}$$

where $a(\alpha_g)$, $b(\alpha_g)$, $c(\alpha_g)$ coefficients determined by the correlation dependences on gas content.

– a volume fraction of water related to the entire flow:

$$\alpha_w = \alpha_{wl}(1 - \alpha_g) \tag{10}$$

Mass consumption of water, oil and associated gas can be determined by known expressions:

$$G_w = \frac{G_{mix}}{{}_{mix}} \; {}_{w} \, {}^{(0)}_{w}$$

$$G_o = \frac{G_{mix}}{{}_{mix}} \; {}_{o} \, {}^{(0)}_{o} \qquad (11)$$

$$G_g = \frac{G_{mix}}{{}_{mix}} \; {}_{g} \, {}^{(0)}_{g}$$

We emphasize that the above methodology for calculating the mass flow rate of phases is applicable under the assumption of equal phase velocities of the water, oil and gas flow.

5. Conclusions

- An intelligent measuring system is proposed - a non-separating multiphase flow meter, which includes the following main components: a gas-liquid flow homogenizer, an infrared fraction meter, a Coriolis flow meter and a physical-mathematical model that relates the phase composition and phase flow to measured parameters.
- An experimental installation has been developed for studying the absorption of NIR radiation in multiphase flows.
- Experimental studies of the absorption of NIR radiation in the flow of GLM (water, oil simulator, air). The dependence of the difference in optical density on the volume fraction of water in the range of the liquid moisture content of 10-90% with a volumetric gas content of the entire stream of 5 and 10% is obtained. The given error in calculating the moisture content according to the generalized correlation equation did not exceed 8%.
- A computational and experimental method is proposed for determining the volume fractions of the phases of oil and gas flows using NIR radiation for various ranges of gas content.

References

[1] Alisov V D and Shalagin N A 2008 et al. *Oil and gas production operator Tutorial* (Ufa: training plant) LLC MPF "SEMS" 200
[2] Boyarov A I 2000 *Handbook of oil production for specialists* (Almetyevsk: Tatneft) 334
[3] Vasilevsky V N and Petrov A I 1989 *Technique and technology for determining the parameters of wells and reservoirs: A reference book of a worker* (Moscow: Nedra) 271
[4] Multiphase flow meters Roxar MPFM 2600 SI Type Description State Registry 47571-11
[5] Multiphase flow meters Vx Spectra SI Type Description State Registry 60560-15
[6] Multiphase flow meters Alpha SI Type Description State Registry 64321-16
[7] Multiphase flow meters Agar MPFM SI Type Description State Registry 65061-16
[8] Rymarenko K V 2010 Hydrodynamic studies and multiphase flow metering: new possibilities and principles of work (using the Vx technology as an example) *Technique and Technologies December* 29-38
[9] Nikolaev A B, Ilyukhin A V, Marsov V I, Zamytskikh P V, Zhuchkov V N and Azovtsev V A 2011 Method for measuring the flow of oil and water mixture by pressure drop using narrowing devices *Automation, telemechanization and communication in the oil industry* **11** 6-11
[10] Sanchez Silva F, Andreussi P and P Di Marco 1991 Total mass flowrate measurement in multiphase flow by means of a Venturi meter Multiphase Production ed. by A P Burns, Elsevier 145-155
[11] Jorge Luiz Goes Oliveira, Júlio César Passos, Ruud Verschaeren and Cees van der Geld 2009 Mass flow rate measurements in gas–liquid flows by means of a venturi or orifice plate coupled

IOP Conf. Series: Materials Science and Engineering **663** (2019) 012002 doi:10.1088/1757-899X/663/1/012002

to a void fraction sensor (*Experimental Thermal and Fluid Science* **33**) Issue 2 January 253-260

[12] He Peixiang, Cees van der Geld, Claudio Alimonti and Julio Cesar Passos 2009 A new method of measuring two-phase mass flow rates in a Venturi *Multiphase Science and Technology* **21** 1–2 157–168

[13] Pushnov A V 2013 Flow measurement of a two-phase flow by a Coriolis flowmeter *Measuring Technology* **4** 45-48

[14] Liu R P, Fuent M J, Henry M P and Duta M D 2001 A neural network to correct mass flow errors caused by two-phase flow in a digital coriolis mass flowmeter *Flow Measurement and Instrumentation* **12** Issue 1 March 53-63

[15] Casimiro R, Henry M, Tombs M, Kroshkin A and Lishchuk A 2014 New multi-phase flow metering technology available for industrial measuring units in the oil and gas industry *Oil Industry* **5** 114-118

[16] Vakulin A A and Shabarov A B 2011 *Diagnostics of thermophysical parameters in oil and gas technologies. Monograph* (Novosibirsk: Publishing House of the Siberian Branch of the RAS) 249

[17] Marlene Bampi, Agnes de P Scheer and Fernandade Castilhos 2013 *Application of near infrared spectroscopy to predict the average droplet size and water content in biodiesel emulsions* Fuel **113** November 2013 546-552

[18] Yong Zhao, Jian Yang, Jin-Qi Wang and Fang-Xiao Gui 2004 High-accuracy low-water-content measurement of crude oil based on a near-infrared spectral absorption method *Optical Engineering* **43** 10 October 2216-2217

[19] Red Eye® water meters Red Eye® 2G and Red Eye® Multiphase SI Type Description. - State Registry 47355-11

[20] Muravyov A V and Otylov E A 2012 *Weatherford Red Eye® water meter family (Red Eye® 2G and Red Eye® Multiphase models) and their use* [NGN] **10** 64–59

[21] RF patent 2669156 Flow hydrometer Zaitsev Evgeny Vyacheslavovich, Vorobyov Vladimir Viktorovich, Nikulin Sergey Gennadevich and Grigoryev Boris Vladimirovich

[22] Brusilovsky A I 2002 *Phase transformations in the development of oil and gas fields* (Moscow: Graal) 575

IOP Conf. Series: Materials Science and Engineering **663** (2019) 012003 doi:10.1088/1757-899X/663/1/012003

Oil displacement by aqueous solutions of surfactants at various temperatures

M Yu Zemenkova [1], O A Kuzina [2,3] and A B Shabarov [2,4]

[1] Industrial University of Tyumen, 38 Volodarskogo str., Tyumen, 625000, Russia
[2] Tyumen State University, 15A Perekopskaya st., Tyumen, 625003, Russia
[3] o.a.kuzina@utmn.ru
[4] kaf_mms@utmn.ru

Abstract. To generalize the experimental data on the displacement of oil by aqueous solutions of surface-active substances (surfactants) at temperatures of 40, 60, 80 °C, we studied the influence of an additional parameter of the ratio of the work of adhesion during the displacement of oil by water to oil adhesion during the displacement of an aqueous solution of surfactant $W = W_w / W$. This parameter characterizes the intensity of mass transfer of oil particles from the surface of the rock using surfactants at different temperatures. The parameter W in combination with the capillary number Ca characterizes the value of residual oil saturation in oil reservoirs. A criterion equation is obtained that allows one to calculate the displacement coefficient from two dimensionless parameters: the capillary number and the relative work of oil adhesion.

1. Introduction

Recently, in connection with the widespread use of surface-active substances (surfactants) to increase oil recovery, studies are increasingly being conducted to reduce residual oil saturation during displacement. [1] Moreover, if the initial oil saturation is determined by the initial geological conditions of the reservoir, the residual oil saturation depends on the surface tension forces arising from the interaction of oil with a displacing agent and other parameters. [2]

Capillary forces, and, consequently, residual oil saturation, depend on the size and type of surfactant micelles, surfactant concentration, and temperature. The decrease in residual oil saturation with increasing temperature can be explained by the weakening of intermolecular interactions of asphaltenes in the volume of structured oils.

Thus, as shown by numerous experimental studies and their analysis, the residual oil saturation depends on the dimensionless capillary number. An extensive review of the results of experimental studies of the effect of capillary number on residual oil saturation during oil displacement by water for core samples from various fields of the world is given in the book by Yu.E. Baturin. [5] However, this parameter alone is not enough to estimate the amount of residual oil saturation. [6]

2. Objects and research methods

The use of an aqueous surfactant solution instead of water as a displacing agent leads to a decrease in residual oil saturation. [1,3] According to the results of laboratory studies on core material, the injection of an aqueous surfactant solution into a porous medium with all relevant reactions leads to a relative decrease in residual oil saturation, which can be up to 30%. [4, 5] This occurs due to a decrease in capillary forces, which are characterized by the capillary number Ca. The form of this number is:

$$Ca = \frac{v \cdot \mu_d}{\sigma}, \tag{1}$$

where v - displacement fluid filtration rate, $m^3/(m^2 \cdot s)$; μ_d – displacement fluid viscosity, Pa·s; σ – surface tension at the oil-displacing fluid interface, N/m.

The capillary number characterizes the ratio between capillary forces, the value of which determines the surface tension at the water-oil interface, the structure of the pore space and the hydrodynamic pressure, which determines the rate of fluid filtration. An analysis of the parameters included in the capillary number formula shows that it is virtually impossible to significantly increase the hydrodynamic forces in a real reservoir. The introduction of surfactants can have a significant effect on capillary forces.

The addition of a surfactant reduces the marginal angles of selective wetting, i.e., increases the wettability of the rock with water. The criterion for the wetting ability of a surfactant is the energy of the interaction of oil with the rock surface, defined as the work of oil adhesion. A decrease in contact angles together with a decrease in interfacial tension entails a significant decrease in the binding energy of oil and the rock surface. The binding energy can be judged by the decrease in the work of adhesion of oil to the surface from the Dupre-Young equation [7]:

$$W = \sigma \cdot (1 - cos\theta) \tag{2}$$

where W - oil adhesion work, J/m^2; σ - interfacial tension at the oil - water phase, N/m; θ - marginal angle of selective wetting, deg.

In the framework of the studies, the surface tension was measured at the interface “oil - aqueous surfactant solution” on an ST-1 stalagmometer by the drop volume method. The results of these studies are given in [8].

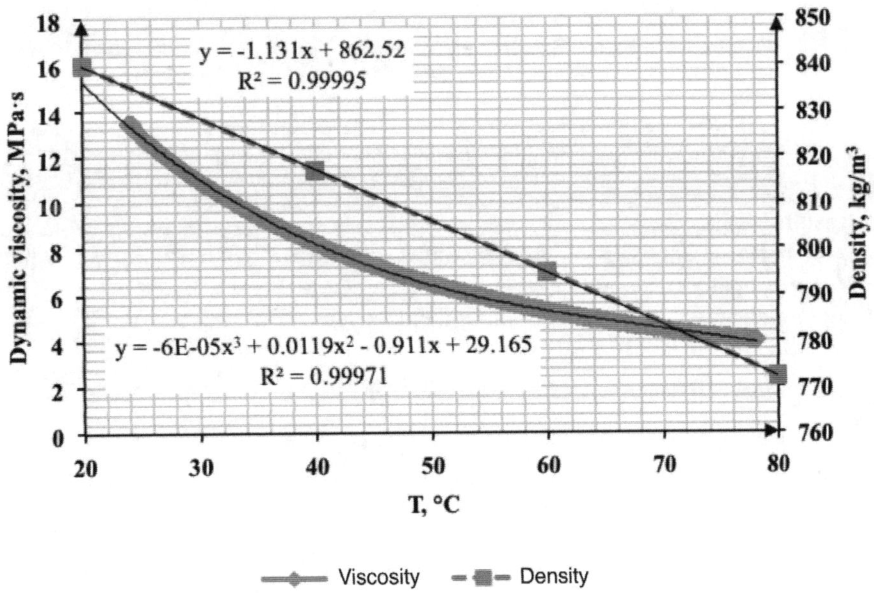

Figure 1. Characteristic dependence of the dynamic viscosity and density of oil on temperature

Filtration studies were carried out on a composite column consisting of three sand cores with similar lithological and porosity and permeability properties. The average open porosity of the samples in water is (22.65 ± 0.2)%; helium gas permeability - (372 ± 20) mD.

An oil with the following characteristics was used: dynamic viscosity at 20 °C μ = 15.2 mPa·s; density at 20 °C ρ = 840 kg/m^3.

A typical form of changes in the dynamic viscosity and density of oil in the temperature range of 20 - 80 °C is shown in Figure 1.

As the aqueous phase, we took aqueous solutions of three surfactants in a reservoir water model. Distilled water with a NaCl salt content of 20 g/L was used as a model of produced water. Surfactants used: synthanol ALM-1, neonol AF 9-12 - belong to the class of nonionic surfactants, sodium lauryl sulfate - anionic surfactant. The concentration of aqueous surfactant solutions was 0.5% of the mass.

The adhesion work *W* was found by the formula (2), where the contact angle was determined by the method of a lying drop. The technique for measuring the contact angles of wetting is based on placing a drop of oil on the bottom surface of the plate suspended horizontally and placed in a cuvette with an aqueous phase. Instead of a core sample, a quartz surface acted as a plate, which models the wetting conditions of the rock pore walls. Using a microscope, the maximum diameter and height of the drop were measured. The diameter of the drop should be from 2 to 5 mm; this ensures that the contact angle will not depend on the diameter. In the case of very small drops, the influence of the surface tension of the liquid itself will be large (spherical drops will form), and in the case of large drops, gravity forces begin to dominate [9, 10]. The contact angle was calculated by the formula:

$$tg\left(\frac{\theta}{2}\right) = \frac{2\cdot h}{d_{max}} \qquad (3)$$

where θ – contact angle, h – drop height, d_{max} – maximum drop diameter.

Filtration was carried out on the PIK-OFP/EP-3 installation, designed to simulate a two-phase flow through a composite column of core samples in reservoir conditions. Filtration experiments were carried out at a mountain pressure of 250 atm, an operating pressure of 30 atm, and a linear filtration rate of 1.7 m/day at three temperatures: 40, 60 and 80 °C.

The oil displacement coefficient was determined by the formula [11]:

$$К_{выт} = \frac{1-S_{*}-S_{p}^{*}}{1-S_{*}}\cdot 100\ \% \qquad (4)$$

where S_{*} – initial residual water saturation; S_{p}^{*} – residual oil saturation.

The value of residual water saturation was determined by centrifugation. [12]

3. Research results

The results of oil displacement experiments for various aqueous surfactant solutions at temperatures of 40, 60, 80 °C, as compared with produced water are shown in Figure 2. The obtained value of the residual oil saturation can be used in determining the dependences of the relative phase permeabilities of oil and water using the methods [13, 14].

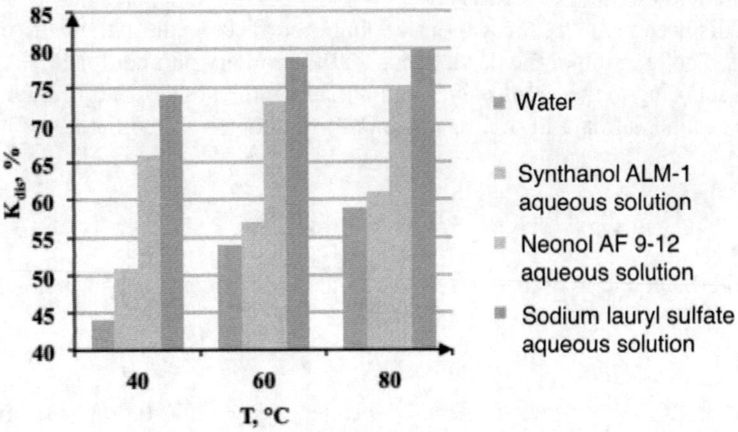

Figure 2. Comparison of oil displacement coefficients with aqueous solutions of various surfactants at different temperatures

Table 1 shows the calculated parameters of the studied system at different temperatures: oil displacement coefficient K_{dis}, surface tension at the oil-aqueous surfactant interface σ, capillary number calculated by the formula (1) Ca, selective contact wetting angles θ, parameter of relative oil adhesion work $\overline{W}$.

Table 1. Values of the found parameters for the investigated three-phase system at various temperatures

T, °C	Displacing agent	K_{dis}, %	σ, mN/m	$Ca\cdot 10^6$	θ, deg	W, mJ/m^2	$\overline{W}$
40	Water	44	36.9	0.41	76	27.9	
60		54	35.8	0.35	70	23.5	1
80		59	35	0.26	47	11.1	
40	Synthanol	51	20	0.75	67	12.2	2.2
60	ALM-1	57	26	0.49	55	11.0	2.1
80		61	31	0.30	44	8.7	1.2
40	Neonol AF 9-12	66	4.8	3.1	57	2.2	12.8
60		73	4.1	3.1	61	2.1	11.1
80		75	4	2.3	43	1.0	10.3
40	Sodium lauryl	74	2.6	5.8	47	0.8	33.8
60	sulfate	79	2.6	4.9	40	0.6	38.7
80		80	2.3	4.1	41	0.5	19.7

Table 1 shows that there is a monotonic dependence of the displacement intensity on the work of adhesion. With increasing temperature, the adhesion work W decreases, due to which the oil recovery increases. Thus, the dimensionless parameter of the relative adhesion work $\overline{W}$ can be used to evaluate oil displacement by aqueous surfactant solutions.

Based on the conducted experimental studies, the dependence of the oil displacement coefficient on two dimensionless parameters was revealed: the capillary number *Ca* and the relative adhesion work $\overline{W}$ in the form:

$$K_{dis} = C \cdot f_1(Ca) \cdot f_2(\overline{W}) \tag{5}$$

where the dimensionless constant $C, f_1(Ca)$ and $f_2(\overline{W})$ are functional dependencies on Ca and $\overline{W}$.

Obviously, when displacement occurs without adding a surfactant, the parameter of the relative adhesion work $\overline{W}$ is 1. The function of the dependence on the capillary number for $\overline{W} = 1$ can be found for specific deposits according to the set of points summarized in the book by Yu.E. Baturin. [5]

We present the experimental data in the logarithmic form adopted in the theory of heat and mass transfer [15,16]:

$$\lg K_{dis} = \lg a_0 + m_1 \cdot \lg Ca \tag{6}$$

In this case, the criterial equation of mass transfer during oil displacement by water has the form:

$$K_{dis} = a_0 \cdot Ca^{m1} \tag{7}$$

As an example, Figure 3 shows the experimental dependence $K_{dis} = K_{dis}(Ca)$ at a temperature of 60 °C in logarithmic coordinates.

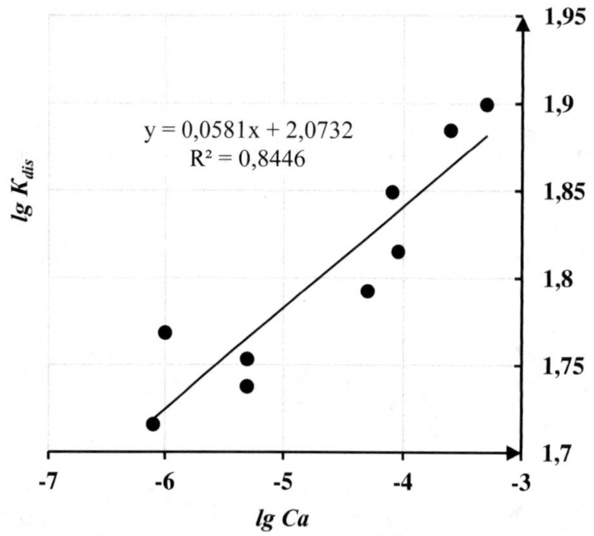

Figure 3. Value of the displacement coefficient depending on the capillary number in the logarithmic coordinates for the Abrams field [5]

Thus, as a function $K_{dis} = K_{dis}(Ca)$ in the logarithmic coordinates, a straight line was taken with approximation reliability of 0.84, which is typical for the criterion equations of mass transfer. [15,16] From the approximation equation, empirical coefficients a_0 and m_1 were found for the studied temperatures.

The general form of the criterion equation for the dependence of the displacement coefficient on dimensionless parameters and temperature is taken in the form:

$$K_{dis} = a_0(T) \cdot Ca^{m_1(T)} \cdot \overline{W}^{m_2}. \tag{8}$$

The value of the coefficient m_2 was found analytically from a system of equations with known data on the displacement coefficient, capillary number, and relative work of adhesion. The found value of the coefficient $m_2 = 0.07$. According to the calculations, it was found that the value of this coefficient does not depend on temperature. While the coefficientsa$_0$ and m_1 are functions of temperature.

According to calculations, the empirical coefficients were:

$a_0(T) = 2600 \cdot T^{-0,748}$;

$m_1(T) = 17,115 \cdot T^{-1,392}$;

$m_2 = 0,07$.

Figure 4 shows the types of surfaces of the displacement coefficients on the parameters of the capillary number Ca and the relative work of oil adhesion $\overline{W}$ for various temperatures.

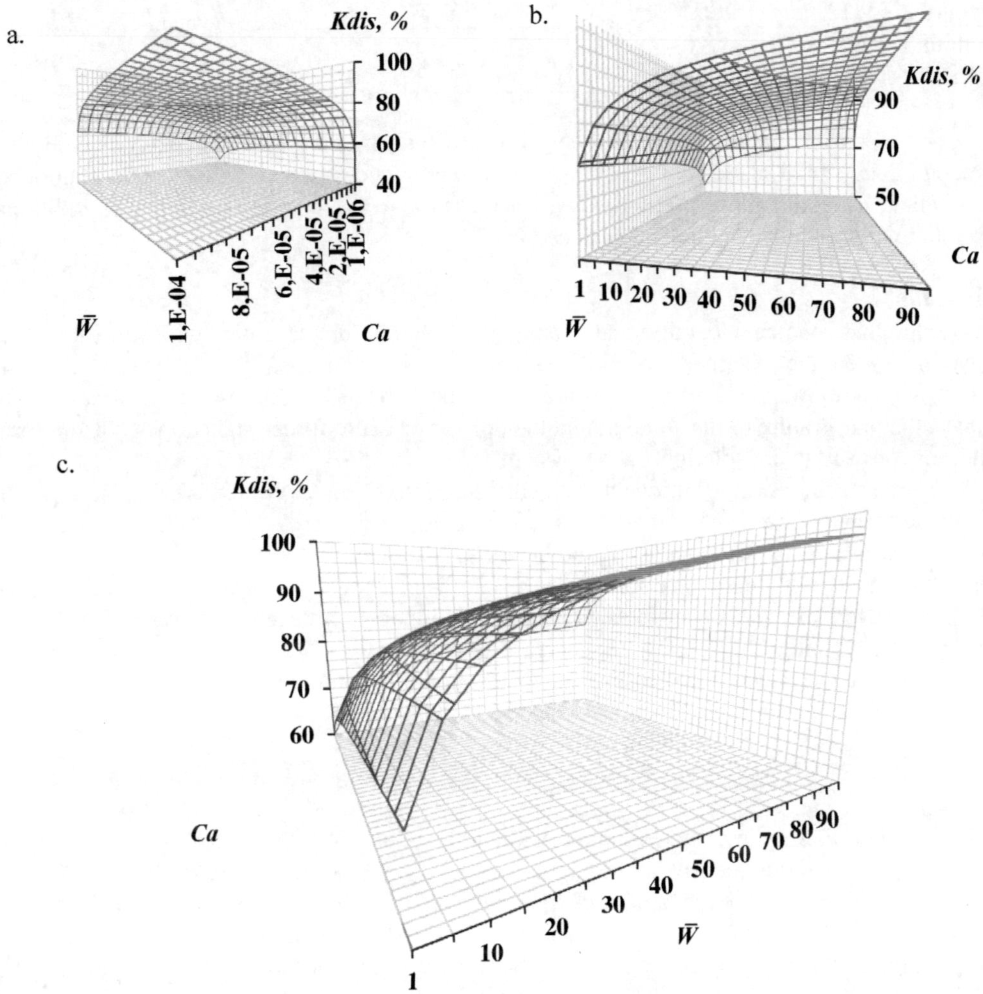

Figure 4. Surfaces of displacement coefficients depending on two dimensionless criteria for temperatures: a – 40 °C; b – 60 °C; c – 80 °C

Table 2 shows the experimental values and values calculated by the formula (8) of the displacement coefficient for various temperatures.

Table 2. Comparison of experimental and calculated values of oil displacement coefficient

Displacing agent	**T = 40 °C**		**T = 60 °C**		**T = 80 °C**	
	$K_{dis.\ exp.}$	$K_{dis.\ calc.}$	$K_{dis.\ exp.}$	$K_{dis.\ calc.}$	$K_{dis.\ exp.}$	$K_{dis.\ calc.}$
Water	44.2	43.7	54.0	53.7	59.2	57.8
Synthanol ALM-1	51.1	49.3	57.4	57.7	61.1	59.1
Neonol AF 9-12	66.3	64.2	73.3	74.0	75.1	74.0
Sodium lauryl sulfate	74.0	73.1	79.7	80.6	80.4	79.1

As can be seen from Table 3, the calculated values within the measurement error coincide with the experimental data, which confirms the adequacy of the obtained criteria equation. Thus, choosing the right surfactant, using the above methodology, it is possible to predict the possibility of additional oil displacement with aqueous solutions of surfactants.

4. Conclusion

1. A procedure has been developed for calculating additional oil displacement with aqueous surfactant solutions, based on the obtained criterion equation for mass transfer based on the experiments performed, which allows predicting the influence of the type of aqueous solution of surface-active substances, the temperature of the solution and the ratio of pressure forces and surface tension forces in the "oil - aqueous surfactant solution - core" system.
2. Experimental studies of oil filtration with aqueous solutions of three different surfactants and without them in the temperature range of 40 - 80 °C were conducted. The increase in oil displacement with the use of additives, as well as with increasing temperature, is shown.
3. It is shown that with increasing temperature, oil displacement by aqueous surfactant solutions increases. However, the use of surfactants is more rational at low temperatures. With increasing temperature, the washing ability of water increases significantly, so the percentage of displacement compared with produced water decreases. So, for example, at a temperature of 40 °C the additional displacement for sodium lauryl sulfate was 68%, and at 80 °C - 35%.

References

[1] Semikhina L P, Shtykov S V, Pashnina A M and Karelin E A 2015 Effect of temperature on the ability of aqueous reagent solutions to wash oil from a solid surface *Bulletin of the Tyumen State University. Physical and mathematical modeling. Oil, gas, energy* **1** 3(3) 39-51

[2] Akhmetova Z R 2016 *Residual oil saturation structuring to substantiate oil recovery technology* National University of Oil and Gas "Gubkin University" Moscow

[3] Moreau P, Morvan M, Rivoal P (Rhodia), Bazin B, Douarche F, Argillier J-F and Tabary R 2010 (IFP) An integrated workflow for chemical EOR pilot design *SPE 129865 presented at the SPE Improved Oil Recovery Symposium held in Tulsa Oklahoma USA* 24-28

[4] Flaaten A A, Quoc Nguyen and Pope G 2008 A systematic laboratory approach to low-cost, high performance chemical flooding *SPE 113469 presented at the SPE/IOR Symposium Tulsa* 19-23

[5] Baturin Yu E 2016 *Design and development of oil and gas and oil fields in Western Siberia* (Surgut: OJSC "Surgutneftegas", Advertising and publishing information center "Oil of the Ob region") **2** 204

[6] Kuzina O A, Semikhina LP and Shabarov A B 2019 Effect of capillary number and adhesion work on oil displacement by aqueous solutions of surfactants *Bulletin of the Tyumen State University. Physical and mathematical modeling. Oil, gas, energy* **5** 3

[7] Babalyan G A 1983 *Development of oil fields using surfactants* (Moscow: Nedra) 216

[8] Grigoriev B V, Vazhenin D A and Kuzina O A 2016 Effect of surfactant concentration of aqueous solutions and temperature on the surface tension coefficient *Bulletin of the Tyumen State University. Physical and mathematical modeling. Oil, gas, energy* **2** 3 35-48.

[9] Aslan S, Najafabadi N F. Firoozabadi A 2016 Non-monotonicity of the contact angle from NaCl and MgCl2 concentrations in two petroleum fluids on atomistically smooth surfaces *Energy Fuels* 30 2858-2864

[10] Lu Y, Najafabadi N F and Firoozabadi A 2019 Effect of Low-concentration of 1-Pentanol on Wettability of Petroleum Fluid-Brine-Rock Systems *Langmuir American Chemical Society*

[11] OST 39-195-86 *Oil. Method for determining the coefficient of oil displacement by water in laboratory conditions*

[12] Kuznetsov A M, Baishev A B and Kuznetsov V V 2010 Determination of initial water saturation and capillary curve by centrifugation *Oil industry* **1** 49-51

[13] Shabarov A B and Shatalov A V 2016 Pressure loss during oil-water mixture flow *Bulletin of the Tyumen State University. Physical and mathematical modeling. Oil, gas, energy* **2** 2 50-72

[14] Shabarov A B, Shatalov A V, Markov P V and Shatalova N V 2018 Methods for determining the functions of relative phase permeability in multiphase filtering problems *Bulletin of the Tyumen State University. Physical and mathematical modeling. Oil, gas, energy* **4** 1 79-109

[15] Isachenko V P, Osipova V A and Sukomel A S 1975 *Heat transfer* (Moscow: Energiya) 488

[16] Lukanin V N, Shatrov M G et al. 2009 *Heat engineering* (Moscow: Vysshaya shkola) 671

IOP Conf. Series: Materials Science and Engineering **663** (2019) 012004 doi:10.1088/1757-899X/663/1/012004

I-res intelligent system for multi-factor risk control of accidents and the technical condition of the main gas pipeline

M Yu Zemenkova [1,2], D D Dubrovin [1], I F Shagbanov [1] and Yu D Zemenkov[1]

[1]Industrial University of Tyumen, 38 Volodarskogo str., Tyumen, 625000, Russia
[2]muzemenkova@mail.ru

Abstract. The paper deals with improving the reliability and safety management system of the linear part of the main gas pipeline. At the Department of Transport of Hydrocarbon Resources, Industrial University of Tyumen, research is being carried out to develop a system for multi-factor risk control of accidents and the technical condition of hydrocarbon transport systems. As an example, an algorithm for monitoring the technical condition and safety of local sections of the pipeline was developed, a software product "I-ReS" (intelligent, reliability, safety) was developed as part of the CI-TREMS system, which allows real-time monitoring of the technical condition and safety of the main gas pipeline.

1. Introduction

In Russia, a system of main pipelines that is unique in length, productivity and safety has been created, operates and is developing. Pipeline transportation is the main method of delivering gas and oil to consumers. According to the annual report of Federal Environmental, Industrial and Nuclear Supervision Service of Russia (Rostechnadzor) for 2017, the total length of the linear part of the main pipelines is more than 257.8 thousand km, of which gas pipelines are 179.5 thousand km. [1] Collaborative intellectual technological reliability and efficiency management of oil and gas systems (CI-TREMS) is a fundamentally new technology of intelligent instrumental neural network engineering control, forecasting and prevention of emergency situations, incidents, accidents, optimization and ensuring the effectiveness of technical solutions in the management of processes of industrial enterprises developed at the Department of Hydrocarbon Resources Transport. CI-TREMS is modular and multi-tasking, flexible and adaptive, based on the theory of neural network programming, the theory of reliability, cybernetics, system analysis, statistics and probability, and others. Currently, the issue of ecology and noosphere culture in the oil and gas industry is relevant and has gained particular importance at facilities from Siberia to the Arctic, and the presence of such systems in hazardous enterprises is a necessity. The theoretical and practical stages of CI-TREMS development provide for individual work with each process facility and specific monitoring and forecasting problems that are implemented within the framework of the modules. The "I-ReS" software product (intelligent, reliability, safety) was developed on the basis of a real gas pipeline and its individual characteristics using the functions provided by c-Mes technology, Asset Management.

Different monitoring and diagnostic kits can be used in different managements of gas or oil pipeline facilities, and to create an integrated monitoring system it is necessary to create different models of periodic monitoring and real-time models. This is due not only to differences in the documents of enterprises but also to the various equipment of the diagnostic means and transport systems of one enterprise.

In modern conditions, there is a large amount of various information and each unit requires a separate methodological study and analysis, and the creation of a separate monitoring module. In the framework of this work, the structure of the enterprise for the facilities of the linear part was analysed and algorithms for analysing reliability and safety were compiled, in addition, algorithms for individual indicators of the technical condition were developed.

In accordance with the Federal Law No. 116-FZ "On Industrial Safety of Hazardous Production Facilities" dated July 21, 1997, gas pipelines are classified as hazardous production facilities - high-risk facilities. Hazard identification in sections of main gas pipelines is associated with an analysis of technological risks. When analyzing technological risks (operational risks), the gas pipeline through which natural gas, which is a hazardous substance, is transported, is identified directly as a source of hazard. The identification of hazards during the operation of the linear part of the main gas pipeline consists in describing the characteristics of the transported product (its properties, state parameters), determining the volume of natural gas in the pipeline sections under consideration, considering possible accident scenarios, their causes, consequences and identifying the most dangerous places. The determination of the hazardous properties of the transported product is related to the characteristics of natural gas, in particular, the type of hazardous substance and the actual pressure and temperature in the section of the main gas pipeline.

2. Materials and methods

The purpose of this work is to improve the reliability management system and develop a methodology and expert system that allows you to quickly monitor the technical condition of gas pipelines in its certain high-risk areas through multifactor control and contributes to the adoption of informed decisions on the implementation of measures ensuring industrial safety requirements.

When performing this work, a number of tasks were solved:

- substantiation of the methodology for multifactor control of the risk of accidents and the technical condition of the main gas pipeline;
- factor and structural analysis of main gas pipeline facilities, in particular, its linear part;
- development of an algorithm for multifactor control of the technical condition and safety of the main gas pipeline;
- development of a software product that allows real-time monitoring of the technical condition and safety of sections of the main pipeline.

The work is aimed at solving the following issues:

- updating data on the main hazards of accidents, including information on the assessment of possible damage to third parties in the event of an accident;

- assessing process data and organizational information for the examination of the technical condition of the facility, including the frequency of diagnosis;

- monitoring the degree of emergency hazard and evaluating the effectiveness of measures to reduce the risk of an accident at a hazardous production facility, including for the effectiveness of industrial safety management systems;

- improvement of operational and maintenance instructions, action plans for localization and elimination of consequences of accidents.

Research in the field of monitoring the reliability and safety of main gas pipeline facilities has been the subject of a large number of works by such famous scientists as A. M. Korolenok, Yu.V. Kolotilov, G.G. Vasiliev, M.V. Sukharev, A.B. Shabarov, E.I. Krapivsky, B.V. Moiseev, M.G. Chuchkalov and others [2-9,11,12,14,15].

Operators of hazardous production facilities maintain pipelines in working condition, monitor their technical condition to ensure the required level of reliability and evaluate operational safety at various levels (Figure 1), however, in some cases, some indicators are replaced by others due to the complexity or impossibility of system monitoring.

IOP Conf. Series: Materials Science and Engineering **663** (2019) 012004 doi:10.1088/1757-899X/663/1/012004

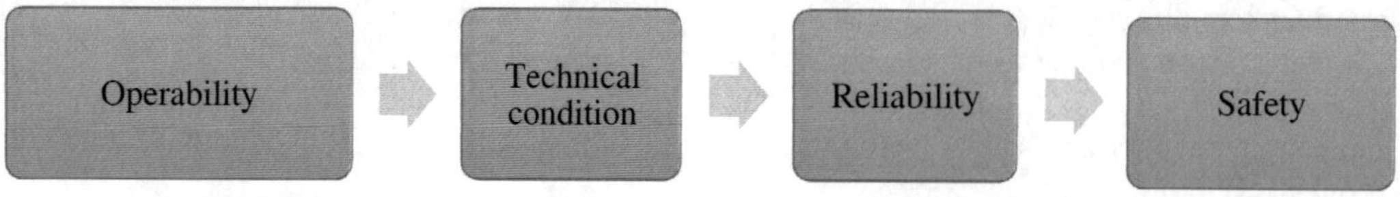

Figure 1. Diagnostic levels while ensuring safe operation of main gas pipelines.

Operators are constantly carrying out measures to ensure the integrity of main pipeline facilities, including various types of diagnostics, routine maintenance, selective and major repairs, and reconstruction of facilities. However, despite this, accidents periodically occur on main pipelines, including fatal ones, the damage from which amounts to millions of rubles.

From Rostekhnador accident reports of 2007 - 2017, 184 accidents occurred on sections of main pipelines, of which more than 70% were attributed to main gas pipelines. Moreover, an analysis of the materials of technical investigations into the causes of accidents shows that more than 60% of incidents occurred on gas pipelines that have been in operation for more than 30 years. The main reasons for other accidents are non-compliance with the procedure of work performance in the gas pipeline protected zone, deviation from design decisions, lack of round checks and inspections, due level of technical inspection, diagnostics of the technical condition of pipelines, as well as current repairs and overhaul.

In the annual reports of Rostekhnadzor, the following main causes of accidents on gas pipelines are given:

- design flaws;
- defects of construction or manufacture;
- pipe metal corrosion (including stress corrosion cracking);
- human error during operation;
- depreciation of equipment;
- natural weather impact;
- mechanical impact.

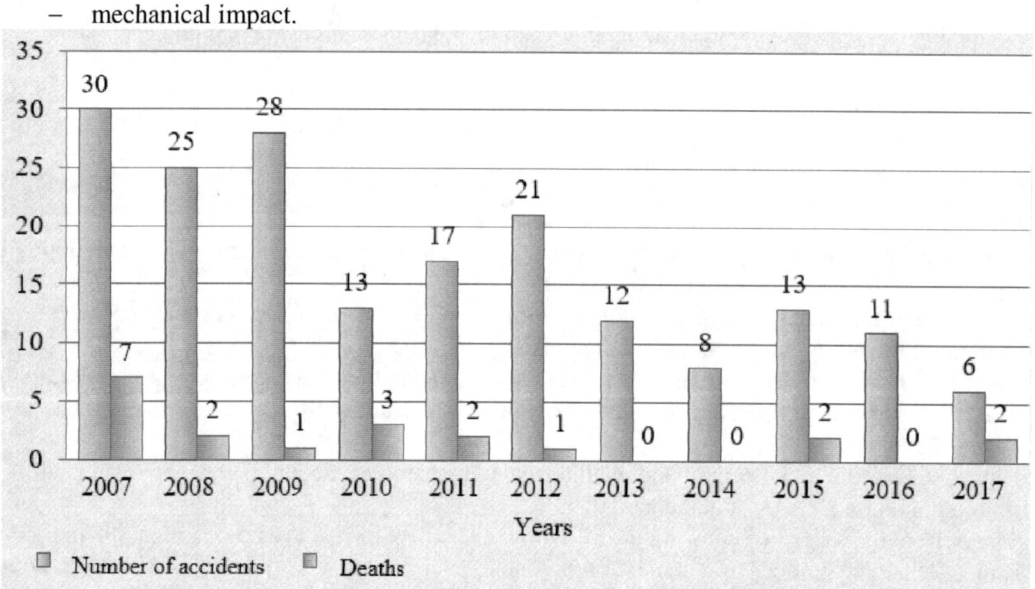

Figure 2. Dynamics of deaths and work-related injuries in 2007 - 2017

The distribution of accidents at the facilities of the main gas pipelines by causes of their occurrence in 2007-2017 is clearly visible in the pie chart (Figure 1.3).

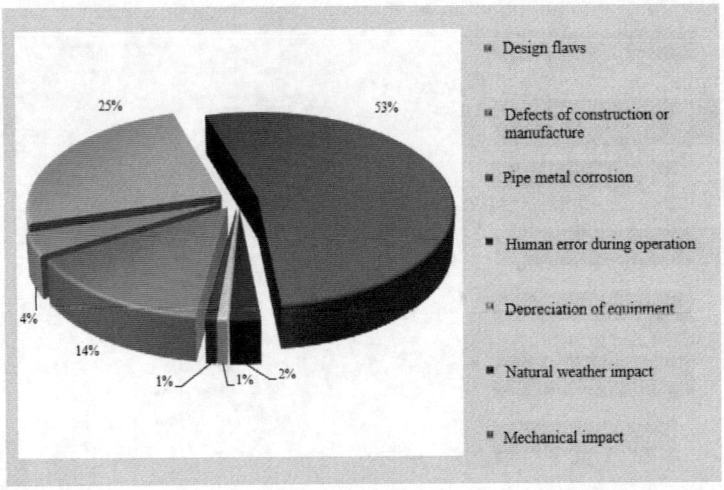

Figure 3. Distribution of accidents at the facilities of the main gas pipelines by causes of their occurrence in 2007-2017

As can be seen from the chart, the main cause of accidents at existing gas pipelines in previous years is corrosion, and there is a growth tendency for such accidents.

It is worth noting that the so-called "industrial safety economy" occupies a special place in risk management. At the same time, it is noted in some works (for example, [11-13]) that limited resources for maintaining the necessary level of industrial safety and risk management are divided into three main areas: safety costs, costs of compensation for damage from accidents, and costs of investing in hazardous production facility safety (Figure 4).

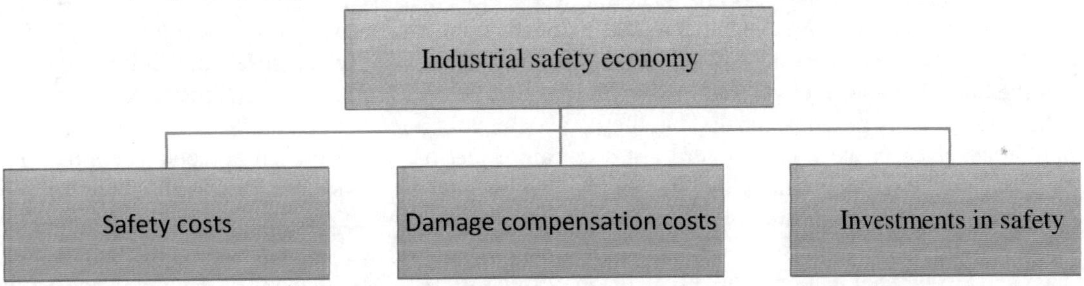

Figure 4. Resource allocation for industrial safety.

In the current documentation in accordance with the requirements of Rostekhnadkhor, there is a classification of possible industry-related events that can occur with a certain degree of probability with a certain technical condition of the object. Prediction issues in such situations remain unresolved [9-12]. Mathematical support issues were published in [7–10]. Mathematical models are provided using a logical-probabilistic mathematical apparatus and the theory of neural networks.

IOP Conf. Series: Materials Science and Engineering **663** (2019) 012004 doi:10.1088/1757-899X/663/1/012004

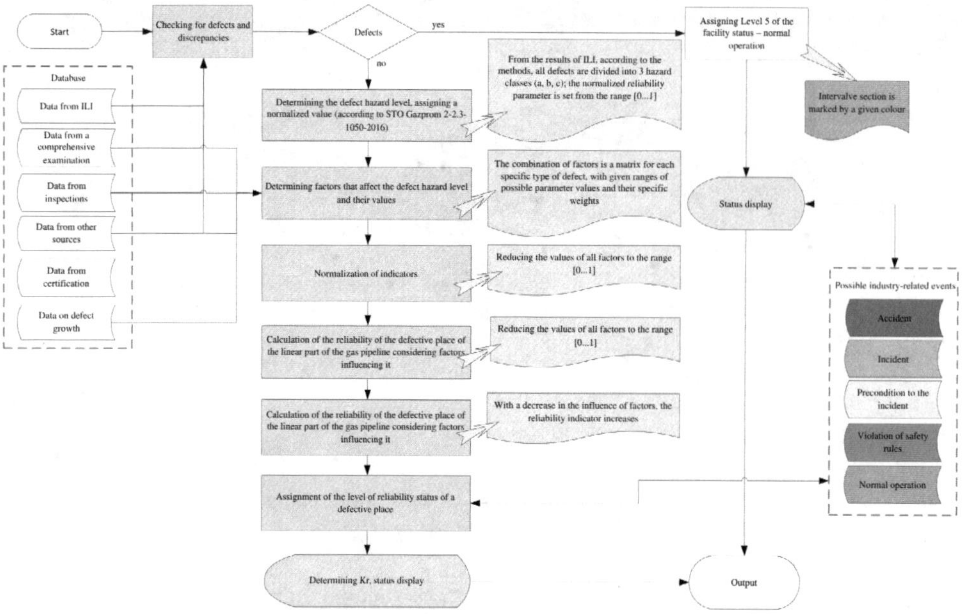

Figure 5. Algorithm for monitoring the reliability of a defective part of a pipeline.

Let us consider an example of the implementation of the concept of monitoring reliability and safety in a real section of a gas pipeline over 30 years old. Conducting diagnostic examinations of the linear part of the main gas pipeline, in particular, in-line flaw detection, provides information on the presence of various kinds of defects in the pipe material, welded joints, as well as geometric shapes and sizes. At the same time, a significant role in the reliability of the information received is played by the data of the questionnaires filled out by the employees of the operator for the entire diagnosed area. It is worth noting that the questionnaires include information on the characteristics of the pipes (diameter, wall thickness, material), category sections, repairs, but they do not contain information and factors that occur in each individual place in the pipeline and have a certain effect on the identified defects. The situation is complicated by the fact that in the area of the pipeline there are long sections with the presence of surrounding objects in hazardous areas.

At present, at the enterprise, most of the information about the technical condition (the results of the diagnostic tests, examinations, etc.) has to be processed by specialists “manually” without the possibility of automatic systematization and processing, as a result of which there is no real-time clear picture of the condition of the hazardous production facility. Therefore, in order to comply with industrial safety requirements, setting priorities in preserving the lives of people and their property requires a completely new approach to monitoring the technical condition, reliability and safety realized through real-time monitoring. To ensure reliable and uninterrupted operation of the linear part of the main gas pipeline, it is necessary to have a system that allows real-time monitoring of the technical condition of the facility using all available information, as well as predicting the state of the system over time.

Based on the foregoing, it is advisable to consider each identified defect from the point of view of its reliability and safety, carry out the ranking of the defect, and also determine its effect on the considered intervalve section and the gas pipeline as a whole. It is proposed to monitor the reliability of each specific defect according to the developed algorithm (Figure 5).

Let us consider in more detail each element of the proposed algorithm.

Initially, a database is needed to carry out any control, monitoring, determination of any indicators. In our case, the database is a collection of information stored in the monitoring system, constantly updated and edited, including:

- data from the results of in-line inspections (defects, their types, characteristics, etc.);

- data from a comprehensive examination (defects, isolation status, state of the ECP, etc.);
- data from diagnostic examinations of underwater crossings (defects, violations of regulatory requirements);
- data from inspections (violation of the design depth, erosion, isolation status, condition of the technical corridor, etc.);
- certification (data on the diameter of the pipeline, wall thickness, categorization of the site, the presence of third-party communications, intersections with roads and railways, the presence of third-party objects in the minimum safe distance zone and information about the characteristics of these objects, etc.);
- data on factors affecting each specific type of defect (it is a matrix of factors affecting each specific type of defect with relative shares of each factor and functions for normalizing the values of factors).

For the purposes of this work, the assumption is made that the pipeline is initially in good condition with reliability indicators corresponding to the "normal operation" range.

At the first stage of work, the defect is processed, and a certain level of reliability is assigned to it.

It is worth noting that in accordance with the requirements of STO Gazprom 2-2.3-1050-2016, all defects are divided into three reliability classes. Moreover, the distribution of defects occurs according to at least three methods, as well as the application of the expert assessment method.

Having determined the type of defect, as well as its reliability, according to the results of the examination, the system proceeds directly to a multivariate analysis of indicators that influence the development of the defect. The matrix of indicators is developed by experimental research and involves the introduction or removal of individual factors for each specific defect.

Since influencing factors belong to different areas and there are different units of measurement, an interpretation of the parameters is necessary. In the framework of this work, this procedure is implemented by normalizing the true values and bringing them to the 0 ... 1 interval according to the normalization formula (linear, sigmoid, probabilistic, logical, etc.), or by expert evaluation.

For example, in the general case, the linear normalization formula has the form:

$$y = \frac{(x - x_{min})(d_2 - d_1)}{x_{max} - x_{min}} + d_1 \tag{1}$$

where: x – the value to be normalized;

$[x_{max}, x_{min}]$ – the x range;

$[d_1, d_2]$ – the interval to which the x value will be reduced.

As an example, let us give an example of the normalization of some factors of the section safety, the higher the value of the normalized rating, the safer the section (Table 1).

Table 1. An example of the normalization of the section safety factors.

Factor	Estimated indicator	True value range		Actual value	Normalized indicator	Normalization method
		min	max			
Pressure	No possibility of sharp pressure increase	yes	no	yes	1.00	Logical
Pipe wall thickness	Relative decrease	0	0.7	0.2	0.29	linear, sigmoid, probabilistic
Stress-strain state of the pipeline	No critical deformation	yes	no	yes	1.00	Logical
Parameters of the electrochemical protection	Compliance with GOST requirements	no	yes	yes	1.00	Logical
Accident rate of the gas pipeline section	Absence of accidents over the previous 10 years	yes	no	no	1.00	Logical
Coefficient of safety pressure	Coefficient value	0	1.2	0.9	0.75	linear, sigmoid, probabilistic

Table 1. Ending.

Term of the external inspection	Term in years	0	50	1	0.02	linear, sigmoid, probabilistic
Insulation coating condition	Integral resistance, Ohm·m^2	0	10000	7000	0.70	linear, sigmoid, probabilistic
Safety by ECP	Compliance with GOST 51164 requirements	yes	no	yes	1.00	Logical
Ground water level	HC mark below the lower pipe generatrix	yes	no	no	0.00	Logical
Corrosive soil aggressiveness	Specific soil resistance	0	50	45	0.90	linear, sigmoid, probabilistic

An example of some windows of the I-ReS software for a real gas pipeline is shown in Figures 6, 7, 8 (the names of the sections were deleted for information security reasons). The software package provides a trend analysis function for characteristics, reliability and safety indicators, probabilistic analysis and ranking of objects for making real-time decisions on maintenance. The program is based on comprehensive certification and complete information about the reliability of the facility.

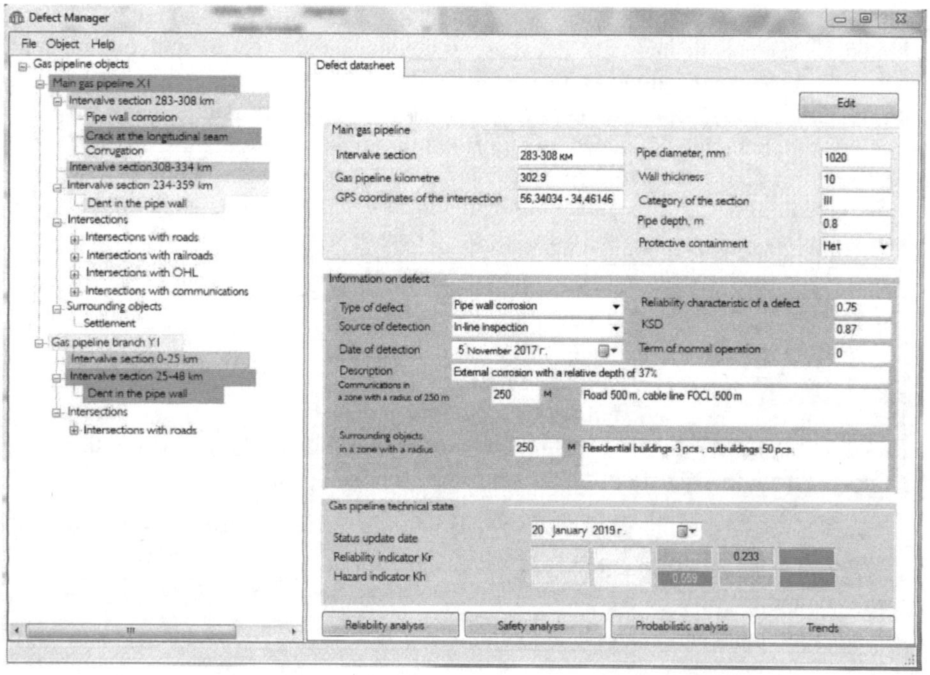

Figure 6. Information processing window (I-ReS).

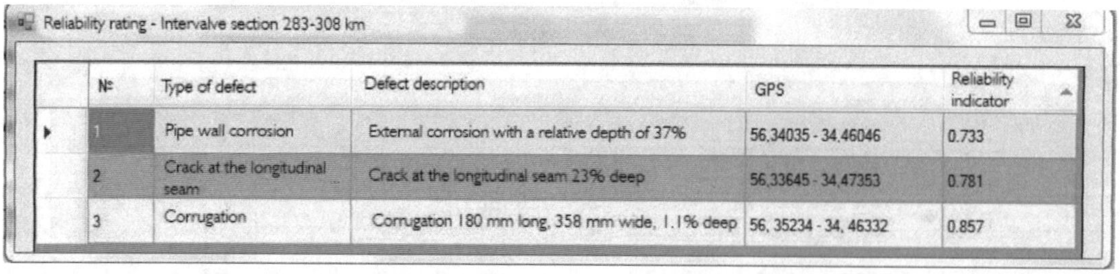

Figure 7. Rating of defects based on reliability assessment (I-ReS).

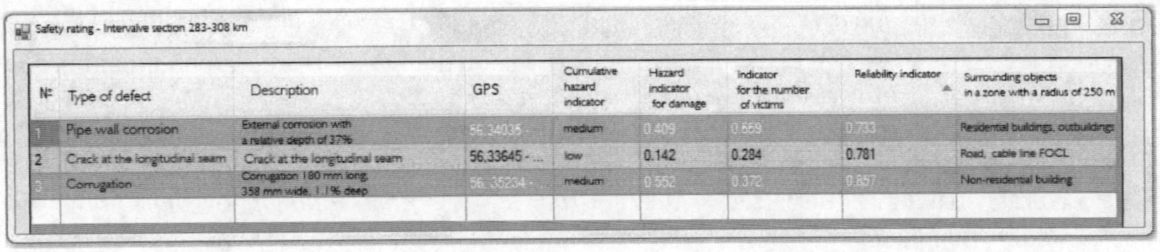

Safety rating - Intervalve section 283-308 km

№	Type of defect	Description	GPS	Cumulative hazard indicator	Hazard indicator for damage	Indicator for the number of victims	Reliability indicator	Surrounding objects in a zone with a radius of 250 m
1	Pipe wall corrosion	External corrosion with a relative depth of 37%	56.34035 - ...	medium	0.409	0.559	0.733	Residential buildings, outbuildings
2	Crack at the longitudinal seam	Crack at the longitudinal seam	56.33645 - ...	low	0.142	0.284	0.781	Road, cable line FOCL
3	Corrugation	Corrugation 180 mm long, 358 mm wide, 1.1% deep	56.35234 - ...	medium	0.552	0.372	0.857	Non-residential building

Figure 8. Rating of defects based on the results of hazard assessment.

3. Conclusion

The methodology of multifactor control of the technical condition of the main gas pipeline was developed as part of the project of an intelligent monitoring system of the Department of Transport of Hydrocarbon Resources, Industrial University of Tyumen. The proposed mathematical models, methods and algorithms using system analysis, probability theory, neural networks, coefficient models and software have been tested in calculating the reliability of Gazprom PJSC facilities, they allow for the correct intelligent monitoring of industry-related events on main gas pipeline sections, and implement a maintenance strategy for actual condition, reliability, safety.

References

[1] Annual report on the activities of the Federal Environmental, Industrial and Nuclear Supervision Service of Russia for 2007-2017 [Electronic resource]. – Availiable at: http://www.gosnadzor.ru/public/ annual_reports. Date of reference 05.09.2018

[2] Korolenok A M 1997 *Technological forecasting of major repairs of main gas pipelines* (Moscow: TsONiK GANG Press) 297

[3] Lisin I U, Ganaga S V, Korolenok A M and Kolotilov Yu V 2019 Logic-simulation modeling of the functioning of energy systems *Oil industry* **2** 94-8

[4] Sukharev M G 2010 *Forecasting methods* (Moscow: MaxPress) 176

[5] Polyakov V A, Shestakov R A and Komarov D N 2018 Integrated parametric method for detecting leaks and unauthorized taps on pipeline *Pipeline transport: theory and practice* **1** (65) 40-2

[6] Krapivsky E I and Pakhotin P A 2013 Substantiation of technologies for remote electromagnetic diagnosis of underwater crossings of oil and gas pipelines *Mining Informational and Analytical Bulletin (scientific and technical journal) (Gorny Informatsionno-Analiticheskiy Byulleten (nauchno-teknicheskii zhurnal)) (GIAB)* **5** 260-4

[7] Zemenkova M Yu 2018 Collaborative intellectual technological reliability real-time management of hydrocarbon transport systems CI-TREMS *Reliability and safety of main pipelines Proc. of the IX International Scientific and Technical Conf.* Ed by V K Lipsky 89-90

[8] Zemenkova M Yu 2017 *System analysis and technological monitoring of reliability and safety during transport and storage of hydrocarbons* (Tyumen: IUT) 270

[9] Zemenkova M Yu 2018 *Methodological support of expert monitoring systems for reliability indicators of hydrocarbon pipeline transport facilities* (Tyumen: IUT) 411

[10] Zemenkova M Yu 2019 *Methods to reduce technological and environmental risks in the transport and storage of hydrocarbons* (Tyumen: IUT) 410

[11] Malyshev A M 2016 Comparative analysis of measures to ensure industrial safety of main oil and gas pipelines *Industrial and construction technologies* **6** (8) 3

[12] Fedorova O B и Chizhevskaya E L 2014 Adapting intellectual property evaluation methods to the region brand evaluation *East J. of Scientific Research* DOI: 10.5829/idosi.mejsr.2014.19.1.12477

[13] Balikaeva M B, Chizhevskaya E L, Ya Grevtseva G, Kotlyarova I O и Volkova M A 2018 Innovative technologies as a means of the development of future engineers' professional

mobility abroad IOP *Conf. Series: Materials Science and Engineering* doi:10.1088/1757-899X/441/1/012007

[14] Gorelik J B, Shabarov A B and Sysoyev Yu S The dynamics of frozen ground melting in the influence zone of two wells *Earth's Cryosphere* **12** (1) 59

[15] Vlasenko V S, Slesarenko V V, Gulkov A N and Lapshin V D 2015 Preparation of arctic oil and gas condensate deposit formation fluid for storage and transport in the form of hydrocarbon hydrate-containing dispersed system *Proc. of the International Offshore and Polar Engineering Conf.* 1863

[16] Safety guide "Guidelines for the classification of industrial events in the field of industrial safety at hazardous production facilities of the oil and gas complex" Approved by Order of the Federal Environmental, Industrial and Nuclear Supervision Service of Russia **29**, January 24, 2018 - Availiable at: http://sudact.ru/law/prikaz-rostekhnadzora-ot-24012018-n-29-ob/rukovodstvo-po-bezopasnosti-metodicheskie-rekomendatsii/ (Date of reference: 1.04.2018)

[17] 1983 *Questions of the mathematical theory of reliability*. Ed by B V Gnedenko. (Moscow: Radio i svyaz) 376

[18] Pevzner L D 2013 *Theory of control systems* (St. Petersburg: Lan) 424

IOP Conf. Series: Materials Science and Engineering **663** (2019) 012005 doi:10.1088/1757-899X/663/1/012005

Analysis of the reliability of structures during operation of gas pipelines on piles in areas of heaving soils

B N Novik[1], S Yu Podorozhnikov[1,2] and M Yu Zemenkova[1]

[1] Industrial University of Tyumen, 38 Volodarskogo str., Tyumen, 625000, Russia
[2] sergey_urevih@mail.ru

Abstract. The routes of the main, field, and process gas pipelines in the Arctic regions of Siberia run through territories with difficult climatic conditions. Along the route, there are areas with heaving soils, which can be used as natural bases only with the use of special anti-heaving measures, since that they are characterized by high mobility, low bearing capacity, uneven settlement and increased compressibility. This raises concerns about a decrease in the bearing capacity of permafrost soils in the territories of Taimyr, Yamal and other regions. The paper discusses various methods of the above-ground laying of pipelines on heaving soils. A pipeline construction option ensuring safer operation is proposed.

1. Introduction

Fields in Western Siberia are located in an area with extremely difficult climatic conditions. Almost the entire territory of the region is characterized by the presence of permafrost soils, which thaw and freeze seasonally, leading to a loss of stability of the supporting structures of gas pipelines.

During the construction of new gas pipelines on such soils, there are many problems and tasks associated with the subsequent safe operation of the gas pipeline.

Frost heaving of the soil is its ability, within the seasonal freezing, under a certain combination of hydrothermal conditions, to increase in volume under the influence of crystallization forces of ice, accompanied by phase transformations of water contained in the soil and pulled additionally to ice crystals. The external manifestation of this property is the uneven elevation of the day surface of the soil through the formation of ice inclusions. [1-5]

There is one more definition of this term: cryogenic frost heaving of soils is the internal volumetric deformation of freezing moist soils, soft ground rocks and soils, which leads to an increase in their volume due to crystallization of water in them, as well as the formation of ice inclusions in the form of lenses, interlayers, polycrystals, etc. The external manifestation of this phenomenon is local uneven uplift of the surface layer of freezing soil, which is subsequently replaced by soil settlement during thawing. [6] Heaving soils, as a rule, are called soils, which, when frozen in conditions of natural occurrence, can increase in volume. [7]

According to the degree of frost heaving, soils are classified by the following criteria:

- particle-size composition;
- natural water content;
- depth of groundwater;
- estimated depth of soil freezing.

IOP Conf. Series: Materials Science and Engineering **663** (2019) 012005 doi:10.1088/1757-899X/663/1/012005

Soils according to this classification are divided into five varieties:

- heavily heaving;
- medium heaving;
- weakly heaving;
- conditionally not heaving;
- not heaving.

Loams, pulverescent sandy loams and pulverescent clays of plastic consistency at a groundwater level in the zone of seasonal freezing or below the standard freezing depth of no more than 0,5 m in sandy loams, and in clay and loams no more than 1 m are among the most heaving frosty soils.

Medium heaving soils include sandy loams, pulverescent sands, clays and loams with natural moisture that exceeds the consistency index of 0,5, with a groundwater level that exceeds the standard freezing depth by no more than 0,6 m in pulverescent sands, in sandy loams - not more than 1 m, in loams - not more than 1,5 m, and in clays - not more than 2 m, in terms of frost heaving.

The group of weakly heaving soils includes pulverescent and fine-grained sands, loams, sandy loams and clays with a stiff texture, and in addition, coarse-grained soils with a pulverescent-clay aggregate at a groundwater level exceeding the freezing depth: in fine-grained and pulverescent sands - no more than 1 m, in sandy loams - no more than 1,5 m, in loams no more than 2 m, in loams (plasticity more than 0,12) - no more than 2,5 m and in clays - no more than 3 m.

2. Materials and methods

The forces of heaving are due to:

- crystallization pressure of water contained in the soil, as well as migration moisture;
- disjoining by films of unfrozen water;
- shrinkage of adjacent soil layers.

Values of normal frost heaving forces, in addition to external pressure and soil properties, depend on:

- soil moisture conductivity;
- compressibility of underlying soil layers. [8, 9]

Values of normal heaving stresses of the soil surface are different from the stresses that develop in the soil mass near the foundations, where these stresses, among other factors, depend on the rigidity of the structures that absorb the frost heaving forces. [10, 11]

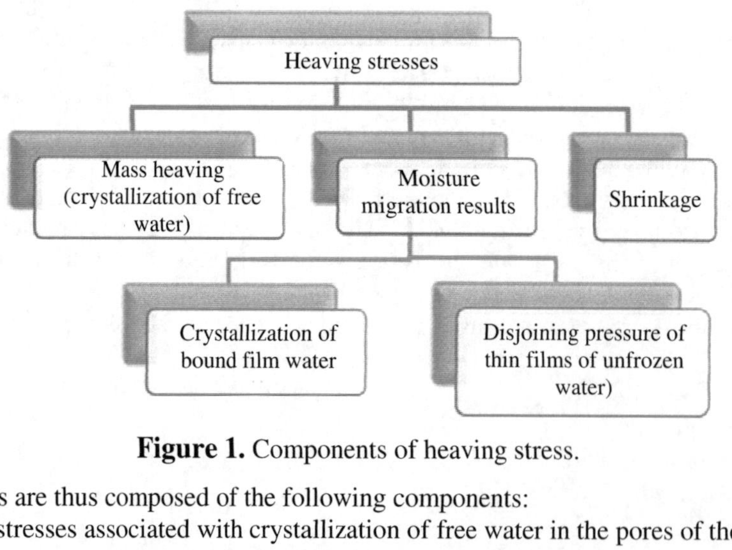

Figure 1. Components of heaving stress.

Heaving stresses are thus composed of the following components:

- mass heaving stresses associated with crystallization of free water in the pores of the soil;

- heaving caused by migratory ice accumulation due to the crystallization of migrating moisture;
- disjoining pressure of non-freezing water films manifested as a result of moisture migration;
- soil shrinkage stresses.

The components of the heaving stresses are schematically shown in Figure 1.

On the one hand, the forces of frost heaving act as a function of the porosity of the soil, and on the other hand, they depend on excess water content, which leads to the appearance of ice, in volume exceeding the pore volume of the soil. [12]

Depending on the geological conditions, the structure of foundations, the features of the "base-pile" structure for supports with thermal stabilization of the soil, various versions of vertical soil thermal stabilizers are used.

Vertical thermal stabilizers can be most widely used to lower the temperature of heaving soils to negative temperatures to increase the bearing capacity of pile supports and prevent buckling of piles.

The installation of vertical thermal stabilizers ensures the formation of a frozen mass of soil near the piles, which ensures the attenuation of vertical movements (Figure 5).

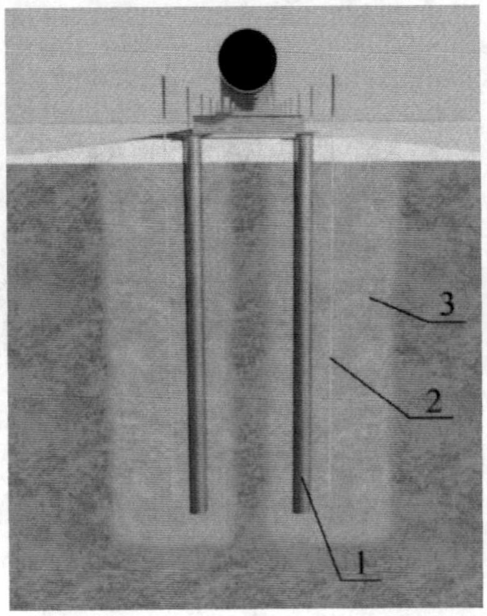

Figure 2. Pile temperature stabilization by means of vertical thermal stabilizers: 1 - pile; 2 - vertical thermal stabilizer; 3 - frozen massif - ice-soil cylinder.

An example of the operation of a gas pipeline in the Arctic regions of Siberia was taken to assess the situation. Permafrost soils are represented by sands with sandy loam interlayers and lenses with pebbles and gravel. The cryogenic texture is massive with small inclusions of ice. The thickness of frozen rocks is from 300 to 500 m, the total water content of sands is at least 30%. Seasonally thawing soils are mainly represented by fine-grained sands, sandy loams and lenses of torn sand. In places, there are thick deposits of peat bogs (in low places) and clay (Table 1).

Frost heaving of tundra soils, together with excess water content, very often leads to a skew of the support crossbar on an operating gas pipeline (Figure 3). The stress-strain states of the pipeline are uneven along the length and with varying deformation values.

Table 1. Characteristics of seasonally thawing soils [14]

Seasonally thawing soil composition	Soil characteristics	Water content, %	Average annual temperatures at the bottom of the layer, °C	Porosity coefficient, e	Liquidity index I_L
Fine-grained, pulverescent, water-saturated sands	Fine-grained, pulverescent, yellow-gray sands; hard frozen, massive cryogenic texture, rarely interbeds of ice up to 0.1 m thick. When thawing, sand becomes flowing	Wc=22-28%	At a depth of 10 m – minus 3.0-4.0 °C. Maximum thawing depth 1.5-2.0 m	0.5	> 1.00

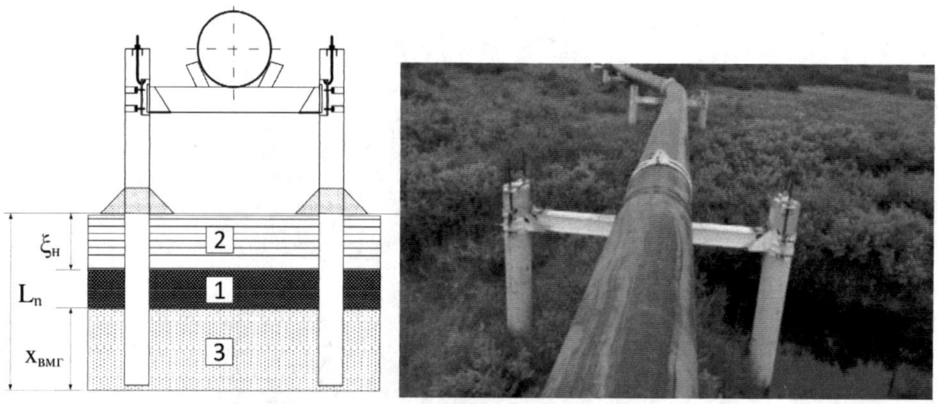

Figure 3. Loaded pile in the freezing soil:

1-thawed soil, 2 - frozen soil, 3 - permafrost soil, $\xi_{н}$ - boundary depth of soil freezing around the pile, X_{perm} – thickness of the permafrost layer, L_{pile} – deepening of the pile, P – support crossbar

Calculation of frost heaving forces for standard gas pipeline supports on heaving soils.

Data for calculation: soil temperature *t= -10°C,* with a freezing soil layer of 1.5 m, the thickness of thawed soil is 0.5 m (based on their maximum thawing depth), the length of the underground part of the pile L_{pile} =4 m, $x_{\text{perm}} = 2.0$ m.

Suppose that the heaving force is maximum and equal to the freezing force (in the freezing layer of thawed soil). Then, the motionlessness condition of piles has the form:

$$F_{frHeav} \leq F_{adhThaw} + F_{frPerm} + N \quad (1)$$

where F_{frHeav} – freezing force (assumed heaving force) of piles with frozen soil, kN; F_{adhThaw} – adhesion force of piles with thawed soil, which keeps the pile from buckling, kN; F_{frPerm} – freezing force of piles with permafrost soil, kN; N –sum of the own weight of the pile and the load on it, kN.

$$F_{frHeav} = (0{,}7 \times R_{af}) \times \xi_{н} \times U \quad (2)$$

$$F_{\text{adhThaw}} = f \times (L_{\text{pile}} - x_{\text{perm}} - \xi_{н}) \times U \quad (3)$$

$$F_{frPerm} = (0{,}7 \times R_{af}) \times x_{perm} \times U \quad (4)$$

where R_{af} – specific strength on the surface of freezing, kPa (Table 2);

f – specific adhesion force of pile material with thawed soil, kPa;

U –circumference of the pile, m;

$(L_{\text{pile}} - x_{\text{perm}} - \xi_{\text{H}})$ - thickness of the layer of thawed soil located below the bottom of the freezing-thawing layer, m;

f - calculated resistance to shear of thawed soil along the pile surface, kPa (Table 3).

Table 2. Calculated resistance to shear on the freezing surface [15]

Soil	Calculated resistance R_{af}, kPa, at soil temperature, °C											
	-0.3	-0.5	-1	-1.5	-2	-2.5	-3	-3.5	-4	-6	-8	-10
Clay	40	60	100	130	150	180	200	230	250	300	340	380
Sand	50	80	130	160	200	230	260	290	330	380	440	500

The value of R_{af} should be multiplied by a factor γ_{af}, depending on the kind of freezing surface.
For hot rolled metal surfaces $\gamma_{af} = 0.7$

Table 3. Specific adhesive force of pile material with thawed soil [15]

Average depth of the soil layer, m	Calculated resistance on the lateral surface of driven and jacked piles and shell piles f, kPa								
	of medium density sands								
	large and medium-large size	small	pulverescent	-	-	-	-	-	-
	clay soils at a yield index I_L, equal to								
	≤0.2	0.3	0.4	0.5	0.6	0.7	0.8	0.9	**1.0**
1	35	23	15	12	8	4	4	3	**2**
2	42	30	21	17	12	7	5	4	**4**
3	48	35	25	20	14	8	7	6	**5**
4	53	38	27	22	16	9	8	7	**5**
5	56	40	29	24	17	10	8	7	**6**
6	58	42	31	25	18	10	8	7	**6**
8	62	44	33	26	19	10	8	7	**6**
10	65	46	34	27	19	10	8	7	**6**
15	72	51	38	28	20	11	8	7	**6**

When calculating, the following conditions should be considered:

1. When determining the calculated resistance of soils on the lateral surface of piles, soil layers should be divided into homogeneous layers with a thickness of not more than 2 m.
2. The values of the calculated resistance of dense sand on the lateral surface of piles should be increased by 30% compared with the values given in the table.
3. The calculated resistance of sandy loam and loam with a coefficient of porosity of 0.5 should be increased by 15% compared with the values given in the table for any values of the yield index.

From (1), (2), (3), (4) it follows:

$$\xi_{\text{H}} = \frac{f \times L_{\text{pile}} - f \times x_{\text{perm}} + 0{,}7 \times \text{R}_{\text{af}} \times x_{\text{perm}} + \frac{N}{U}}{0{,}7 \times \text{R}_{\text{af}} + f} \quad (5)$$

- critical freezing depth of each pile individually

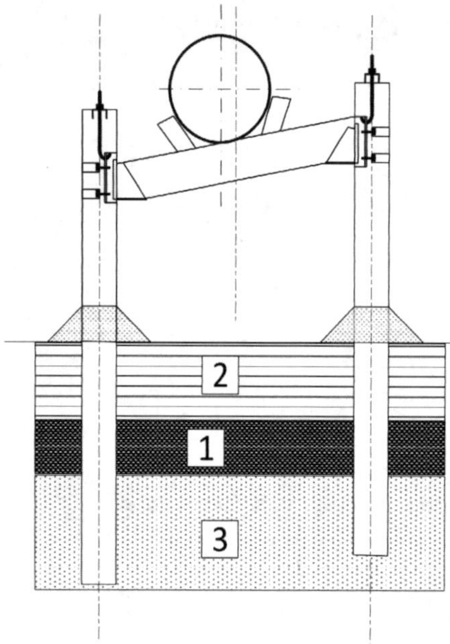

Figure 4. Stress-strain state of a gas pipeline during pile heaving

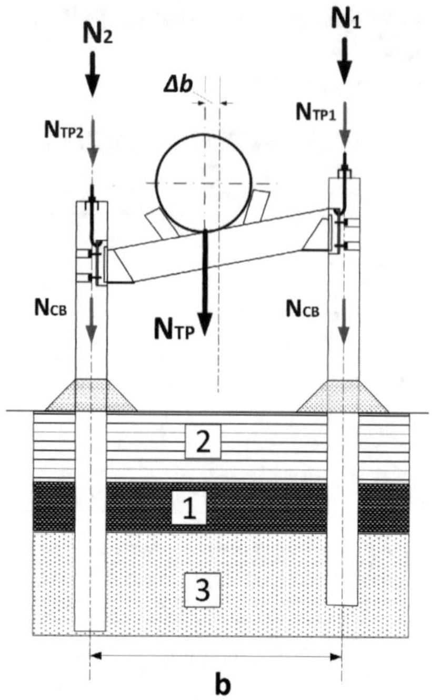

Figure 5. Scheme of stress-strain state

Humidity, structure, composition, density and temperature of the soil affect the specific forces R_{af} and f included in the formula (4). In turn, the critical freezing depth for each of the piles will also change (Figure 4) since even at small distances the soil parameters may differ (shrub, snowdrift, uneven solar radiation, uneven load distribution from the gas pipeline due to its transverse displacement along the crossbar).

Let us denote (Figure 5) the distance between the centers of the two piles on which the crossbar rests as $\boldsymbol{b}$, the displacement of the lower generatrix of the pipeline relative to the center of the crossbar as $\Delta\boldsymbol{b}$, the load on the entire support from the weight of the pipeline as $\boldsymbol{N_{PL}}$, the weight of the pile as $\boldsymbol{N_{PILE}}$, the load on the pile from which the pipeline has shifted as $\boldsymbol{N_1}$, the load on the pile towards which the pipeline has shifted as $\boldsymbol{N_2}$. The load on each of the piles is equal to the sum of the weight of the pile $\boldsymbol{N_{PILE}}$ and the load from the part of the weight of the pipeline $\boldsymbol{N_{PL}}$, falling on this pile ($\boldsymbol{N_{PL1}}$ and $\boldsymbol{N_{PL2}}$), $\boldsymbol{N_{PL}} = \boldsymbol{N_{PL1}} + \boldsymbol{N_{PL2}}$:

$$N_1 = N_{PL1} + N_{PILE} = N_{PL} \times (\frac{1}{2} - \frac{\Delta b}{b}) + N_{PILE} \tag{6}$$

$N_2 = N_{PL2} + N_{PILE} = N_{PL} \times (\frac{1}{2} + \frac{\Delta b}{b}) + N_{PILE}$.

The gas pipeline with a diameter of 720 mm, thickness s = 10 mm, shifted 0.2 m with a distance between the centers of the piles of 2.2 m.

$N_{PL} = q_m \times l_{between\ supports}$, (in our case$l_{between\ supports}$=15 m) (7)

from (7) $N_{PL} = 175\ {}^{kg}/_{m} \times 15\ \text{m} \times 9.8\ {}^{m}/_{s^2} = 25.73$ kN.

Piles are made of steel pipes with a diameter of 325 mm and a wall thickness of 10 mm. The length of the above-ground part of the pile L_{ag} = 3.5 m.

Hence, the weight of the pile from (7)

$N_{PILE} = 77{,}7\ {}^{кг}/_{м} \times 3{,}5\ м \times 9{,}8\ {}^{м}/_{с^2} = 2{,}7\ кН$.

From (6) $N_{PL1} = 10{,}5$ кН , $N_{PL2} = 15{,}2$ кН.

$N_1 = 10{,}5 + 2{,}7 = 13{,}2$ кН $N_2 = 15{,}2 + 2{,}7 = 17{,}9$ кН.
$f = 3 \times 1{,}15 = 3{,}45\ kPa$ (Table 3), $R_{af} = 500$ kPa (Table 2).
By (5)

$$\xi_{н1} = \frac{f \times L_{pile} - f \times x_{perm} + 0{,}7 \times \mathrm{R_{af}} \times x_{perm} + \frac{N_1}{U}}{0{,}7 \times \mathrm{R_{af}} + f} = \frac{3{,}45 \times 4{,}0 - 3{,}45 \times 2{,}0 + 0{,}7 \times 500 \times 2{,}0 + \frac{13{,}2}{3{,}14 \times 0{,}325}}{0{,}7 \times 500 + 3{,}45} = 2{,}04\ \mathrm{m}$$

$$\xi_{н2} = \frac{f \times L_{pile} - f \times x_{perm} + 0{,}7 \times \mathrm{R_{af}} \times x_{perm} + \frac{N_2}{U}}{0{,}7 \times \mathrm{R_{af}} + f} = \frac{3{,}45 \times 4{,}0 - 3{,}45 \times 2{,}0 + 0{,}7 \times 500 \times 2{,}0 + \frac{17{,}9}{3{,}14 \times 0{,}325}}{0{,}7 \times 500 + 3{,}45} = 2{,}05\ \mathrm{m}$$

Using the module of frost heaving of the soil $m_f = 3.5$ [5] with a particle size of 0.15 mm, we find the annual increment of the difference in the amount of buckling of piles:

$$\Delta = m_f\,(|\xi_{н1} - \xi_{н2}|) = 3{,}5 \times 0{,}01 = 0{,}035\ \text{м}\ [6] \tag{8}$$

It follows from this that for:
$\xi_{н1} = \xi_{crit} = 2.04$ m and $\xi_{н2} = \xi_{crit} = 2{,}05$ м over 1 year, the difference in the height of the supports can reach 3.5 cm; it was assumed that the heaving force is maximum and equal to the freezing force (in the freezing layer of thawed soil).

According to observations during seasonal thawing and freezing in different areas, the data on the actual buckling of piles of gas pipeline supports differ from these calculations, which confirms the heterogeneity of the lithological composition, dispersion, water content and strength of the soil. In many areas, there was an increased depth of drilling wells for piles - up to 6 meters. It does not always provide a positive result.

Figure 6. Structural defect

It is also necessary to take into account the moments of forces during the freely movable displacement of the pipeline in compensatory, longitudinally movable, especially high-altitude sections. These moments (Figure 6), together with the thawing and freezing of seasonally thawing soil, lead to structural defects in the supports.

Recently, there has been a steady increase in climate and permafrost soils temperature, which means there are well-founded fears about a decrease in the bearing capacity of permafrost soils in the territories of Taimyr, Yamal and other regions. [19]

At the moment, supports with soil thermal stabilization (STS) are very efficient in operating on permafrost soils (Figure 5). [20,21]

In comparison with STS, soil modules (Figures 7, 8) [21, 22] do not have sufficient load-bearing capacity to ensure safer operation of the above-ground gas pipeline. The essence of the method of laying from soil modules is that a canvas made of Dornit synthetic non-woven material is laid on the soil surface, and soil modules of the GP-1500 type are installed on it and filled with soil. The soil is compacted, after which the height-adjustable support of the conductor is removed. A protective sand layer with a thickness of 50-100 mm is poured onto the module, on which, in turn, a thermally insulated pipeline is laid. The soil module is a cellular structure filled with local, including icy and silty clay soils. Soil modules are made of the estimated width from tapes of technical fabric with various cell sizes. The necessary bearing capacity of the soil base when taking into account static and dynamic loads (pipeline weight, snow cover and thermal insulation) is achieved by changing the size of the cells.

Figure 7. Supports with soil thermal stabilization

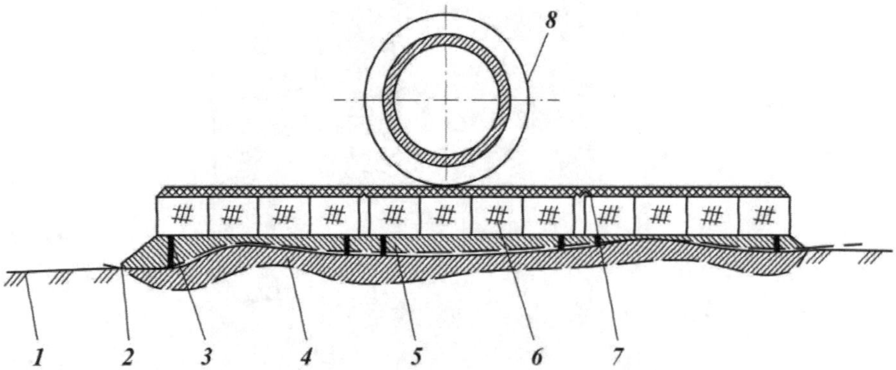

Figure 8. Scheme of laying a gas pipeline on soil modules: 1 - surface of weakly bearing soil; 2 - synthetic non-woven canvas material; 3 - conductor support, adjustable in height; 4 - near-surface compacted soil layer; 5 - soil compacted between non-woven synthetic material and the bottom end of the soil module; 6 - soil module; 7 - protective layer of sand; 8 - pipeline

3. Conclusion

There is a need for a qualitative review of permafrost construction methods. When choosing the optimal support structure, it is necessary to be guided by its versatility for different permafrost sections, with different soil characteristics. The relatively small amount of construction work, which, as a rule, on its own destroying the tundra cover of the Arctic, is one of the many and important reasons for premature thawing of soils, will also contribute to lower costs.

References

[1] Kiselev M F 2002 *Recommendations for the design of bases and foundations on heaving soils* (Moscow: Research Institute of Bases and Underground Structures of the USSR Gosstroy) 250

[2] Snezhko O V 1997 Deformations of culverts and their causes in the Western section of the Amur Railway *Proceedings of the Far Eastern State Transport University* (Habarovsk) Issue 11 134

[3] Belyaev K I 1983 Prevention of heaving soil deformations *Construction industry* No. 4

[4] Tsytovich N A and Sumgin M I 1987 *Fundamentals of frozen soil mechanics* (Moscow: USSR Academy of Sciences Press) 207

[5] Kiselev M F 1971 *Measures against the deformation of buildings and structures from the action of the forces of frosty bulging foundations* (Moscow: Stroyizdat) 123

[6] Dalmatov B I and Lastochkin V S 1978 *Installation of gas pipelines in heaving soils* (Leningrad: Nedra) 200

[7] *Handbook on the design of bases and foundations on heaving soils* 1979 (Moscow: Stroyizdat) 183

[8] Kudryavtsev S A 2003 Influence of migration moisture on the process of frost heaving of seasonally freezing soils [Electronic resource] *Internet magazine "Reconstruction of cities and geotechnical construction"* **7** 233-40 (URL: http://www.georec.spb.ru/joumals/07/files/pdf/0407018.pdf (date of reference: 17.01.2019))

[9] Puskov V I 1993 Power effects of frost heaving of soils on the foundations of structures and methods for their calculation *Siberian transport university* (Novosibirsk; Moscow) 40

[10] Medvedev O P 1992 *The work of pyramidal-prismatic piles in dusty clay soils* (Krasnoyarsk; Perm) 20

[11] Michalowski R A 1993 Constitutive model of saturated soils for frost heave simulations *Gold Regions Science and Technology* **22** 47-63

[12] Nevzorov A L 2000 *Foundations on seasonally freezing soils* (Moscow: ASV Publishing House) 152

[13] GOST 25100-2011. Soils. Classification. Moscow 2013

[14] SP 25.13330-2012. Bases and foundations on permafrost soils. Moscow 2013

[15] SP 24.13330.2011. Pile foundations. Moscow 2011

[16] PNIIIS Gosstroy of the USSR. Recommendations for accounting and prevention of deformations and forces of frost heaving of soils 1986 (Moscow: Moscow Printing House) **4**

[17] Marakhtanov V P 2013 Cryogenic deformations of the pile base of pipelines *Pipeline transport: theory and practice* **5** (39) 18-22

[18] Streletsky D A, Shiklomanov N I and Grebenets V I 2012 Change in the bearing capacity of frozen soils due to climate warming in the north of West Siberia *Earth's Cryosphere* (Novosibirsk: Academic Publishing House "Geo") **16** 22-32

[19] Lisin Yu V, Soshchenko A E, Surikov V I, Pavlov V V and Zotov M Yu 2014 Technical solutions for the methods of laying the Zapolyarye oil pipeline - Purpe OPS (Moscow: Pipeline Transport Institute, LLC) **1** 24-8

[20] Gamzaev R G and Kronik Z A 2016 Systems of soil thermal stabilization during construction in the permafrost zone *Materials of the fifth conference of Russian geocryologists* (Sumy: Universitetskaya kniga (University Book) Publishing and Trading House LLC) **1** 245-52

[21] Vasiliev G G, Lezhnev M A and Gainulin E N 2011 Laying pipelines on permafrost soils using

soil modules *Science and technology of pipeline transport of oil and gas* (Moscow: Pipeline Transport Institute, LLC) **3** 12-7

[22] Goryainov Yu A, Volodchenkova O Yu, Shamukaeva A A and Leonovich I A 2014 Promising technologies for the construction of a pipeline on weakly bearing soils *Proceedings of the National University of Oil and Gas "Gubkin University"* (Moscow: National University of Oil and Gas "Gubkin University") **4** 48-59

IOP Conf. Series: Materials Science and Engineering **663** (2019) 012006 doi:10.1088/1757-899X/663/1/012006

Neural network monitoring and predictive control of the reliability and safety of gas distribution networks using deep learning algorithms

M Yu Zemenkova[1,2]
[1] Industrial University of Tyumen, 38 Volodarskogo str., Tyumen, 625000, Russia
[2]muzemenkova@mail.com

Abstract. The work deals with the development of an algorithmic, mathematical complex based on neural networks of deep learning to assess and predict the reliability and safety of gas distribution equipment based on diagnostic data. The article describes the features of creating a predictive control system within the framework of the CI-TREMS complex. The work describes the algorithms and presents an example of the implementation of the system on one of the nodes of the gas distribution network.

1. Introduction

The relevance of the problem of forecasting undesirable events at gas supply facilities is confirmed by system-related accidents at facilities [1-4], especially at residential facilities, where annually the number of victims is measured in dozens.

Researchers at the Department of Hydrocarbon Resources Transport developed a complex of collaborative intellectual technological reliability and efficiency management of oil and gas systems (CI-TREMS) - a fundamentally new technology of intelligent instrumental neural network engineering control, forecasting and prevention of emergency situations, incidents, accidents, optimization and ensuring the effectiveness of technical solutions in the management of processes of industrial enterprises. [4,5,9-12]

The process of creating analytical and predictive models and methods of the system is associated with the following steps:

1) analysis of the flow scheme of the facility, parameters and characteristics, determination of production features of the project and control of parameters and modes;

2) creation of a diagnostic architecture that allows evaluating the technical state, reliability and safety of the facility and the reliability of technological processes;

3) substantiation of the methodological support of the predictive control system;

4) software implementation of models and creation of a software package;

5) testing and updating the system at a specific production facility.

The CI-TREMS system is modular and multi-tasking, flexible and adaptive, based on the theory of neural network programming, the theory of reliability, cybernetics, system analysis, statistics and probability, and others.

The main objectives of the system are forecasting and preventing emergency situations.

For this, researchers at IUT have developed methodological support using the apparatus of the probability theory and statistics, graph models, neural network technologies and machine learning to evaluate the indicators of technical state, reliability and risk analysis of gas supply system facilities.

IOP Conf. Series: Materials Science and Engineering **663** (2019) 012006 doi:10.1088/1757-899X/663/1/012006

2. Problem

One of the important and relevant tasks of the operation of the industrial fuel and energy complex is the creation of a monitoring system for gas networks.

When implementing a production facility management system for real-time monitoring, prevention of incidents and accidents at gas supply facilities and increasing the efficiency of decision-making by personnel, it becomes possible to increase the safety of the system.

Operational experience allows us to identify the main causes of accidents at gas distribution networks and stations: deterioration of equipment; poor quality of maintenance and repair; pipe metal corrosion; defects of construction; external influences. Gas distribution networks themselves are a multi-level complex technical system. At each level, a complex of analytical studies of a technological and mathematical nature is required, as well as the creation of a diagnostic tool kit.

Figure 1. Explosion of the gas distribution network of a residential building [1,2]

Figure 2. Destruction of gas pipelines of various levels

3. Materials and methods

The system-related problem of ensuring the reliability and safety of transport and storage of hydrocarbons in real-time is very time-consuming and multifaceted, requires the use of various methods and models.

Let us consider an example of a predictive system based on deep learning neural networks.

The advantages of using a deep learning neural network include the following:

- ability to handle significant amounts of data;
- ability to learn automatically according to the operational information provided;
- automatic detection of statistically significant factors and their combinations and elimination

The main reasons for reducing the reliability of GDS is the presence of various operational defects, they are varied and operational services conduct extensive work to prevent and eliminate them. The most common defects detected are gas leaks (Figure 3) at the nodes. This defect manifests itself as a result of leaks in threaded and flange joints and worn out mating surfaces of insignificant factors.

Leakage-related defects pose a direct threat to human life and health, and increase the risk of accidents and incidents at hazardous production facilities. One of the main causes of emergencies, as well as the decommissioning of individual sections or entire piping systems, is corrosion. Identified foci of corrosion in the technological cuts of the insulation also provide information on the status of the technological pipelines of the gas distribution system.

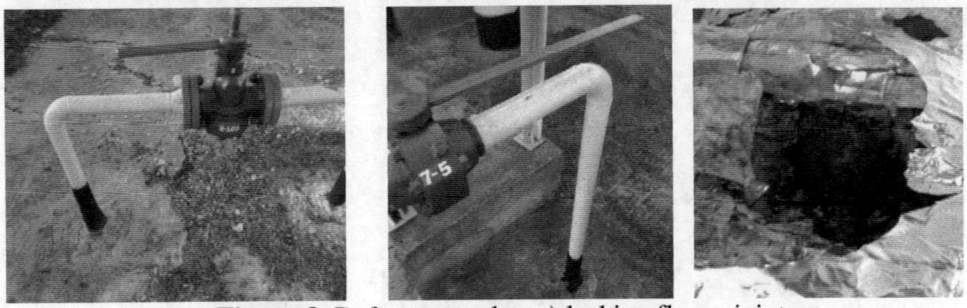

Figure 3. Defect examples: a) leaking flange joint; b) nonstandard element; c) foci of pipeline corrosion

Table 1 – Factors determining the technical state of the gas distribution network (example)

No. of factor	Name of factor (presence of fact, process)	Instrumental diagnostic method; normative document
1.	Corrosion defects	Diagnosis [16]
2.	Stress-strain state level	Diagnosis [16]
3.	Aging of the metal of the pipe, leading to a decrease in strength properties and embrittlement of the metal	Diagnosis [16,17,18]
4.	Weld defects	Diagnosis [16,19]
5.	Loss of tightness (including gas leakage)	Diagnosis [16], monitoring
6.	Electrochemical protection mode and service life	Diagnosis [16], monitoring
7.	Poor condition of the insulation coating	Diagnosis [16]
8.	Electrical contact with the supporting structures	Diagnosis [16]
9.	Use of non-standard elements	Diagnosis [16]
10.	No contact between pipes and supports	Diagnosis [16]
11.	Thinning elements (tee)	Diagnosis, ultrasound monitoring [16]
12.	Thinning elements (branch)	Diagnosis, ultrasound monitoring [16]
13.	Thinning elements (coil)	Diagnosis, ultrasound monitoring [16]
14.	Mechanical damage	Diagnosis [16]
15.	Change of geometrical sizes	Diagnosis [16], visual and measuring control
16.	Vibration	Diagnosis [20], visual and measuring control

An analysis of production experience and regulatory documentation made it possible to determine a list of groups of factors that determine the reliability of gas distribution stations.

As an example, a 16-factor model of a neural network for implementation in the Python environment is taken as sufficient and necessary for a high-quality diagnosis of the node. Data processing is carried out by three neural networks: the first determines the fact of the incident, the second - the fact of the accident, the third allows you to assess the condition of the object by five levels of danger.

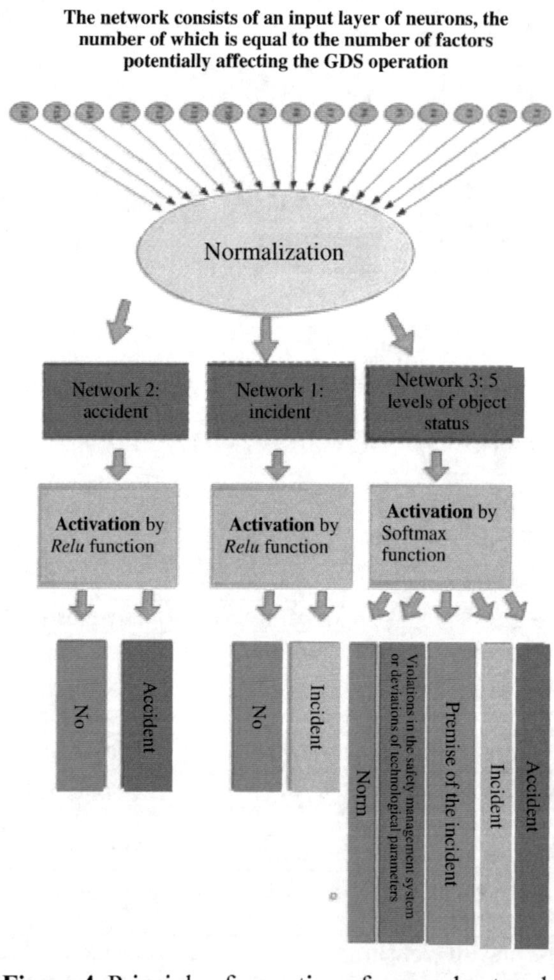

Figure 4. Principle of operation of a neural network

IOP Conf. Series: Materials Science and Engineering **663** (2019) 012006 doi:10.1088/1757-899X/663/1/012006

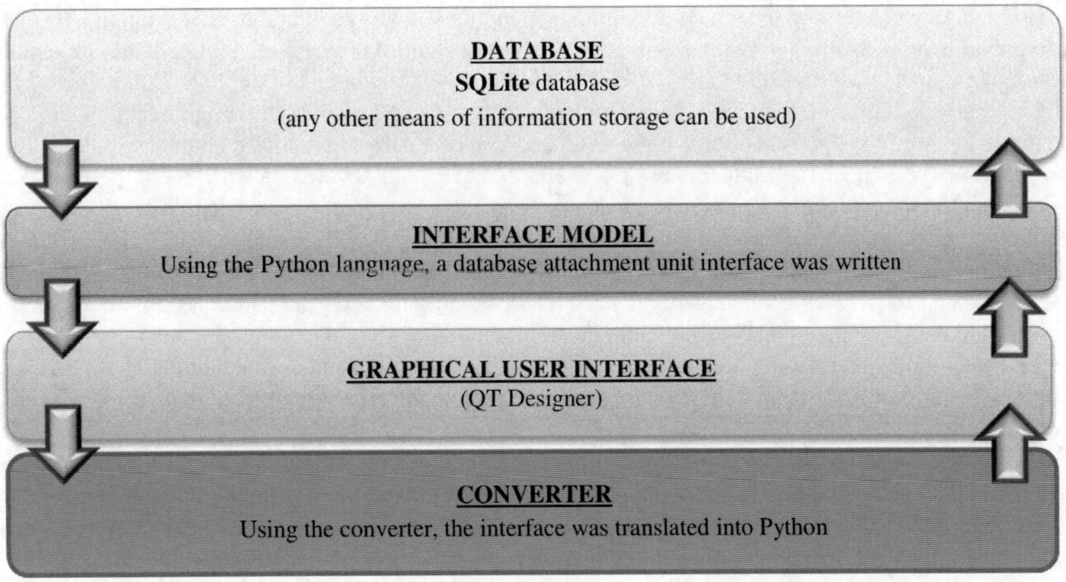

Figure 5. Basic algorithm for the software implementation of a deep neural network

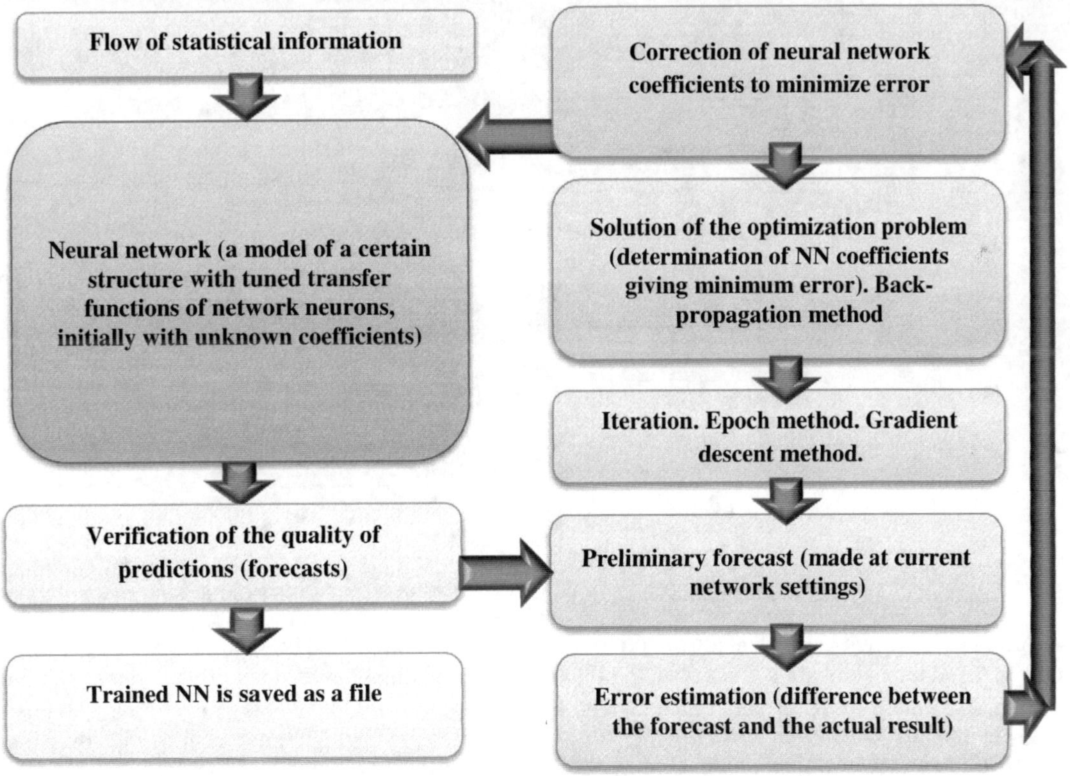

Figure 6. Neural network learning algorithm

As an example, based on the analysis of statistics on breakdowns and accidents, it is enough to single out several levels - from X - normal operation to 1 - accident, or binary - 2 levels of possible states (accident and incident), since they are associated with interruptions in GDS operation. It is assumed that GDS operation is controlled by at least 16 groups of factors (hereinafter in the program code - Factor 1, Factor 2 ... Factor 16. Their number is not crucial and can be easily adjusted. The database can be. Each combination (vector) of the input factors can be associated with several hazard levels: If there are 5 states, then the output layer will have 5 neurons (activated by the softmax function). We will create two three-layer models (one for breakdowns and one for accidents) based on model data. The number of neurons in the first layer is equal to the number of factors (16). The number of neurons of the second layer is taken from 32 to 64 to take into account possible pair and triple interactions of factors. A prerequisite is the import of TensorFlow and the Keras module.

To monitor the learning process, the model displays the minimum information on the learning progress to the console, but often receiving more detailed information in graphical form would be preferable.

To do this, the fit model training method provides a series of information that can be analyzed. We will create the corresponding function and in the future, we will use it on trained models. The function displays the results of training the model in terms of accuracy and error (the first and second window of the graph, respectively) in the form of a graph. According to the presented error graphs, it is possible to determine the optimal number of epochs or the volume of tasks. In our case, the optimal indicator was 5-8 epochs of learning.

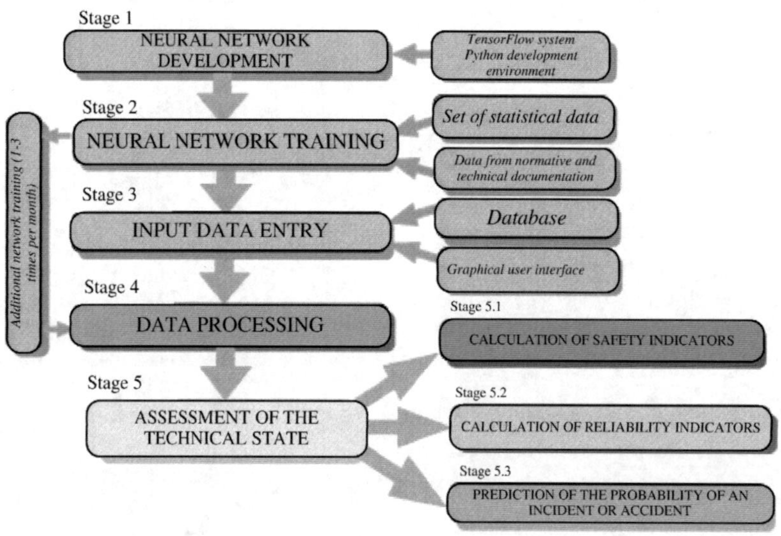

Figure 7. Program operation algorithm and data streams

A model is created and layers are added to it (input - 16 neurons, hidden - 64 and 16, output - 2). The input and hidden layers are activated by the relu function, the output layer by softmax. Next, the model is compiled. Since Keras provides several optimizers, the most effective one was chosen - stochastic gradient (default parameters, in accordance with the recommendation). The model is trained over seven epochs with a task volume of 32. According to the results, an object for monitoring is presented - history1, which is visualized. An example of visualization graphs for estimating prediction error is presented below.

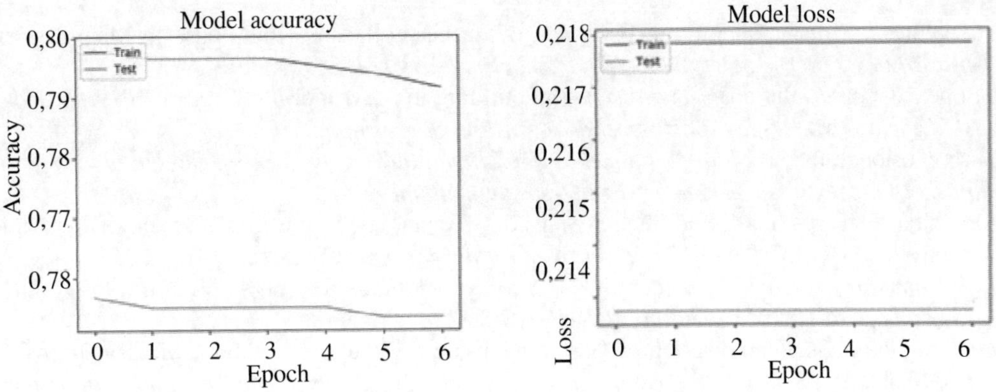

Figure 8. Model learning indicators

It can be seen that the accuracy of the trained binary model is very high and amounts to about 90%, which is a good indicator. Models are tested similarly for more states and epochs.

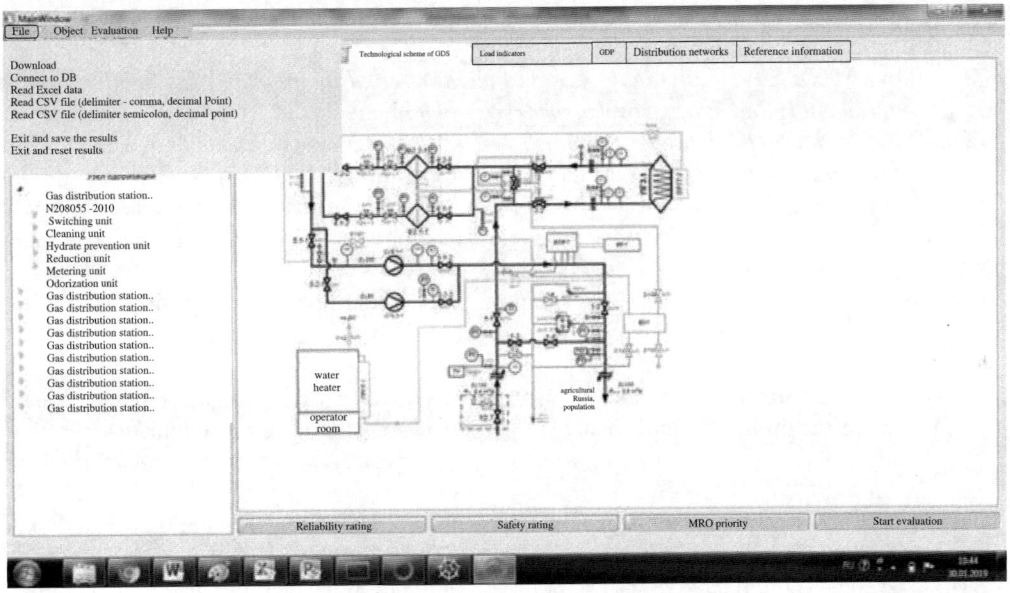

Figure 9. Program interface (by nodes)

4. Conclusion

The scientific novelty lies in the development of algorithms and provisions of the state assessment method based on a neural network with deep learning. The use of gas supply in the control system of a production facility for operational control, prevention of incidents and accidents on the basis of continuous analysis of the technical state will improve the safety and reliability of the systems. The use of stochastic models to create neural networks can improve the efficiency of predicting.

The practical value lies in the use of gas supply facilities in the control system of the production facility for operational control, prevention of incidents and accidents based on continuous analysis of the technical state. The presented software package was tested at the production facilities of PJSC Gazprom during the assessment and forecasting of the state and risk analysis of gas supply system facilities.

References

[1] Emergencies Ministry announced the threat of further collapse of the house in Magnitogorsk. *https://www.rbc.ru/society/31/12/2018/5c2a10cd9a794745269c533c*

[2] SK opened a case after a fire in a residential building in Stavropol *https://rg.ru/2019/05/06/reg-skfo/zheleznovodsk-delo-posle-pozhara-v-zhilom-dome.html*

[3] Gas explosion in Izhevsk *http://izhlife.ru/criminality/83097-vzryv-gaza-v-izhevske-obvinyaemyy-poshel-na-prestuplenie-iz-nepriyazni-k-sosedyam.html*

[4] Kurushina V, Kurushina E and Shabarov A 2016 Technology Audit: Assessment of Innovative Portfolio MATEC *Web of Conferences XV International Conference "Topical Problems of Architecture, Civil Engineering, Energy Efficiency and Ecology – 2016"* https://doi.org/10.1051/matecconf/20167307028

[5] Modeling the Phase Composition of Gas Condensate in Pipelines Dudin S M, Zemenkov Y D and Shabarov A B *IOP Conference Series: Materials Science and Engineering* https://iopscience.iop.org/article/10.1088/1757-899X/154/1/012010 doi:10.1088/issn.1757-899X

[6] Goodfellow I, Bengio Y, Courville A 2018 *Deep learning* (Moscow: DMK Press) 652

[7] Tariq Rashid 2018 *Make your own neural network* (SPb: LLC "Dialectics") 272

[8] Haykin Simon *Neural networks* 2018 (Moscow: LLC "I.D. Williams") 1104

[9] Zemenkova M, Zemenkov Y, Gladenko A and Podorozhnikov S 2016 Estimation of Emissions during Monitoring of Pipelines in the Dynamic Mode of Operation DOI 10.1051/matecconf/20168604053

[10] Zemenkova M Yu 2017 *System analysis and technological monitoring of reliability and safety during transport and storage of hydrocarbons* (Tyumen: IUT) 270

[11] Zemenkova M Yu *Methodological support of expert monitoring systems for reliability indicators of hydrocarbon pipeline transport facilities* 2018 (Tyumen: IUT) 411

[12] M Yu Zemenkova, M A Glukh, I F Shagbanov and D D Dubrovin 2019 Predictive control of the reliability of hydrocarbon distribution systems using neural networks with deep learning *Oil and gas terminal* ***17*** (Tyumen: IUT)

[13] Fedorova O B and Chizhevskaya E L. Adapting intellectual property evaluation methods to the region brand evaluation 2014 Middle *East Journal of Scientific Research* DOI: 10.5829/idosi.mejsr.2014.19.1.12477

[14] Balikaeva M B, Chizhevskaya E L, Ya Grevtseva G, Kotlyarova I O and Volkova M A 2018 Innovative technologies as a means of the development of future engineers' professional mobility abroad IOP *Conference Series: Materials Science and Engineering doi:10.1088/1757-899X/441/1/012007*

[15] Lessons learned from accidents. Federal Service for Ecological, Technological and Nuclear Supervision. URL: *http://www.gosnadzor.ru/industrial/oil/lessons/*

[16] STO Gazprom RD 1.10-098-2004 Methods for the comprehensive diagnosis of pipelines and piping of technological equipment of gas distribution stations of gas pipelines *https://meganorm.ru/Index2/1/4293841/4293841820.htm*

[17] GOST 22761-77. Metals and alloys. Brinell method of measuring hardness with portable hardness testers of static action URL: *http://docs.cntd.ru/document/gost-22761-77*

[18] GOST 9012-59 (ISO 410-82, ISO 6506-81) Metals. Brinell hardness measurement method URL: *http://docs.cntd.ru/document/1200004695*

[19] GOST R 55724-2013 Non-destructive testing. Welded joints. Ultrasound techniques URL: *http://docs.cntd.ru/document/1200107569*

[20] STO Gazprom 2-2.3-328-2009 Assessment of the technical condition and term of safe operation of the technological pipelines of compressor stations. URL: *https://meganorm.ru/Data1/58/58891/index.htm*

Calculation of thermodynamic characteristics of gas flows in concentrated fire vortices

A G Obukhov [1,2]

[1] Industrial University of Tyumen, 38 Volodarskogo str., Tyumen, 625000, Russia
[2] obuhovag@tyuiu.ru

Abstract. The article presents the results of numerical simulation of gas flows in free fire vortices. The fundamental possibility of physical modeling of the occurrence of concentrated fire vortices is presented in a series of experimental studies carried out at the Joint Institute for High Temperatures of the Russian Academy of Sciences. In the model of a compressible continuous medium for the complete system of Navier-Stokes equations, an initial-boundary-value problem is proposed that describes complex three-dimensional unsteady flows of viscous compressible heat-conducting gas in ascending swirling heat fluxes. Using explicit difference schemes and the proposed initial-boundary conditions, approximate solutions of the Navier-Stokes system of equations are constructed and the thermodynamic characteristics of three-dimensional unsteady gas flows are numerically determined. The calculation results showed that in the process of formation of fiery vortices, it is formally possible to distinguish several stages. The calculations of the thermodynamic characteristics of gas flows during heating of the underlying surface by several local sources showed that the selected mathematical model under the corresponding initial and boundary conditions allows numerical experiments to describe the arising of complex unsteady three-dimensional flows.

1. Introduction

The possibility of physical modeling of free fire vortex structures in a laboratory without special swirling devices was demonstrated in experimental work [1]. To generate concentrated vortices, a simple installation was used, in which 19 tablets of fuel were located on a metal surface. During the experiment, the tablets were set on fire, and during their combustion, generation of fire vortices was observed, the height of which significantly exceeded the height of the flame above each of the burning tablets. A sign of the appearance of fire vortex structures was the appearance of helical trajectories in heated particles of combustion products.

The first fire vortex structures begin to form in 3 minutes, and the last ones - in 12 minutes after setting fire to the tablets. Complete combustion of tablets occurs in 15–17 minutes. The formation of fire vortices occurs in the center of the fuel area and on its periphery. The number of vortex structures observed is up to 15 in one experiment. The lifetime of the vast majority of generated vortices is from 1 to 5 seconds. The highest height of the fire vortex structures reaches 0.7 meters, and their maximum diameter is 0.05 meters. The process of formation of a fire vortex is accompanied by a noticeable radial air flow from all directions to the base of the main central convective flow and the accompanying swirling of the flow relative to the vertical axis due to the presence of velocity and temperature gradients. The flow of combustion products in a fire vortex consists of a combination of relatively thin vortex filaments rotating and interacting with each other.

Thus, the possibility of physically simulating fire vortices without the use of mechanical swirling devices is shown. Generated fire vortices are laboratory analogues of fire tornadoes that occur during large forest fires, fires on large areas of agricultural land and in urban areas with dense buildings, in timber processing enterprises.

The numerical calculation of the thermodynamic characteristics of complex gas flows arising in the described experiments is the goal of this work. The presented results continue and develop the formulated direction of numerical simulation of unsteady three-dimensional flows of compressible viscous heat-conducting gas in swirling heat flows and fire vortices [2-10].

2. Mathematical model. Initial and boundary conditions

To describe complex flows of compressible viscous heat-conducting gas with dissipative properties of viscosity and thermal conductivity, we use the complete system of Navier-Stokes equations, which, in dimensionless variables and taking into account the action of gravity and Coriolis in vector form, has the form [11-14]:

$$\begin{cases} \rho_t + \boldsymbol{V}\cdot\nabla\rho + \rho\,\mathrm{div}\boldsymbol{V} = 0, \\ \boldsymbol{V}_t + (\boldsymbol{V}\cdot\nabla)\boldsymbol{V} + \frac{T}{\gamma\rho}\nabla\rho + \frac{1}{\gamma}\nabla T = \boldsymbol{g} - 2\boldsymbol{\Omega}\times\boldsymbol{V} + \frac{\mu_0}{\rho}\left[\frac{1}{4}\nabla(\mathrm{div}\boldsymbol{V}) + \frac{3}{4}\Delta\boldsymbol{V}\right], \\ T_t + \boldsymbol{V}\cdot\nabla T + (\gamma-1)T\,\mathrm{div}\boldsymbol{V} = \frac{\kappa_0}{\rho}\Delta T + \frac{\mu_0\gamma(\gamma-1)}{2\rho}\Big\{\Big[\big(u_x - v_y\big)^2 + \\ +(u_x - w_z)^2 + \big(v_y - w_z\big)^2\Big] + \frac{3}{2}\Big[\big(u_y + v_x\big)^2 + (u_z + w_x)^2 + \big(v_z + w_y\big)^2\Big]\Big\}, \end{cases} \tag{1}$$

where the values of dimensionless viscosity and thermal conductivity coefficients are: $\mu_0 = 0.001$, $\kappa_0 \approx 1.46\mu_0$. In system (1): t – time; x, y, z – Cartesian coordinates; ρ – gas density; $\boldsymbol{V} = (u, v, w)$ – gas velocity vector with projections on the corresponding Cartesian axes; T – gas temperature; $\boldsymbol{g} = (0, 0, -g)$ – gravity acceleration vector, and $g = const > 0$; $-2\boldsymbol{\Omega}\times\boldsymbol{V} = (av - bw, -au, bu)$ – Coriolis force acceleration vector, where $a = 2\Omega sin\psi$, $b = 2\Omega cos\psi$, $\Omega = |\boldsymbol{\Omega}|$; $\boldsymbol{\Omega}$ – Earth angular velocity vector; ψ – latitude of O – origin of the Cartesian coordinate system $Oxyz$, rotating with the Earth.

The initial conditions when describing flows of a compressible viscous heat-conducting gas at constant values of the coefficients of viscosity and thermal conductivity are functions that determine the exact solution [15] of system (1):

$$u = 0,\ v = 0,\ w = 0,\ T_0(z) = 1 - kz\,,$$

$$k = \frac{lx_{00}}{T_{00}},\ l = 0.0065\,\frac{K}{м},\ x_{00} = 1\,м\,,\ T_{00} = 288^oK, \tag{2}$$

$$\rho_0(z) = (1 - kz)^{\nu - 1},\quad \nu = \frac{\gamma g}{k} = const > 0. \tag{3}$$

The calculation area is a cube with unit lengths of sides $x_0 = y_0 = z_0 = 1$.

The boundary conditions and methods for calculating the gas-dynamic characteristics on the faces of the calculation area are presented as follows [16].

The density values on the lateral faces of the cube $x = 0$, $x = x_0$, $y = 0$, $y = y_0$ and on the top face $z = z_0$ are calculated from the continuity condition, that is, the density values on the lateral faces are calculated by linear interpolation of the values at the two nearest nodes within the calculation area.

The symmetry condition is set for the density on the bottom face $z = 0$. This means that the density values on the bottom face are calculated from the condition of their zero derivative normal to the face.

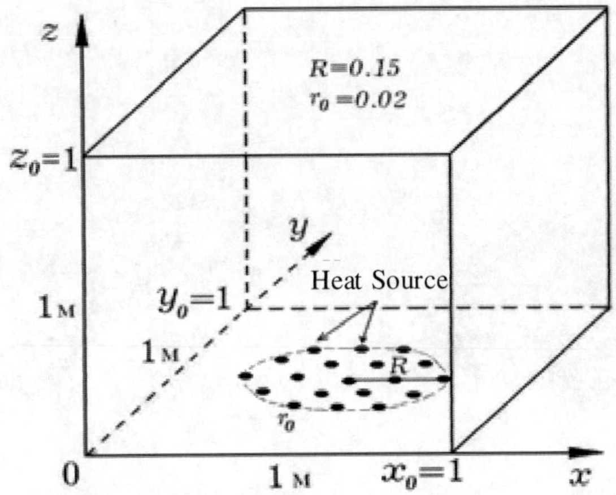

Figure 1. Calculation area in the form of a cube

On the lateral faces and the top face of the calculation area, continuity conditions are imposed on the values of all components of the velocity vectors, i.e., the values of all velocity components on the boundary are calculated by linear interpolation from the internal nodes of the calculation area.

On the bottom face, the conditions of non-leakage are set for the velocity components. The third velocity component is equal to zero$w|_{z=0}$=0, and the first and second components of the velocity vector are determined from the symmetry condition, that is, they are calculated from the condition that their derivatives are equal to zero along the normal to the face.

The temperature on the lateral faces and the top face is calculated from the condition of continuity.

The gradual heating of nineteen regions of the bottom face to a temperature of 300 °C is set by a functional dependence on time and coordinates

$$T(x,y,t) = 1 + T^{*}\left(1 - e^{-10t}\right)\sum_{l=1}^{19} e^{-\frac{(x-x_l)^2+(y-y_l)^2}{r_0^2}}, \quad (4)$$

where $T^* = 0.99$ – coefficient determining the addition to the temperature value taken as a unit scale value, $r_0 = 0.02$ – dimensionless values of heating radii, x_l, y_l, $l = 1 \div 19$ – coordinates of the centers of the heating sources of the bottom plane of the calculation area located in the center of the bottom face of the calculation area inside the circle of radius $R = 0.15$.

Thus, the boundary conditions set simulate the appearance of unsteady three-dimensional flows of a compressible gas with the properties of viscosity and thermal conductivity initiated by heating by nineteen heat sources. Boundary conditions suggest the possibility of free movement of gas through all faces of the calculation area, except for the bottom face. The scaled dimensional values of density, velocity, distance, and time in the calculations were taken as: $\rho_{00} = 1.29\ kg/m^3$, $u_{00} = 333\ m/s$, $x_{00} = 1$m, $t_{00} = \frac{x_{00}}{u_{00}} = 0.003\ s$. Discrete steps in three coordinates $\Delta x = \Delta y = 0.005$, $\Delta z = 0.05$, a time step $\Delta t = 0.001$.

3. Calculation results

Figures 2-13 show typical results of calculations of the thermodynamic characteristics of the air flow that appears when the bottom plane is heated for a height of 0.1 m at different points in time: 10; 20; 30; 40 seconds. The axes and are the numbers of nodes of the computational grid.

Figures 2-5 show the calculated temperature distributions at the indicated time points. It can be seen from the figures that at the beginning of heating, the temperature has elevated values only in the areas where the heat sources are located. Over time, areas with elevated temperatures expand horizontally and combine with each other.

IOP Conf. Series: Materials Science and Engineering **663** (2019) 012007 doi:10.1088/1757-899X/663/1/012007

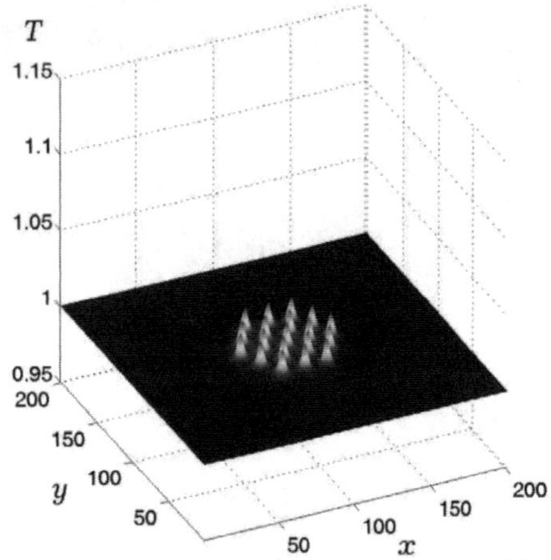

Figure 2. Temperature if $t_1 = 10$ s

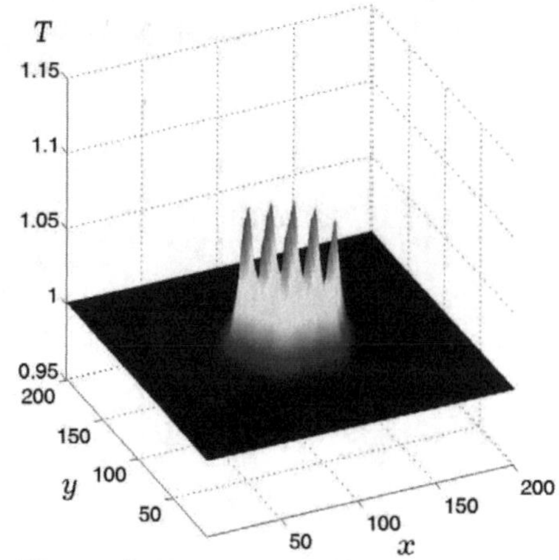

Figure 3. Temperature if $t_2 = 20$ s

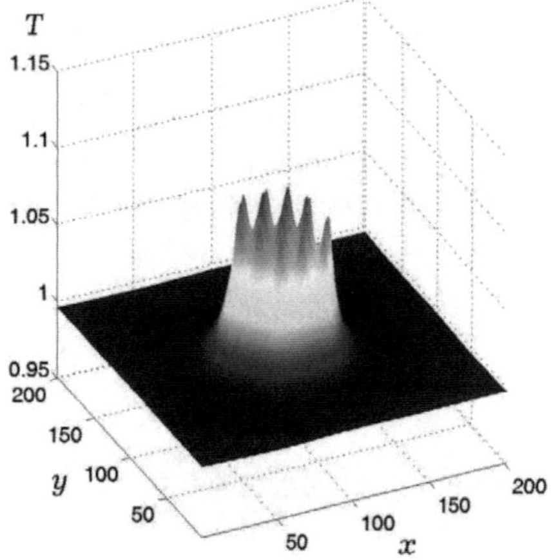

Figure 4. Temperature if $t_3 = 30$ s

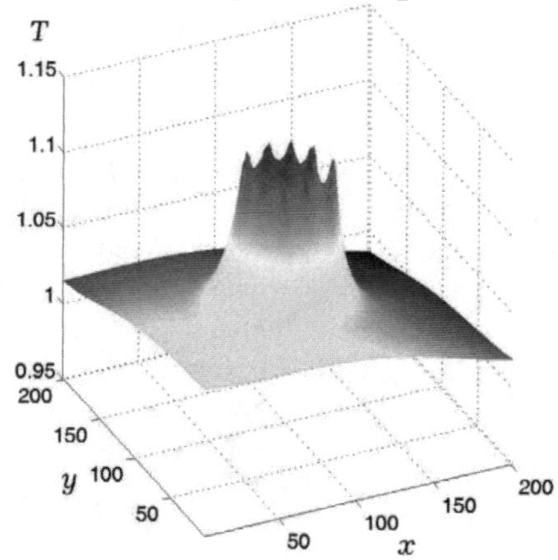

Figure 5. Temperature if $t_4 = 40$ s

The peripheral temperature values at the beginning of heating correspond to a unit scale value of 288°K (Figures 2, 3). Over time, their decrease is observed (Figure 4), and then an uneven increase (Figure 5).

Figures 6–9 show graphs of calculated gas densities for the same corresponding time instants.

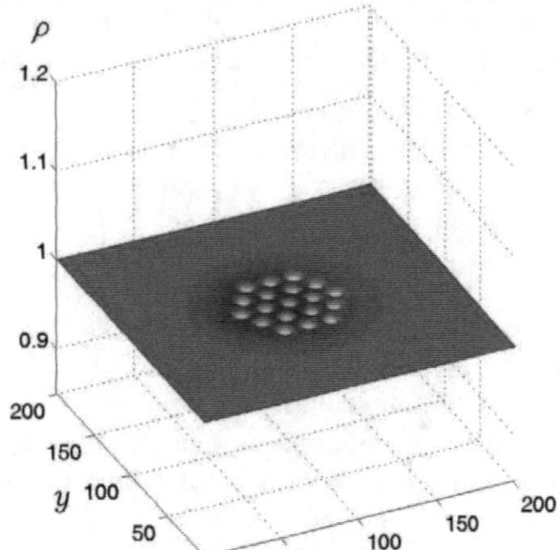

Figure 6. Density if $t_1 = 10$ s

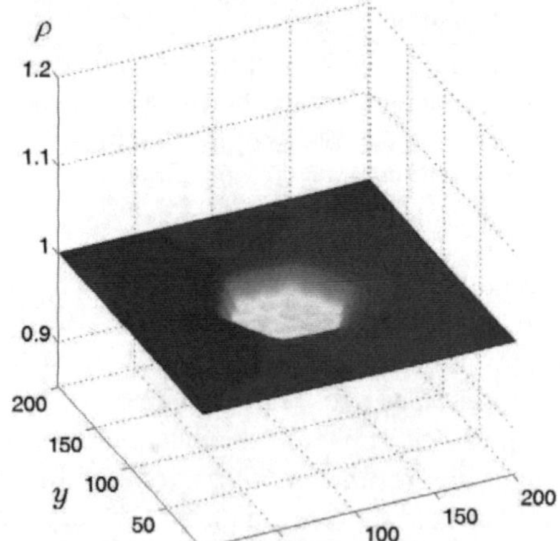

Figure 7. Density if $t_2 = 20$ s

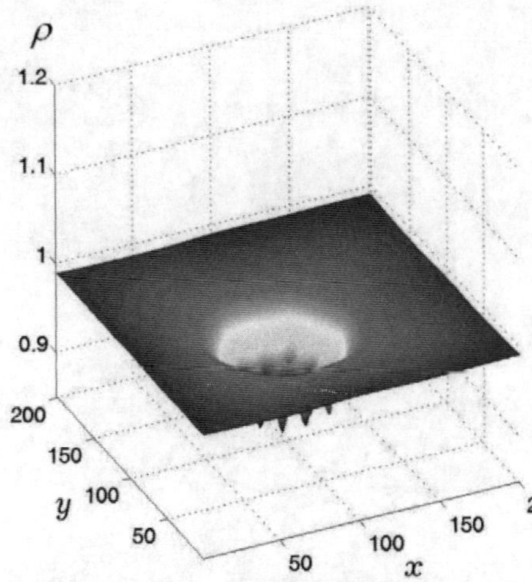

Figure 8. Density if $t_3 = 30$ s

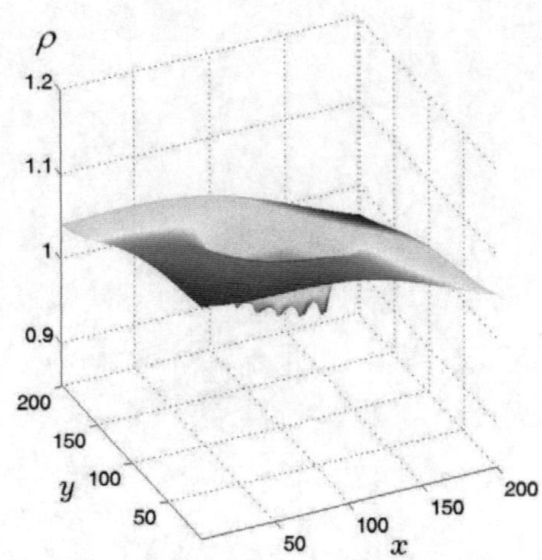

Figure 9. Density if $t_4 = 40$ s

It can be seen from the figures that, at the initial moments of time, the gas density has reduced values in the form of funnels localized in the heating regions (Figure 6). Regions with reduced density values quickly combine into one (Figure 7), and over time this region expands along the radii to the peripheral regions (Figure 8). The calculation results indicate that the peripheral density values, like the temperature, undergo uneven changes at the initial time moments. For example, the values of the gas density at time $t_4 = 40$s near the beginning of a rectangular coordinate system are higher than in the opposite part of the calculation area (Figure 9). However, this relationship between density values varies both in time and in coordinates.

Figures 10-13 show the results of pressure calculations for the above time moments.

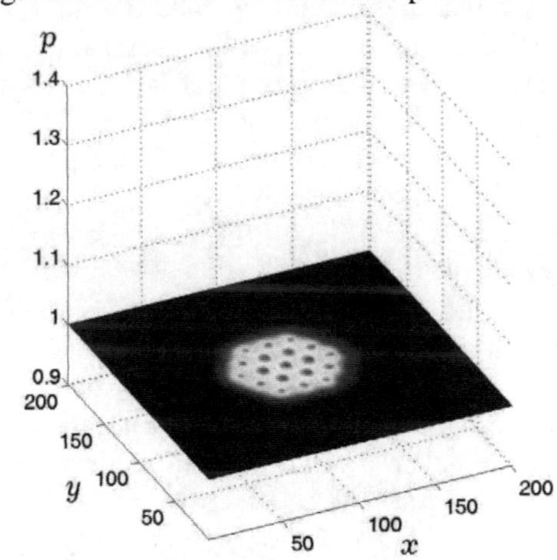

Figure 10. Pressure if $t_1 = 10$ s

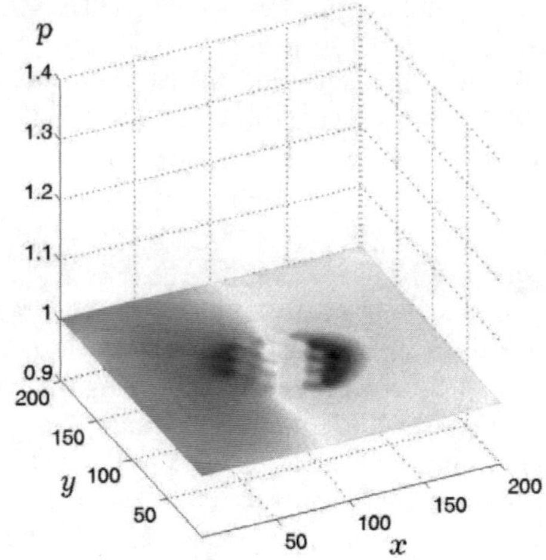

Figure 11. Pressure if $t_2 = 20$ s

IOP Conf. Series: Materials Science and Engineering **663** (2019) 012007 doi:10.1088/1757-899X/663/1/012007

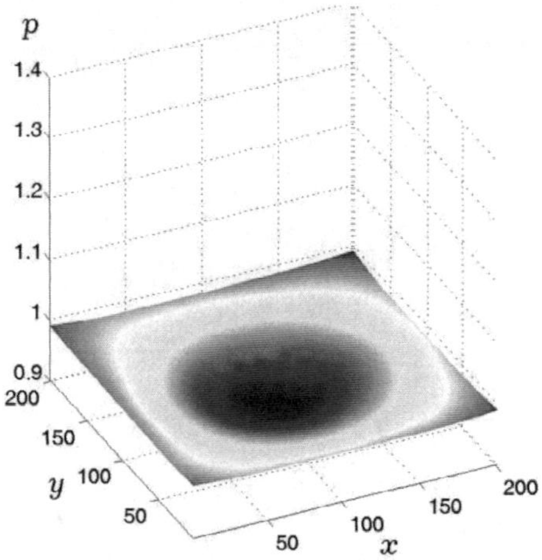

Figure 12. Pressure if $t_3 = 30$ s

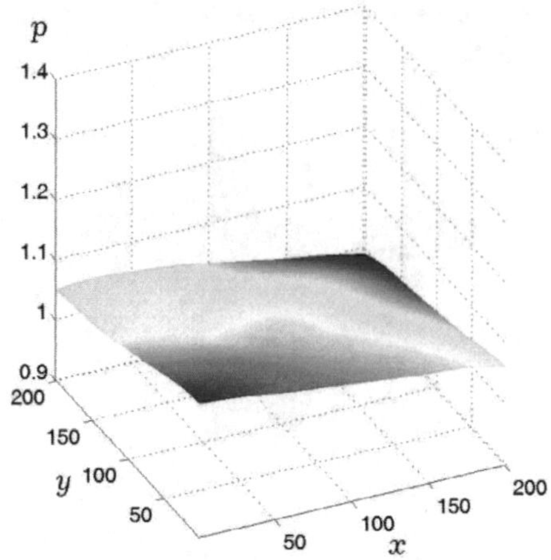

Figure 13. Pressure if $t_4 = 40$ s

The calculations revealed the presence of local pressure maxima located in the places of the heating sources (Figure 10) at the initial stage of heating. Over time, annular, gradually expanding regions with reduced pressure values form around each of the local areas of increased pressure. In addition, at the initial stage of heating, the peripheral pressure values correspond to the initial values of the steady distribution. During heating, the peripheral pressure begins to change, forming a region of reduced pressure in the center of the calculation area (Figure 12). Then, by analogy with changes in temperature and density, an uneven change in pressure is observed in the horizontal plane (Figure 13).

The calculations of the thermodynamic characteristics of air flows during heating of the underlying surface by several local sources showed that the selected mathematical model — the complete system of Navier-Stokes equations — under the corresponding initial and boundary conditions allows us to perform numerical experiments to describe the arising complex unsteady three-dimensional flows. The local changes in temperature, density, and pressure that are revealed in the process of calculations, which change in time, lead to corresponding changes in the gas flow, both in horizontal and in vertical planes. It should be noted that the fine structure observed in the calculations in changes in temperature, density, and pressure manifests itself more clearly near the bottom face of the calculation area.

4. Conclusions

The results of the calculations indicate a complex structure of emerging flows, their pronounced unsteadiness. All thermodynamic characteristics of flows at arbitrary time instants are calculated.

The calculations showed that during the formation of fire vortices, several characteristic stages can be distinguished, the successive change of which leads to the formation of a common large thermal vortex from small vortex formations due to the influx of external air receiving a positive swirl under the action of the Coriolis force.

The complex and dynamic structure of the flow is due to the fact that the compressibility, viscosity and thermal conductivity of the medium, as well as gravity and Coriolis forces, simultaneously affect the physical process. All this leads to the appearance of nonzero pressure gradients in the local parts of the flow. The calculated patterns of thermal gas-dynamic flows in many elements coincide with the characteristic features of the flows observed in experiments.

Acknowledgements

The research was supported by the Ministry of Education and Science of the Russian Federation (project No. 1.4539.2017/8.9).

References

[1] Varaksin A Yu, Romash M E, Kopeytsev V N and Горбачев M A 2014 On the possible generation of fire vortices without the use of forced spin *Proceedings of the Academy of Sciences* **456** No. 2 159–61

[2] Obukhov A G and Barannikova D D 2014 Features of the gas flow in the initial stage of formation of a thermal ascending swirling flow *News of higher educational institutions. Oil and gas* **6** 65-70

[3] Bautin S P, Abdubakova L V, Barannikova D D, Kazachinsky A O, Krutova I Yu, Mezentsev A V, Obukhov A G and Sorokina E M 2015 Mathematical and experimental modeling of ascending swirling flows *Proceedings of the XI All-Russian Conference on Fundamental Problems of Theoretical and Applied Mechanics* (Kazan) 378-80

[4] Bautin S P, Obukhov A G and Barannikova D D 2018 Numerical Simulation of Fire Vortices with Consideration of Gravity and Coriolis Forces *High Temperature* **56** No.2 241-46

[5] Bautin S P, Krutova I Yu and Obukhov A G 2015 Twisting of a Fire Vortex Subject to Gravity and Coriolis Forces *High Temperature* **53** No. 6 921–3

[6] Barannikova D D and Obukhov A G 2017 Mathematical numerical simulation of fire vortices *Bulletin of the Tyumen State University. Physical and mathematical modeling. Oil, gas, energy* **3** 60-71

[7] Barannikova D D and Obukhov AG 2017 Calculation of the thermodynamic characteristics of an upward swirling flow with five heat sources *News of higher educational institutions. Oil and gas* **3** 100-6

[8] Barannikova D D and Obukhov A G 2017 Mathematical modeling and calculation of gas-dynamic characteristics of free thermal air vortices *News of higher educational institutions. Oil and gas* **4** 93-8

[9] Barannikova D D, Borzykh V E and Obukhov A G 2018Numerical simulation of fire vortex *International Conference "Transport and Storage of Hydrocarbons" IOP Conference Series: Materials Science and Engineering* **357** 012040. doi:10.1088/1757-899X/357/1/012040

[10] Bautin S P, Barannikova D D and Obukhov A G 2018 Numerical Simulation of Fire Vortices with Consideration of Gravity and Coriolis *High Temperature* **56** 2 229–33

[11] Bautin S P 1987 Representation of solutions of the Navier-Stokes system of equations in the vicinity of the contact characteristic *Applied Mathematics and Mechanics* **51** issue 4 574–84

[12] Bautin S P and Obukhov A G *Mathematical modeling of destructive atmospheric vortices* (Novosibirsk: Nauka) 152

[13] Bautin S P, Krutova I Yu, Obukhov A G and Bautin K V 2013 *Destructive atmospheric vortices: theorems, calculations, experiments* (Ekaterinburg: Ural State University of Railway Transport Press) 215

[14] Bautin S P, Deryabin S L, Krutova I Yu and Obukhov A G 2017 *Destructive atmospheric vortices and the rotation of the Earth around its axis* (Ekaterinburg: Ural State University of Railway Transport Press) 336

[15] Bautin S P and Obukhov A G 2013 One exact stationary solution to the system of equations of gas dynamics *News of higher educational institutions. Oil and gas* **4** 81–6

[16] Bautin S P and Obukhov A G 2013 On a type of boundary conditions in the calculation of three-dimensional unsteady flows of compressible viscous heat-conducting *News of higher educational institutions. Oil and gas* **5** 55–63

IOP Conf. Series: Materials Science and Engineering **663** (2019) 012008 doi:10.1088/1757-899X/663/1/012008

Calculation of gas velocities in concentrated fire vortices

A G Obukhov [1,2]

[1] Industrial University of Tyumen, 38 Volodarskogo str., Tyumen, 625000, Russia
[2] obuhovag@tyuiu.ru

Abstract. The article presents the results of numerical modeling of the generation of free fire vortices in laboratory conditions without the use of special swirling devices. The fundamental possibility of physical modeling of the occurrence of concentrated fire vortices is described in a series of experimental studies conducted under the supervision of A. Yu. Varaksin, corresponding member of the RAS. In the model of compressible continuous medium for the complete system of Navier-Stokes equations, an initial-boundary-value problem is proposed that describes complex three-dimensional unsteady flows of viscous compressible heat-conducting gas in ascending swirling heat flows. Using explicit difference schemes and the proposed initial-boundary conditions, approximate solutions of the complete system of Navier-Stokes equations are constructed and the velocity characteristics of three-dimensional unsteady gas flows initiated by local heating of the underlying surface by nineteen heat sources are numerically determined.

1. Introduction
The possibility of physical modeling of free fire vortex structures in a laboratory without special swirling devices was demonstrated in experimental work [1]. The following setup was used to excite concentrated vortex structures. Nineteen tablets of hexamine were placed on the underlying metal surface, forming a hexagon inscribed in a circle with a diameter of 300 mm. During the experiment, the tablets were set on fire, and during their combustion the appearance of a fire vortex was detected, the height of which significantly exceeded the height of the flame above each burning tablet. The appearance of helical paths in the particles of combustion products was a sign of the generation of fire vortex structures. In one experiment, up to 15 vortex structures arise. Moreover, most vortices function on average up to 5 seconds. The height of the vortex fire structures is 0.7 meters, and their diameter is 5 centimeters.

The formation of fire vortices occurs with a significant radial influx of ambient air to the main central convective flow and swirling of the flow relative to the vertical axis. The flows of combustion products in the fire vortex consist of a complex combination of thin vortex filaments that rotate and interact with each other.

The numerical simulation of the generation of free fire vortices in a laboratory without the use of forced swirl devices and the calculation of the velocity characteristics of gas flows in such fire vortices is the goal of this work. The results of numerical studies presented in this paper continue and develop the previously formulated direction of numerical investigation of three-dimensional unsteady flows of a heat-conducting viscous gas in swirling heat flows and fire vortices.

2. Mathematical model
For the numerical modeling of complex air flows as a compressible continuous medium with dissipative properties of viscosity and thermal conductivity, we use the complete system of Navier-Stokes

equations, which, in dimensionless variables and taking into account the action of gravity and Coriolis in vector form, has the form [2-13]:

$$\begin{cases} \rho_t + \mathbf{V}\cdot\nabla\rho + \rho\,\mathrm{div}\,\mathbf{V} = 0, \\ \mathbf{V}_t + (\mathbf{V}\cdot\nabla)\mathbf{V} + \dfrac{T}{\gamma\rho}\nabla\rho + \dfrac{1}{\gamma}\nabla T = \mathbf{g} - 2\boldsymbol{\Omega}\times\mathbf{V} + \dfrac{\mu_0}{\rho}\left[\dfrac{1}{4}\nabla(\mathrm{div}\,\mathbf{V}) + \dfrac{3}{4}\Delta\mathbf{V}\right], \\ T_t + \mathbf{V}\cdot\nabla T + (\gamma-1)T\,\mathrm{div}\,\mathbf{V} = \dfrac{\kappa_0}{\rho}\Delta T + \dfrac{\mu_0\gamma(\gamma-1)}{2\rho}\Big\{\big[(u_x - v_y)^2 + \\ +(u_x - w_z)^2 + (v_y - w_z)^2\big] + \dfrac{3}{2}\big[(u_y + v_x)^2 + (u_z + w_x)^2 + (v_z + w_y)^2\big]\Big\}, \end{cases} \quad (1)$$

where $\mu_0 = 0.001$ and $\kappa_0 \approx 1.46\mu_0$ – values of constant dimensionless viscosity and thermal conductivity coefficients.

In system (1): t – time; x, y, z – Cartesian coordinates; ρ – gas density; $\mathbf{V} = (u, v, w)$ – gas velocity vector with projections on the corresponding Cartesian axes; T – gas temperature; $\mathbf{g} = (0, 0, -g)$ – gravity acceleration vector, and $g = const > 0$; $-2\boldsymbol{\Omega}\times\mathbf{V} = (av - bw, -au, bu)$ – Coriolis force acceleration vector, where $a = 2\Omega\sin\psi$, $b = 2\Omega\cos\psi$, $\Omega = |\boldsymbol{\Omega}|$; $\boldsymbol{\Omega}$ – Earth angular velocity vector; ψ – latitude of O – origin of the Cartesian coordinate system $Oxyz$, rotating with the Earth.

The initial conditions when describing flows of a compressible viscous heat-conducting gas at constant values of the coefficients of viscosity and thermal conductivity are functions that determine the exact solution [14] of system (1):

$$u = 0,\ v = 0,\ w = 0,\ T_0(z) = 1 - kz,$$

$$k = \frac{lx_{00}}{T_{00}},\quad l = 0.0065\,\frac{K}{\text{м}},\quad x_{00} = 1\,\text{м},\quad T_{00} = 288\,{}^\circ K, \quad (2)$$

$$\rho_0(z) = (1 - kz)^{\nu-1},\qquad \nu = \frac{\gamma g}{k} = const > 0. \quad (3)$$

The calculations were carried out in the area, which is a cube with sides $x_0 = y_0 = z_0 = 1$, и and shown in Figure 1.

The conditions at the boundaries of the calculation area, as well as methods for calculating the gas-dynamic characteristics on its faces, are described in detail in [15].

The density values on the lateral faces of the cube $x = 0$, $x = x_0$, $y = 0$, $y = y_0$ and on the top face $z = z_0$ are calculated from the continuity condition, that is, the density values on the lateral faces are calculated by linear interpolation of the values at the two nearest nodes within the calculation area domain.

IOP Conf. Series: Materials Science and Engineering **663** (2019) 012008 doi:10.1088/1757-899X/663/1/012008

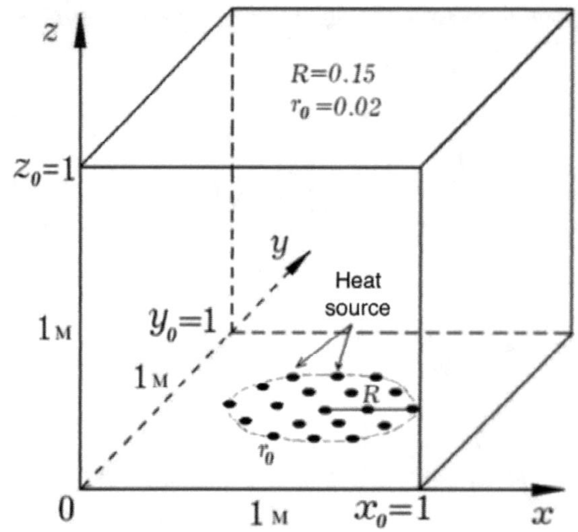

Figure 1. Cubic area for calculations

The symmetry condition is set for the density on the bottom face $z = 0$. This means that the density values on the bottom face are calculated from the condition of their zero derivative normal to the face.

On the lateral faces and the top face, the velocity values obey the continuity conditions and are calculated by linear interpolation of the values at the two nearest nodes inside the calculation area.

On the bottom face, the conditions of non-leakage are set for the velocities. The vertical velocity is equal to zero $w\big|_{z=0} = 0$, and the other two components are calculated from the condition of symmetry and are calculated from the condition of their zero derivative normal to the face.

The temperature on the lateral faces and the top face is calculated from the condition of continuity.

The gradual heating of nineteen regions of the bottom face to a temperature of 300 °C is set by a functional dependence on time and coordinates

$$T(x,y,t) = 1 + T^{*}\left(1 - e^{-10t}\right)\sum_{l=1}^{19} e^{-\frac{(x-x_l)^2+(y-y_l)^2}{r_0^2}}, \qquad (4)$$

where $T^{*} = 0.99$; $r_0 = 0.02$ – values of the radii of the heating areas; x_l, y_l, $l = 1 \div 19$ – coordinates of the centers of heat sources. The boundary conditions set in this way simulate the flow of a compressible gas initiated by heating of the bottom plane by nineteen heat sources. Boundary conditions imply the possibility of intersection of gas of all faces of the calculation area, apart from the bottom one. The following scale values were used in the calculations: $\rho_{00} = 1.29$ kg/m³, $u_{00} = 333$ m/s, $x_{00} = 1$ m, $t_{00} = x_{00} / u_{00} = 0.003$ s. Discrete steps in three coordinates $\Delta x = \Delta y = 0.005$, $\Delta z = 0.05$, a discrete time step $\Delta t = 0.001$.

3. Calculation results

Next, an analysis of the velocities of complex gas flows arising when heating the lower surface by nineteen heat sources. Figures 2-5 show the results of calculations of the first component U of the gas flow rate during heating of the bottom face by nineteen sources. The results for a height of 0.1 m for various time moments are presented. The abscissa and ordinate axes show the numbers of nodes of the computational grid.

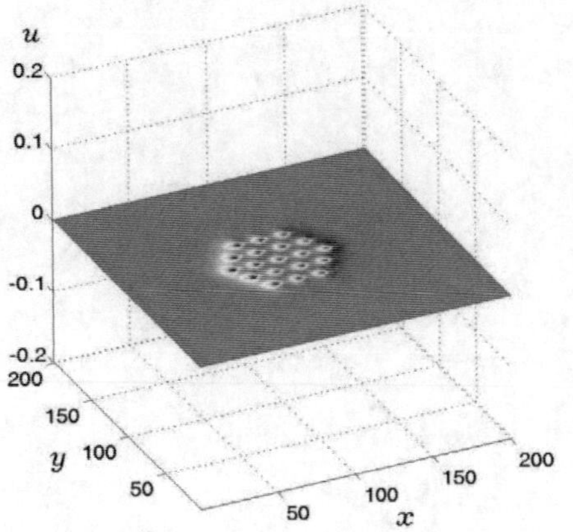

Figure 2. Velocity U if $t_1 = 10$ s

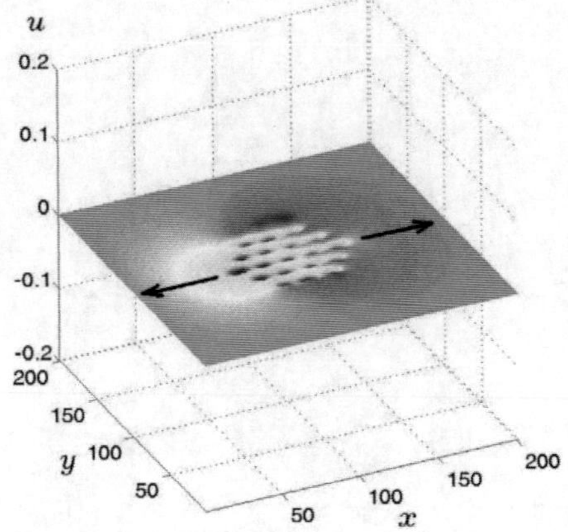

Figure 3. Velocity U if $t_2 = 20$ s

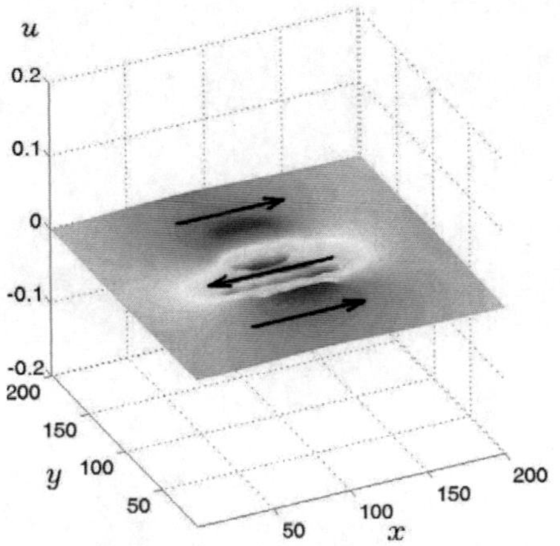

Figure 4. Velocity U if $t_3 = 30$ s

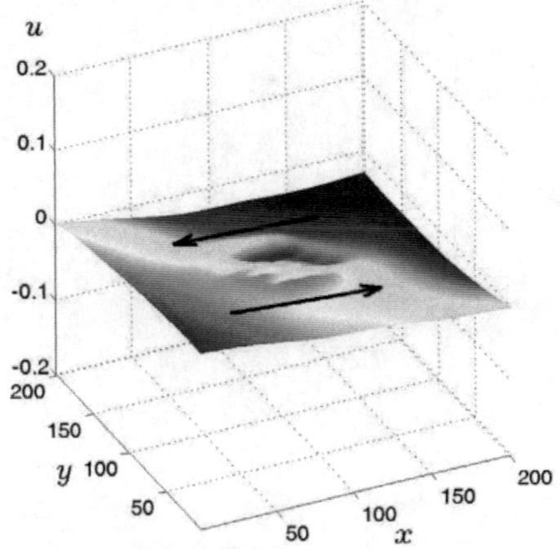

Figure 5. Velocity U if $t_4 = 40$ s

According to the results of the calculations of the first velocity component, several phases of the development of thermal vortex structures can be distinguished. The first phase is characterized by the appearance of multiple small vortices with the opposite direction of swirling near each of the nineteen heat sources (Figure 2). The yellow and red colors in the figure correspond to a positive gas flow velocity (along the Ox axis), and the blue color corresponds to a negative gas flow velocity (against the Ox axis). The second phase is characterized by the superposition of a larger gas flow diverging from the heating region onto a smaller vortex structure (Figure 3). With the onset of the third phase, two vortices arise with opposite directions of rotation (Figure 4). The vortex closest to the observer has a positive swirling direction, and the vortex closest to the observer - negative. And finally, the calculations show the presence of another stage in the development of the fire vortex. At this stage, due to the intense influx of external air and the action of the Coriolis force on it, the positive swirling direction of the entire flow gradually prevails (Figure 5). As a result, the fire vortex observed in the experiments is formed. The marked stages of the formation of a fire vortex are indistinct in time. However, the distinctive features of each of these stages formally make it possible to arrange them in a certain chronological sequence.

The behavior of the second velocity component V is similar to the behavior of the first one described above. Graphic dependences on the coordinates of this component for the same time moments are shown in Figures 6-9.

IOP Conf. Series: Materials Science and Engineering **663** (2019) 012008 doi:10.1088/1757-899X/663/1/012008

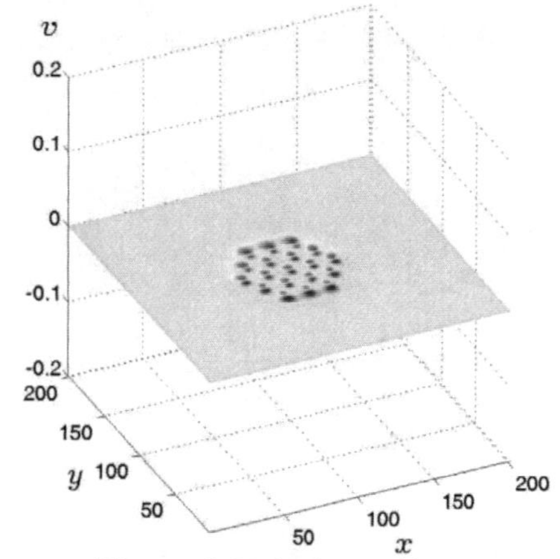

Figure 6. Velocity V if $t_1 = 10$ s

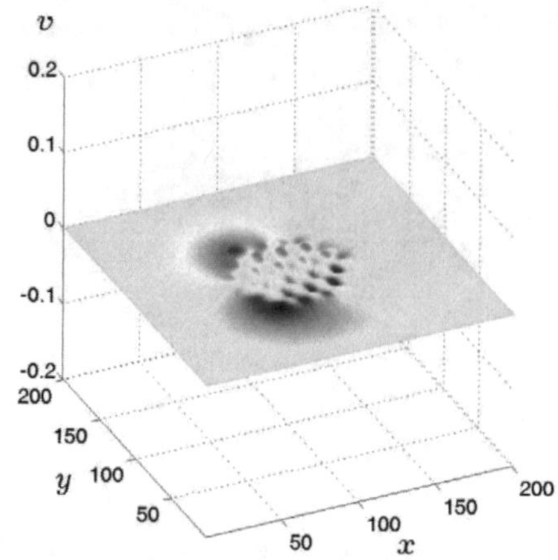

Figure 7. Velocity V if $t_2 = 20$ s

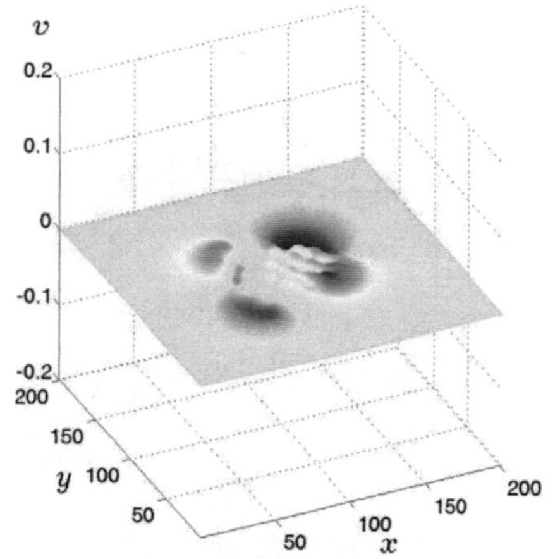

Figure 8. Velocity V if $t_3 = 30$ s

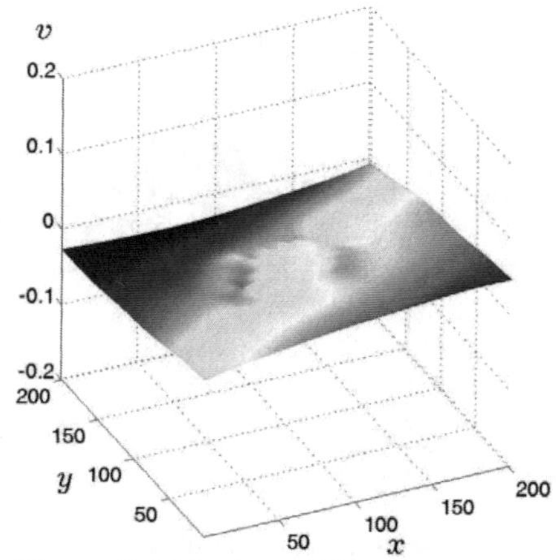

Figure 9. Velocity V if $t_4 = 40$ s

Figures 10-11 show the third velocity component W for time moments $t_1 = 10$ s and $t_4 = 40$ s for a height of 0.1 m from the bottom surface.

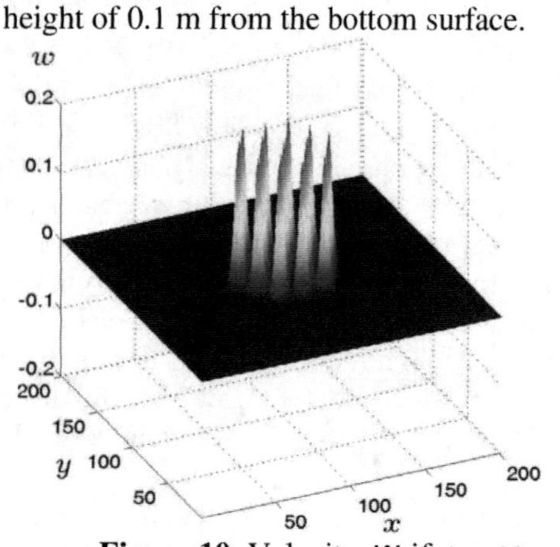

Figure 10. Velocity W if $t_1 = 10$ s

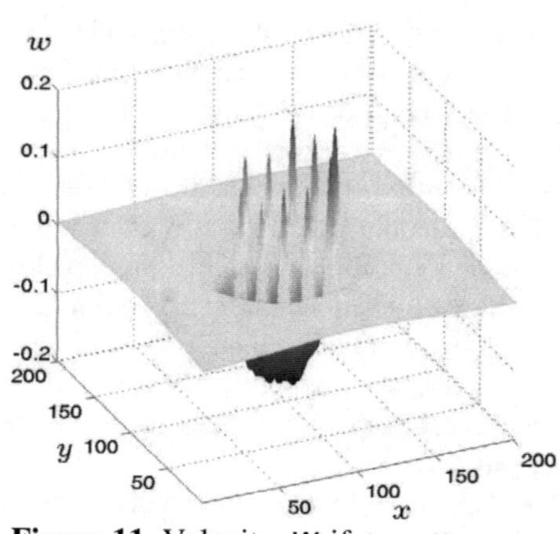

Figure 11. Velocity W if $t_4 = 40$ s

Figure 10 demonstrates that by the time moment $t_1 = 10$ s, local vertical air flows are formed above the heat sources. Maximum velocities of 33 m/s correspond to the location of the sources, and when moving away from them they decrease to zero. In the space located near the heat sources, the vertical velocity has negative values corresponding to the downward movement of the gas flow.

4. Conclusions

The results of the calculations indicate a complex structure of emerging flows, their pronounced unsteadiness. The velocity characteristics of flows at arbitrary time instants are calculated.

The calculations showed that during the formation of fire vortices, several characteristic stages can be distinguished, the successive change of which leads to the formation of a common large thermal vortex from small vortex formations. Such a process occurs due to the influx of external air receiving a positive swirl under the action of the Coriolis force. Due to a substantial increase in the swirl velocity, a reduction in size occurs, followed by the disintegration of the common thermal vortex into smaller ones. The expansion of gas from the calculation area actually ends the life cycle of one fire vortex, and then the formation of a new fire vortex begins.

The calculations showed that with these parameters, the lifetime of one fire swirling stream is 1 minute.

Acknowledgements

The research was supported by the Ministry of Education and Science of the Russian Federation (project No. 1.4539.2017/8.9).

References

[1] Varaksin A Yu, Romash M E, Kopeytsev V N and Горбачев M A 2014 On the possible generation of fire vortices without the use of forced spin *Proceedings of the Academy of Sciences* **456** No. 2 159–61

[2] Bautin S P 1987 Representation of solutions of the Navier-Stokes system of equations in the vicinity of the contact characteristic *Applied Mathematics and Mechanics* **51** issue 4 574–84

[3] Bautin S P and Obukhov A G *Mathematical modeling of destructive atmospheric vortices* (Novosibirsk: Nauka) 152

[4] Bautin S P, Krutova I Yu, Obukhov A G and Bautin K V 2013 *Destructive atmospheric vortices: theorems, calculations, experiments* (Ekaterinburg: Ural State University of Railway Transport Press) 215

[5] Bautin S P, Deryabin S L, Krutova I Yu and Obukhov A G 2017 *Destructive atmospheric vortices and the rotation of the Earth around its axis* (Ekaterinburg: Ural State University of Railway Transport Press) 336

[6] Barannikova D D and Obukhov A G 2017 Calculation of velocity characteristics of an upward swirling flow with five heat sources *Fundamental research* **9**-1 14-8

[7] Barannikova D D and Obukhov A G 2017 Mathematical numerical simulation of fire vortices *Bulletin of the Tyumen State University. Physical and mathematical modeling. Oil, gas, energy* **3** 2 60-71

[8] Barannikova D D and Obukhov A G 2017 Mathematical modeling and calculation of gas-dynamic characteristics of free thermal air vortices *News of higher educational institutions. Oil and gas* **4** 93-8

[9] Barannikova D D, Borzykh V E and Obukhov A G 2018 Numerical simulation of fire vortex *International Conference "Transport and Storage of Hydrocarbons". IOP Conference Series: Materials Science and Engineering* **357** 012040. doi:10.1088/1757-899X/357/1/012040.

[10] Bautin S P, Obukhov A G and Barannikova D D 2018 Numerical Simulation of Fire Vortices with Consideration of Gravity and Coriolis Forces *High Temperature* **56** 2 229–33.

[11] Bautin S P, Krutova I Yu and Obukhov A G 2015 Twisting of a Fire Vortex Subject to Gravity and Coriolis Forces *High Temperature* **53** 6 921–3

[12] Barannikova D D and Obukhov A G 2017 Calculation of velocity characteristics of an upward swirling flow with five heat sources *Fundamental research* **9-1** 14-8

[13] Barannikova D D and Obukhov A G 2017 Mathematical modeling and calculation of gas-dynamic characteristics of free thermal air flows *News of higher educational institutions. Oil and gas* **4** 93-8

[14] Bautin S P and Obukhov A G 2013 One exact stationary solution to the system of equations of gas dynamics *News of higher educational institutions. Oil and gas* **4** 81–6

[15] Bautin S P and Obukhov A G 2013 On a type of boundary conditions in the calculation of three-dimensional unsteady flows of compressible viscous heat-conducting *News of higher educational institutions. Oil and gas* **5** 55–63

IOP Conf. Series: Materials Science and Engineering **663** (2019) 012009 doi:10.1088/1757-899X/663/1/012009

Modelling the movement of a cultivating shovel in the ground

S P Pirogov[1,2], D A Cherentsov[1],
[1] Industrial University of Tyumen, 38 Volodarskogo str., Tyumen, 625000, Russia
[2]Northern Trans-Ural State Agrarian University, Tyumen, Russia

Abstract. The paper presents the results of the calculation of the forces acting on the cultivating shovel when moving in the ground. For research, the finite element method implemented in the ANSYS program was used. The tasks of constructing a grid model of the ground were solved, the horizontal component of the force of the impact of the ground when moving at different speeds was determined.

1.Introduction

For trenching, various devices with cultivating shovels (CSs) are used. In the proposed design, as a working body, a flexible tubular element is used, at the end of which a CS is attached.

The change in pressure in the inner cavity of the rack causes the cross-sections to deform and the free end with the cultivating shovel moves. In the process of cultivating the soil, the working body is influenced by the resistance forces of the soil of a variable nature, which causes rack vibrations. The flow of working fluid into the cavity of the rack under variable pressure leads to oscillatory movements of a CS with a certain amplitude and frequency, which depend on the parameters of the supplied pressure. Calculation of vibration parameters of flexible tubular springs is given in [1-5].

2. Formulation of the problem

To ensure trouble-free operation of the cultivator, it is necessary to solve the following problems:

1. Building a grid model of the CS movement in the ground.
2. Determination of the horizontal component of the force R of the soil impact on the shovel, depending on the speed of movement.

3.Theory

To simulate the movement of a CS in the soil, we assume the following - the treated soil will be considered as a continuous medium. The movement of an object in a continuous medium can be equivalently replaced by “flowing around” a body at rest with a continuous medium at the same speed; this approach will allow the use of ANSYS Fluent.

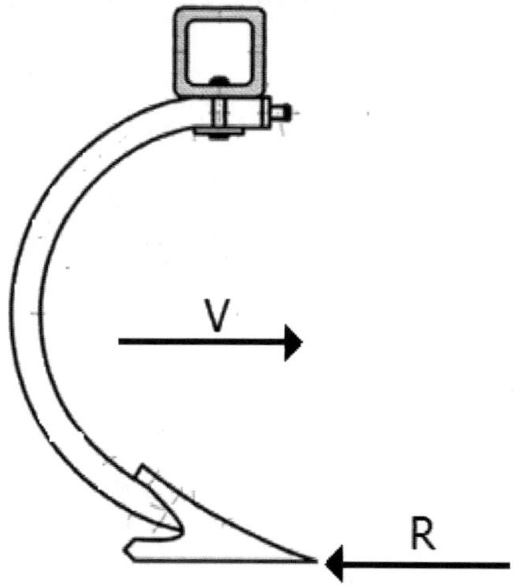

Figure 1. Cultivator working body.

To determine the relationship between the horizontal component R and the velocity V of the cultivating shovel (CS), we use the dependence:

$$R = C_x F \frac{\rho V^2}{2} \tag{1}$$

here F is the mid-sectional area — the largest cross-sectional area; ρ is the density of the continuous medium (of the soil under consideration), V is the CS velocity in the soil, Cx is a dimensionless coefficient characterized by the geometric component and the rate of flowing around with the continuous medium (i.e., it is a function of velocity).

So, Cx and the horizontal component R directly depends on the design of the CS, in this paper, we consider the influence of the most primitive design below but evaluate the influence of the grid model and the size of its elements, as well as models of the behaviour of a continuous medium.

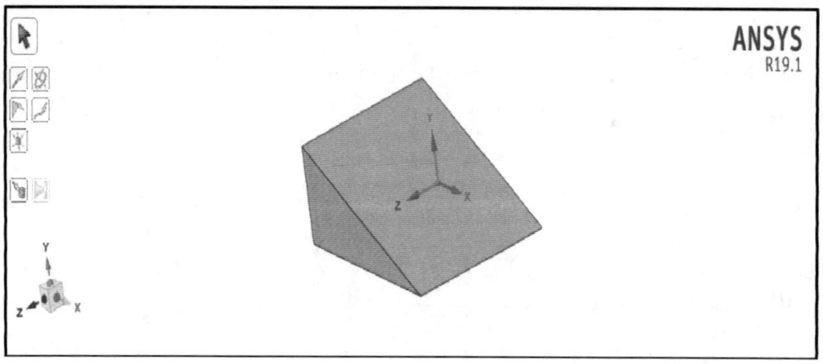

Figure 2. Schematic diagram of the cultivating shovel

4. Experiment results

The CS is made of steel, geometric characteristics: length - 500 mm, height - 200 mm, width - 400 mm.

Hydrodynamic calculation was carried out in toolbox - fluent.

All hydrodynamic problems require an iterative approach to solving, we consider the change in the value of the drag coefficient Cx depending on iterations and on the velocity of flowing around with a continuous medium for various models of continuous medium behavior.

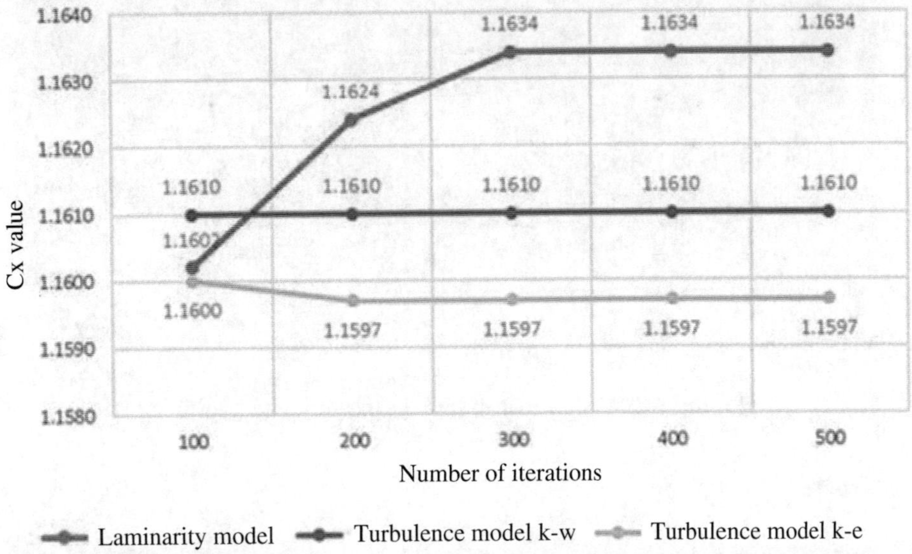

Figure 3. Results of the evaluation of the influence of the number of iterations

The calculation results showed that to obtain reliable results, iterations may be limited to 300.

The distribution of velocities in the simulated volume is shown in Figure 4, and the effect of the movement velocity on the resistance force is shown in Figure 6.

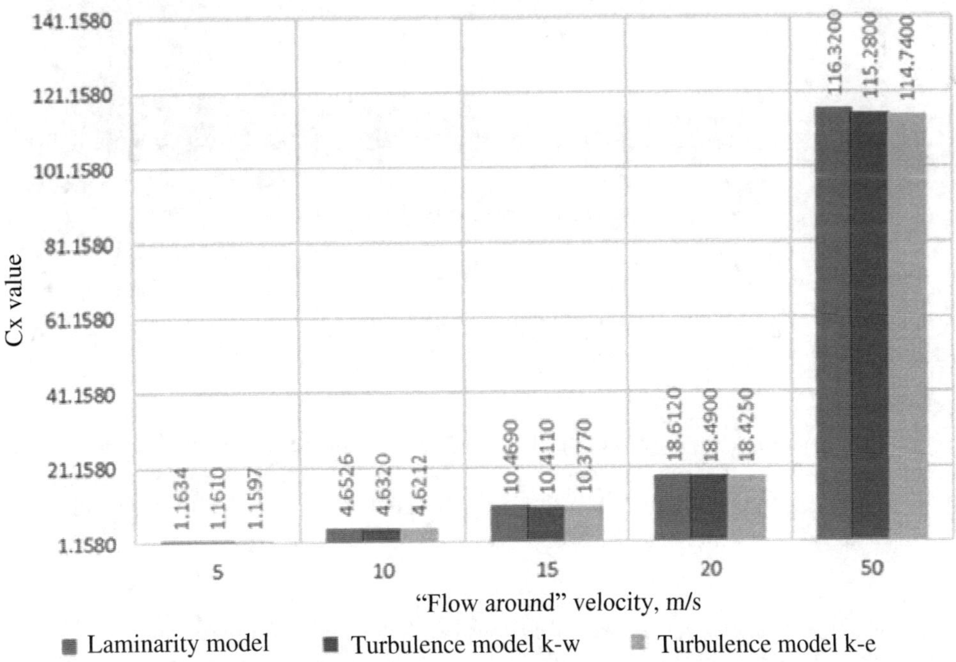

Figure 4. Results of the evaluation of the effect of flow velocity

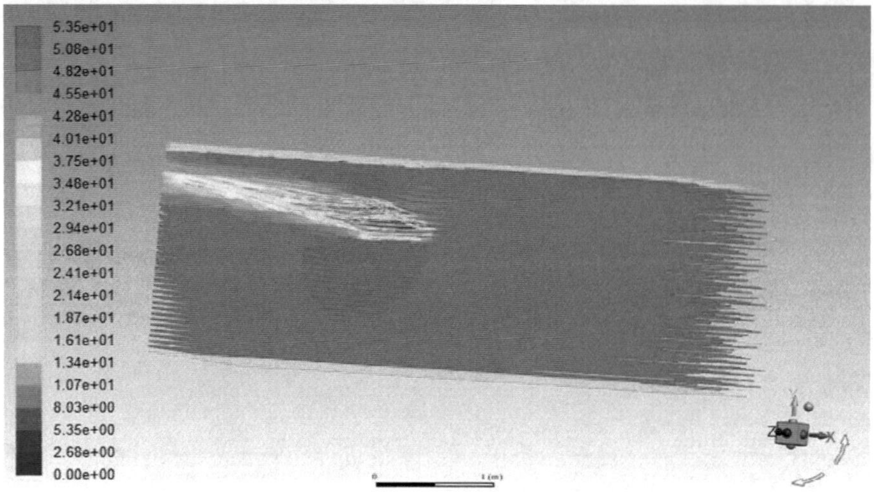

Figure 5. Velocity distribution in the simulated volume

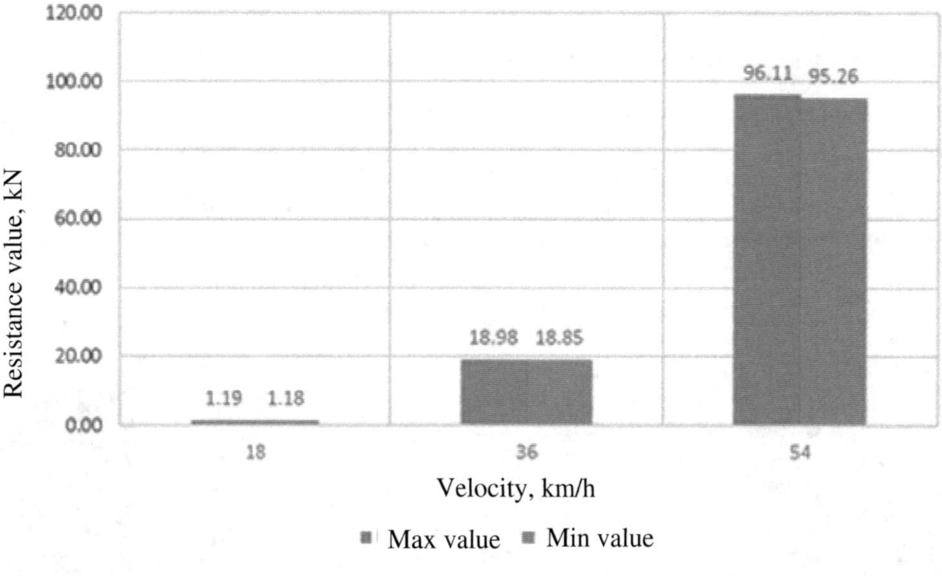

Figure 6. The influence of the movement velocity of the cultivating shovel on the resistance

5. Conclusions

The proposed methodology for determining the resistance of soil to the CS movement allows one to calculate the permissible velocity of its movement with a fixed geometric dimension of the element to which it is attached. More accurate modelling of the design of the CS will refine the value of the movement velocity.

References

[1] Pirogov S P 2009 *Manometric tubular springs* (St. Petersburg: Nedra) 276
[2] Pirogov S P, Cherentsov D A and Chuba A Yu 2015 *Fluctuations of manometric tubular springs* (Tyumen: TSOGU) 95

[3] Pirogov S P, Cherentsov D A and Chuba A Yu 2018 *Tubular spring mechanisms* (Tyumen:IUT) 78

[4] Pirogov S P and Chuba A Yu 2017 Application of manometric tubular springs in agricultural machines *Agro-food policy in Russia* **9** (69) 82-8

[5] Andreeva L E 1962 *Elastic elements of devices* (Moscow: Mashgiz) 456

[6] Akselrad E L 1976 *Flexible shells* (Moscow: Nauka) 376

[7] Cherentsov D A 2014 Determination of generalized resistance forces in second-order Lagrange equations describing the oscillations of manometric springs in the presence of viscous friction forces *Natural and technical sciences* **4** (72) 12-5

[8] Cherentsov D A, Chuba A Yu and Pirogov S P 2013 Model of oscillations of a manometric tubular spring in a viscous fluid *In the collection: Problems of the functioning of transport systems Materials of the All-Russian Scientific and Practical Conference of Students, Graduate Students and Young Scientists, dedicated to the 50th anniversary of the founding of the Tyumen Industrial Institute* 373-7

[9] Cherentsov D A, Pirogov S P and Dorofeev S M 2014 A mathematical model of a manometric spring in a viscous medium *Bulletin of the Tyumen State University. Physical and mathematical modelling. Oil, gas, energy* **7** 234-41

[10] Cherentsov D A, Pirogov S P, Dorofeev S M and Cherentsova S A 2017 Study of damped oscillations of manometric springs with a rigid tip *News of higher educational institutions. Oil and gas* **1** 116-20

[11] Pirogov S P 2016 Foundations of the Design of Vibration-Resistant *Measurement Techniques.* Volume **59**, Issue 8, November, 845–9.

[12] Pirogov S P *2018* Oscillations of manometric tubular springs with rigid end S P Pirogov, D A Cherentsov, S M Dorofeev and Y S Ryabova *IOP Conf. Series: Materials Science and Engineering* ***357*** *012030 doi:10.1088/1757-899X/357/1/012030*

[13] S P Pirogov, A Y Chuba and D A Cherentsov *2018* Effect of section shape on frequencies of natural oscillations of tubular springs *IOP Conf. Series: Materials Science and Engineering* ***357*** *012032 doi:10.1088/1757-899X/357/1/012032*

[14] Pirogov S P and Cherentsov D A 2016 Theoretical foundations of the design of vibration-resistant manometers *Measurement Techniques* **59** 8 845-9

[15] Pirogov S P, Cherentsov D A and Gulyaev B A 2016 Prospects of applying vibration-resistant pressure gauges in the oil and gas industry *IOP Conference Series: Materials Science and Engineering "International Scientific-Practical Conference of Students, Graduate Students and Young Scientists "Transport and Storage of Hydrocarbons""* C 012013

IOP Conf. Series: Materials Science and Engineering **663** (2019) 012010 doi:10.1088/1757-899X/663/1/012010

Planning the activities of the utility vehicles department in the structure of a joint stock company

E L Chizhevskaya[1,2]

[1]Industrial University of Tyumen, 38 Volodarskogo str., Tyumen, 625000, Russia
[2]chizel76@yandex.ru

Abstract. Planning the activities of a structural unit is one of the sources for finding reserves to improve the efficiency of the company as a whole. The correctness of the methods used, the completeness of the source data, the software used is an incomplete list of factors affecting the reliability of the plans of the company's departments, which in turn make up a plan of the organization's activity as a whole. The article provides an overview of the planning process for the Utility Vehicles Department, considers planning tools, the possibility of increasing the efficiency of the process and achieving planned indicators.

1. Introduction

Planning the enterprise's activity is associated with the selection of the optimal procedure for the formation of development plans for structural units included in its structure. In this case, planning should take into account the specifics of the activities of structural units and their contribution to the implementation of the main technological processes and key financial and economic indicators.

The object of planning the Utility Vehicles Department (UVD) is the production and economic activity that it carries out on orders and instructions of the Company. A characteristic feature of the motor transport company is that the main production process - the transportation of goods - is carried out outside its territory, and consists of work on the line. On the territory of the enterprise, only auxiliary processes are carried out related to ensuring the readiness of the rolling stock for transportation. [4]

2. Materials and methods

For the effective organization of the UVD's activities, it is necessary to receive timely information on the progress of transportation, consumption of fuel and lubricants and spare parts, volumes of goods transported. To build and study models of management objects in their temporal relationship, it is advisable to identify the main stages of the transportation process (Figure 1).

The level of management of the production process is the most important factor determining the level of production efficiency. Special management requirements are presented in the organization of road transport. The production process of motor transportation should organically integrate the production processes of customers, directly linking operations into a single cycle from the moment the need for information arises to the receipt of products, including the satisfaction of needs not directly related to specific material objects. [1] Road transport management is a fairly complex integrated system that includes bodies, personnel and equipment of the department.

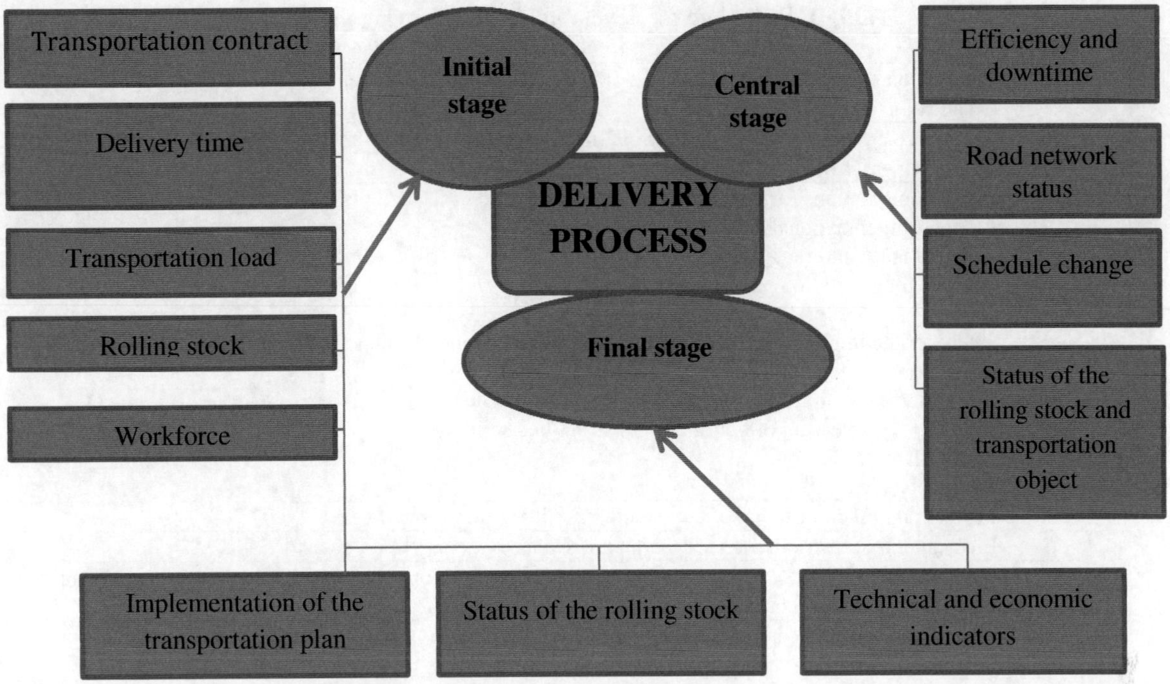

Figure 1. Transportation process model [3]

The following distinctive features are characteristic of the UVD as a management object.:

1) dependence of the UVD functioning on the technological processes of customers, as well as the impact of the transportation process on the economic results of their activities;
2) dependence of the activity of the elements of the organizational structure and management effectiveness on external conditions;
3) dynamism and stochasticity due to the simultaneous influence of many factors, some of which have elements of randomness.

Planning for the UVD is preceded by an analysis of its activities. The formation and implementation of the plan is a continuous process, which can be graphically represented by the following scheme (Figure 2). [2] Under the guidance of the head of the department, the UVD plan is developed with the large-scale and active participation of the entire team in identifying reserves, developing measures for their use and preparing the draft plan as a whole. [9]

The development of the structural unit plan is carried out in stages. At the first stage, a draft plan for the next year is designed taking into account possible additional instructions of the Company. This project is compiled only according to the main aggregated indicators determining the upcoming traffic volume, the need for supplies of the rolling stock and the level of volumes of transport services. Then follows the refinement and approval of the indicators of the plan (Table 1).

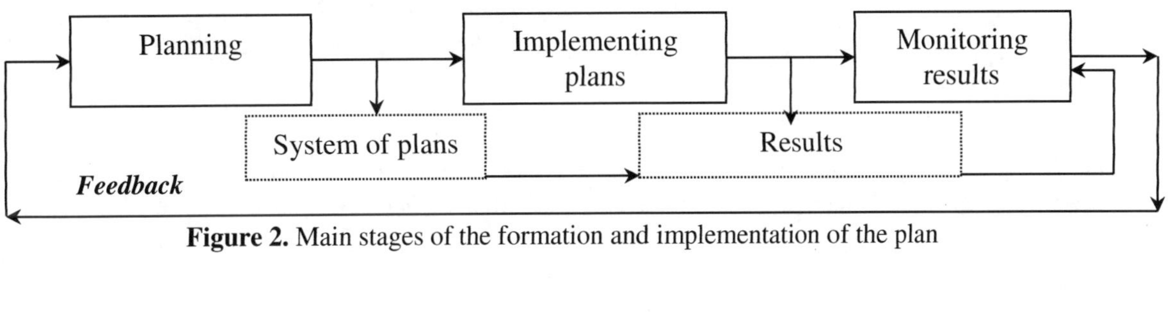

Figure 2. Main stages of the formation and implementation of the plan

Table 1. Procedure for developing and approving the UVD plan

Stage	Stage Content
1	Preparation of preliminary proposals of the structural unit for the planned year (based on the analysis of indicators of the reporting year)
2	Bringing the control numbers of OJSC for the structural unit for the planned year
3	Issuing an order by the UVD supervisor to develop a plan
4	Preparation of plans for areas of activities: - Plan of organizational and technical activities; - Maintenance and repair plan; - Goods transportation plan; - Rolling stock operation plan; - Plan for the implementation of new vehicles and advanced technology; - Labor and wage plan; - Procurement plan; - Plan for the overhaul of buildings and structures; - Cost plan; - Financial plan.
5	Formation of the draft plan of the structural unit for the planned year
6	Presentation of the draft plan for the planned year to the parent company
7	Continuation of work on consideration and generalization of proposals aimed at improving the efficiency of the structural unit
8	Obtaining an approved plan for the planned year (by main indicators) by the structural unit from the OJSC
9	Refining and detailing the previously developed draft plan of the structural unit for the planned year
10	Approval of the plan for the planned year by the head of the department and bringing it to the shops and columns
11	Notification of the OJSC on approval of the plan and its main indicators
12	Implementation of previously developed measures aimed at ensuring the implementation of the structural unit plan

The first stage begins in April-May of the current year preceding the planned one. On the basis of a thorough analysis of the economic activity of the previous period, the progress of the plan for the past period of the current year, the planned changes in the size and structure of the rolling stock, the needs of the clientele served and other factors, the UVD prepares preliminary proposals on the possible increase in traffic for the next year and presents them to the Company. [3] The latter, based on the control figures and the generalization of the preliminary proposals of the enterprise, brings the control figures for a limited circle of the most important indicators to its structural unit, which, taking them into account, carries out the first stage of the development of the plan.

The head of the department issues an order, which regulates the procedure and terms for developing the plan, appoints responsible persons for individual sections of the plan. At the same time, special attention is paid to the active involvement of workers, engineers and technicians, inventors and rationalizers in the development of proposals related to the quickest elimination of bottlenecks existing in the enterprise and ensuring an increase in the efficiency of the enterprise's economic activity.

The development of the plan is carried out under the direct supervision of the head of the planning and economic department. The organization's economic service, organizing this work, engages technical, operational, and other services in it, ensuring thorough and coordinated preparation of all sections of the plan. The first stage of this work should be completed by early July so that, based on the development of the plan, the enterprise has the opportunity to present to the Company its draft plan for the established indicators for the coming year at the time. [15] At the same time, the company continues to collect, review and generalize proposals related to the implementation in the planning period of organizational and technical measures aimed at improving the efficiency of the enterprise.

The draft plan for the next year submitted by the UVD is reviewed, summarized and reflected in the unified plan of the Company. At the beginning of October, the plan for the main quantitative and qualitative indicators is brought by the Company to the UVD as a binding document.

On this basis, and taking into account additionally identified reserves, the next stage of work is carried out, which consists in refining and detailing the previously developed draft plan. This work should be carried out within a month and completed no later than November so that the UVD head has the opportunity to approve the plan in advance, and the planning and economic service can bring the approved tasks to the workshops, transport columns, sections, brigades and prepare events with other services ensuring the successful implementation of the plan of the structural unit, including the organization of systematic and effective control over its implementation.

Upon completion of the work, the head of the UVD is obliged to inform the Company about the completion of the development and approval of the plan and provide a certificate of the main indicators of this plan. This will enable the Company to take into account the obligations undertaken by the structural unit itself in the generalized material. [8, 14]

The UVD plan includes various sections, each of which has its own indicators that reveal their content and place in the plan (Table 2).

Table 2. Key plan indicators communicated to the unit

№	Key plan indicators group	Name of indicators
1	Cargo transportation	- total estimated volume of traffic in tons indicating the list of clients served; - cargo turnover
2	Operation of rolling stock	- truck fleet utilization rate; - loaded mileage proportion; - time on duty
3	Maintenance	- number of technical services; - total labor input for maintenance and repair work; - coefficient of technical readiness; - vehicle downtime in repair
4	Labor	- payroll plan; - employee plan; - training plan; - workforce social development plan; - working efficiency
5	Finance	- total value of transport services in monetary terms; - payments to the budget; - budget appropriations, financing of the parent company; - reducing the cost of road transport
6	Capital investments	- level of capital investments; - commissioning of fixed assets and production facilities; - production development fund generation standards
7	Introduction of new vehicles and equipment	implementation of new technological processes, complex mechanization and automation
8	Procurement	volume of supplies of rolling stock, tires, fuel, lubricants and other material and technical resources

Planning for the UVD begins with a transportation plan that defines the volume and structure of the transportation. This planning stage serves as the basis for the development of all other parts of the plan.

Initial data for drawing up a plan for the transportation of goods are the applications of enterprises, which indicate the types of goods, points of departure and destination, the volume of transportation in tons and ton-kilometers with distribution by quarters. It is important to note that the high level of reality of planned indicators is determined by their recalculation when changing external factors. [5]

Customers draw up applications for the transportation of goods in accordance with the production plan. The volume of traffic is determined by the main types of cargo characteristic of the company.

When summarizing and analyzing applications, the possibility of centralizing transportation is primarily considered. This is preceded by a detailed study of transportation from points that have a large volume of exported cargo to various structural units. Then the volume of traffic, cargo turnover, the condition of roads, access roads and loading platforms, the equipment of loading and unloading means, the mode of operation of consignors and consignees are determined. [13] After processing the data and agreeing on the organization of transportation with the heads of enterprises, they draw up a list of shippers from whose warehouses in the planning year centralized delivery of goods will be carried out. The centralized transportation plan is an integral part of the overall transportation plan for the planned period. After developing a plan for the centralized transportation of goods, the remaining applications are examined, checked, summarized, and as a result, a general plan for the transportation of goods by the enterprise is developed, which can be specified when it is completed. If it is impossible to quantify the transported goods during transportation, vehicles that operate at the hourly rate are allocated. These services are planned in auto hours.

The developed plan for the transportation of goods for the planning period is compared with the carrying capacity of the vehicle fleet. For this, the balances of the carrying capacity of the vehicle fleet are compiled according to the classes of goods, that is, the security of transportation of goods by various types of rolling stock is revealed. As a result, it can be established that not all cargo groups by transportation classes are provided with sufficient carrying capacities. In this case, measures are taken to increase the productivity of rolling stock by improving performance. If this does not provide the necessary carrying capacity, it is necessary to raise the question of either increasing the number of vehicles in the fleet or partially removing the volume of cargo transported. Draft transportation plans for the planned year, developed in the indicated sequence, by type of cargo and consignors are sent to the Company. The annual transportation plan is approved quarterly. Based on the approved annual and quarterly cargo transportation plans, the UVD enters into agreements with a fixed clientele and develops detailed monthly transportation plans, for the preparation of which they receive detailed orders from shippers on the nomenclature of goods indicating the points of departure and destination of goods. Shippers draw up monthly requests in accordance with the volume of traffic planned in annual and quarterly plans. [7, 12]

Monthly plans agreed with shippers and approved by the UVD are the basis for daily planning. Shift-daily plans determine the volume of traffic, the number of rolling stock at the facilities, and are used to develop schedules for the arrival of rolling stock at loading, unloading and vehicle routes.

The constituent parts of the transport process are operational, storage and loading and unloading operations, the organization of which largely determines the downtime of vehicles under loading and unloading operations and the safety of cargo, and therefore, rolling stock productivity and transportation costs.

It should be noted that the centralized method of organizing transportation is the most progressive compared to the decentralized one. The decentralized method of organizing transportation is old and ineffective. With it, cars are allocated to cargo owners who carry out loading and unloading operations.

With centralized transportation, the entire system for organizing loading and unloading and transportation of goods by road changes. The shipper, the UVD and consignees are involved in the transport process. At the same time, the duties are distributed between them as follows: loading of goods onto vehicles at bases and warehouses is carried out by the forces and means of the consignor; cargo transportation and forwarding are carried out by the UVD; unloading of goods from vehicles is carried out by forces and means of consignees. [10, 11]

The centralization of transportation provides the possibility of using heavy trucks in a wide range. The documentation for dispatch and receipt of goods and payment for transportation is greatly simplified since this is done by one organization.

The existing shortcomings in the operation of technological transport are largely due, primarily, to the imperfection of the planning system. This includes a low level of information in the management and communication system both within it and in relation to other systems, insecurity or low level of use of various technical means, insufficient coverage of planning activities, etc. These shortcomings lead to violations, disruption of transportation plans, cargo delivery schedules and other organizational failures both in the service unit and in the parent company. [6]

However, existing developments in the planning of performance indicators of the Utility Vehicles Department in practice do not ensure the achievement of predicted values of the performance indicators of its activities. This is due to the main drawback of research in the field of improving the planning of the transportation process, namely the lack of a systematic approach to the problem under study, to its consideration. As a result, many of the connections and functions of the planning system were not properly implemented in practice, and in the studies performed, they were not given due attention, some parts of the system were completely missed. This leads to insufficient efficiency of the system of operational planning and operation of transport. Thus, the great importance of the transportation planning system in the implementation of the transport process does not correspond to the level of implementation of the system. Ensuring the effectiveness of the system requires a change in attitude towards it, considering it as a whole, and taking into account all links.

3. Conclusion

Based on the foregoing, we can distinguish the following most important areas of improving the planning of activities of the structural unit of the enterprise:

- strengthening planning discipline, enhancing the validity of plans based on technical and economic calculations for all its positions and drawing up a multivariate plan for choosing the optimal plan;
- expanding the use of modern means of automation of the collection, processing and transmission of information;
- strengthening the planning and economic service, improving the skills and responsibilities of economists;
- organizing continuous monitoring of the implementation of the plan and the prompt elimination of the reasons preventing the achievement of the goals.

When developing the plan of the structural unit, special attention should be paid to the feasibility study of each indicator based on data characterizing the presence and condition of the rolling stock, the structure and organization of transportation, and the plan for improving the vehicles.

To determine the carrying capacity, the average number of cars that the UVD will have in the planning period is initially calculated. To calculate the average number of cars in the UVD, first of all, the number of days of stay at the enterprise for incoming and outgoing vehicles is determined.

Then, on the basis of data on the number of rolling stock of each type for the planning period and the estimated level of its productivity, in accordance with the operational indicators adopted in the plan, a production program for the operation of road transport is developed, which is the basis for compiling all other sections of the structural unit plan.

The annual plan of the enterprise is built on an information basis, in which a special place is occupied by the regulatory and reference base.

References

[1] Antipiev V N, Zemenkov Yu D, Malyushin N A et al 2001 *Operation of main oil pipelines. Pipeline transportation of oil* (Omsk)

[2] Vasiliev G G, Gulkov A N, Zemenkov Yu D and Prokhorov A D 2008 *Operation of gas industry equipment and facilities* (Vologda) **2**

[3] Dudin S M, Shipovalov A N, Podorozhnikov S Yu and Zemenkov Yu D 2011 Prospects for reducing energy costs for hydrocarbon transport *News of higher educational institutions. Oil and gas* **2** 65-9

[4] Zemenkov Yu D, Kurushina V A, Vylegzhanina A O, Barmenkova V V and Khairullina N G 2013 *Modern management and innovative, scientific and technical policy* (St. Louis)

[5] Zemenkov Yu D et al 2007 *History and development prospects of the oil and gas industry and the fuel and energy complex* (St. Petersburg)

[6] Zemenkova M, Zemenkov Y, Gladenko A and Podorozhnikov S 2016 Estimation of Emis-sions during Monitoring of Pipelines in the Dynamic Mode of Operation *MATEC Web of Conferences DOI: 10.1051/matecconf/20168604053*

[7] Fedorova O B and Lenkova O B 2011 The technology of choosing tools for financing innovative solutions in oil production *Management in Russia and abroad* **6** 23-8

[8] Fedorova O B, Chizhevskaya E L and Plenkin V M 2010 *Financial aspects of enterprises and organizations* (Tyumen: TSOGU) 164

[9] Chizhevskaya E L and Musalaev G M 2014 Features of the activities of small enterprises in the oil and gas industry *Current problems and prospects of regional and industrial development: a collection of scientific papers* (Tyumen: TSOGU) 135-41

[10] Chizhevskaya E L and Kot A D 2016 Program for the growth of the technical level of production of a production company *Journal of Economy and Entrepreneurship* **1** (part 2) 536-9

[11] Plenkina V V, Tonysheva L L, Andronova I V and Deberdieva E M 2000 *Functional management technology in complex business systems* (Tyumen: TSOGU) 194

[12] Andronova I V 2005 *Forecasting the development of economic systems in oil and gas production* (Tyumen: Vector Buk) 214

[13] Andronova I V 2003 Systematic representation of complex oil producing structures: theoretical aspect *Management: theory and practice* **3** 13

[14] Yakunina O G 2011 Possibilities of reorganizational transformations in the construction of oil and gas facilities *Current problems and prospects of regional and industrial development: a collection of scientific papers* (Tyumen: TSOGU) 364-9

[15] Andronova I V and Osinovskaya I V 2011 Making consistent decisions in the context of organizational transformations in oil production *Current problems and prospects of regional and industrial development: a collection of scientific papers* (Tyumen: TSOGU) 177-82

IOP Conf. Series: Materials Science and Engineering **663** (2019) 012011 doi:10.1088/1757-899X/663/1/012011

Modelling the stress-strain state of tubular springs in the ANSYS software package

S P Pirogov [1,2], D A Cherentsov[1]

[1] Industrial University of Tyumen, 38 Volodarskogo str., Tyumen, 625000, Russia
[2]Northern Trans-Ural State Agrarian University, Tyumen, Russia

Abstract. The paper presents the results of calculating the stresses and strains of manometric tubular springs under loading by internal pressure and external force. For research, the finite element method implemented in the ANSYS program was used. The problems of constructing a grid model of the tubular element were solved, the displacements of the free end and stresses from internal pressure were determined, and the maximum permissible pressure value was found.

1. Introduction

A manometric tubular spring is a curved non-circular tube. Under the influence of internal pressure, the tube straightens, and its end experiences displacement.

Tubular springs were originally proposed as elastic sensing elements of manometric devices, and now they are widely used as power elements of various mechanisms (brakes, manipulators, agricultural machines). [1-3]

2. Formulation of the problem

Figure 1 shows a diagram of a manometric tubular spring.

Many researchers study manometric tubular springs. The main calculation methods were energy methods [4] and the theory of flexible shells [5-8].

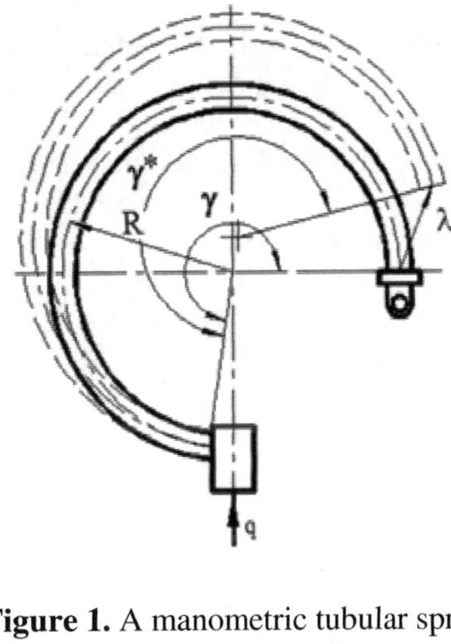

Figure 1. A manometric tubular spring.

Significant drawbacks of all methods were that they described the stress-strain state of the tubes without taking into account the end sections, which strongly affect the distribution of stresses. In addition, only small displacements were considered, that is, the geometrical nonlinearity arising from sufficiently large displacements of the tube end was not taken into account. Also, these methods did not allow studying the conditions of stability loss.

To eliminate these drawbacks, it was proposed to simulate the stress-strain state of tubular springs by the finite element method in the ANSYS software package. The application of this method also allows one to determine the natural frequencies of vibrations.

3. Theory

The accuracy of the calculations will directly depend on the quality of the grid model of the structure under consideration. A rigid constraint (attachment point) and a tip are constructed by default using the Sweep method, and in order to construct the grid model of a manometric spring, it is necessary to determine the best grid method (Tetrahedrons or Sweep) and the minimum size of the elements, which ensures the correct solution without the loss of accuracy.

Tetrahedrons method **Sweep method**

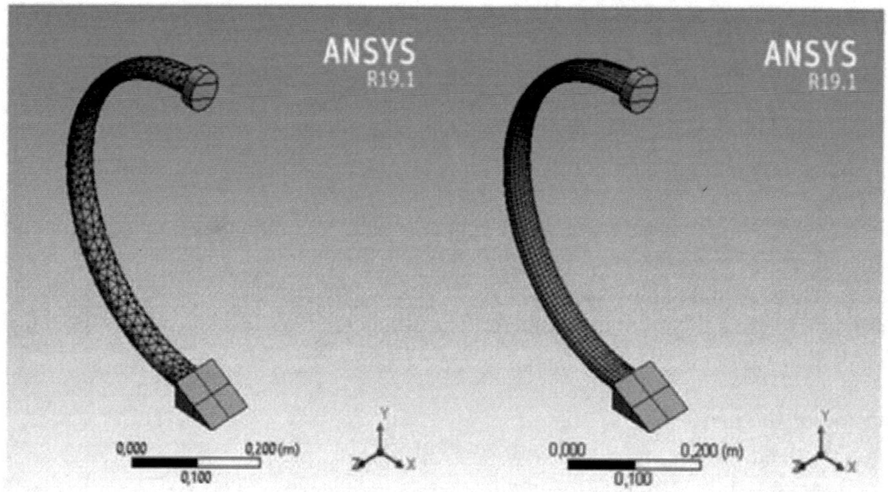

Figure 2. Building a grid model.

A tube with the following geometric characteristics was taken as a sample: the central angle is 180 degrees, the radius of curvature is 500 mm, the semi-major axis of the cross-section is 25 mm, the semi-minor axis of the cross-section is 12.5 mm, the wall thickness is 2.5 mm, the material is 36NHTY steel.

To assess the influence of methods for building the grid and element sizes, we determined the displacement of the free end of the tube under the influence of a horizontal force of 1000 N on the free end of the tube. Grid models are shown in Figure 2.

The calculations were made in toolbox - Static Structural, the results are presented in Figure 3.

Analyzing the obtained results, we can conclude that with a decrease in the size of the elements of the grid model of the tube, the values of the free end displacement tend to a certain limit. A smoother solution is obtained using the Sweep method; the minimum element size at which deviation of the calculation results is less than 0.5% is 5 mm.

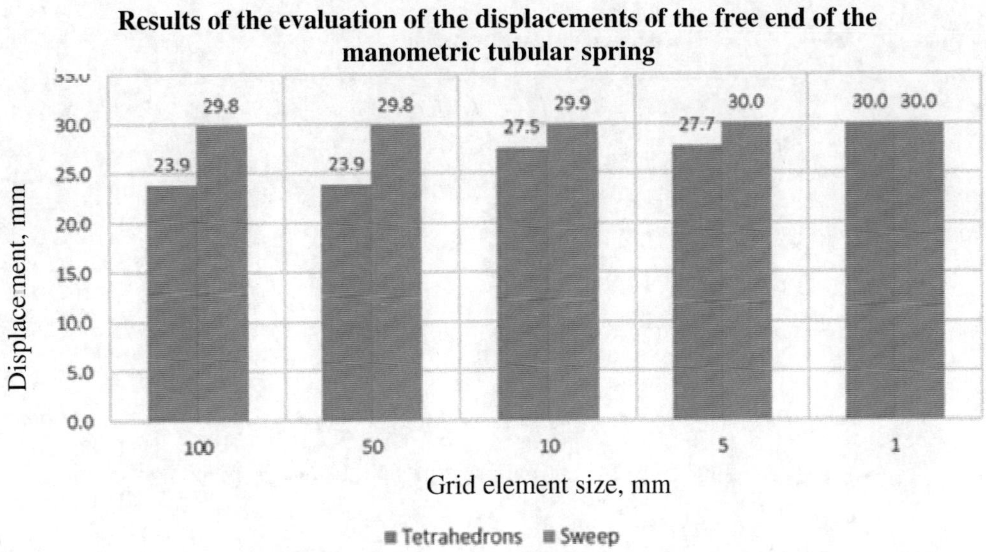

Figure 3. Results of the evaluation of the displacements of the free end of the tube.

4. Experiment results

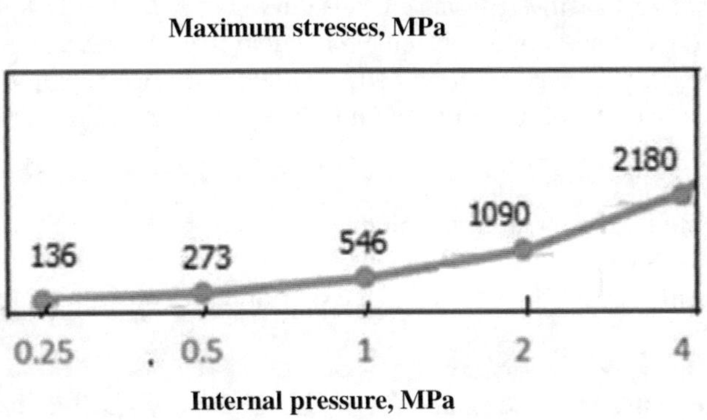

Figure 4. Influence of internal pressure on maximum stresses.

Figures 4 and 5 show the results of computational experiments to determine the maximum stresses and displacements of the tube end.

We also set and solved the task to determine the maximum value of the horizontal force R applied to the end of the tube, at which a loss of stability is observed.

The calculations were made in toolbox – Eigenvalue Buckling.

The calculation results showed that the loss of stability will occur at a force value of 12.8 kN. The maximum stresses, as well as the violation of the integrity of the manometric tubular spring, are observed at the base of the rigid restraint.

IOP Conf. Series: Materials Science and Engineering **663** (2019) 012011 doi:10.1088/1757-899X/663/1/012011

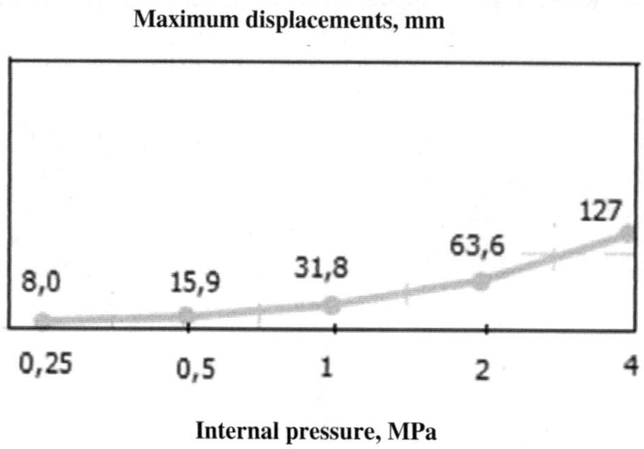

Figure 5. Influence of internal pressure on tube end displacements.

5. Discussion

When performing mathematical calculations of a tubular spring, it was found that the best way to build a grid and the minimum size of the elements, providing the correct solution without loss of accuracy, is 1 mm. Analysis of the stress-strain state under the action of an external force of 1000 N showed that the movement of the end of the tube will be 30 mm. For stability loss, the required horizontal force is at least 12800 N. This confirms a sufficient margin of safety.

An analysis of the dependences of displacements and stresses on internal pressure shows that at the initial stage of loading, at pressures up to 4 MPa, a linear increase in stresses and displacements is observed; at high pressures, a deviation from the linear dependence occurs.

6. Conclusions

The study of stresses and strains by the finite element method in the ANSYS software package showed that this method can be successfully used to determine the stresses, displacements, and stability of manometric tubular springs under the influence of external forces and internal pressure.

References

[1] Pirogov S P, Cherentsov D A and Chuba A Yu 2018 *Tubular spring mechanisms* (Tyumen: IUT) 78

[2] Pirogov S P and Chuba A Yu 2017 Application of manometric tubular springs in agricultural machines *Agro-food policy in Russia* **9** (69) 82-8

[3] Andreeva L E 1962 *Elastic elements of devices* (Moscow: Mashgiz) 456

[4] Akselrad E L 1976 *Flexible shells* (Moscow: Nauka) 376

[5] Pirogov S P 2009 *Manometric tubular springs* (St. Petersburg: Nedra) 276

[6] Pirogov S P, Cherentsov D A and Chuba A Yu 2015 *Fluctuations of manometric tubular springs* (Tyumen: TSOGU) 95

[7] Cherentsov D A, Chuba A Yu and Pirogov S P 2013 Model of oscillations of a manometric tubular spring in a viscous fluid *In the collection: Problems of the functioning of transport systems Materials of the All-Russian Scientific and Practical Conference of Students, Graduate Students and Young Scientists, dedicated to the 50th anniversary of the founding of the Tyumen Industrial Institute* 373-7

[8] Cherentsov D A, Pirogov S P and Dorofeev S M 2014 A mathematical model of a manometric spring in a viscous medium *Bulletin of the Tyumen State University. Physical and mathematical modelling. Oil, gas, energy* **7** 234-41

[9] Cherentsov D A, Pirogov S P, Dorofeev S M and Cherentsova S A 2017 Study of damped oscillations of manometric springs with a rigid tip *News of higher educational institutions. Oil and gas* **1** 116-20
[10] Cherentsov D A, Pirogov S P, Zemenkov Yu D and Chuba A Yu 2015 Results of the calculation of the attenuation parameters of manometric tubular springs *News of higher educational institutions. Oil and gas* **1** 70-3
[11] S P Pirogov, A Y Chuba and D A Cherentsov Effect of section shape on frequencies of natural oscillations of tubular springs 2018 *IOP Conf. Series: Materials Science and Engineering* 357 012032 doi:10.1088/1757-899X/357/1/012032
[12] Pirogov S P and Cherentsov D A 2016 Theoretical foundations of the design of vibration-resistant manometers *Measurement Techniques* **59** 8. 845-9
[13] Pirogov S P, Cherentsov D A and Gulyaev B A 2016 Prospects of applying vibration-resistant pressure gauges in the oil and gas industry *IOP Conference Series: Materials Science and Engineering "International Scientific-Practical Conference of Students, Graduate Students and Young Scientists "Transport and Storage of Hydrocarbons""* 012013
[14] Pirogov S P and Cherentsov D A 2016 Foundations of the Design of Vibration-Resistant Manometers *Measurement Techniques* **59** 8 845–9
[15] Pirogov S P, Cherentsov D A, Dorofeev S M and Ryabova Y S Oscillations of manometric tubular springs with rigid end *IOP Conf. Series: Materials Science and Engineering* 357 (2018) 012030 doi:10.1088/1757-899X/357/1/012030

IOP Conf. Series: Materials Science and Engineering **663** (2019) 012012 doi:10.1088/1757-899X/663/1/012012

Modeling the heat transfer processes in the pipe-soil system

V V Golik [1,4], Yu D Zemenkov [1], A A Gladenko[2] and I V Seroshtanov[3]

[1] Industrial University of Tyumen, 38 Volodarskogo str., Tyumen, 625000, Russia
[2] Omsk State Technical University, 11 Mira Ave., Omsk, 644050, Russia
[3] Gazprom Transgaz Yugorsk, 15 Mira str., Yugorsk, 628260, Russia
[4] darkpaint@mail.ru

Abstract. The paper presents a mathematical model and methodology for conducting thermotechnical calculations of buried multilayer pipelines. A section of the existing oil pipeline running under difficult geological conditions is considered. The methodology for calculating thermal processes and the main results for this section of the pipeline are described, and halos of thawing of permafrost are modeled on the basis of a mathematical model. Modeling was carried out using several options of insulating material with a constant internal diameter of the pipeline and pumping mode. Modeling was carried out in an all-purpose software system of finite element modeling - ANSYS and a TPS module developed by the authors for conducting thermotechnical calculations.

1. Introduction

The study of the force effect of frost heaving of soils on an existing pipeline is associated with freezing of soils - a complex process that depends on heat and mass transfer in frozen and thawed soil, the nature of the phase transition and chemical transformations, on a possible change in the stress-strain state of the soil massif (thawed and frozen) with subsequent structural change. Such multifactorial nature of the freezing process, the dependence of its course on many different physical and chemical parameters should be taken into account with an adequate mathematical description of the phenomenon under study. [1]

Even more difficult is the study of the mechanism of freezing of soils in the presence of thermal influence from a buried pipeline, which is explained by both a “deterioration” of the process geometry and a change in the energy carrier temperature along the route.

2. Materials and methods

The sections of the main pipelines are divided (according to the temperature of the transported energy carrier) as follows:

1) hot (the energy carrier temperature is positive at any time of the year);
2) warm (the average annual product temperature is positive, but may be negative for some time);
3) cold (the average annual energy product temperature is negative, but may be positive for some time).

The first include the entire length of oil pipelines since at low temperatures there is a significant increase in the viscosity of the oil, which requires large energy costs when pumping it.

Gas pipelines belong to the second or third type, with the exception of relatively short sections downstream compressor stations, which can be attributed to the first type. [2,3]

In general, under certain conditions, a main pipeline can have all three types of the indicated sections.

The thermal impact of buried pipelines on the dynamics of seasonal freezing of soils is determined by a combination of average annual $\overline{t_{zh}}$, maximum $t_{zh.max}$ and minimum $t_{zh.min}$ temperatures of the transported product (Figure 1).

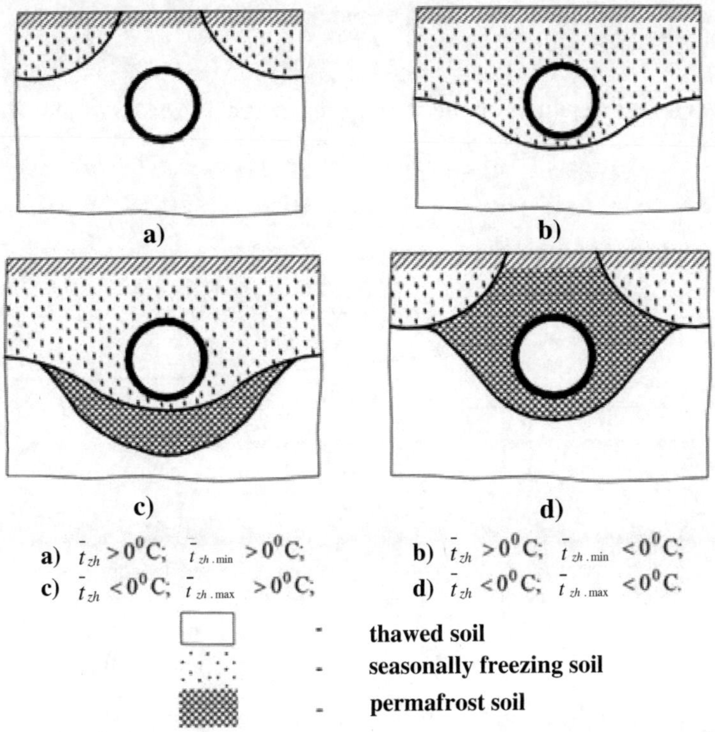

Figure 1. Thermal interaction of underground pipelines with soils

As noted above, the formation of the temperature field of freezing soil is determined by a seasonal decrease in the ambient temperature and the thermal impact of the pipeline. In other words, the task set can be solved within the framework of a joint study of heat exchange processes inside the pipeline and in freezing soil with the conjugation of temperature and heat flux in the "pipe-soil" system. [4,5]

The scheme of soil freezing along the pipeline route is shown in Figure 2, the air temperature $t_N(\tau)$ is negative, i.e. $t_N(\tau) \leq 0$. Heat removal from frozen soil to the outside air (through the snow cover) is carried out by convection with a heat transfer coefficient α_N. The depth of the pipeline laying to its axis is H_0, the mass flow rate G_{zh} of the energy carrier and its temperature t'_{zh} at the inlet of the pipeline are considered set. The following physical and thermophysical parameters of the transported product are also assumed to be known: density ρ_{zh}, kinematic viscosity ν_{zh}, thermal conductivity λ_{zh}, specific heat c_{zh} and thermal diffusivity a_{zh} . [1]

In addition to the laying depth H_0, the defining geometric parameters of the process include the internal D_{vn} and external D_n diameters of the pipeline, as well as the insulation thickness of δ. If the insulation thickness practically does not affect the outer diameter of the pipeline when calculating its force interaction with the soil, then when calculating the heat transfer coefficient from the energy carrier to the soil, it makes a significant contribution due to the low thermal conductivity of the insulation. The thermal conductivity of the pipeline material and insulation are equal to λ and λ_{iz}, respectively, the thermal conductivity of thawed λ_{Th} and frozen λ_F soils, as well as their volumetric heat capacities c_{Th} and c_F are determined using SP 25.13330.2012 based on data on the density of the soil skeleton and its moisture. [6]

The average energy carrier temperature depends on the z coordinate, measured from the inlet of the pipeline and the current time τ (in fractions of a month):

$$t_{zh} = t_{zh}(z;\tau). \tag{1}$$

The temperature fields of both thawed and frozen soils are determined by the coordinates y (vertical), x (horizontal), z (along the length of the pipeline) and time τ:

$$t_{Th} = t_{Th}(x;y;z;\tau) \text{ и } t_F = t_F(x;y;z;\tau). \tag{2}$$

The dependences for temperatures on the internal and external surfaces of the pipeline look similar:

$$t_{1tr} = t_{1tr}(x;y;z;\tau);\ t_{2tr} = t_{2tr}(x;y;z;\tau). \tag{3}$$

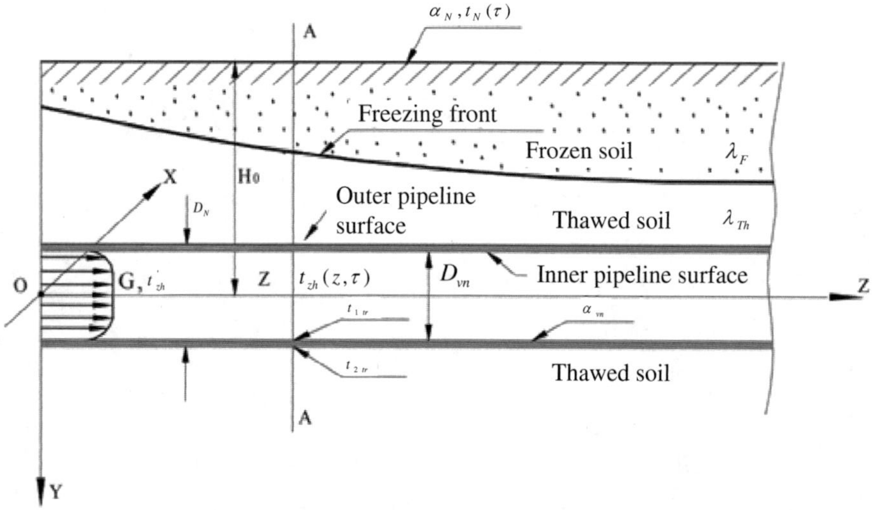

Figure 2. Soil freezing along the pipeline route during the cold period ($t_N(\tau) \leq 0$ °C)

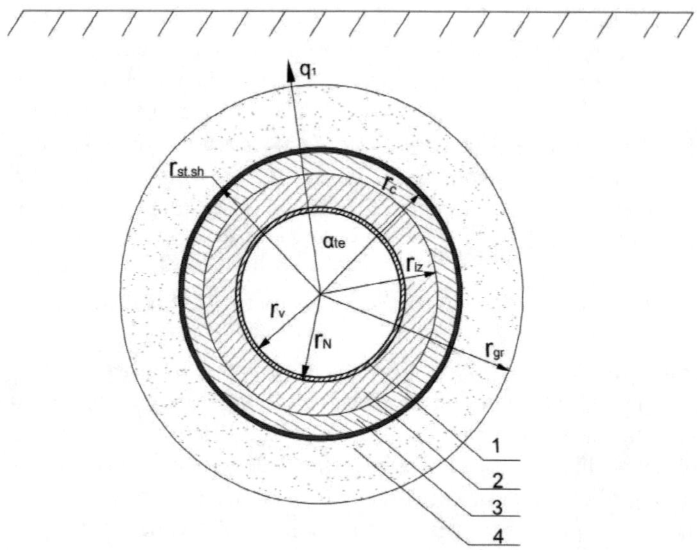

Figure 3. Calculation model of thermal interaction of an underground oil pipeline in permafrost soils: 1. steel pipe; 2. PU foam thermal insulation; 3. nano-modified concrete coating; 4 permafrost soil.

The most complete mathematical model of the processes of freezing and thawing of moist soils in the framework of a continuous medium is the Stefan problem. [7] The presence of moving interface between zones with different aggregate states of pore moisture in the soil, the law of movement of which is not known in advance, puts the Stefan problem on a par with the most complex nonlinear problems of mathematical physics. Given the above conditions, the calculation model of the heat distribution process around the underground oil pipeline is shown in Figure 3.

A mathematical model of the thermal interaction of a pipeline of radius r laid to a depth H in the soil describes the heat exchange in the soil [8]:

$$c_{ef}\frac{\partial t}{\partial \tau} = \frac{\partial}{\partial x}\left(\lambda\frac{\partial}{\partial x}\right) + \frac{\partial}{\partial y}\left(\lambda\frac{\partial}{\partial y}\right), \tag{4}$$

where t – soil temperature;

τ – calculation time.

Assuming that the phase transition of pore soil moisture occurs in a certain temperature range Δt_i, the effective heat capacity C_{ef}:

$$c_{ef} = \begin{cases} c_{Th}, & \text{at } t > t_f + \Delta t; \\ \rho_x c - \chi\gamma\frac{dW(t)}{d\tau}, & \text{at } t \in [t_f - \Delta t, t_f + \Delta t,], \\ c_F, & at\ t > t_f - \Delta t, \end{cases} \tag{5}$$

where t_f – moisture phase transition temperature;

χ – latent heat of the moisture phase transition of;

W – unfrozen moisture content;

γ – soil skeleton specific gravity;

ρ_x – coefficient taking into account the density of the soil during the subsidence of the pipeline.

The indices T and M correspond to thawed and frozen soil conditions:

$$\lambda = \begin{cases} \lambda_{Th}, \text{при } t > t_f; \\ \lambda_F, \text{при } t \le t_f. \end{cases} \tag{6}$$

If the temperature on the surface of the pipe T_{tr} is set, then for a period of 5 years the steady-state field T_n of the soil temperature is calculated:

$$T(x,y,\tau)\Big|_{y>0}^{\tau=0} = T_n. \tag{7}$$

On the surface of the ground, the equality of heat fluxes is set:

$$\left.\frac{\partial t}{\partial y}\right|_{\substack{y=0 \\ y<x<\infty}} = \frac{h_{sr}}{\lambda_{gr}}\left[t(x,y,\tau)\Big|_{\substack{y=0 \\ y<x<\infty}} - t_{sr}\right], \tag{8}$$

where t_{sr} – air temperature;

λ_{gr} – surface thermal conductivity coefficient;

h_{sr} – heat transfer coefficient of the soil surface.

The derived relation of symmetry with respect to the OY axis is the condition

$$\left.\frac{\partial t}{\partial x}\right|_{x=0}. \tag{9}$$

Boundary conditions at infinity:

$$\left.\frac{\partial t}{\partial x}\right|_{x\to\infty} = 0; \quad \left.\frac{\partial t}{\partial x}\right|_{\substack{x=0 \\ 0<y<\infty}} = 0; \quad \left.\frac{\partial t}{\partial y}\right|_{y\to\infty} = 0. \tag{10}$$

In this statement, the shock-capturing method with the introduction of effective heat capacity makes it possible to solve one general equation for the entire studied area.

However, such an approach is more applicable to one-dimensional problems and encounters great computational difficulties already in the transition to two-dimensional temperature fields. An additional complicating factor is the fact that under certain conditions it is possible to split the freezing front, which makes the approach proposed above unacceptable. [9]

In this case, we restrict ourselves to the conditions:

$$\lambda = \begin{cases} \lambda_{Th}, \text{при } t > t_f, \\ \lambda_F, \text{при } t \le t_f. \end{cases} ; \quad c_{\text{эф}} = \begin{cases} c_{Th}, & \text{при } t > t_f; \\ c_F\,, & \text{при } t \le t_f. \end{cases} \tag{11}$$

We transform the basic equation by introducing a new value - thermal diffusivity a:

$$a = \frac{\lambda}{c \cdot \rho}, \tag{12}$$

$$\frac{\partial t}{\partial \tau} = a\left(\frac{\partial^2 t}{\partial x^2} + \frac{\partial^2 t}{\partial y^2}\right). \tag{13}$$

We expand the differential equation in a Taylor series. Let

$$\frac{\partial t}{\partial \tau} = \frac{t_i^{j+1} - t_i^j}{\Delta\tau}, \tag{14}$$

then

$$\frac{\partial^2 t}{\partial x^2} = \frac{t_{x+1}^j + t_{x-1}^j - 2t_x^j}{(\Delta x)^2}, \quad \frac{\partial^2 t}{\partial y^2} = \frac{t_{y+1}^j + t_{y-1}^j - 2t_y^j}{(\Delta y)^2}. \tag{15}$$

Thus, expressing t_i^{j+1}, we obtain a numerical method for solving the equation (1):

$$t_i^{j+1} = t_i^j + a \cdot \Delta\tau \cdot \left(\frac{t_{x+1}^j + t_{x-1}^j - 2t_x^j}{(\Delta x)^2} + \frac{t_{y+1}^j + t_{y-1}^j - 2t_y^j}{(\Delta y)^2}\right). \tag{16}$$

The software implementation of this model was carried out in Visual Basic for Applications (VBA) and ANSYS.

Initial calculations of the mathematical model were carried out in VBA, boundary conditions were set corresponding to the information obtained during the examination of the Yarudeyskoye oil field pipeline (Figure 4). In order to increase the accuracy of the calculation, an approximation is made of the function of the temperature change on the soil surface using the MATLAB software package (Figure 5). [10]

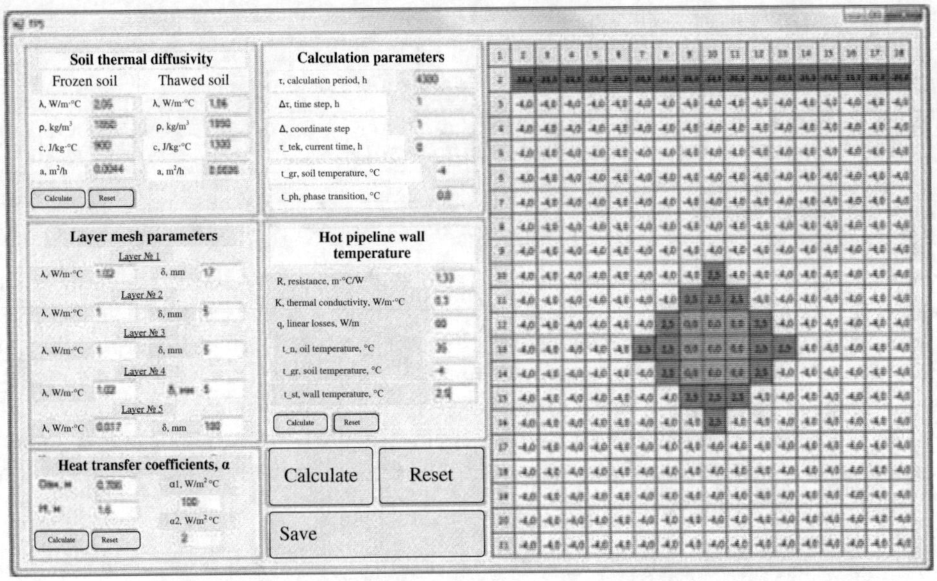

Figure 4. Entering initial conditions in the program interface

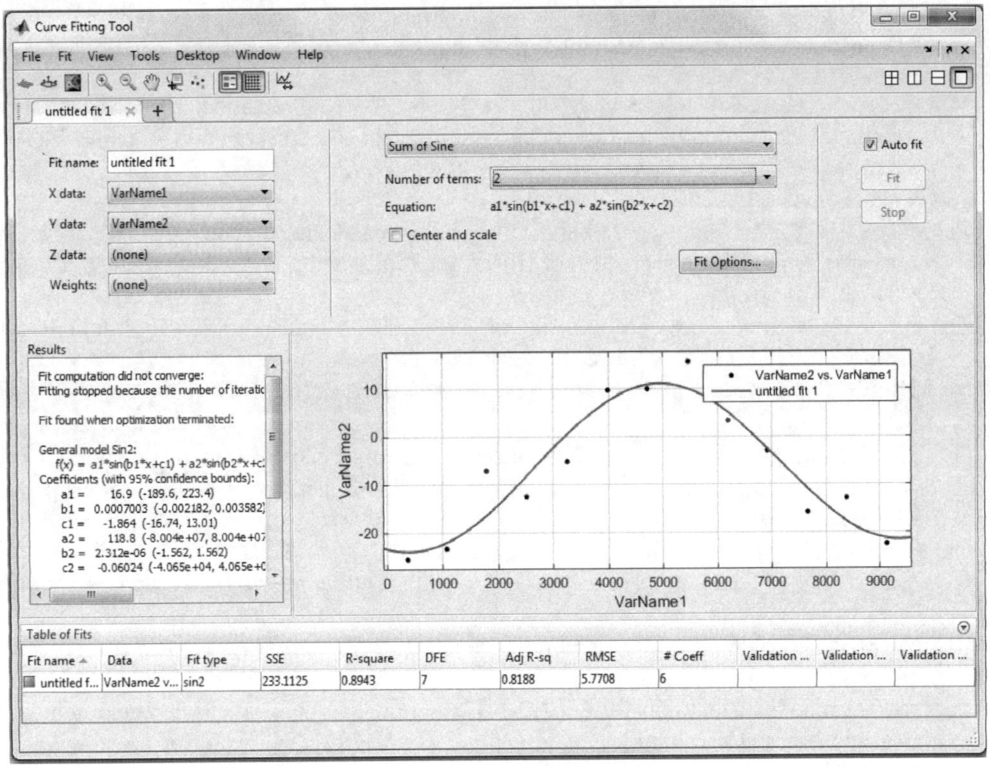

Figure 5. Approximation of the function of the ambient temperature in MATLAB

Table 1 presents the average monthly ambient temperature according to the weather station WMO ID 23256 in Tazovsky region.

Table 1. Average monthly ambient temperature in °C

Jan 2017	Feb 2017	Mar 2017	Apr 2017	May 2017	Jun 2017	Jul 2017	Aug 2017	Sep 2017	Oct 2017	Nov 2017	Dec 2017	Jan 2018
-25.6	-23.3	-7.1	-12.5	-5.2	9.8	10	15.7	3.3	-3	-15.8	-12.9	-22.5

3. Conclusion

The developed mathematical model includes a large number of variables, including coefficients developed by the team of authors under the guidance of Professor B.V. Moiseev, which allows for the most accurate description of the processes occurring in the "pipe-soil" system over a sufficiently long period of time. When conducting mathematical modeling according to the developed methodology, results were obtained that suggest that it is advisable to use multilayer pipelines in the Arctic. The developed TPS program based on the above mathematical model will allow design engineers at the design stage to have complete data on the behavior of the future structure for many years to come.

References

[1] Under general ed. of Grigoriev V A and Zorin V M 1982 *Heat and mass transfer. Thermotechnical experiment. Reference book* (Moscow: Energoizdat) 512

[2] Zemenkov Yu D, Moiseev B V, Bogatenkov Yu V, Nalobin N V and Dudin S M 2015 *Energy-technological complexes in the design and operation of transport and storage facilities of hydrocarbon raw materials* Under general ed. of Moiseev B V (Tyumen: TSOGU, TyumGASU) 256

[3] Moiseev B V, Kushakova N P and Nalobin N V 2002 Mathematical modeling of unsteady temperature interaction of heat conductors and seasonally freezing soils *Composite building materials: Collection of scientific papers of the international scientific-practical conference.* (Penza: PenzGASA) 261-4

[4] Zemenkova M Y, Shastunova U, Shabarov A, Kislitsyn A and Shuvaev A 2018 Physical and mathematical modeling of process of frozen ground thawing under hot tank *IOP Conference Series: Materials Science and Engineering*

[5] Zemenkova M Methodology of Monitoring of Hydrocarbon Transportation Hydraulic Reliability 2018 *Advances in Intelligent Systems and Computing*

[6] SP 25.13330.2012 *Bases and foundations on permafrost soils. Updated edition of SNiP 2.02.04-88* (with changes 1, 2)

[7] Golik V V, Moiseev B V, Gulkova S G and Zemenkov Yu D 2018 Mathematic simulation of the effect of a buried oil pipeline on permafrost soils *IOP Conference Series: Materials Science and Engineering*

[8] GOST 25100-2011 *Ground. Classification* (Amended)

[9] GOST 26263-84 *Ground. Laboratory method for determining the thermal conductivity of frozen grounds*

[10] RD-75.180.00-KTN-198-09 2009 *Unified technological calculations of facilities of main oil pipelines and oil product pipelines*

[11] Fedorova O B and Chizhevskaya E L 2014 Adapting intellectual property evaluation methods to the region brand evaluation *Middle - East Journal of Scientific Research* DOI: 10.5829/idosi.mejsr.2014.19.1.12477

[12] Balikaeva M B, Chizhevskaya E L, Ya Grevtseva G, Kotlyarova I O and Volkova M A 2018 Innovative technologies as a means of the development of future engineers' professional mobility abroad *IOP Conference Series: Materials Science and Engineering doi:10.1088/1757-899X/441/1/012007*

[13] Chizhevskaya E L, Fedorova O B and Kot A D 2013 Regional branding as factor of achievement of competitiveness of the territory *World Applied Sciences Journal*

[14] Stepanov O, Moiseev B, Chekardovskiy M, Aksenov B and Shapoval A 2017 Physical and mathematical conditions of non-stationary thermal conditions of the underground air channels *International Journal of Applied Engineering Research*

[15] Moiseev B, Zemenkov Y, Nalobin N and Dudin S 2017 Thermal calculations of underground oil pipelines *MATEC Web of Conferences*

IOP Conf. Series: Materials Science and Engineering **663** (2019) 012013 doi:10.1088/1757-899X/663/1/012013

Temperature condition of a stopped underground oil pipeline

A U Yakupov[1], K S Voronin[1] and D A Cherentsov[1]

[1]Industrial University of Tyumen, 38 Volodarskogo str., Tyumen, 625000, Russia

Abstract. Carrying out thermal calculations of "hot" oil pipelines, through which highly viscous and high pour point oils are transported, is one of the main ones. Determining the cooling rate of oil in a stopped oil pipeline is of great practical importance. The cooling time is used to calculate the safe stop time of the "hot" oil pipeline. The time at which there will be no freezing of the "hot" oil pipeline, and the station pressure is enough to overcome the shear stresses that arise when cooling highly viscous and high pour point oil. The methods of finding the cooling time of oil in a stopped pipeline are considered, which allow determining the time of a safe stop of the pipeline. Installed thermal stabilizers designed for cooling the soil can affect the cooling rate of the "hot" oil pipeline.

1. Introduction

Thermal calculation of the pipeline is an important task of pipeline transport to ensure reliability and safety, especially when laying in the Far North. [1,8,10-13] Since the majority of oil pipelines in Russia run in severe climatic conditions, the calculation of the rate of cooling of oil in the pipeline determines a safer time to stop it. With a long shutdown, highly viscous petroleum products can freeze, which will lead to significant economic costs. Thus, the correct calculation of the cooling time of oil and oil products allows one to determine the period during which the resumption of pumping will not have complications.

2. Object of study

Currently, there are several methods for determining the cooling time of oil in the pipeline. The thermal calculation of safe oil pipeline shutdown time, described in RD 39-30-139-79, was developed by Vadim Ivanovich Chernikin, Doctor of Technical Sciences, Professor. [2]

In this method, the heating process of the soil-pipeline system is determined to a greater extent by the rate of heating of the soil surrounding the pipeline. In turn, the heat of the oil product is neglected, since it is small relative to the heat accumulated in the soil. The soil surrounding the pipeline is considered to be isotropic with invariable physical constants. Thus, the process of cooling the oil product is due to the cooling rate of the soil.

The cooling process is characterized by the cooling rate or dimensionless temperature, determined by the following formula:

$$\theta = 1 - \frac{E_i(-h_0^2/R_2^2 \cdot F_0) - E_i(-1/4F_0)}{2\,ln(2h_0/R_2)}, \quad (1)$$

where E_i – sign of integral exponential function;

h_0 – pipeline laying depth, along the axis, m;

R_2 – outer pipeline radius, m;

F_0 – Fourier parameter calculated by the formula:

$$F_0 = \frac{a' \cdot T_{cool}}{R_2^2}, \quad (2)$$

where a' – thermal diffusivity, m²/s;

T_{cool} – duration of the pipe cooling process, s.

At $h_0/R_2^2 \geq 4$ the value $E_i(-h_0^2/R_2^2 \cdot F_0)$ almost tends to zero. With this assumption, the formula will take the following form:

$$\partial = 1 + \frac{E_i(-1/4F_0)}{2\, ln(2h_0/R_2)} \quad (3)$$

When the pipeline cools down, the process of changing the temperature of the oil product along its length is described by the dependence:

$$t = t_0 + \Theta(t_{in} - t_{fin})e^{-Shu\frac{l}{L}}, \quad (4)$$

where t_0 – ambient temperature around the pipeline;

t_{in} – initial fluid temperature;

t_{fin} – final fluid temperature;

Shu – Shukhov's parameter;

l – current section length;

L – full section length.

If during a stop $T_{ост}$=0, then $\Theta = 1$ and we have Shukhov's law for the distribution of oil temperature along the length of the pipeline for stationary operation.

The value of the heat transfer coefficient from the pipe to the ground during cooling is described by the formula:

$$a_2 = \frac{a_{2\infty}}{1+4F_0}, \quad (5)$$

where $a_{2\infty}$ – heat transfer coefficient from the pipe to the ground in the outer surface of the pipeline to the environment during a stationary heat transfer mode, determined by the formula:

$$a_{2\infty} = \frac{2\lambda_{cr}}{D_2\, ln\frac{4h_0}{D_2}}, \quad (6)$$

where λ_{cr} – critical value of the thermal conductivity coefficient.

This formula is used only for $h_0/D_2 \geq 3$. With a smaller relative depth of laying, as well as with a depth of backfill of the pipe less than 0.7 m, it is necessary to take into account the thermal resistance at the soil - air boundary.

Another method is described in the book by Alexander Leonidovich Yastrebov "Engineering communications on permafrost soils." [3] When calculating the cooling time of the liquid in the pipeline, it was assumed that heat loss does not depend on the flow rate, since friction heat affects only at speeds greater than 2 m/s. The cooling time of the liquid during the underground laying of the pipeline will be calculated by the formula:

$$\tau_{ОСТ} = \frac{d_1^2 \gamma c}{8\lambda_T} ln\left(\frac{2h}{d_2} + \sqrt{\frac{4h^2}{d_2^2} - 1}\right) ln\frac{\lambda_T t_{in} - \lambda_F t_s}{\lambda_T t_{fin} - \lambda_F t_s}, \quad (7)$$

where d_1 – internal diameter of the pipeline, m;

d_2 – outside diameter of the pipeline, m;

λ_T – coefficient of thermal conductivity of thawed soil, kcal/h·deg;

λ_F – coefficient of thermal conductivity of frozen soil, kcal/h·deg;

t_s – soil temperature at the laying depth of the pipe axis, assuming that the pipe is absent, deg;

h – pipe laying depth from the surface, m;

γ – fluid volumetric weight, kg/m³;

c – specific heat of liquid, kcal/kg ·deg;
t_{in} – initial fluid temperature before cooling begins, deg;
t_{fin} – final fluid temperature, deg.

In the case when the pipeline has been working for a long time with a constant flow rate, a thawed zone will form around the pipe, and the adjacent frozen soil will be substantially warmed up. The cooling time of the liquid can be calculated using this formula:

$$\tau_{\text{ОСТ}} = -\frac{\rho G_W d_2^2}{8\lambda_F t_s}\left\{\left[\frac{(d_2+2\xi)^2}{d_2^2} - 1\right] ln\frac{6{,}8h}{d_2} - -\frac{(d_2+2\xi)^2}{d_2^2} ln\frac{(d_2+2\xi)^2}{d_2^2}\right\}, \quad (8)$$

where ρ=80 kcal/kg - heat of melting ice;
ξ – thickness of thawed soil layer around the pipe, m;
G_w – weight of water and ice per unit volume of soil, kg/m^3, determined by the formula:

$$G_W = \frac{W}{1+W}\gamma_0, \quad (9)$$

where W – soil water content in fractions of a unit; γ_0 – volumetric weight of frozen soil, kg/m^3. In formula (8), the remaining notations are the same as in formula (7).

Formula (8) should be used when operating the pipeline for more than 5 years, provided that most of this time it works at a constant flow rate. For the initial period of operation of the pipeline or during its work with interruptions, formula (7) should be applied.

The considered basic methods for calculating the safe shutdown time of an underground oil pipeline have significant drawbacks. There are errors in the calculations using formula (1) with the results of experiments conducted on a pilot plant. Attempts have been made to compare the calculation results by the analytical formula with the results of numerical modeling [4], where differences were also obtained in solutions. The author of [4] obtained the estimated time intervals for safe shutdown of the pipeline which are less than the time allotted according to the plan for the elimination of possible accidents for facilities of the main pipeline, which provides for an increase in time by 30-50% depending on the complexity of the route.

It is necessary to refine the calculation formula to determine the time for safe shutdown of an underground oil pipeline. We will consider taking into account additional factors and parameters that may affect the cooling time of an underground oil pipeline.

Reliability of pipelines is laid at the design stage during calculations. Improving the calculation methods which will increase the accuracy of the results will ensure the reliability and safety of pipelines laid in the harsh climatic conditions of the Far North and the Arctic.

Oil product cooling in underground pipelines depends on the temperature of the soil around it. A "hot" stopped oil pipeline exchanges heat with the surrounding soil, which has not the best heat-conducting properties, and is close to insulators in terms of thermal conductivity [7]. The cooling process will take place due to the difference in temperature between the soil and the pipeline.

Installed thermal stabilizers designed to cool the soil to the design value [6] and maintain this state throughout the entire life cycle should affect the cooling rate of oil in the pipeline. As the heat stabilizers used for these purposes contribute to the additional cooling that the soil receives during the active period of operation — in winter [5]. The rate of hydrodynamic processes inside the device is much higher than the rate of temperature redistribution in the soil. Therefore, at negative atmospheric temperature, the thermal stabilizer will intensively cool the soil under the pipeline, which will affect the process and the cooling rate of oil.

3. Methods of study

In this work, a numerical experiment using modern mathematical modeling tools is carried out. Numerical modeling of the process of cooling oil in an underground oil pipeline was carried out using modern mathematical modeling tools.

The authors of the article proposed and implemented a calculation scheme: a model of a stopped oil pipeline with heated oil inside and a thermal stabilizer installed next to the oil pipeline in a two-dimensional setting. According to the problem condition, the soil was set homogeneous, therefore, to optimize the calculations, the model was divided into two symmetric parts and the symmetry condition was set on the border in the Model tab. The model is shown in Figure 1.

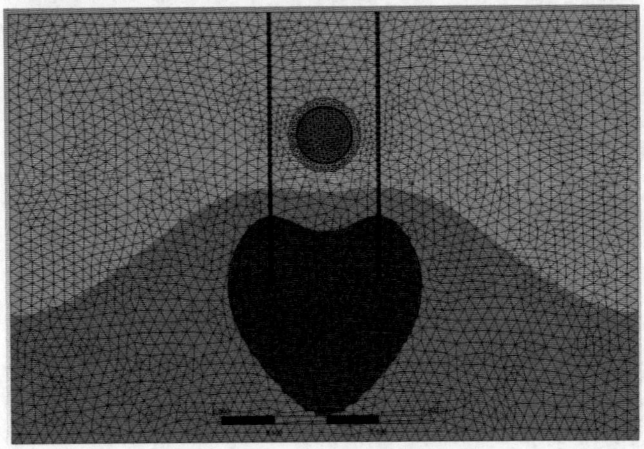

Figure 1. Oil pipeline model

The boundary conditions, based on the statement of the problem, the Neumann conditions (zero heat flux), Heat Flow command, are set at the boundaries of the simulated soil mass from below and from the side.

$$-\lambda\left(\frac{\partial T}{\partial \bar{n}}\right)_w = q_w(x, y, z, t) = 0 \tag{10}$$

where w– normal to the surface of the body.

On the surface, a boundary condition is set - the interaction of soil with the atmosphere - taking into account convective heat exchange with the atmosphere, the Heat Flow command.

$$-\lambda\left(\frac{\partial T}{\partial \bar{n}}\right)_w = q_w(x, y, z, t) \tag{11}$$

Also, the freezing power of the thermal stabilizer taking into account fluctuations in ambient temperature will be taken into account. We set the boundary condition on the outer wall of the evaporator, which is in contact with the soil according to the formula from [4]

$$\frac{a_c \cdot S_c}{S_{ev}} \cdot (t_a - t_g) = -\lambda_f \cdot \left(\frac{\partial t}{\partial r}\right)_{R1} \tag{12}$$

where a_c – heat transfer coefficient from the condenser part;
S_c – condenser area;
S_{ev} – evaporator area;
t_a – air temperature;
t_g - temperature on the outer wall of the evaporator;
λ_f – soil thermal conductivity;
t – soil temperature depending on the radial coordinate r and time τ.

4. Results

When determining the safe shutdown time, the main step is to determine the cooling time of the oil product in the underground pipeline. Existing methods do not allow taking into account all parameters that may affect the cooling time. Oil product cooling in underground pipelines depends on the temperature of the soil around it. The stopped oil pipeline exchanges heat with the surrounding soil, which has not the best heat-conducting properties, and is close to insulators in terms of the coefficient of thermal conductivity. The cooling process will take place due to the difference in temperature between the soil and the pipeline. It was discovered that the installed thermal stabilizers designed to prevent thawing of permafrost soil can affect the cooling time. During operation, soil thermal stabilizers freeze the soil adjacent to the underground oil pipeline, which receives a negative temperature and creates an additional temperature difference. Using the proposed formulas by the author and carrying out numerical modeling, using modern mathematical modeling tools, the process of cooling oil in an underground oil pipeline, it is possible to assess the degree of influence of the established thermal stabilizers on the cooling time of the oil product.

References

[1] Aliev R A et al 1988 *Pipeline transport of oil and gas* (Moscow: Nedra) 368

[2] Chernikin V I 1958 *Pumping of viscous and high pour point oils* (Moscow: Gostoptekhizdat) 168

[3] Yastrebov A L 1972 *Engineering communications on permafrost soils* (Leningrad: Building Literature Publishing House) 175

[4] Suleymanov V A 2018 Estimation of the time for safe shutdown of the pipeline through which high pour point oil is pumped *Scientific and technical collection · NEWS OF GAS SCIENCE* **2** 36–43

[5] Gorelik Ya B and Seleznev A A 2016 On the efficiency of fins of vertical two-phase naturally convective cooling devices for construction on frozen soils *Earth's Cryosphere* (Tyumen: Earth's Cryosphere Institute) **XX** 2 78-89

[6] Holubec I and Eng P 2008 Flat loop thermosyphon foundations in warm permafrost *Prepared for Government of the NT Asset Management Division Public Works and Services and Climate Change Vulnerability Assessment Canadian Council of Professional Engineers*

[7] Cheng X et al 2009 *Numerical simulation of a buried hot crude oil pipeline during shutdown* 73-82

[8] Voronin K S et al 2018 Modernization of the individual device for temperature stabilization of the soil *IOP Conf. Ser. Mater. Sci. Eng.* **445** 1 012011

[9] Yakupov A U 2017 Individual seasonally operated cooling devices *West Siberian Oil and Gas Congress a collection of scientific papers of the XI International Scientific and Technical Congress of the Student Branch of the Society of Petroleum Engineers - Society of Petroleum Engineers (SPE)* 139-40

[10] Markov E V, Pulnikov S A, Sysoev Y S 2018 Methodology for calculating the safe stop time of underground pipeline with high pour point oil *International Journal of Civil Engineering and Technology* **9** 8 1699-705

[11] Markov E V, Pulnikov S A and Uporov I A 2016 Thermal calculation of buildings on permafrost soils. Optimization of the methodology for choosing the dimensions of the design area *Science and technology of pipeline transportation of oil and oil products* **4** (24) 48-53

[12] Dudin S, Bahmat G, Zemenkov Y, Voronin K and Shipovalov A 2017 Strategy for monitoring and ensuring safe operation of Russian gas transportation systems *MATEC Web of Conferences* https://doi.org/10.1051/matecconf/201710606004

[13] Zemenkova M, Zemenkov Y, Gladenko A and Podorozhnikov S 2016 Estimation of Emissions during Monitoring of Pipelines in the Dynamic Mode of Operation *MATEC Web of Conferences DOI: 10.1051/matecconf/20168604053*

[14] Dudin S M, Zemenkov Y D, Maier A V and Shabarov A B 2016 Research and Design of Thermophysical Gas-Liquid Mixture Parameters in Product Pipelines *IOP Conference Series: Materials Science and Engineering doi:10.1088/1757-899X/154/1/012021*

[15] Fedorova O B and Chizhevskaya E L 2014 Adapting intellectual property evaluation methods to the region brand evaluation Middle - East Journal of Scientific Research DOI: 10.5829/idosi.mejsr.2014.19.1.12477

[16] Balikaeva M B, Chizhevskaya E L, Ya Grevtseva G, Kotlyarova I O and Volkova M A 2018 Innovative technologies as a means of the development of future engineers' professional mobility abroad *IOP Conference Series: Materials Science and Engineering doi:10.1088/1757-899X/441/1/012007*

IOP Conf. Series: Materials Science and Engineering **663** (2019) 012014 doi:10.1088/1757-899X/663/1/012014

Improving the production activities of a unit of an industry-specific company

E L Chizhevskaya[1,2] **and O B Fedorova**[1]

[1] Industrial University of Tyumen, 38 Volodarskogo str., Tyumen, 625000, Russia
[2] chizel76@yandex.ru

Abstract. The goal of finding ways to optimize the production and economic activities of enterprises is usually cost optimization. In this case, the solution of the indicated problem should occur without compromising the organization of production processes of the enterprise units. In this regard, the most urgent is the search for possibilities to increase the efficiency of the use of resources available to the economic entity. The article presents the results of an analysis of indicators of production activity of a unit of an industry-specific company, considers the areas of searching for and utilization of reserves for the growth of production organization efficiency.

1. Introduction

An integral part of the strategy to increase the efficiency of the core business of any production company is to ensure cost control and the desire to its constant reduction. At the same time, the cost-cutting program implemented by the enterprise is based on measures that allow the use of internal reserves to reduce the cost of materials, fuel and energy, increase the efficiency of the use and operation of fixed assets.

The object of the study, the purpose of which was to search for opportunities for increased efficiency and, as a result, production, financial and economic indicators of the enterprise, was the production structural unit of one of the leading Russian oil companies.

2. Materials and methods

An important condition for the smooth operation of the enterprise is the full security of the need for material resources with sources of coverage [1, 3]. Analysis of material and technical supply showed incomplete implementation of the plan for the supply of material resources, delays in supplies. The organization of transport services revealed a lack of vehicles. These shortcomings became the reasons for unproductive time during the repair of wells.

When analyzing the organization of repair work, unproductive time was revealed, i.e. downtime for reasons such as waiting for transport, lack of material and technical resources and pipes.

With a significant distance of oil production areas from the Central pipe base, transporting pipes over long distances becomes technically and economically inexpedient. In these cases, in the immediate vicinity of promising and existing production areas, it is possible to create a rolling repair shop or a pipe repair site. At the enterprise under study there is no such workshop and tubing repair, as indicated above, is carried out at the pipe base. This leads to some disadvantages, such as high transportation costs, unproductive loss of time, including due to untimely delivery of pipes, poor-quality repair, etc. [7]

The number of well accidents has increased for various reasons, including because of poor-quality pipe repair, their use without rejection, poor quality of pipe metal, careless handling of pipes during loading, unloading, transportation and storage.

Figure 1 shows a scheme of reserves to increase production efficiency [2, 4].

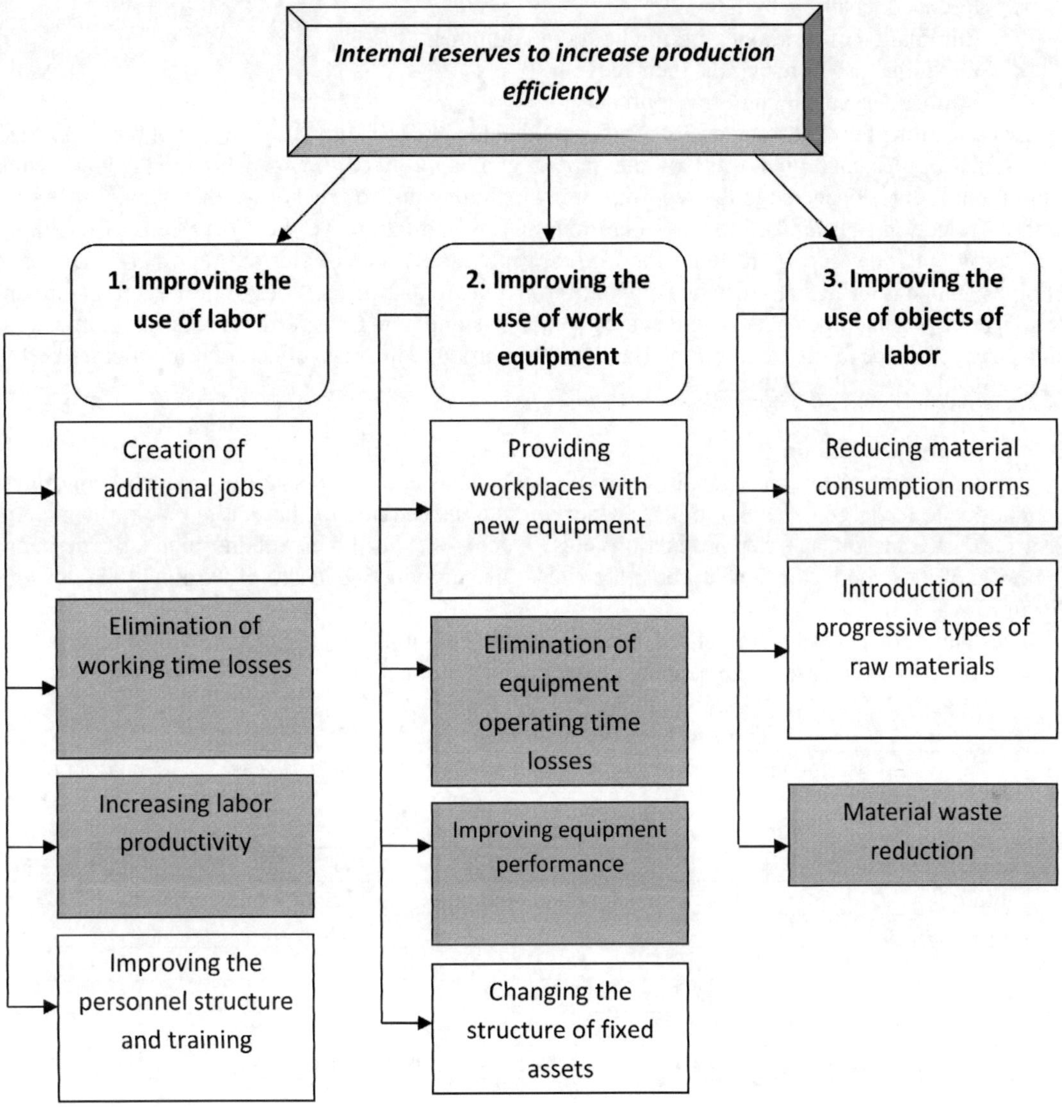

Figure 1. Reserves for increasing production efficiency

Based on the analysis of the effectiveness of the study object and the analysis of the organization of auxiliary production of the enterprise, it is possible to identify the most important reserves for increasing the efficiency of production activities in the following areas:

- use of material and technical base;
- use of labor in time and technical parameters;
- introduction and use of advanced technology and technical equipment of production;
- increase in production organization efficiency.

To increase the efficiency of the production activities of the structural unit, which consists in increasing the volume of production at minimal cost, based on the identified reserves and following the classification of areas for increasing the efficiency of production activities [9, 15], the following areas can be proposed in which it is necessary to develop measures to increase the efficiency of production activities:

1. Reduced repairs.
2. Elimination of losses of working hours of equipment.
3. Economic use of materials, their recycling.
4. Rational organization of transport services.

Investigation of the causes of accidents is carried out by the reliability department of the Engineering-Economic Implementation Center of the parent company. According to their data, 94 downhole equipment failures occurred in the reporting year, including due to breakdown of tubing, due to its use without rejection, substandard repair, operation and transportation. To meet the needs of drilling and production, the oil company includes the Central Pipe Base, which carries out repair and maintenance of pipes, which is located at a distance of 90 km from the production facilities of the structural unit under study. Transporting pipes over such distances requires significant transportation costs and time. With a significant distance of the Central Pipe Base from oil production areas, transportation of pipes becomes economically inexpedient. [8, 14].

3. Results and Discussion

Analysis of the use of transport by type of service showed a large share of its load being internal freight due to the large volume of pipe transportation to the central pipe base, as a result of which there is a lack of transport at other production sites. Incomplete fulfillment of the plan of procurement, violation of its smooth flow also became the causes of unproductive losses of working time during the repair of wells.

These and other reasons influenced the increase in the number and duration of well repairs (Table 1), which led to a decrease in the potential volume of oil production.

Table 1. Data on the number and duration of repairs

Indicator	Previous year	Reported year		Deviation from	
		plan	fact	previous year	plan
Total repairs including:	1932	1984	2002	+70	+18
- routine	1681	1722	1737	+56	+15
- major	251	262	265	+14	+3
Average duration of 1 repair, hour:					
- routine	83.1	-	85.6	+2.5	-
- major	601.3	-	608.8	+7.5	-
Overhaul period, days	433	439	435	+2	-4

In the reported year, the total number of repairs increased both in comparison with the previous one and in comparison with the plan. The average duration of one repair has also increased. As a result, the overhaul period of the wells was reduced by 4 days.

When analyzing the number of repairs, it is very important to find out the reasons for the deviations of the actual number of repairs from the planned one, since the increase in the duration of repairs is the cause of the decline in oil production [5, 6]. So, in the reported year, the number of routine and major repairs increased compared with the plan by 15 and 3, respectively. An increase in the number of routine and major repairs compared to the plan indicates unplanned downhole equipment failures and accidents, as planned underground repairs are made to prevent them.

A comparative analysis of the balance of working hours of well workover for the study period indicates a change in the ratio of production and non-production time. In the previous year, productive time amounted to 94.26% of the total time structure, in the reporting year it decreased slightly and amounted to 93.11%. The rest of the time in the balance of workover for the reporting year - 6.89% - was unproductive time, i.e. downtime. Table 2 shows the distribution of crews' downtime by causes of downtime.

Table 2. Causes for the downtime of workover crews

Causes of downtime	Previous year		Reported year		Deviation	
	Total hours	Share, %	Total hours	Share, %	hours	%
Weather conditions	4609	27.7	9385	43.9	+4776	236.8
Lack of fuel and lubricants	18	0.1	44	0.2	+26	244.4
Waiting for materials	1857	11.1	2381	11.1	+524	128.2
Elevator repair	1790	10.7	1334	6.2	-456	74.5
lack of special vehicles and equipment	3667	22.0	2869	13.5	-798	78.2
No kill fluid	496	3.0	509	2.4	+13	102.6
No vehicles for moving	1286	7.7	1085	5.1	-201	84.4
Waiting for Geophysicists	246	1.5	321	1.5	+75	130.5
No shift transport	0	0	5	0	+5	-
Waiting for technical operations	89	0.5	76	0.4	-13	85.4
Waiting for pipes	678	4.1	758	3.6	+80	55.9
No electricity	140	0.8	156	0.7	+16	111.4
Repair of equipment, tools	621	3.7	594	2.8	-27	95.7
Understaffing	351	2.1	50	0.2	-301	14.2
Stop of Regional Surveying Company's work	838	5.0	1798	8.4	+960	214.6
TOTAL	16686	100	21365	100	4679	128.0

As can be seen from the Table, non-productive time in the reporting year increased by 4679 hours compared to the previous year. The main reason for this increase is the increase in downtime due to weather conditions, which amounted to 9385 hours in the reporting year versus 4609 in the previous one. Downtime due to lack of materials also increased (by 524 hours), but their share in the structure did not change. Indicators such as elevator repair and lack of special machinery and equipment for moving dropped and the decrease in downtime for these reasons amounted to 456, 798 and 201 hours, respectively. An indicator such as waiting for pipes increased by 80 hours, and downtime due to understaffing decreased by 301 hours.

4. Conclusion

The foregoing has become the basis for developing a program to improve the efficiency of production activities of the company. In order to improve production and economic activities, it is advisable to implement the following measures (Figure 2) [10, 11-13]:

1. Opening of a workshop for repair of tubing in the pre-production workshop with the use of the new Tuboskope Vetko flaw detection unit during repair. The Tuboskope Vetko unit is a fully computerized inspection installation for monitoring and tubing. This is a fundamentally new modern equipment, which yielded positive results when tested and used in a similar profile enterprise. The unit is designed to conduct research of tubing and pipe products made of ferromagnetic material, with the aim of identifying transversely oriented defects, corrosion pits and metal loss of the pipe wall, as well as monitoring the grade of steel. The production line with the Tuboskope Vetko unit allows repairing one tubing up to 2-3 times, which increases the useful life of tubing 1.5-2 times and improves the quality of their repair. Pipe inspection allows reducing the number of repairs to 9% per year due to a more

accurate determination of the suitability of the pipe for further operation and the quality of repairs. The practice of restoring damaged parts and assemblies instead of acquiring new ones.

2. Justification and introduction of the required number of missing pieces of equipment to reduce downtime due to lack of transport.

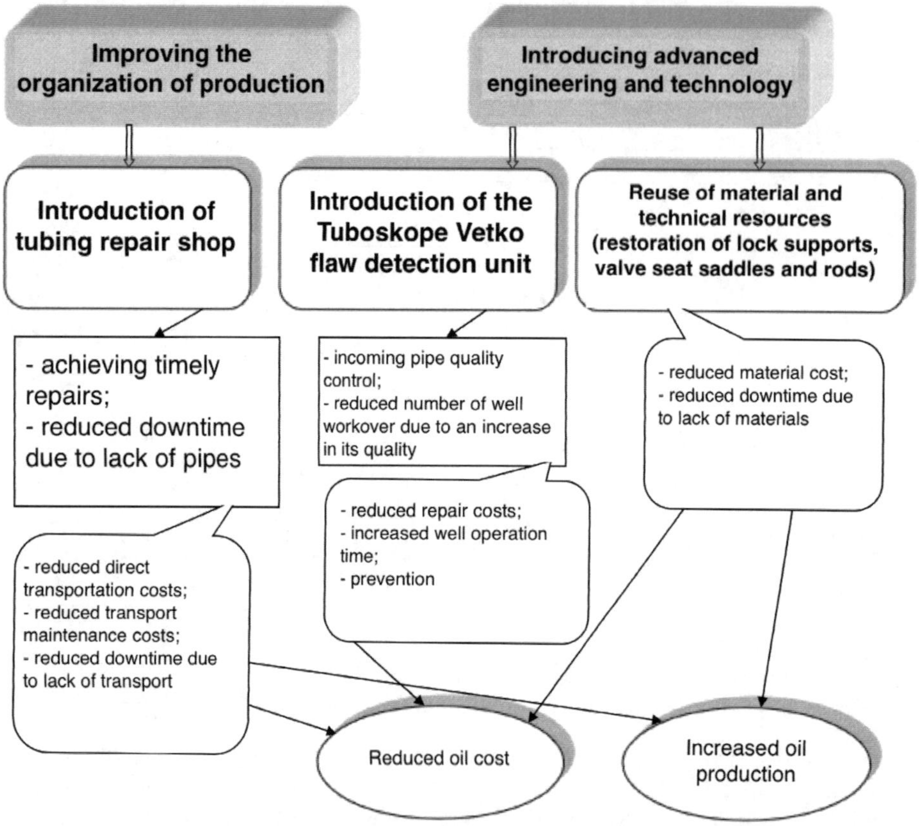

Figure 2. Areas for improving the production activities of the oil company unit

The proposed program of measures includes significant changes in the organization of production activities, namely, the rejection of the services of the Central Pipe Base for the repair and maintenance of tubing and involves the creation of its own pipe repair shop, the purchase of a repair line with flaw detection to make pipe repairs more efficient. An effect is expected from the installation in the form of a reduction in the number of repairs due to a reduction in accident rate and unplanned failures of downhole equipment.

The organization of its own tubing repair shop will require additional costs for its creation and maintenance, but the refusal of the Central Pipe Base services frees up funds that should cover the costs of maintaining the tubing repair shop. In addition, cost savings are expected due to a reduction in transportation costs in terms of transporting pipes to the Central Pipe Base.

The proposed measure to save material and technical resources involves specific material values, which with modern technology can be restored and repaired. These are the lock supports, valve seat saddles and rods. Previously, they were not restored, which required large material costs for the acquisition of new ones. The introduction of the practice of restoration of these resources will allow significant savings in reducing material costs. In addition, this measure will reduce downtime due to the lack of these material resources.

To implement this program, the values of the cost of fixed assets and the number of personnel of the enterprise should increase by 0.4% and 0.5%, respectively.

The calculations showed the effectiveness of the implementation of these measures. The expected effect should be expressed in a reduction in material and transportation costs, in a reduction in the number of repairs, which in turn will save costs for repairs and increase the operating time of wells.

For the comparison base in the calculations, the actual values of the indicators of the reported year were taken with constant indicators of the number of wells in the existing stock and the average production rate of wells. Implementation of the proposed program to improve the production and economic activities of the enterprise unit will allow increasing oil production by 10 thousand tons. The cost of 1 ton of gross oil under these conditions should decrease by 3.9%, which in the current economic conditions is the most important indicator of the effectiveness of the proposed measures.

Thus, the analysis of the production activities of the studied structural unit of the oil company showed that it has sufficient production potential, and with the implementation of the proposed measures will achieve higher performance indicators.

References

[1] Kurushina V and Zemenkov Y 2014 Innovative cyclical development of the Russian pipeline system *WIT Transactions on Ecology and the Environment* DOI: 10.2495/EQ140822

[2] Levitin R E and Zemenkov Yu D 2016 The new approach to accurately determination of oil tanks evaporation *Neftyanoe Khozyaystvo Oil Industry* OIJ-2016-01-110-114-RU

[3] Dudin S M, Zemenkov Y D and Shabarov A B 2016 Modeling the Phase Composition of Gas Condensate in Pipelines *IOP Conference Series: Materials Science and Engineering*

[4] Dudin S M, Zemenkov Y D, Maier A V and Shabarov A B 2016 Research and Design of Thermophysical Gas-Liquid Mixture Parameters in Product Pipelines *IOP Confer-ence Series: Materials Science and Engineering* DOI: 10.1088/1757-899X/154/1/012021

[5] Dudin S M, Shipovalov A N, Podorozhnikov S Yu and Zemenkov Yu D 2011 Prospects for reducing energy costs for hydrocarbon transport *News of higher educational institutions. Oil and gas* **2** 65-9

[6] Zemenkov Yu D, Kurushina V A, Vylegzhanina A O, Barmenkova V V and Khairullina N G 2013 *Modern management and innovative, scientific and technical policy* (St. Louis)

[7] Zemenkova M, Zemenkov Y, Gladenko A and Podorozhnikov S 2016 *Estimation of Emis-sions during Monitoring of Pipelines in the Dynamic Mode of Operation* MATEC Web of Conferences DOI: 10.1051/matecconf/20168604053

[8] Fedorova O B and Lenkova O B 2011 The technology of choosing tools for financing innovative solutions in oil production *Management in Russia and abroad* **6** 23-8

[9] Fedorova O B, Chizhevskaya E L and Plenkin V M 2010 *Financial aspects of enterprises and organizations* (Tyumen: TSOGU) 164

[10] Chizhevskaya E L and Musalaev G M 2014 Features of the activities of small enterprises in the oil and gas industry *Current problems and prospects of regional and industrial development: a collection of scientific papers* (Tyumen: TSOGU) 135-41

[11] Chizhevskaya E L and Kot A D 2016 Program for the growth of the technical level of production of a production company 206 *Journal of Economy and Entrepreneurship* **1** part2 536-9

[12] Dudin S, Bahmat G, Zemenkov Y, Voronin K and Shipovalov A 2017 *Strategy for monitoring and ensuring safe operation of Russian gas transportation systems* MATEC Web of Conferences https://doi.org/10.1051/matecconf/201710606004

[13] Voronin K S and et al 2018 *Modernization of the individual device for temperature stabilization of the soil* IOP Conf. Ser. Mater. Sci. Eng. **445** 1 012011

[14] Tonysheva L L and Nazmutdinova EV 2010 *Formation of an industry development strategy: theoretical and implementation aspects* (Tyumen: TSOGU) 143

[15] Tonysheva L L and Chumlyakova D V 2016 Corporate social responsibility: the principles and the process of integration into the system of strategic management *Asian social science* **9** 115-2

IOP Conf. Series: Materials Science and Engineering **663** (2019) 012015 doi:10.1088/1757-899X/663/1/012015

Technology of road construction and laying of pipelines in marshy and waterlogged areas using pontoon modules

A V Ryabkov [1,2]

[1] Industrial University of Tyumen, 38 Volodarskogo str., Tyumen, 625000, Russia
[2] an10091980@gmail.com

Abstract. The development of environmentally friendly technology for the construction of non-rigid roads and systems for laying pipelines on block elements from composite pontoon modules in swampy and waterlogged areas for the Far North and Siberia is presented.

1. Introduction

A characteristic feature of the development of hydrocarbon resources in the vast swampy and waterlogged areas of Western Siberia and especially the Far North is the consideration and interaction of the environment with the pipeline system, long-distance and line service roads.

Problems arise at the first stage of route selection: how to connect points A and B, where the route will pass, what obstacles it will cross, how to assess the most hard-to-pass sections of the route, how to link the route with existing and planned future pipeline systems, existing and planned auto and railways, how to link them together, etc. [1]

It is known that in conditions of high swampiness, the creation and maintenance of line service roads on the territory of Western Siberia, including the regions of the Far North, in operating condition requires enormous costs.

The studies conducted by the authors in the field, as well as the analysis of the results of space diagnostics of the territories of Western Siberia and the Far North, showed that more than 48% of the territory is covered by type 3 swamps, 23% by water crossings, and the rest by permafrost soils with deep seasonal freezing (>3.0m) [5].

2. Problem description

Under these conditions, when searching for the optimal route of pipelines and line service roads while avoiding the above obstacles, the route lengths are overestimated by 43%, thereby creating complex configurations of the oil and gas pipeline network, which in turn additionally factorially increases the failure of the entire system. Based on the studies of failure diagnostics in the main oil and gas pipelines of Western Siberia and the Far North, diagrams are constructed where these situations are clearly shown (Figure 1).

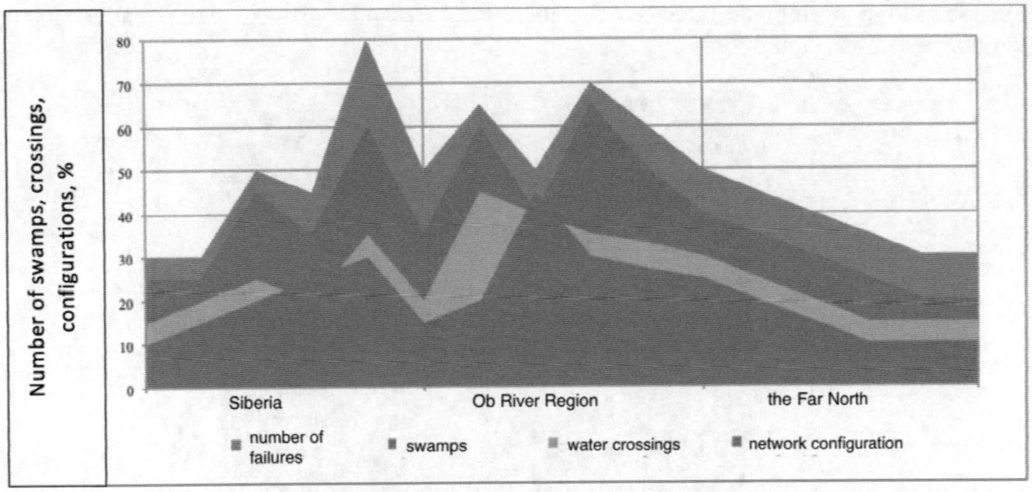

Figure 1. Distribution of oil and gas pipelines failures in Siberia, Ob River Region and the Far North

Analysis of the distribution of failures at the main oil and gas pipelines showed that the number of accidents of the linear part largely depends on the degree of swampiness of the territory, on water crossings and configuration.

Therefore, when laying pipelines with low bearing capacity of type 3 swamps underground, in waterlogged areas, permafrost soils and in soils with deep seasonal freezing, significant movements of pipelines occur, with a loss of longitudinal stability and their emergence to the surface, with the formation of arches and their configuration (Figure 2).

Figure 2. Formation of an arch with the emergence of the pipeline to the surface

All this changes the stress state of the pipelines and leads to an overstatement of stresses by 3–4 times relative to the normative ones [2, 8]. These system parameters make it necessary to adjust the design schemes and strength calculations to determine the stress-strain state and select appropriate materials, due to loads and impacts based on the limit ones.

In the process of heavy traffic of heavy construction equipment and vehicles along temporary line service roads with elevated soil moisture, sharp daily temperature fluctuations (up to 30°), various types of deep gauges (internal, external, deep, surface) appear. At the same time, the layers of pavement, subgrade, roads become unstable, since their physicochemical properties do not meet the requirements for shear resistance, strength and other indicators that change during the operation of line service roads and the pipeline itself [4, 10]. This leads to the fact that the equipment is drowning, falling through, getting stuck in a swamp and often remains there forever, which leads to significant economic and environmental costs (Figure 3).

Figure 3. Work of excavators in the winter and summer periods on swampy and waterlogged areas

According to our proposal, for the first time in the design of the pipeline and line service roads of a new type, the above-mentioned shortcomings can be reduced or completely eliminated.

To this end, in order to create reliable and long-term integrated road structures and oil and gas pipelines, a fundamentally new method for the construction and operation of pipeline systems is proposed. This method will greatly complement traditional construction methods with new progressive technological properties.

3. Materials and methods

The essence of the method is as follows. A profile of a complex structure constructed from composite materials (pontoons for bedding) shown in Figure 4 with rutting guides will provide increased reliability of the line service road, buoyancy, smooth running of a self-propelled lifting and loading equipment designed as a "centipede", which has a number of significant advantages in the construction and operation of the entire complex, as shown in Figure 5.

Figure 4. Pontoon pipe section

Figure 5. Laying a pipeline with an automated pipe layer

On the pontoons, made in the form of a module of composite materials and having inside the power frame assembled in the form of a guy, two rutting guides of the channel type are mounted, facing the shelves up. The gauge is so large that construction and repair equipment (the "centipede" pipe-laying machine) and automobile transport can pass along these guides. The number of modules is determined on the basis of the bearing capacity of the module itself and the load that will be applied to the pontoon base (constant load from the mass of the pipeline and temporary load from the mass of the automated pipe layer during the construction, repair, and maintenance of the pipeline).

Pontoons in the longitudinal and transverse directions are fastened together by composite cables. Longitudinal cables are fixed at fixed distances in the land part with “anchors” installed in wells or small buildings, with hydraulic lifts to maintain a constant level of the road and pipeline in the pontoon system [6, 9].

The cables passing along the side and bottom of the pontoons enter the well through special holes and are wound onto the drums using an engine with a gearbox (Figures 6 and 7).

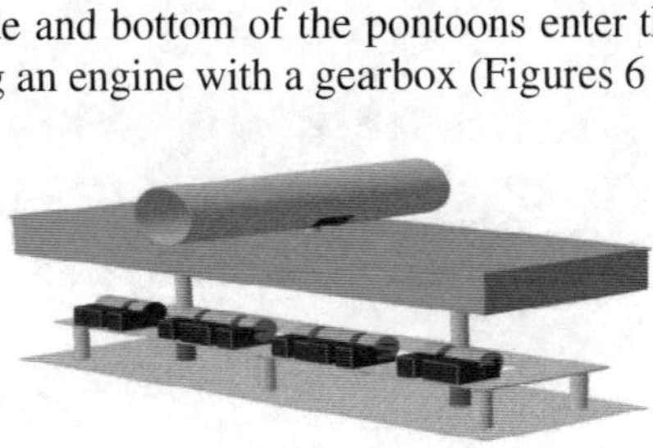

Figure 6. Technological well with the automatic cable control system

Figure 7. General view of a technological well with cables

The technological well is delivered to the route in the form of a one-piece, fully equipped unit of modular type. When installing the technological well unit (TWU), it is necessary to ensure the integrity of the structure, to prevent water, dirt, etc. from getting inside the unit.

In parallel with the installation of the TWU in the design position, anchors must be installed. Shot anchors, which can be buried in the underlying soil up to 5 m, regardless of the thickness of the peat layer, were adopted as anchors. The anchors are immersed in the ground with the help of special installations on caterpillar tracks with the increased off-road performance [7, 11].

The following is an example of a completed project for the construction of a pontoon-type roadbed made of modules using a "centipede" connected in the form of a continuous guy for driving equipment and laying the pipeline, as shown in Figure 8 a, b, c, d.

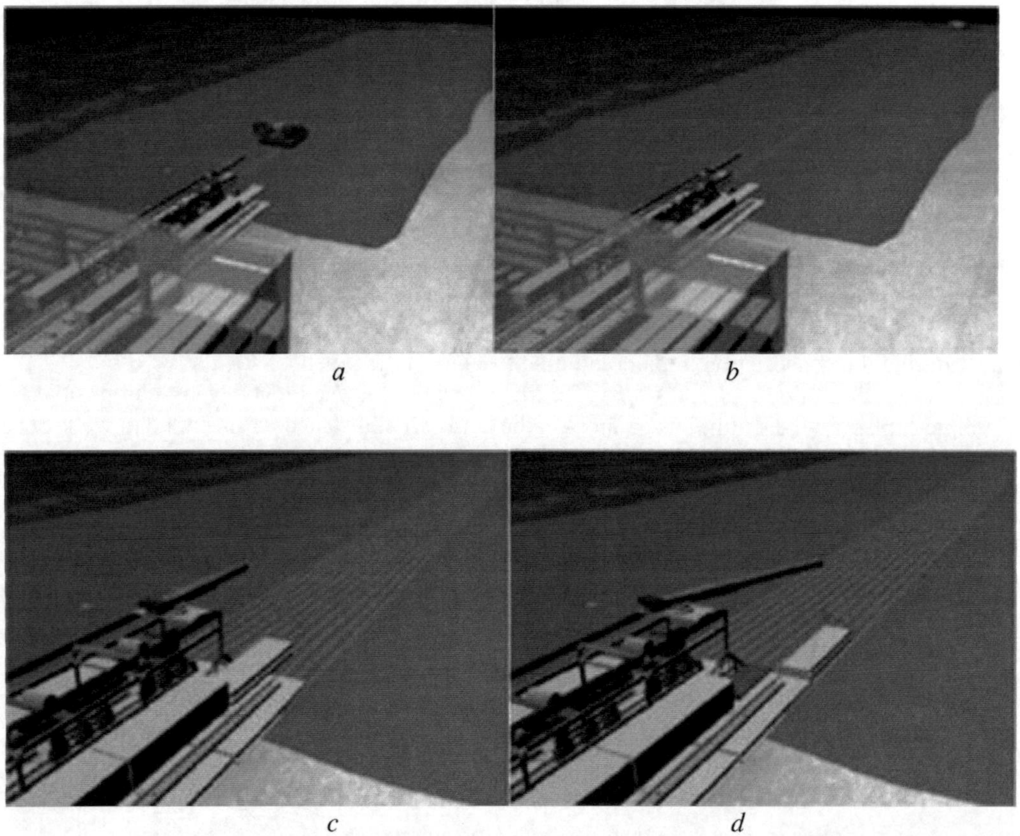

Figure 8. Construction of a roadbed from composite pontoon modules: *a* - laying a cable through an obstacle using a ground-effect vehicle, *b* - cable fastening in a well, *c* - laying the pontoon module on cables, *d* - fixing the pontoon module on cables.

On the pontoons, made in the form of a module of composite materials and having inside the power frame assembled in the form of a guy, two rutting guides of the channel type are mounted, facing the shelves up. The gauge is so large that construction and repair equipment (the "centipede" pipe-laying machine) and, if necessary, automobile transport can pass along these guides. [12,13].

As you can see, this system completely eliminates the use of automated pipe layers and excavators for digging trenches.

All this creates the conditions for a new technology of laying roads from new structural materials and pipelines on bedding elements by the ground method, tangible savings in fuel and resources when laying such a pipeline route and almost eliminates overburden work. The profile of the earth's surface, due to the peculiarities of the formation of the relief by natural influences, is much closer to curves with a variable radius, which gives the best fit of such a route into the terrain [14,15].

4. Conclusion

An analysis of the research work carried out in recent years on the adoption of an innovative concept for the design, construction and operation of roads using new technological systems, as well as maintenance and repair of pipelines, indicates the complexity of the problem, the optimal solution of which is possible only with the use of complex mathematical methods for design analysis, modern materials, computer technology for managing innovative transport systems working on alternative types of energy. It should be noted that some tasks were solved at a sufficient scientific level and currently have high scientific and practical value.

References

[1] SNiP 2.05.06-85* *Main pipelines*

[2] SNiP 2.06.04-82* *Loads and impacts on hydraulic structures (wave, ice and from ships)*

[3] SNiP 3.04.03-85 Protection of building structures and structures from corrosion

[4] Merdanov Sh M 2006 *Snowy roads: research, construction, construction technology, mechanized complexes* (Tyumen: TSOGU)160

[5] Ryabkov A V, Ivanov V A and Zakuraev A F *Development of a new technology for laying pipelines on composite pontoon modules in Siberia and the Far North* (Tyumen: Tyumen Printing House) 391

[6] Ryabkov A V, Stafeeva N.A, Ivanov V A and Zakuraev A F 2018 A comprehensive solution to the problem of all-weather construction of roads and pipelines on universal composite pontoon modules *IOP Conference Series: Materials Science and Engineering*

[7] Ryabkov A V, Shipovalov A N and Seroshtanov I V 2015 Year-round construction of pipelines on pontoon modules in the regions of Siberia and the Far North *Business magazine "Neftegaz.ru"* [11 – 12] 30 – 3

[8] Dudin S M, Shipovalov A N, Podorozhnikov S M and Zemenkov Yu D Prospects for reducing energy costs for hydrocarbon transport *News of higher educational institutions Oil and gas* **2** 65-9

[9] Zemenkova M Yu, Babichev D A and Zemenkov Yu D 2007 Methods of system analysis in solving problems of managing complex technical systems *Oil and gas business* **2** 22

[10] Zemenkov Yu D, Pimnev A L and Kurushina E V 2016 System of controlling the reliability of hydraulic machinery in oil and gas facilities *IOP Conference Series: Materials Science and Engineering* **127** (1) 012055

[11] Shantarin V D, Bezzubtseva N A and Zemenkov Y D 2018 Noospheric ecological imperative in culture of technocratic society *IOP conference series: materials science and engineering* 012001

[12] Vasiliev G G, Gulkov A N, Zemenkov Yu D and Prokhorov A D 2008 *Operation of gas industry equipment and facilities* (Vologda: Publishing house "Infra-Engineering") **2** 624

IOP Conf. Series: Materials Science and Engineering **663** (2019) 012015 doi:10.1088/1757-899X/663/1/012015

[13] Karnaukhov N N, Merdanov Sh M, Shefer V V and Ivanov A A 2005 *Operation of machinery in construction* (Tyumen: TSOGU) 393

[14] Igoshin M E, Vorontsov V V, Kraev A N and Ushakov A E 2014 Numerical modeling of a constructive solution to strengthen the foundation and embankment of the roadbed *Actual problems of construction, ecology and energy conservation in Western* Siberia (Tyumen: TyumGASU) 21-8

[15] Kurushina V and Zemenkov Y 2014 Innovative cyclical development of the Russian pipeline system *WIT TRANSACTIONS ON ECOLOGY AND THE ENVIRONMENT* **190** 881-8

IOP Conf. Series: Materials Science and Engineering **663** (2019) 012016 doi:10.1088/1757-899X/663/1/012016

Analysis of the development of BS10 oil reservoir at the Savuyskoye field

E V Ogudova[1,2] , Z S Aleskerova[1,3]

[1]Industrial University of Tyumen, 38 Volodarskogo str., Tyumen, 625000, Russia
[2]ogudovaev@tyuiu.ru
[3]aleskerovazs@tyuiu.ru

Abstract. Based on the analysis of oil production at the BS10 formation of the Savuiskoye field, the article provides the calculation and justification of the necessary filter permeability which limits sand production in the well during operation of wells with weakly cemented unstable reservoirs.

1. Introduction

The Savuiskoye gas and oil field is located in the eastern part of the Surgut oil and gas region of the Central Ob region of the West Siberian oil and gas province. The commercial oil content was established in the sediments of the Sortymsky Formation of the Upper Valanginian age of the Lower Cretaceous System (BS10 formation).

The BS10 reservoir is elongated in a Northwest direction and occupies most of the licence area. The deposit is fully developed by production wells. BS10 reservoir reserves are represented by two categories - A and C2. Initial oil reserves (category A) of the BS10 formation, listed on the State balance sheet, are: geological - 109 942 thousand tons, recoverable - 50 523 thousand tons, oil recovery factor - 0.460.

The object is at the last stage of development. We are interested in the issue of its development in order to determine the residual reserves.

2. Materials

Since the beginning of development, 48 897 thousand tons of oil have been produced at the field. With a current water cut of 96.3%, the oil recovery factor is 0.393 of the approved recoverable reserves of the BC1 categories, and the production of initial recoverable reserves is 94.2%.

The BS10 formation provided 99.7% of the extracted oil (48 764 thousand tons). The production of initial recoverable reserves for the BS10 facility is 96.5% with a water cut of 96.6%, the current recovery factor is 0.444.

In the US2 formation, the degree of reserves development is low, the production of initial recoverable reserves is 9.8%.

Three sites of experimental program were organized at the US2 formation. Thus, the immediate prospect of further development of the field is associated with the low-productive formation US2.

3. Methods

As of January 01, 2018, studies on monitoring the development of oil reserves in the BS10 reservoir by production logging methods were carried out in 403 wells, including 285 production, 112 injection and 6 control wells. Based on the results of studies of production and injection wells, the parameters of the formation development were determined:

- working thickness coefficient - the ratio of the working thickness to the effective oil-saturated perforated formation thickness (determined by temperature logging and flow metering); - coefficient of secondary completion - the ratio of the effective perforated thickness to the total effective oil-saturated formation thickness;
- coefficient of sweep efficiency - the ratio of the working thickness to the entire effective oil-saturated formation thickness;
- waterflood thickness ratio - the ratio of the waterflood thickness to the entire effective formation thickness.

In production wells, 81% of the effective oil-saturated thickness of the formation was perforated. Incomplete secondary completion of the formation is associated with the lack of reliable impermeable sections at the OWC level and the presence in the bottom of the formation of individual productive lenses that were not tapped by perforation (well No. 357, etc.). Fluid is recovered throughout the entire effective thickness. The coefficient of the working thickness of the reservoir is 1.15. The sweep coefficient over the thickness of the BS10 formation varies in the range from 0.64 to 1.0 with an average value of 0.94. The effective oil-saturated thickness of the BS10 formation in the studied wells varies from 1.8 m (well No. 702) to 26.2 m (well No. 316) and averages 11.8 m.

The current oil-saturated thickness of the BS10 formation (as of January 01, 2018) varies from 0.7 m (well No. 738) to 9.4 m (well No. 563) and averages 4.1 m. The source of watering the production of wells is the passage of the front of the injected water and pulling of formation water to the perforation interval from the bottom water-saturated part of the formation in the marginal zone.

The state of the current water flooding of the BS10 reservoir thickness is shown in Figures 6-9, plotted for different sections of the reservoir. The development of oil reserves takes place over almost the entire oil-saturated thickness with the leading flooding of the bottom part of the thickness by the water injected. The work does not involve individual low-permeable intervals of the reservoir and non-perforated lenses. The high stratification and variability of the thickness of the reservoir over the area of the reservoir is also reflected in the sweep coefficient, which, as noted above, varies from 0.64 to 1.0.

The results of testing the control wells by carbon-oxygen (C/O) logging show that the development process covers almost the entire thickness of the BS10 formation. The highest displacement coefficient is mainly observed in the bottom of the formation.

In control wells No. 817K, 818K, 820K, the data of carbon-oxygen (C/O) logging show a decrease in the oil saturation coefficient over the entire oil-saturated thickness. In July - October 2003, the BS10 formation in these wells was perforated. An inflow of oil with water was obtained in the wells, with a water cut of 80 - 93%.

High water cut of well production is confirmed by the results of C/O logging.

According to the results of C/O logging as of September 6, 2017, the BS10 formation in the control well No. 813K was oil saturated and in October 2017 it was perforated. For two months, the well operated with practically anhydrous products (water cut - 3.3%). As of September 2006, 51.8 thousand tons of oil was produced from it with a water cut of 80%. At the time of the secondary completion of the BS10 formation in well No. 813K, its surrounding wells No. 318, 387, 506, worked with a water cut of 91% - 98%. That is, in the area of these wells, a "water finger" formed and without tapping the BS10 formation in well No. 813K, it would be difficult to extract the bypassed oil formed in its area.

In exploratory well No. 4251P, drilled in November 1998, the weighted average oil saturation coefficient of the BS10 formation was 47.3%. Nearby production wells No. 270, 594, 595 at that moment were operating with a water cut of 90% - 93%.

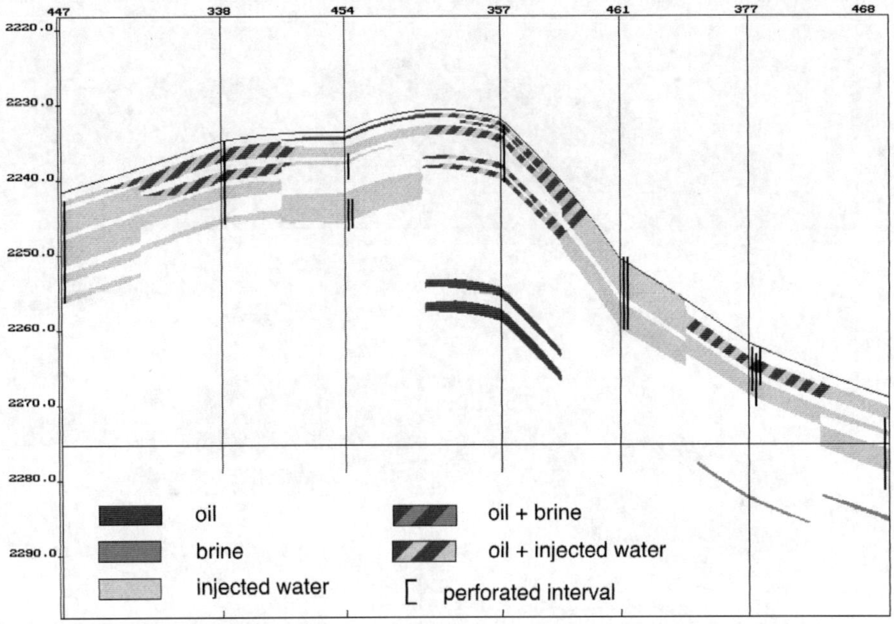

Figure 1. The production profile of the reservoir BS10 along the line of wells №№ 447-468

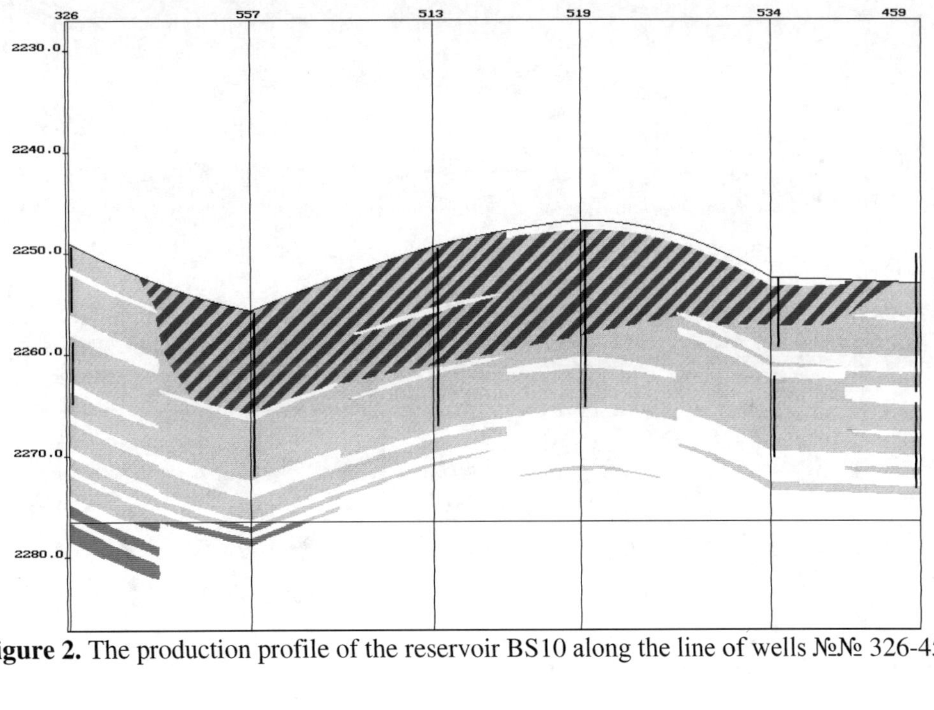

Figure 2. The production profile of the reservoir BS10 along the line of wells №№ 326-459

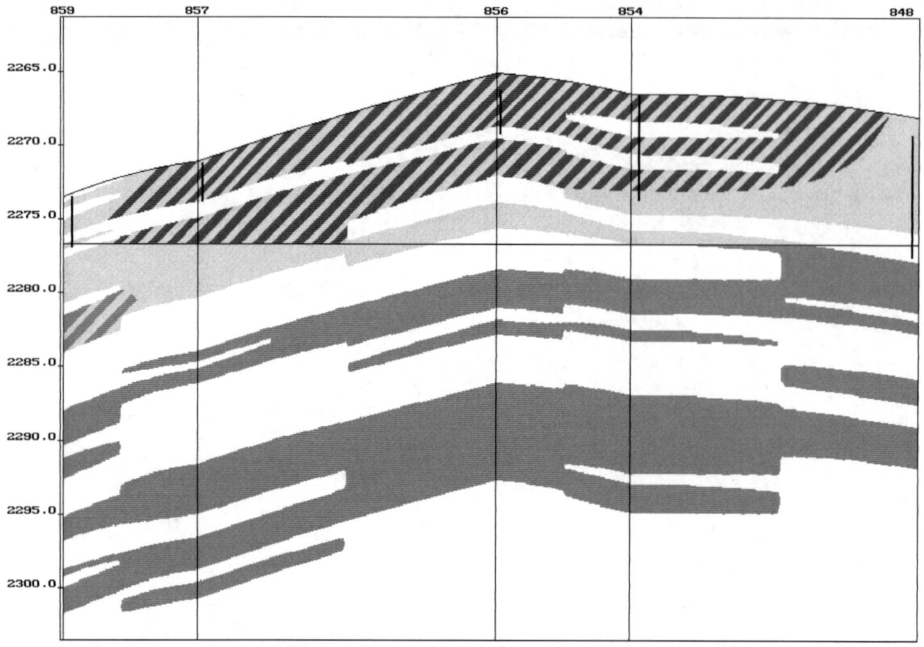

Figure 3. The production profile of the reservoir BS10 along the line of wells №№ 859-848

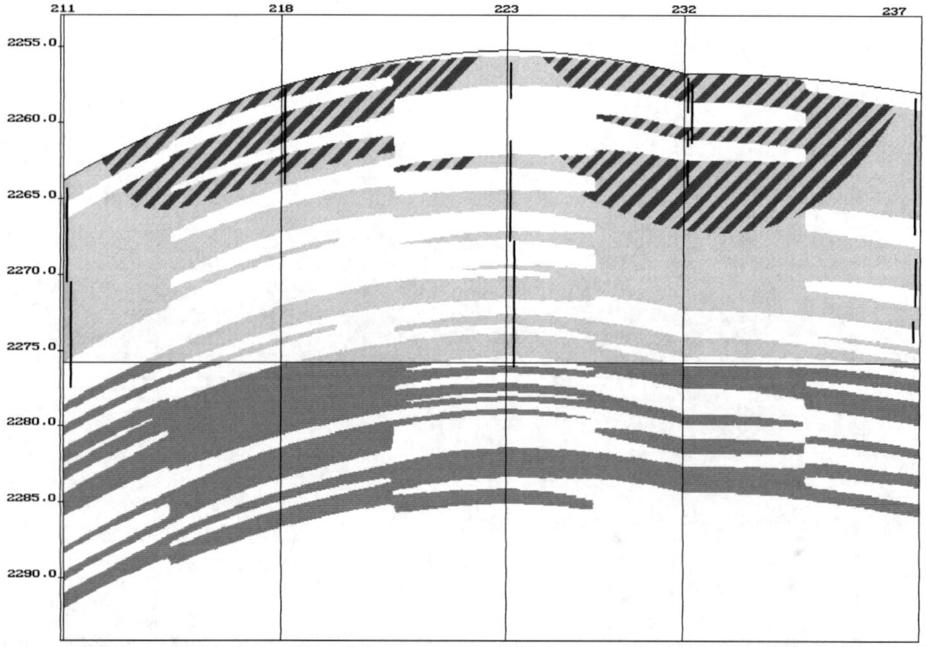

Figure 4. The production profile of the reservoir BS10 along the line of wells №№ 211-237

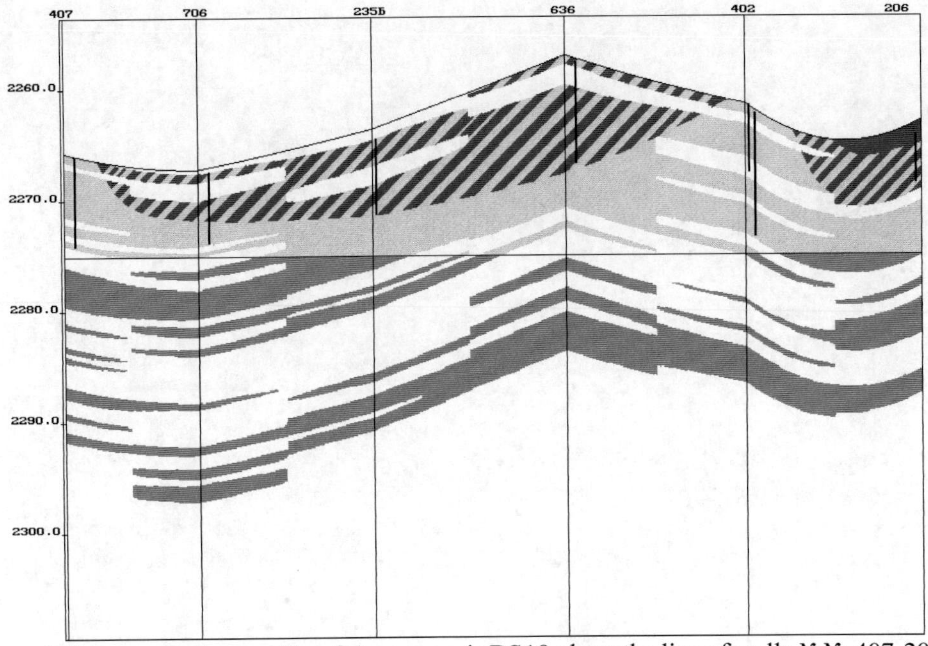

Figure 5. The production profile of the reservoir BS10 along the line of wells №№ 407-206

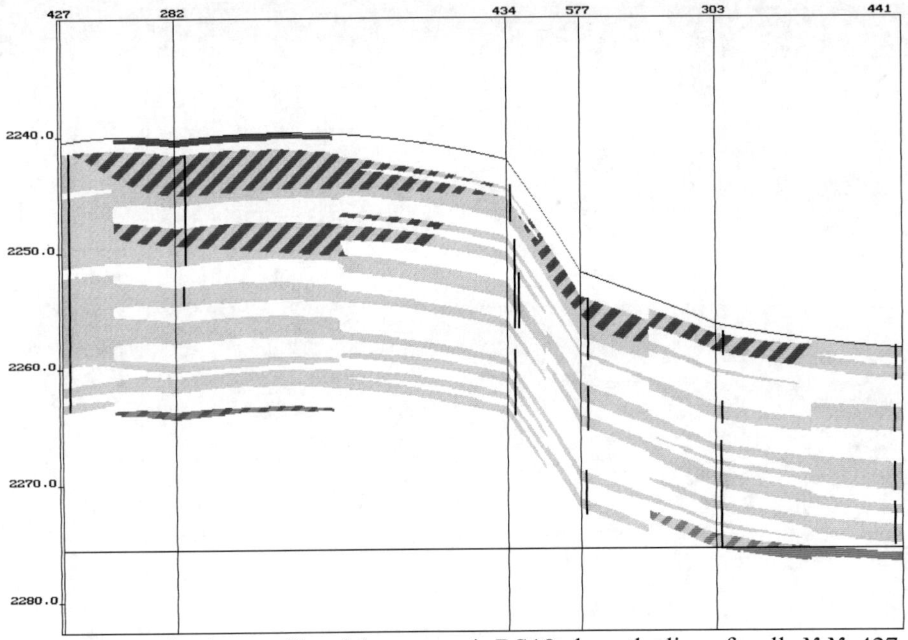

Figure 6. The production profile of the reservoir BS10 along the line of wells №№ 427-441

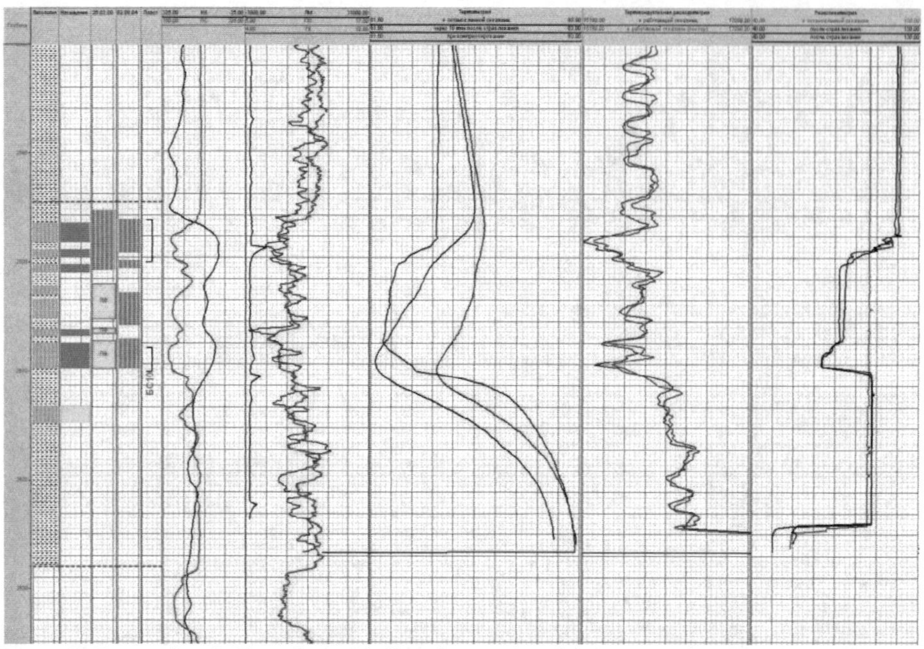

Figure 7. Example of allocation of inflow intervals of the reservoir BS10 for well № 507

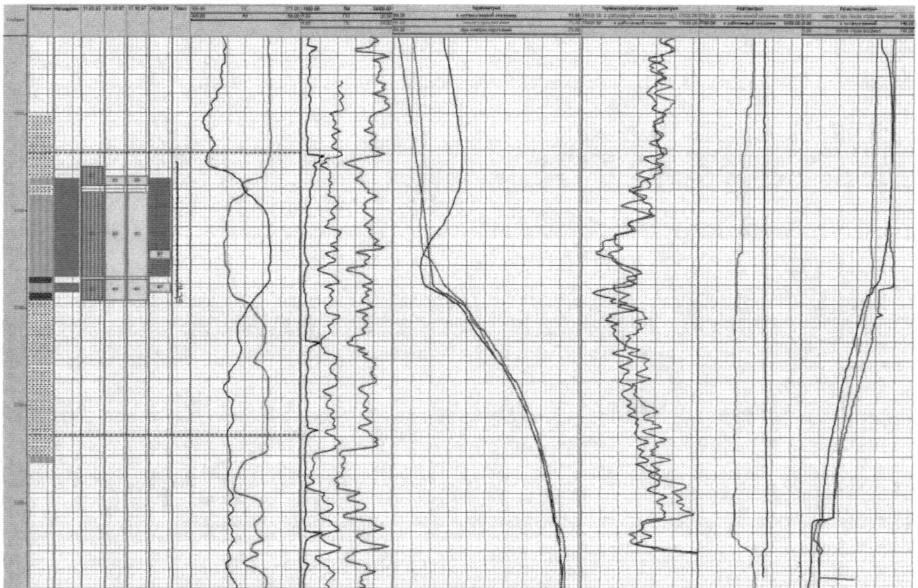

Figure 8. Example of allocation of inflow intervals of the reservoir BS10 for well № 517

The cumulative oil production in well No. 270 amounted to 377 thousand tons, and in wells No. 594, 595, respectively, 66.4 thousand tons and 36.2 thousand tons. In the area of the above wells, a certain "water finger" formed.

The reservoir receives the injected water within the effective oil saturated thickness. Perforation tapped 91% of the oil-saturated thickness of the reservoir. The coefficient of working thickness varies in the range from 0.15 to 1.0 with an average value of 0.9. 82% of the oil-saturated thickness of the reservoir is involved in the displacement process.

IOP Conf. Series: Materials Science and Engineering **663** (2019) 012016 doi:10.1088/1757-899X/663/1/012016

In general, the development of oil reserves in the BS10 formation is uneven both in thickness and in area.

Based on the results of studies of production, injection and control wells, a map of the current oil-saturated thicknesses of the BS10 formation is constructed. The average oil-saturated thickness of the formation varies from 0.7 - 9.4 m with an average thickness of 4.1 m.

The maximum current oil saturated thicknesses are identified in the southwestern part of the reservoir. Residual oil reserves are concentrated in the roofing of the reservoir and non-perforated lenses.

The areas with the highest density of oil reserves are promising for sidetracking.

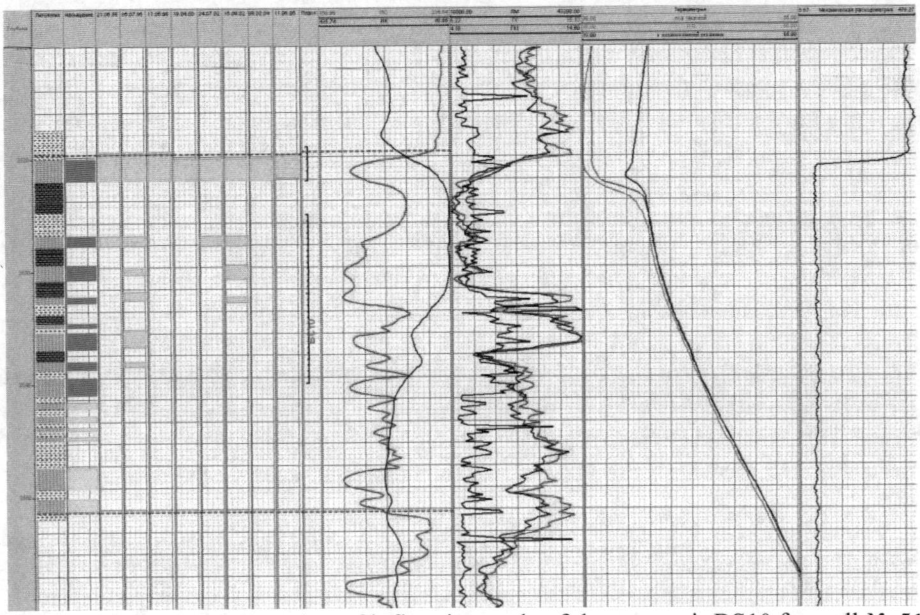

Figure 9. Example of allocation of inflow intervals of the reservoir BS10 for well № 517

4. Conclusions

The main development target at the Savuiskoye field is BS10, both in terms of geological reserves - 88% of the total in the field and in cumulative oil production - 99.7%. The BS10 formation is at its final, fourth stage of development. As of January 01, 2018, 48763 thousand tons of oil were recovered, which is 96.5% of the initial recoverable reserves, with a water cut of 96.5%. Oil production in 2017 amounted to 303.5 thousand tons at a rate of recovery from initial recoverable reserves of 0.6%. The current oil recovery factor is 0.444. The implemented system of the BS10 reservoir development is a three-row system of well location dividing the reservoir into 13 production blocks.

The maximum values of both current and cumulative oil production are characterized by blocks 8-12, in which a significant part of the area is a water-free oil zone. The development of reserves for the blocks is uneven. Current oil recovery factors of the blocks vary in the range 0.140 - 0.632 with a change in water cut of 98.5% - 95.5%.

The nature of watering in all blocks is identical - a rapid increase in produced water. However, current oil recovery factors differ by 2–3 times. By the magnitude of the current oil recovery factor and the intensity of pumping, operational blocks can be combined into three groups. Group 1 (operational blocks 5, 6, 8, 9, 12). The degree of pumping of the blocks of this group is 1.712 - 2.013 of the oil-saturated volume of the formation. The blocks are characterized by maximum values of the current oil recovery factor (0.632 - 0.481). The maximum oil recovery factor (0.632) was obtained in block 12 with a maximum degree of pumping of 2.013. Group 2 (operational blocks 4, 7, 10, 11). With a change in

IOP Conf. Series: Materials Science and Engineering **663** (2019) 012016 doi:10.1088/1757-899X/663/1/012016

the degree of pumping in the range of 1.416 - 1.545, the current oil recovery factor for the blocks varies from 0.379 to 0.447. Group 3 (operational blocks 1, 2, 3, 13). All blocks of this group (except block 13) are located in the water-oil zone. The smallest value of the current recovery factor is obtained for the regional block 1 - 0.140. The maximum oil recovery factor in this group does not exceed 0.304 (block 2). In recent years, a characteristic feature of the process of developing the reserves of the BS10 formation is an increase in the pumping rate, which in turn at the late stage of the development of the formation sharply increases the value of the current WOR. With a close degree of water cut of all blocks (about 96%), the degree of development of reserves of the selected groups of blocks differs significantly.

The difference in the intensity of development of reserves for these groups of blocks is noted at all stages. When the pumping factor reaches 0.5, they differ by a factor of 1.5–3, while the difference in the values of the accumulated WOR is 3-4 times. An increase in oil recovery factor with an increase in pumping from 0.5 to 1.0 of the pore volume in the blocks of group 3 is almost two times lower than the growth rate for the blocks of group 1. The separation of the blocks according to the intensity of development of reserves is largely due to the share of water-oil zone in the block area, the reserves and the number of active wells. The current water cut of extracted products by blocks varies from 95.5% to 98.5%.

The effective oil-saturated thickness of the BS10 formation in the studied wells varies from 1.8 m to 26.2 m and averages 11.8 m. The current oil-saturated thickness of the BS10 formation (as of January 01, 2018) varies from 0.7 m to 9.4 m and averages 4.1 m. The source of watering the production of wells is the passage of the front of the injected water and pulling of the formation water to the perforation interval from the bottom water-saturated part of the formation in the marginal zone.

The figures show examples of the allocation of working and water flood intervals of the reservoir in production wells according to well logging data. The development of oil reserves takes place over almost the entire oil-saturated thickness with the leading flooding of the bottom part of the thickness by the water injected. The work does not involve individual low-permeable intervals of the reservoir and non-perforated lenses.

The high stratification and variability of the thickness of the reservoir over the area of the reservoir is also reflected in the sweep coefficient, which, as noted above, varies from 0.64 to 1.0. The results of testing the control wells by carbon-oxygen (C/O) logging show that the development process covers almost the entire thickness of the BS10 formation. The highest displacement coefficient is mainly observed in the bottom of the formation. In general, the development of oil reserves in the BS10 formation is uneven both in thickness and in area. The maximum current oil saturated thicknesses are identified in the south-western part of the reservoir.

References

[1] Addition to the development project Savuiskoye field 2010 FE SurgutNIPIneft Tyumen

[2] Guidelines for the implementation of diploma projects for students of specialty 130503 “Development and operation of oil and gas fields” 2010 TSOGU Tyumen

[3] Methodical recommendations on design of development of oil and gas-oil fields 2009 NP NAEN Moscow

[4] Muromtsev V S 1984 *“Electrometric Geology of sand bodies-lithological traps of oil and gas”* Leningrad “Nedra”

[5] Rozhkov G F 1978 *“Geological interpretation of grain-size parameters according to the fractional sieve analysis. Granulometric analysis in Geology”* Moscow GI an USSR

[6] Petersil'e V I Poroskun V I Yatsenko G G 2003 *"Guidelines for calculation of geological reserves of oil and gas volumetric method"* Moscow-Tver VNIGNI SPC “Tvergeofizika”

[7] Khamatdinov R T Enikeeva V H Velizhanin V A Zhuravlev B K and others 1989 *"Guidelines for conducting neutron and gamma-ray logging in oil and gas wells equipment recovery and processing of measurement results"* Kalinin “Soyuzpromgeofizika”

[8] Methodical guide on hydrodynamic, field-geophysical and physico-chemical methods of oil field

IOP Conf. Series: Materials Science and Engineering **663** (2019) 012016 doi:10.1088/1757-899X/663/1/012016

development control 1991 Moscow VNII
[9] Sekachev V N 1994 *"Analysis of the development of the largest oil fields, and the CIS countries"* Moscow VNIIONG
[10] Ivanov M M 1976 *"Dynamics of mining of oil deposits"* Moscow "Nedra"
[11] Lysenko V O Greifer V I 2005 *"Rational development of oil fields"* Moscow "Nedra"

IOP Conf. Series: Materials Science and Engineering **663** (2019) 012017 doi:10.1088/1757-899X/663/1/012017

Environmental safety estimation of drill cuttings using a composition mixture based on zeolite

A I Agoshkov, M O Tretyakova, I V Moskovaia and T A Brusentsova

Engineering school, Far Eastern Federal University, 8, Sukhanova Str., Vladivostok, 690091, Russia

E-mail: bgdtsdvfu@mail.ru

Abstract. Drill cuttings is a complex mixture of spent drilling mud with various additives having different chemical properties and drilled solids, and which significantly pollutes the environment when it is buried. This article proposes and studies a zeolite-based composite mixture that converts drill cuttings into environmentally friendly soil. As a result of the experiment, samples of model drill cuttings and soil with different holding times are obtained. A sequential extraction method is used to determine the degree of metal desorption in a laboratory study of samples. The analysis consists of an estimation of metals content (gross form and extracts in 3 extractants), the content of petroleum products and pH value. Concentrations of metals in both gross and mobile form in the processed drill cuttings have significantly decreased in comparison with the original drill cuttings, and do not exceed standard values. Also, oil concentration in the soil is lower than in the drill cuttings. pH reaction has decreased from alkaline to neutral. The proposed zeolite-containing composite mixture allows implementing drill cuttings as a new, sustainable in the environment, useful product.

1. Introduction

Oil production is a complex process consisting of several basic technological operations, such as exploration, well drilling, oil production itself and transportation. Each of these operations is accompanied by a colossal negative impact on the environment.

One of the main problems is the formation of significant amounts of toxic drill cuttings (DC). On average, DC consists of 30–45 % of drilled solids (particles of clay and sand), 30–45 % of drilling mud, and 10–20 % of groundwater and oil [1]. All drilling muds (DM) can be divided into 3 groups: water, hydrocarbon and synthetic. The most dangerous for the environment are hydrocarbon-based DM [2–3], an important component of which is a dispersion medium [4–5] represented by diesel fuel which has a high phytotoxicity [6]. The following are used as additives for DM: corrosion inhibitors, anti-foam additives, emulsifiers, flocculants, lubricating compounds, SAA, dispersing agents, thickeners, weighting agents, etc. [7]. Thus, DM is a multicomponent liquid, and the mutual influence of the components on each other further enhances the toxic effect of each of them.

“Lebedinskoe” oil field in northern Sakhalin, Russia was selected for the study. Drilling invert-emulsion solution based on hydrocarbon with the addition of weighting agent, thickener and stabilizer is used for drilling.

A variety of methods, such as thermal, physical, biological, and physico-chemical methods are used to process DC [8–10]. In world practice, the most commonly used thermal method is incineration, which is inefficient for the removal of heavy metals, as well as when the content of petroleum products in DC

IOP Conf. Series: Materials Science and Engineering **663** (2019) 012017 doi:10.1088/1757-899X/663/1/012017

is more than 16 % [9, 11]. This process is accompanied by the release of pollutants, and the resulting burned-out soil has not found further application. The physical methods are based on the processes of mixing and physical separation. Physical methods are not applicable to DC, if they contain nonvolatile and dense resins and asphaltenes [11]. The most environmentally friendly physical method is "reinjection" – pumping DC into the inner annulus or into a specially drilled well. The disadvantage is the presence of a receiving layer, waterproof layers above and below the receiving layer, there is a risk of emergency situations. Biological methods involve the addition of microorganisms to waste [10]. Application area is limited to the conditions: range of biological products activity, ambient temperature, acidity, aerobic conditions. Physico-chemical methods are the processing of cuttings with reagents that cause a change in the physico-chemical properties of the waste and allow getting new materials. Adding materials for sorption or coagulation neutralization has become widespread due to the simplicity of the method and relative universality, as well as the possibility of recycling the product [10, 11].

Many previously published works that describe DC processing methods using composite mixtures involve the use of binders such as portland cement, hydrated lime, gypsum, ash, and most are associated with the subsequent disposal of the resulting product or use as a basis for construction [12–15]. For example, a mixture of sand, peat, drywall, and clay-gypsum has been proposed to reduce the toxicity of DC [12]. However, there is no component in this mixture that promotes the transfer of heavy metals into insoluble form and their retention in the sediment, a low level of stability of solidified wastes. There is also a mixture, manure or peat, mineral soil and fertile soil layer, and besides the technology of using this mixture involves drying DC in the sun until the moisture level reaches 8–14 % [16], what is applicable not in all climatic conditions. The authors propose a technology with the preparation of 2 mixtures for different purposes - for the lower recultivation layer and for the biological reclamation stage with different percentages of sand or sandy loam, peat, neutralizing additives (aluminosilicate, coal sorbent, "Neftedestruktor", "Centrin" biological products, mineral, organic, humic fertilizers) of gypsum or phosphogypsum [13]. However, low mechanical qualities, such as strength and elasticity, limit the use of the resulting product in construction, and the use of a large amount of peat is not justified. Also a large selection of additional neutralizing additives with the lack of data on the unique effectiveness of the proposed composite materials is presented.

This paper discusses the use of zeolite as an additive in the composition of mixture. Ion-exchange properties of zeolites are widely used for the treatment of wastewater contaminated with heavy metal salts and petroleum products. The effeciency of zeolite sorption properties with respect to heavy metals and petroleum products has been studied in the works [17–21]. Zeolites were used to treat wastewater from such heavy metals as Fe, Cu, Zn, Al, Mn, Cr, Cd and Pb. The work [22] presents the study results of zeolites sorption capacity with respect to petroleum products, which averaged 2.5 mg/g.

The purpose of this work is to determine the effectiveness of the developed zeolite-based composite mixture for neutralizing DC. For this, immobilization of metals and hydrocarbons in the resulting soil mixture is estimated, as well as the compliance of these indicators with regulatory requirements.

2. Materials and research methods

2.1. Research objects

Drill cuttings. The drilling invert emulsion solution was taken as a prototype for the preparation of a model hydrocarbon drilling mud used for drilling at "Lebedinskoe" oil field, Sakhalin, Russia. The ratio of hydrocarbon phase / aqueous phase was 70:30. The hydrocarbon phase composition included: diesel fuel, Okhinskaya oil, emulsifier, stabilizer, inhibitor / structurant, the aqueous phase composition: fresh water and weighting agent.

The rock that corresponds to the drilled rock of "Lebedinskoe" oil field in the interval from 120 to 2000 m along the wellbore – sandstones, sands, clays in the ratio of 60:20:20 was chosen for the preparation of model DC (Fig. 1 (a)).

Man-made soil. The developed zeolite-based composite mixture was added to DC to obtain samples of man-made soil, and additionally containing peat, sand, gypsum and organic-mineral fertilizer. Zeolite content in the mixture was 15 %. The resulting mixture was kept for 2 months. Sampling from the obtained sample was carried out 2 times - after 1 month and after 2 months, respectively (Fig. 1 (b), (c)).

(a) (b) (c)

Figure 1. Resulting samples: (a) – DC, (b) – soil after 1 month of ageing treatment, (c) – soil after 2 month of ageing treatment

2.2. Research methods

Extract preparation. 3 extracts were prepared using extractants to determine the content of various forms of metals: distilled water, ammonium acetate buffer (AAB) with 4.5 pH, 1%-th ethylendiaminetetraacetic acid solution (EDTA) in ammonium acetate buffer with 4.5 pH. Carbon tetrachloride was used as an extractant to determine the total oil content in samples.

Sample preparation. Sample DC and 2 samples of man-made soil were dried on filter paper under an extract for 1 week at room temperature to an air-dry state, and then were sifted on1 mm filter. Part of the dried samples was used to determine gross content of metals, the other part – for extraction in extracts.

Sample analysis. Dried samples were analyzed by X-ray fluorescence method on Innov X-X 50 analyzer to determine gross metal content. The peculiarity of the method is the possibility of simultaneous analysis of qualitative composition and quantitative content of elements in complex multicomponent mixtures.

Extracted samples were analyzed on atomic absorption spectrophotometer AA-6800 "SHIMADZU". For this, a sample weighing 10 g was selected from each sample, the total number of samples was 9 (3 for each form of extract). Then each sample was diluted with a liquid corresponding to extract shape. The extraction time on a mixing device was 30 minutes, after which the samples were settled for 12 hours. Then the samples were filtered on "blue ribbon" filters. Sample solutions were sprayed into the acetylene-air flame of the high-temperature burner of the atomizer, and then changes in the corresponding absorption signals were recorded.

Samples weighing 0.5 g, which were filled with carbon tetrachloride, then stirred for 1 hour were taken to determine the mass fraction of petroleum products. The extract obtained was filtered through a "blue ribbon" paper filter then extraction followed by filtration was repeated 2 more times with new portions of carbon tetrachloride. All extracts were combined into a measuring cylinder and the total volume was recorded. The mass fraction of petroleum products was determined by the method of IR-spectrometry on the analyzer of AN-2 oil bitumen content.

The pH was determined by a potentiometric method in aqueous extracts using a Multi 340i fluid analyzer.

3. Results and discussion

3.1 Environmental hazard estimation for metals

Metal concentrations were determined for the obtained DC and soil samples (Table 1).

Table 1. Metal concentrations in samples (gross form)

Element	Concentration, mg/kg			MAC/APC, mg/kg
	Model DC	Soil after 1 month	Soil after 2 months	
Fe^{2+}, Fe^{3+}	25203 ± 127	12265 ± 89	12238 ± 86	
Pb^{2+}	43 ± 6	18 ± 4	18 ± 3	32
Sr^{2+}	132 ± 3	65 ± 3	62 ± 2	
Zr^{4+}	183 ± 2	92 ± 2	91 ± 1	
K^{+}	25987 ± 322	22760 ± 202	22634 ± 200	
Ca^{2+}	46668 ± 284	29842 ± 197	29801 ± 194	
Cr^{3+}	55 ± 3	24 ± 3	23 ± 3	100
Mn^{2+}	253 ± 5	200 ± 4	200 ± 4	1500
Cd^{2+}	12 ± 1	5 ± 1	5 ± 2	1
Cu^{2+}	46 ± 28	23 ± 6	23 ± 6	66
Zn^{2+}	47	25	25	110

The results showed that due to DC dilution with composite materials, the concentration of metals decreased by almost 2 times. The exceptions are K, Mn and Ca due to the addition of K+ ions in the composition of fertilizer and peat, as well as the processes of substitution of K+, Mn2+, Ca2+ ions in zeolite for metals from DC. MAC and/or APC over lead, copper and cadmium are exceeded in DC. And in the obtained product, only the concentrations of cadmium are exceeded due to the high background content of this metal in the soils of Sakhalin [23].

Since the resulting soil will be placed in the soil, we have analyzed the research results [24–25] and selected those metals that are characterized by the highest degree of danger: Cd, Cr, Pb, Cu, Fe. Metals are contained in several forms in soils, but the most dangerous for the environment are mobile or exchangeable forms of elements, which are extracted from the soil with extractants that mimic to a certain extent the dissolving effect on the soil particles of natural waters and plants. Extracts with the following extractants are used to characterize the group content of metals in soils in addition to the determination of water-soluble element forms [26]:

- ammonium acetate buffer with 4.5 pH – to extract exchangeable and soluble in weak acid forms, which characterizes the current stock of the element in the soil;
- 1%-th EDTA solution in ammonium acetate buffer (pH 4.5), which allows, in addition to the exchangeable and soluble in weak acid forms, extracting elements from organic complexes and determining mobile heavy metal compounds;
- 1N HCl (HNO_3) – to extract the elements making up the amorphous compounds, which characterizes the entire potential stock of the element in the soil.

In order to determine the degree of environmental impact, the first two extracts are of the greatest interest, which allow determining the elements that are prone to ion exchange and are available to plants and living organisms. Thus, we have determined the concentrations of selected most dangerous metals in 3 different exposures to assess the environmental hazards of DC and soil (Table 2).

Having analyzed the study results of aqueous extracts (1), it has been determined that the decrease in water soluble forms due to DC treatment with composite mixture is more pronounced in Fe and Pb. In general, there is a positive trend – the concentration of iron decreased by 105 times, lead – 70 times, cadmium – 1.7 times, copper – 4.5 times, chromium – 1.2 times. The selectivity is explained by the sorption process of ions by zeolite pores. The larger ion radius with the same charge, the better it is sorbed, and the greater ion charge, the stronger ion is attracted to the oppositely charged surface of the zeolite [27]. This is also due to the interaction with humic acids – Fe forms intraspheric complexes, which are more durable [28]. Complex compounds of humic acids with Pb ions are much more stable than complexes with Cd ions [29]. The release of Cd ions can be through the molecular topological effect [30]. Thus, a decrease in desorption of metals from the obtained soil is observed in the aqueous extract, in comparison with DC.

Table 2. Concentrations of hazardous metals in different exposures

Element	Drill cuttings				Man-made soil. Exposure -1 month				Man-made soil. Exposure -1 months			
	1	2		3	1	2		3	1	2		3
		Fact	MAC			Fact	MAC			Fact	MAC	
Fe	14.543	943.32		1437.52	0.2745	494.34		950.5	0.1375	117.25		565.22
Cd	0.047	0.4		0.455	0.044	0.315		0.34	0.028	0.27		0.29
Cu	0.3	2.115	3.0	2.415	0.0805	2.045	3.0	2.145	0.067	1.275	3.0	1.91
Pb	5.1	8.355	6.0	9.715	0.121	5.23	6.0	5.82	0.0725	1.33	6.0	4.595
Cr	1.182	9.725	6.0	10.24	1.103	7.88	6.0	9.455	1.024	5.515	6.0	6.305

1 - Water soluble form.
2 - Exchange and soluble form in weak acids.
3 - Exchange + complex form.

Comparing the experiment results in all forms of extracts (1)–(3), we see a tendency to increase the degree of metals extraction from aqueous extract (1) to 1 % EDTA extract in AAB with pH of 4.5 (3), which is consistent with study results of other authors. In general, the desorption of metals under the influence of an acidic medium and extractants in DC is significantly higher than in the resulting soil, which indicates the retention of metals strength in the soil due to their sorption by the zeolite and fixation in pores under the influence of gypsum. Over time, the concentration extracts of recoverable metals also decrease, which indicates the rationality of keeping the resulting soil for 2 months.

Thus, a decrease in the ion-exchange capacity of the obtained soil, a decrease in the concentrations of metals in the extracts for 2 months, as well as the absence of MAC exceedances, give grounds for assuming that this material is safe for the environment.

3.2 Environmental hazard estimation for petroleum products

We have determined the concentrations of petroleum products (gross form) for the obtained samples. (Table 3).

Table 3. Concentrations of petroleum products in samples (gross form).

Concentration, mg/kg		
Model DC	Soil after 1 month	Soil after 2 months
59897	23166	11742

The concentration of petroleum products in the soil after 1 month of exposure decreased by 2.5 times compared with the original DC due to dilution and destruction processes. Concentrations decreased by 2 times in the soil after 2 months of exposure, which is also associated with the processes of petroleum products destruction.

Due to the absence of petroleum products in soils approved by MAC or APC, the obtained data were compared with the project of regulatory and technical document "Standards of permissible residual content of oil and products of its transformation in the soil (SPRCOP) after carrying out recultivation and other restoration work in Sakhalin Region". The resulting soil can be used for most types of soil, both organic-mineral and mineral, industrial and forestry soils.

3.3 PH value

Additionally, pH of aqueous extract of samples has been determined (Figure 2).

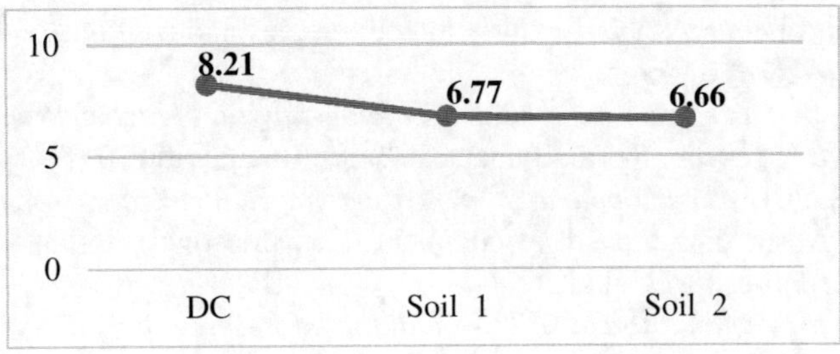

Figure 2. pH value of aqueous extract of samples

It was established that after a month of soil ageing treatment pH value decreased from slightly alkaline to neutral. This may be due to dilution with DC by peat and the addition of humic acids to the mixture, with reactions of dissociation of hydroxyl groups of the zeolite and gypsum mixture.

4. Conclusion

In this paper, we can conclude that a zeolite-based composition mixture can be used to process DC with a view to reduce its environmental hazard. As a result of the experiment, samples of man-made soil with different exposure times were obtained under laboratory conditions.

The concentrations of metals (gross form) in the soil do not exceed standard values, with the exception of cadmium, the background concentrations of which are exceeded throughout Sakhalin. Concentrations of mobile forms of metals that can be available to plants and living organisms after adding a composition mixture based on zeolite, have significantly decreased and do not also exceed standard values. A number of metals according to the degree of their retention in man-made soil are as follows: Fe>Cu>Cd>Pb>Cr.

Also, the concentration of oil products in the resulting soil has decreased, which allows using it in soils of various types. Neutralization of pH also has a positive effect on soil properties.

Thus, the environmental estimation of process using zeolite-containing DC composite mixture allows us to conclude about the possibility of its safe reuse and the application of this technology in general when handling drilling waste.

5. Acknowledgments

The work was supported by British Petroleum Scholarship Program.

References

[1] Batalin B S and Nechaeva A E 2013 Utilization of drill cuttings processing into materials for construction purposes *Master's J.* **2** 148–52

[2] Islamov Kh M 2011 Geoecological safety of chemicals use for the treatment of drilling fluids *Geology, Geography and Global Ecology* **3** 174–9

[3] Iagafarova G G, Mavliutov M R and Barakhnina V B 1998 Biotechnological method of disposal of oil cuttings and drilling waste *Mountain Bull.* **4** 43–6

[4] Petrov V G, Kharaldina E A and Shumilova M A 2011 Integrated disposal and disposal of drilling waste *Bull. of Udmurt University* **4–2** 77–9

[5] Genoveseva M 2008 Bioremediation of benzene, toluene, ethylbenzene, xylenes-contaminated soil: a biopile pilot experiment *J. Appl. Microbiol* **105(5)** 1694–702

[6] Wieczorek D, Marchut-Mikolajczyk O and Bielecki S 2012 Phytotests as tools for monitoring the bioremediation process of soil contaminated with diesel oil *Biotechnol. Lett.* **93(4)** 431–9

[7] Classification of drilling solids 2009 *Appendix to J. Oil and Gas Technol.* (Moscow: Fuel and Energy)

[8] Mishunina A S 2014 Methods of drilling waste disposal *Problems of development of hydrocarbon and ore mineral deposits* **1** 125–8

[9] Perevalov S N and Ivleva A A 2013 Current technologies and methods of drilling waste disposal *Int. Res. J.* **11–1(18)** 63–6

[10] *Information and technical reference book on the best available technologies* 2016 Disposal and disposal of waste (except thermally neutralization (incineration)) (Moscow: BAT Bureau)

[11] Shpinkova M S 2014 *Development of a method for neutralization of oil waste of different composition* Abstr. dis. ... cand. of chemical sci. (Moscow: Russian State University of Oil and Gas named after I M Gubkin)

[12] *New technology for reclamation of soil through the processing of drilling waste in the extraction of oil and gas* 2015 Explanatory note (sci. background) (Moscow: FSBU UralRI "Ecology")

[13] Koltsov I N, Mitrofanov N G and Petukhova V S 2013 *Sludge-soil soil mixture (versions) for recultivation of disturbed lands and a method of quarrying of quarries and disturbed lands* Pat. 2491135 Russian Federation

[14] Tsipper A A et al 2016 *Composite building material "Gumikom"* Pat. 2575950 Russian Federation

[15] Mokrousova M A and Glushankova I S 2015 Remediation of drill cuttings and oil-contaminated soils using humic preparations *Transport. Transport facilities. Ecology* **2** 57–72

[16] Burlaka N V, Burlaka I V and Burlaka V A 2010 *Method of disposal of waste drill cuttings* Pat. 2379137 Russian Federation

[17] Valieva I R and Nefedov V A 2012 Sorption of copper and zinc from waste and surface waters by zeolitic and zeolite-montmorillonite rocks of the Urals *Fan-science* **6** 57–9

[18] Vatin N I and Chechevichkin V N 2013 Application of clinoptilolite type zeolites for the purification of natural waters *Engineering and Construction J.* **2** 81–8

[19] Veisgeim A S, Nazarenko O B and Zarubina R F 2012 Removal of iron from borehole water on the filter with loading from Badinsky zeolite *Siberian J. of Sci.* **4(5)** 23–9

[20] Nazarenko O B, Zarubina R F et al 2011 Application of Sahapta zeolite to improve the quality of drinking water *Bull. of Tomsk Polytechnic University* **3** 28–32

[21] Chichina S B and Ovsiannikova O O 2012 Evaluation of the effectiveness of cleaning oil-contaminated waters using coke and zeolite *Omsk Sci. Bull.* **114** 207–11

[22] Obuzdina M V 2010 Natural and modified zeolites as adsorbents of petroleum products from industrial wastewater *Proc. of IrSTU* **4(44)** 104–10

[23] Ministry of Natural Resources and Environmental Protection of Sakhalin Region 2016 *Report on the environmental situation and environmental protection of Sakhalin region in 2015* (Yuzhno-Sakhalinsk: LLC Eikon)

[24] Crommentuijn T, Sijm D and Bruijn J 2000 Maximum Permissible Concentrations and Negligible Concentrations for metals, taking background concentrations into account *J. of Environmental Management* **2** 121–43

[25] Vodianitskii Iu N 2011 On hazardous heavy metals/metalloids in soils *Bulletin of the Soil Institute named after V V Dokuchaev* **68** 56–82

[26] Titova V I, Dabakhova E V and Dabakhov M V 2011 *Agro-and biochemical methods for studying the state of ecosystems: textbook for universities* (N Novgorod: VVASS Publ. House)

[27] Lukhneva O L, Chikin A Iu and Proidakova O A 2010 Nature of interaction of metal ions with mineral sorbents in the process of purification of wastewater from electroplating plants *Proc. of IrSTU* **5(45)** 194–7

[28] Popov A I 2004 *Humic substances: properties, structure, education* (St. Petersburg: St. Petersburg University Publ. House)

[29] Stevenson F J 1976 Stability constants of Cu2+, Pb2+, and Cd2+ complexes with humic acids *Soil Sci. Soc. Amer.* **40(5)**

[30] Liu C and Huang P M 1999 Ionic Strength and Cadmium Effects on Surface Features of Humic Acid *Understanding Humic Substances. Advanced Methods, Properties and Applications* **247** 87–99

IOP Conf. Series: Materials Science and Engineering **663** (2019) 012018 doi:10.1088/1757-899X/663/1/012018

Concept of eliminating past environmental damage in the area of mining industry of the North Caucasus

I D Alborov[1], F G Tedeeva[1], F Kh Gutsaev[1], E A Gridnev[1], M M Tebloev[2], T F Tsgoev[1] and A K Berkaeva[3]

[1]North Caucasian Institute of Mining and Metallurgy (State Technological University), 44, Nikolaeva Str., Vladikavkaz, 362021, Republic of North Ossetia-Alania, Russia
[2]North Ossetian State Medical Academy, 40, Pushkinskaia St., Vladikavkaz, 362019, Republic of North Ossetia-Alania, Russia
[3]North Ossetian State Pedagogical Institute

E-mail: ekoskgmi@rambler.ru

Abstract. The article presents the results of many years research on the surface infrastructure influence of mining industry of the North Caucasus on biosphere components in its area of activity. Considering that industrial mining of ores on the eastern slope of Caucasus Mountains (Sadonskoe silver-zinc deposit) began in the middle of the nineteenth century by Belgian industrialists and continues to the present. The accumulation of heavy and toxic metals in the soil horizon is shown (2.5–3.5 MAC), creating a real threat to health and livelihoods of the population living here. Fragmentary variability of landscapes, leading to the depletion of biological diversity, up to the disappearance of its various types has been revealed. According to the intensity of dust emission and the share of negative sources participation in the overall balance of dust pollution of the area atmosphere, it has been found that man-made deposits of ore processing waste are potential foci of catastrophic processes in the destruction of fencing dams by mountain streams, as they are located on terraced areas of mountain valleys. It is analytically established that safety, both general and environmental, can be achieved only by eliminating these sources from potentially dangerous places.

1. Ecological state of the research object

A specific characteristic of the Caucasus mountain ecosystems is their isolation, isolation with a strongly dissected relief, high gustiness of mountain-valley winds due to the proximity of zones with eternal snow, low barometric pressure of the atmosphere, etc. Air and water flows, solid runoff and gravitational structural anomalies are directed into the system, the most reduced relief areas, valleys of the gorges, where water, solid streams are accumulated, and local currents are largely regulated, and in general correspond to the hydraulic network. In turn, watercourses are a dendrated (branched multi-order) system in which the entire runoff is ultimately integrated into the trunk structure - the main river valley. By tracking the upstream of main rivers (Ardon, Fiagdon, Baksan, Urup, Dzhodzhora, Terek) variations in the content of pollutant substance, one can reach the tributary in each of the above watercourses, and the content of technogenesis ingredients rises above the mouth of the river and sometimes higher due to the release of the ore-bearing deposits into the river bed [1].

IOP Conf. Series: Materials Science and Engineering **663** (2019) 012018 doi:10.1088/1757-899X/663/1/012018

2. Research subject and results

Large-scale sampling of bottom sediments ensures prompt screening of the hydraulic network elements into contaminated and background. In the future, with this technology of processing information data in the absence of significant changes in distribution of man-made load on the hydrosphere, background areas can be excluded from research, many times reducing their volume.

Having established a network of monitoring reference stations, it is possible to have data on the distribution of pollutants that adequately reflect the level and nature of chemical pollution of water flows of any valley and territories of the Greater Caucasus mountain ecosystem. These stations will be located in the mouths of different orders streams, in the valleys of which there are sources of relevant pollution [4-6].

The transition from conceptual modeling to a formalized automated model is associated with the accumulation of a significant amount of quantitative information and systematization of the objects cited earlier, content of observations, inclusion of elements of air basin studies, materials of hydrochemical sampling and biological studies of soil and vegetation into the model. A new form for the choice and selection of criteria for assessing the environment state, parameters of permissible and normalized loads on ecosystem objects, especially in the absence of the developed MAC has been proposed. Integrated environmental load on natural environment of a mountain region is determined by formula 1.

$$Э_Н = \int\left(\sum_{Н=1}^{n} Х_Н \cdot \sum_{Н=1}^{n} Ф_Н \cdot \sum_{Н=1}^{n} Б_Н\right) \tag{1}$$

where $\sum_{Н=1}^{n} Х_Н \cdot \sum_{Н=1}^{n} Ф_Н \cdot \sum_{Н=1}^{n} Б_Н$ – indicators characterizing chemical, physical and biological deviations in the environment associated with activities of the natural-technical system (mining complex), c.u.;

$\sum_{Н=1}^{n} Х_Н$ – Х1, Х2, Х3, ... Хn – parameters of chemical elements and phenomena affecting pollution and destruction of the environment (according to geochemical soil testing), c.u.;

$\sum_{Н=1}^{n} Ф_Н$ – Ф1, Ф2, Ф3, ... Фn – parameters of physical processes associated with activities of the mining and industrial complex include noise, vibration, magnetic, thermal and other forms of environmental impact, c.u.;

$\sum_{Н=1}^{n} Б_Н$ – Б1, Б2, Б3, ... Бn – indicators of unnatural biological processes caused by activities of the mining and industrial complex in the region, housing and communal, animal husbandry, etc., c.u.

Private models are developed for each element of a mountain object; subsequent similar objects can be combined with similar parameters.

The general model of sustainable development of ecosystem includes forms of the considered object negative impact on the hydrosphere, lithosphere and atmosphere through private indicators - on the integral environmental estimation of environmental management efficiency. Environmental load on the natural man-made system environment-infrastructure-mining-society is determined by formula 1 and compared with the normative (if available), and if it is not available, the comparison is made according to the preservation of biosphere basic properties (self-regulation and self-purification) and the acceptability of economic activity intensification or its weakening in the region. Monitoring studies conducted in the zone of influence of the natural-technical system environment-infrastructure-mining-society in the conditions of Sadonsky Lead-Zinc Combine, Tyrnyauz Tungsten-Molybdenum Combine, Urupsk MPC, Kvaisinsky Mine Directorate have found that the content of heavy metals in soils exceeds sanitary norms two or more times. Gross contents of main ore elements along vertical section of the soil to a depth of 1 m, with an interval of approbation of 0.2 m, in the area of Sadonsky LZC activity, showed a stable, sometimes quite sharp drop in the content of heavy metals with depth, which unequivocally confirms the man-made nature of the detected anomaly.

IOP Conf. Series: Materials Science and Engineering **663** (2019) 012018 doi:10.1088/1757-899X/663/1/012018

3. Obtained results and their scientific and practical value for the economy

It is established that maximum content of lead and zinc is confined to upper layers of the soil cut. The gross distribution of lead and zinc, relative to their maximum permissible concentrations, is shown in Fig. 1, which confirms man-made origin of lithosphere soil pollution.

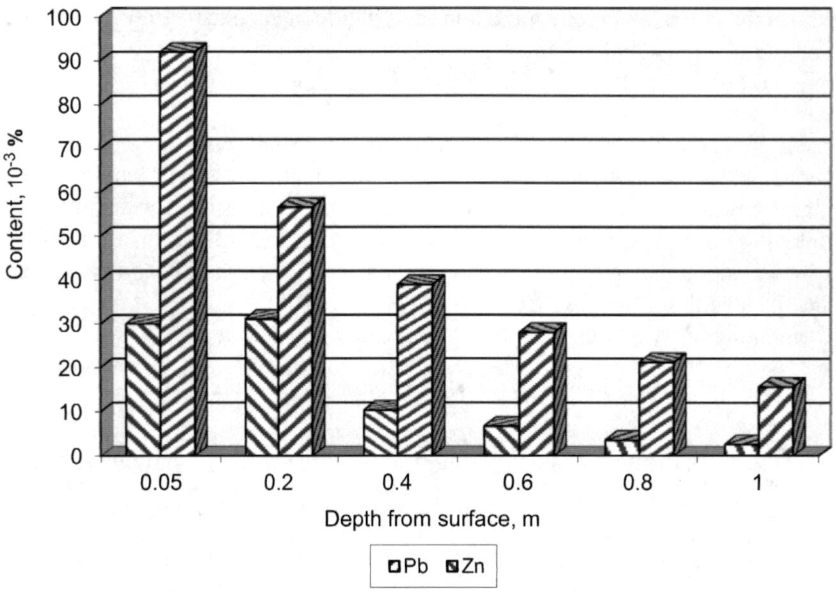

Figure 1. Maximum content of lead and zinc in different layers of the soil cut

Block diagram of the model for the development of man-made environmental loads, including human environment, is shown in Fig. 2.

When a natural-technical system functions, the level of its impact on individual components of the biosphere depends on the integral index of entire system. When any parameter goes beyond the acceptable level, it is necessary to use effective means of neutralizing or reducing the effect of the disturbing factor to restore its normal functioning.

The intensification problem of production, with the continuous growth of requirements for its greening, is typical for all industrialized countries. In general, an integral part of enterprises high efficiency should be a high efficiency of environmental management, which refers to the environmental and socio-economic impact of the natural resources use and the exploitation of natural environment [3].

In this regard, the authors propose to estimate production impact on the environment by its intensification degree, which will make it possible to normalize emissions not by gross volumes, but by its specific values for final products.

Environmental efficiency indicator of $K_Э$ in mining enterprises (2):

$$K_Э = \frac{\sum_{i=1} П_{iVoz} + \sum_{i=1} П_{iVod} + \sum_{i=1} П_{iPochv}}{V} \tag{2}$$

where $П_{iVoz}$, $П_{iVod}$, $П_{iPochv}$ – amount of payments for air, water and soil pollution, respectively, c.u.; V – annual ore production, c. u.

Such a standard characterizes the efficiency of environmental management at a particular enterprise, and also makes it possible to judge environmentally acceptable products of different enterprises from the aggregate of harmful effects on biosphere components, determines technical level of production.

IOP Conf. Series: Materials Science and Engineering **663** (2019) 012018 doi:10.1088/1757-899X/663/1/012018

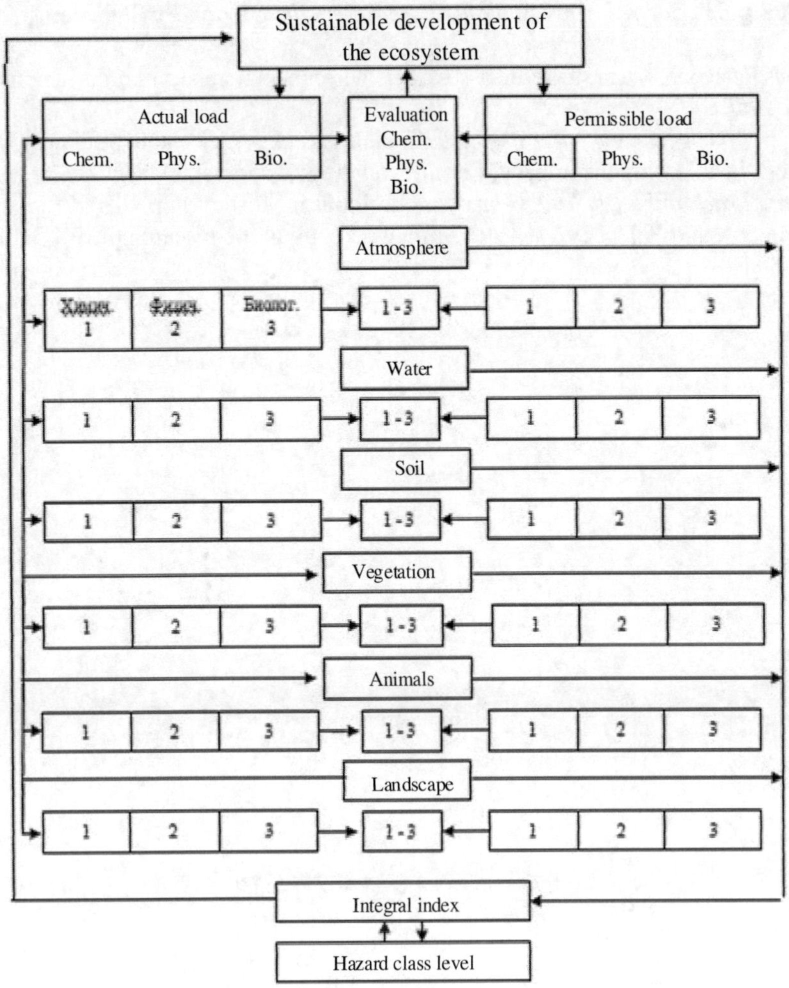

Figure 2. Block diagram of a mining region model

4. Conclusions

1. Accumulation of heavy and toxic metals of technogenic origin in the soil horizon (2.5–3.5 MAC) creates a real threat to health and livelihoods of the population living in this area.

2. Fragmentary variability of landscapes leading to the depletion of biological diversity, up to the disappearance of its various species is noted.

3. Potential foci of development of catastrophic processes remain coastal areas enclosing landfill sites for dams enrichment, since they are located in an area of increased risk when mountain rivers are poured.

4. Proposed indicator of environmental efficiency allows us to judge the environmentally acceptable products of different enterprises by the totality of harmful effects on biosphere components, at the same time determines technical level of production.

References

[1] Integrated environmental monitoring of "Mountain Valley" ecosystem (on the example of the valley of Fiagdon river) 1993 *Report on RW SUE ERPC "Ecology", b. 1* (Vladikavkaz) 139 p

[2] Estimation of ecological and geochemical situation in the area of tailing dump of Sadonsky Lead-Zinc Combine MSU *Report 1990–1991* (Moscow) 120 p

[3] Monin A S and Losev K S 1999 *Ecological Encyclopedic Dictionary* (Publ. house Noosphere) 930 p

[4] Guidelines for geochemical studies to assess the environmental impact of projected mining enterprises 1986 *MMEE* 100 p

[5] Chaplygin N N, Papichev V I, Blizniuk G I, Proshliakov A A and Stoliarov D O 1998 New methods in assessing the impact of mining on the environment *Mining sci. at the turn of the 21st century, Proc. of the 1997 Int. Conf.* (Yekaterinburg; Moscow) pp 468–77

[6] Chaplygin N N and Papichev V I 1997 Mining ecology in the research of IPKON RAS *Mountain Bull.* **5** 86–92

Ecological situation in the area of non-ferrous metals mining on the slopes of the Eastern Caucasus

I D Alborov[1], F G Tedeeva[1], O G Burdzieva[2], E A Gridnev[1], A P Glazov[1], F M Dzeboeva[1] and S D Vaneev[1]

[1]North Caucasian Institute of Mining and Metallurgy (State Technological University), 44, Nikolaeva Str., Vladikavkaz, 362021, Russia
[2]Geophysical Institute of Vladikavkaz Scientific Center RAS, 93a, Markov Str., Vladikavkaz, 362002, Russia

E-mail: ekoskgmi@rambler.ru

Abstract. The article provides information on the history of the development of Sadon ore field, Tyrnyauz tungsten-molybdenum and Urupsk deposits. A description of landscape changes occurred in the process of mining and processing of non-ferrous metal ores in places of mining enterprises dislocation is given. It has been shown that soil pollution in the area of mining facilities activity with heavy and toxic trace elements has mineralization of mining origin. It has been established that in the areas of dislocation of existing mountain objects, landscapes and soils, which are an integral part of ecosystems, have a pronounced zonation, each of which has certain components that influence the formation of ecological situation. Low levels of lead and zinc have been found in seasonal vegetables (potato tubers) and high, exceeding MAC – in perennial fruit crops (apples, pears). It is noted that only the use of scientifically based measures based on real positive results, taking into account geo-environmental factors, can affect efficient environmental management under the considered conditions.

1. Characteristics of the zone of non-ferrous metals mining in the North Caucasus Sadon ore-bearing region

The deposits of polymetallic ores of North Ossetia are localized both in foundation rocks and in Jurassic volcanic-sedimentary deposits. The most significant in terms of reserves of polymetallic ore are Sadono-Unal ore field, represented by Sadon, Zgid, Arkhon, Oktiabrskoe, Kholst deposits, etc. Venous deposits, the main ore minerals are sphalerite, galena, pyrite, pyrrhotine, chalcopyrite, secondary – arsenopyrite and marcasite. The main useful components of ores are lead and zinc; the associated ones are silver, copper, cadmium, and indium. The distribution of the main ore elements is extremely uneven, the lead content in the ores varies from 0.42 to 12.8 %, zinc from 1.87 to 26.2 %.

Fiagdon ore field is represented by deposits of Levoberezhnoe, Kadatskoe, Khanikamskoe, Kakadurskoe and etc., located in thick sandy shale strata of Jurassic in the form of brecciation and carbonatization zones. Brecciation zones carry galena, sphalerite, pyrite, arsenopyrite, chalcopyrite; the satellite elements are tin, silver, gold, cadmium, bismuth.

The area of activity of the Sadon mines is located in the mountainous part of North Ossetia and is characterized by medium-high-altitude relief with absolute elevations of 1.800–2.800 m and relative elevations up to 800–1.600 m. Mountains are sharply divided by numerous rivers and their tributaries into a number of narrow and long gorges of various directions. The longest gorges have submeridional directions and are formed by Ardon, Urukh, Fiagdon rivers – the main tributaries of Terek river. The climatic regime and nature of vegetation in gorges is variable and closely related to the relief. The relief also determines local distribution of thermal regime, humidification regime, and air flow direction.

The valleys of Ardon and Fiagdon rivers belong to th “Solar valleys” with a characteristic microclimate. By the number of days with solar activity, this zone does not concede to Swiss Alps. The average temperature of the warmest month of July in the area of Unal and Fiagdon villages is + 16 °C and the coldest in January is –2 °C. Annual precipitation is 300–400 mm with the highest loss in summer and autumn. Winters are with little snow with periodic snow cover about 10 cm. The area is characterized by strong valley winds up the gorge during the day and back at night.

There are two main types of soil within the valleys: wormwood-grass and slope mountain-forest-meadow; brown podzolized and humus-carbonate soils. Agriculture has a major fruit-vegetable and livestock trend in the area. Apple-pear orchards are widely developed here; they have not only of local, but also wide industrial significance. Potato and corn are the most common among vegetable and grain crops.

In the transverse profile of the Ardon river valley is asymmetrical. The left bank is steep, bounded by rock outcrops of sedimentary rocks of the Middle Jurassic and landslide sediments of Kion-Khokh Mountain. Transcam road connecting Transcaucasia with the North Caucasus passes through it. The right bank is represented by a series of multi-level accumulative basement river terraces. All terraces are composed of boulder-pebble sediments with a gravel-sandy aggregate and covered with a layer of loam, often loess-like. The upper soil layer is represented by black soil with a thickness of 0.2–0.3 to 0.4 m in the area of N. Unal village, Unal tailing dump, commissioned in 1984 is on the lower terrace near the bridge across Ardon river. It is located in the floodplain of the left bank of Ardon river, 0.5 km north of N. Unal and 10–12 km north-east of Mizur concentrating mill. The tailings dump beds are pebbles of Ardon river, the right board is separated by a concrete dam from the riverbed. The absolute mark of of the floodplain river bottom is 873.3 m; N. Unal village is 900 m above sea level. Refinement tailings of Mizur concentrating mill are fed to the tailing dumps. The entire infrastructure of Sadon Lead-Zinc Combine, including residential zone, is located in the narrow valley of Ardon. The bedrock and geological exploration waste by motor transport are simultaneously removed and stored in the floodplain of Ardon. Fiagdon tailing dump has been functioning since 1970. It is located in a narrow, canyon-shaped gorge of Khanikomdon river, a right tributary of Fiagdon river at an altitude of 1250 m above sea level. It has a sharply elongated shape, bounded by rocky sandy-rocky and limestone outcrops of the Jurassic and Cretaceous on both sides. Fiagdon tailing dump is separated by a high bulk dam from the river valley; and it is located about 600 m downstream of Khanikomdon river from Old Dzuarikau village. The village is located in the floodplain of a narrow valley of the brook; a dirt road between Khanikom gallery and Fiagdon concentrating mill passes along the left slope of 400–500 m. Tails were fed by pipeline hydrotransport to the tailing dumps by two lines.

2. Tyrnyauz tungsten-molybdenum deposit

The field is worn out by the underground mine “Molibden” and single-breasted quarries of “Mukulanskii” and “Vysotnyi”.

Mine production facilities are located at various levels: concentrating mill – 1350 m, amenity complex and main haulage adit – 2004 m, main scarn – 2 600 m, “Mukulanskii” quarry – 2700 m, “Vysotnyi” 3300 m. The town of Tyrnyauz occupies floodplain terraces of Baksan river and its tributaries Sakashili-Su and Kamyk-Su at elevations of 1250–1300 m from sea level. The area is characterized by a strong dissection, narrow deep gorges, rocky slopes, large amplitudes of relative heights. The relief features are also clearly defined traces of glacier activity, which gave the river valleys a trough-shaped form, known as troughs. The activity of glaciers is rocky ridges on the banks of rivers. There are powerful terraces composed of loose-detrital material resulting from the erosion-accumulative activity of glaciers in areas of river valleys, especially in the mouths of rivers flowing into Baksan river, as well as in Baksan gorge. The formation of loose-detrital material on the bottom and on the slopes of river valleys is explained by the weathering of rocks. The reasons for the formation of terraces and loose cones of loose-detrital material are mudslides in most watercourses of the region.

The vegetation cover of the region is of zonal in nature. Floodplains of Baksan river valley are covered with grass and bushes. Woody vegetation appears higher on the slopes of Baksan gorge. First deciduous trees are dominated, with a height giving way to pine forest. Areas of alpine meadows with xerophytic vegetation – mosses, lichens, and alpine grasses are above the forest zone.

Baksan river and its tributaries in the upper part of the basin are mountain-type watercourses. The main base of the water regime is the spring-summer flood, during which, even during rain floods, rivers carry suspended and sediment. Loose-detrital material is piled up on the slopes of gorges in many places: taluses, debris cones, etc. Formation of air pollution sources is associated with the wind regime of the area and heat and moisture factor.

The wind regime of the area is determined by orographic conditions and temperature conditions defined by the location proximity of eternal snows-glaciers, in accordance with which local mountain-valley winds dominate, characterized by diurnal variability. During the day, the wind blows from the bottom up – the valley wind; at night – in the opposite direction – the mountain wind.

In the area of the mine, strong side winds are observed from the Greater Caucasus Mountain Range side, descending into the valley of Baksan gorge. Winds have the character of mountain dryers. The environment is polluted by dust particles from their places of dislocation – debris on terraces, dumps of overburden rocks of "Mukulanskii" and "Vysotnyi" quarries; the accumulation of geological work dumps, etc. [3].

The wind regime is characterized by average annual repetition in the directions of Table 1 and average annual wind speeds in the directions of Table 2.

Table 1. Average annual repetition of winds in directions, %

N	NE	E	SE	S	SW	W	NW
22.8	43.8	2.2	1.7	12.2	15.5	1.0	0.8

Table 2. Average annual wind speed in directions, m/s

N	NE	E	SE	S	SW	W	N-W
4.7	4.8	2.0	2.1	6.7	4.7	1.1	0.9

3. Characteristics of mining enterprise activities

Mountainous provinces are ecosystems specific for mountainous areas, characterized by their significant isolation in terms of water and air flows distribution, the concentration of slope demolition in thalwegs, and mosaic microclimatic conditions. Anthropogenic load of mountain valleys determines the degree and nature of pollution not only of the valley, but also of flat landscapes in their mouth parts.

The current level of anthropogenic pressure on mountain valleys is very high: the distribution of load across different valleys of the North Caucasus is extremely uneven. The most tense environmental situation has developed in the valleys of Fiagdon, Ardon, Baksan and Urup rivers, where mining complexes with a developed surface infrastructure operate.

4. Soil horizon contamination with heavy metals

Studies conducted at Geological Faculty of Moscow State University have found that the content and toxic metals in the soil are in the area of activity in which regularities are located, and the content of metals in thickness depends on the depth of soil horizon, which is evidenced by negative impact of mine activity on the extraction and processing of non-ferrous metal ores.

The ingress of metals into the soil occurs in the process of applying environmentally harmful and hazardous technological processes of production: transportation of ore from mines in dump trucks to the processing complex; concentrate on metallurgical processing; removal of dust fractions from the surface of the tailing dumps and landfills of sub-standard ores; incomplete cleaning of mine runoff and wastewater from the processing plants; pollution of the atmospheric basin by polluted air removed from the mines; leaching of ore minerals in warehouses, dumps, sludges, etc.

The most dangerous environmental pollutants are mercury, cadmium, lead, arsenic, chromium, copper, nickel, zinc, iron, sulfur, etc. Fluent trace elements are considered to be of particular ecological, biological and health significance.

The behavior of trace elements in the soil is very different, the residence time of pollutants in the soil is incomparably longer than in other parts of the biosphere, and its contamination with especially heavy metals is almost eternal.

According to CMGE laboratory, lead, zinc, copper, cobalt, strontium, molybdenum, cadmium, silver, bismuth, arsenic, selenium, manganese, chromium, nickel, and thallium are present in the soils of the Sadon industrial region.

Metals that accumulate in soils are slowly assimilated by plants, and also as a result of erosion and soil deflation, form dust pollution of the atmosphere.

In respect that the ecosystem under consideration includes continuous industrial development of non-ferrous metals, the applied technology allows using only a small part (first percent) of the extracted rock mass. The rest is accumulated on the surface in the form of dumps and sludge, which are potential sources of environmental pollution by a wide range of chemical elements.

The presence of residential areas located in the zone of influence of the mining complex, as well as agricultural development sites with the cultivation of fruit and vegetable crops, while simultaneously occupying the population living there with livestock raising creates a real danger of environmentally safe living in this area [4].

Research has established that the content of the main ore elements in soils in the territory of Nizhny Unal village is: lead – 0.15 %, zinc – 0.4 %, copper – 0.1 %, silver – $30 \cdot 10^{-5}$ % and they are confined to soil humus horizon. Low levels of lead and zinc have been found in seasonal vegetables (potato tubers) and high, exceeding MAC – in perennial fruit crops (apples, pears) [2].

The areas and intensity of soil contamination by ore elements exceeding MAC in the area of Nizhny Unal village are given in Table 3.

The background contents of ore elements correspond to average values of alluvium-deluvial formations for North Ossetia in general [3].

Landscapes and soils being an integral part of the ecological systems under consideration are distinguished by a great variety. High-altitude zonality of landscapes, forming the corresponding climatic zones, each of which has certain components (Table 4) is pronounced in the areas of dislocation of existing mountain objects (Sadon mines, "Molibden" and "Urupsk" mines).

Table 3. Contamination area and the content of ore elements (mg/kg) in the soils of Nizhny Unal village

Chemicalel ements	Cf	MAC_s	Contamination area %	C on contamination area	Drift, MSC_s	C_{max}	Drift, MAC_s	$q_{e/02m}$
Pb	47.5	200	45	460	2.3	1500	7.5	45
Zn	100	400	50	1200	3.0	2000	5.0	119
Cu	27	100	Dotted	100	1.0	100	1.0	–
Ag	0.05	5	Dotted	2	1.0	2	1.0	–

Glacial-nival zone is located above the snow line and is an area of snow accumulation, which turns into firn and glacier ice. Intensive physical weathering is developed in this zone, as a result, detrital material is formed.

The soils are thin, chirping, representing the formation stage of mountain-meadow soils in the subnival belt.

Table 4. Zonality of landscapes

Zones	Belt	height of floodplain from sea level, m	Note (mine)

Glacial-nival		More 3500	
Mountain-meadow	subnival -alpine	3000–3500	"Molibden"
	alpine	2400–3000	
	subalpine	1000–2500	"Sadon"
Mountain-steppe		800–2000	"Sadon"
Mountain-forest	Pine and birch forests	1600–2400	"Sadon"
	Deciduous forests	750–1600	"Urupsk"

Erosion processes and especially chemical weathering play a large role in the alpine and subalpine belts [4–6]. Here, the slopes of the ridges are dissected by deep valleys, the bottom of which lies at an altitude of 1800–2000 m. The bottom of basins and the lower parts of slopes are accumulation place of detrital material, and with mountain-long winds its dust-like fractions serve as a steady source of atmospheric dusting.

5. Conclusions

1. It is established that ecological state of the environment depends on the interaction and location of pollution sources and destruction objects.

2. It is shown that soil pollution in the activity area of mining facilities with heavy and toxic trace elements has mineralization of mine origin.

3. It is shown that a pronounced zonality, each of which has certain components that influence the formation of ecological situation takes place in dislocation areas of existing mountain objects of landscapes and soils, which are an integral part of ecosystems.

4. It is established that on the territory of N. Unal village the content of the main ore elements in soils amounts to: lead – 0.15 %, zinc – 0.4 %, copper – 0.1 %, silver – $30 \cdot 10^{-5}$ % and they are confined to the humus horizon of the soil.

5. Low levels of lead and zinc are found in seasonal vegetables (potato tubers) and high, exceeding MAC – in perennial fruit crops (apples, pears).

References

[1] To the 125th anniversary of the Sadon Lead-Zinc mines 1968 *Issues of improving mining production (to the 125th anniversary of the Sadon mines)* (Ordzhonikidze) 3–16

[2] *Assessment of the ecological geochemical situation in the area of the Sadon Lead Zinc Combine* MSU Report for 1990–1991 (Moscow) 120 p

[3] Trubetskoi K N 1999 Resource-saving technologies and their role in ecology and rational environmental management during the development of mineral resources *Theses of sci. and techn. Conf. Environmental problems of mining production* (Moscow: IACGO) pp 3–4

[4] Alborov I D, Zaalishvili V B, Tedeeva F G et al 2013 *Environmental risk, principles of environmental assessment and public health* (Vladikavkaz) 347 p

[5] Chaplygin N N and Papichev V I 1997 Mountain ecology in IPKON studies *Mountain Bull.***5** 87–92

[6] Alborov I D, Tedeeva F G, Burdzieva O G et al 2017 Ecological aspects of functioning of the ecosphere with the mining and processing component *Col. of works "All-Rus. Congress of Ecologists"* (Grozny) pp 19–23

IOP Conf. Series: Materials Science and Engineering **663** (2019) 012020 doi:10.1088/1757-899X/663/1/012020

Pathways to sustainable development of urban areas in the Southern Urals

N G Blagovidova and N V Iudina

Urban Planning Department, Moscow Institute of Architecture, 11/4, Rozhdestvenka

E-mail: nablago7@yandex.ru

Abstract. The paper considers retrospective development of the mountain region in the Southern Urals as a special industrial and natural landscape monument of Russia. Following global trends, the authors are aware of both high significance and great vulnerability of mountain areas. The Southern Urals is considered as the region with a tendency to stagnation. The historical background of the problems is highlighted, and ecological status and risks of the region are analyzed on the example of a network of small and medium-sized cities of Zlatoust district, which contributes to solution of the problems. The cities are united by modern specialization, infrastructure, socio-economic situation and natural heritage typical of the entire Ural region, which enables generalization of the principles for both the Urals and other mountain areas. Their special environmental features require complex measures aimed to restore the integrity and the value paradigm. It is proposed to create a cluster based on the potentials, which offers specific directions for the development of a group settlement and functional autonomy. In this regard, it seems necessary to eliminate past ecological damage and to minimize adverse factors that violate environmental conditions. The urban ecology strategy is proposed to resolve contradictions in the development of resource-rich and historically identical areas of the Southern Urals.

1. Introduction

The ecosystem of our planet has come to the bifurcation point of its development, when people's need for all types of resources, including natural, recreational and cultural, significantly exceeds the capacity and capabilities of natural and anthropogenic systems. In 1992, the United Nations Conference on Environment and Development (UNCED) in Rio de Janeiro adopted Program 21, the 13th chapter of which is devoted to preservation and rational use of mountain areas that belong to fragile ecosystems [1]. Being one of the main sources of fuel and organic resources, and tourist activity, mountain areas are under tremendous pressure. This situation is particularly relevant for one of the largest and most developed regions of mining industry – the Southern Urals, which currently faces serious development contradictions.

The study aims to consider the prospects for the balanced development of mountain areas through the example of small cities in Zlatoust mountain region of the Southern Urals in Russia.

Objectives of the study are to:

- identify the features of historical development of the Southern Urals;
- assess and characterize the current status of the Ural region in terms of environmental, cultural and social aspects;
- identify and assess the resource potential of the Southern Urals;

- suggest prospects for sustainable development of mountain areas through the example of Zlatoust region, including such cities as Karabash, Miass, Zlatoust, Kusa, Satka, Chebarkul, Ozersk, and Kyshtym.

2. Retrospective of the development of mountain areas in the Southern Urals

In the 18th century, resource-rich mountain regions were attractive for development and urbanization when the state was particularly in need of mining and industrial production. Ural region and its eastern and southern regions were first to develop. New cities were founded around factories built by such entrepreneurs and industrialists as A.N. Demidov, V.N. Tatishchev, V.I. Gennin, and the Stroganovs [2]. The Stroganovs contributed greatly to the formation of metallurgical production and built 14 factories that formed a resettlement framework. In 1734, which is indicated by L.E. Joffe as one of the turning points in the history of development of the Urals, the region amounted to 40 factories. By 1767, the number of factories had increased to a hundred [3]. The enterprises became the basis for cities-factories, which exhibited a high level of socio-economic development and comfort. As a rule, the main square with the temple forms the central urban space between the working settlement and the factory. The unity of the city ensemble was achieved by a common style of buildings [3]. Therefore, until now the historical center of the cities of the Southern Urals is of great historical and cultural significance.

The introduction of advanced technologies resulted in a variety of industries, and different types of resources provision and transport accessibility, which formed a dense frame of settlements. However, settlements still used vast amounts of available natural resources. The surrounding forests played a special role in the factory functioning. The use of forests for construction and fuel supply often led to devastating fires, so that in the 19th century forests were taken under protection and became the basis for creation of subsequently protected areas (specially protected natural areas).

The Urals started to develop again during the Second World War, when 667 factories of the European part of Russia were evacuated to this region, as well as 18 million people, who settled mainly in mining settlements. Thus, eleven of these settlements were granted city status [3]. Until 1990, the population of the Urals grew steadily in all major cities and settlements, which increased production and its impact on the environment.

2.1 Zlatoust district

Zlatoust mountain district located at the foot of the Bolshoy Taganay ridge, the area abundant in lakes and small rivers, plays an important role in terms of a special symbiosis of natural areas and increased industrial activity, The city of Zlatoust was founded in 1754 at the foot of the Kosotour mountain, at the confluence of the Ai and Tesma rivers. In the early 19th century the city became the center of Zlatoust mountain district, including the Zlatoust, Satka, Kusinsk, Artinsk and Miass factories. The city occupied the intermountain valley and extended more than 17 km from west to east [Figure 1 (a)]. Zlatoust was focused mainly on ferrous metallurgy and iron smelting, which factories were located on flat areas, whereas settlements occupied less-polluted mountain areas. To date, the city has retained a production function supported by educational activities (the branch of Chelyabinsk Technical University is located in Zlatoust). The city exhibits an ensemble of factories of the first half of the 19th century, and monuments of religious architecture and civil engineering. Zlatoust is adjacent to the protected nature monuments such as the Akhmatov Mineral Lance and the Taganai National Park (Response Ridge).

The city of Kusa, the regional center of Chelyabinsk region, is located on the western slope of the Southern Urals, 244 km west of Chelyabinsk on the Kusa river, not far from its confluence with the Ay river [Figure 1 (b)]. The city's foundation was directly related to the construction of ironworks that produced cast iron shells and high-quality art castings [4]. In 1943, Kusa was granted city status, and now it focuses on mechanical engineering and production of precision technical stones. Logging, iron ore and marble are underway in the area.

Satka is a city of regional subordination, the regional center of Chelyabinsk region. Satka is located on the slope of the Southern Urals to the west of Chelyabinsk. The city's foundation in 1785 was due to the Troitsk-Satka ironworks built by S.G. and A.S. Stroganovs [Figure 1 (c)]. Satka was granted city

IOP Conf. Series: Materials Science and Engineering **663** (2019) 012020 doi:10.1088/1757-899X/663/1/012020

status in 1937 [5]. Preserved low-rise wooden and stone buildings, as well as a regional museum can be found in the eastern part of the city that is higher in elevation. At the same time, a modern center with cultural institutions is located in the western part, lower in elevation. Currently, Satka is a center for mining and processing of magnesia. There are factories for the production of refractory products and cast iron. The branch of Magnitogorsk Institute of Mining and Metallurgy is located in the city.

The city of Miass is located in the Southern Urals at the foot of the Ilmen mountains, on the Miass river. The area is rich in copper, so the city' foundation in 1773 was associated with the construction of copper factory. In 1797, the richest gold mines were discovered near Miass, where about 40 pounds of gold were mined annually [6]. In 1926, Miass acquired city status. Further development was due to the automobile factory evacuated from Moscow, which contributed to the increased production of electronics and equipment [7]. Chelyabinsk State Technical University also played a significant role in the industrial development of the city. The historical and cultural potential of Miass is determined by the historical regular planning of the old part of the city, numerous wooden buildings of the 19th–early 20th centuries. A unique resource of the environment is the neighborhood with the Ilmensky Reserve, which includes a mineralogical museum.

(a) (b) (c)

Figure 1. Historic photos of cities of Zlatoust district (a) Zlatoust; (b) Kusa; (c) Satka

The historical framework of Zlatoust district is complemented by nineteenth-century cities founded on the basis of fortresses: Karabash, Chebarkul, Kyshtym, as well as by the autonomous territorial entity – the cities of Ozersk, Chelyabinsk-40 and Chelyabinsk-65. In particular, Karabash founded in 1822 as a settlement of gold miners included settlements of three factories, of which only copper factory has been operating to date [8]. The city has preserved the heritage of industrial architecture, has a regional museum and is adjacent to the scenic areas of Karabash Mountain and a network of lakes. The city of Chebarkul founded in 1736 as an outpost of the South Ural Cossacks comprises a unique landscape monument of the Krutik Peninsula, as well as resort areas (including Yelovoye), sanatoriums, recreation centers and the Kisegach mud resort.

2.2 Resource potential of Zlatoust district

To date, the considered cities occupy an economically advantageous position near railways, highways, and waterways (federal highway M-5–E-30) in the vicinity of the regional center. The development of cities is ensured by the orientation towards the production and mining complex and the existing infrastructure, which in some cases has been added by new residential areas and public service enterprises, including those in the recreational sector. The branches of technical universities located in the cities show a focus on innovative technologies. The increased attention paid to the heritage is supported by museums with their numerous collections.

On the other hand, the development of urban areas is closely related to the formation of specially protected natural areas. In Chelyabinsk region, Zyuratkul (86700 ha) and Taganai (56400 hectares) National Parks established between 1991 and 1993 are of high significance. In 1920, Ilmensky reserve was created near Miass (34380 hectares). The filtration and health-improving function of forests should be considered in conditions of high pollution of the Urals.

IOP Conf. Series: Materials Science and Engineering **663** (2019) 012020 doi:10.1088/1757-899X/663/1/012020

However, all these cities are single-industry [10]. The population number has been decreasing since 1990 [Table 1], which indicates negative trends in urban development. Manufacturing in small and medium-sized cities is not able to compete with that in large urban centers (Chelyabinsk), which means that cities become vulnerable and can lose their identity. The situation is aggravated by a worsening ecological situation, urban areas extended towards natural territories, and outdated production technologies, which represent a huge load on the environment.

Table 1. Dynamics of the population of cities in the South Ural[a]

City	1931 [a]	1979 [b]	1992 [c]	2017 [d]
Zlatoust	75 000	197 660	207 800	167 978
Miass	20 100	150 179	172 400	151 856
Satka	16 900	46 191	51 000	42 214
Chebarkul	3 700	45 845	51 400	39 914
Kyshtym	26 700	39 830	42 700 [a]	37 480
Kusa	13 300	21 926	22 800	17 368
Karabash	8 300	17 619	16 400	11 718
Ozersk	–	–	89 200 [e]	79 265

[a] Data of the People's Encyclopedia of Cities and Regions of Russia My City [Electronic resource] URL: http://www.mojgorod.ru/cheljab_obl/index.html (Appeal Date 03/24/2019)
[b] Data from the 1979 All-Union Population Census. Urban population of the RSFSR, its territorial units, urban settlements and urban areas by gender *Demoscope Weekly* [Electronic resource] URL: http://www.demoscope.ru/weekly/ssp/rus79_reg2.php (Date applications 24.03.2019)
[c] Data of the Cities of Russia: encyclopedia / Ch. ed. Gm Lappo. M: Great Russian Encyclopedia, 1994–559 pp.
[d] Data of the Federal State Statistics Service ROSSTAT Population of the Russian Federation by municipalities as of January 1, 2017 [Electronic resource] URL: www.gks.ru/free_doc/doc_2017/bul_dr/mun_obr2017.rar (Date of circulation 24.03.2019)
[e] Data obtained in 1996

Thus, we can argue about a special position of the small cities of Zlatoust district. The similarity of potentials and proximity reveal common problems, including the systematic disturbance of natural ecosystems.

3. Current ecological status of the Southern Urals

The retrospective of the development of the Southern Urals reveals underlying causes of the current situation. Natural resources of mountain areas and their natural potential became the basis for the establishment of unique mining and metallurgical settlements but led to a total disruption of the natural ecosystem. Thus, the above-described areas exhibit an excessive mining complex as a result of regional specialization. In this system, the lack of local structures and common regional unity forces preservation of single-type production in a great number of cities, that provides meager economic effect and causes extensive damage to the environment. Distributed production belongs to the dirtiest production: metallurgy, oil production and mining. Energy generation requires burning of fossil fuel without by-product filtration, usage of hydropower including damming, installation of dams and hydraulic structures, which leads to the imbalanced hydrographic system. Suffering small rivers are of fundamental importance for the general condition of the hydrographic system and the biocenotic integrity of the region. The condition of small rivers has a direct impact on groundwater and cleanliness of large rivers and lakes.

The current situation virtually excludes the cultural and recreational potential of the region, since it undermines its foundation – the landscape environment. The mechanical destruction of soil cover and the formation of disturbed lands take place in the region, that is, the loss of the initial soil integrity and

IOP Conf. Series: Materials Science and Engineering **663** (2019) 012020 doi:10.1088/1757-899X/663/1/012020

the formation of man-made relief [10]. According to Rosreestr (Federal Service for State Registration, Cadastre and Cartography) as of January 1, 2016, in 2015 the area of disturbed lands was 1037 hectares, and for the first time in six years it dropped below 20800 hectares. The greatest violations can be observed in mining, exploration and construction areas [11]. The spread of cities towards natural territories reduces the most important ecological resource of the forest. For example, in Miass, part of the forest was preserved between the constructions of a new multi-storey complex in order to implement some pseudo-ecological ideas. As a result, these areas separated from the general system and deprived of proper care seem deplorable.

Disturbance and pollution of natural ecosystems significantly worsen environmental conditions, affecting, among other things, people's health. According to the Ecological Atlas of Russia, diseases of the respiratory tract, nervous system and cancer are more often recorded in mountain areas. In addition, people living in mountain areas have lower life expectancy compared to average.

According to experts, ecosystems are capable of self-healing, but this will require hundreds and thousands of years of anthropogenic non-interference [11]. Nevertheless, the authors argue that the possible solutions to the problems of the region do not imply the withdrawal of industry – the main town-forming component of the Southern Urals. These are the strategies for eco-city development, including the formation of local agglomerations of mixed specialization.

Thus, it can be concluded that the adverse effect of industry on the surrounding landscape and, accordingly, on the health of people living in these areas is so great that it requires "transition from the ecological paradigm based on pollution and environmental control to the paradigm based on foresight, assessment and prevention of ecological damage" [11]. "Land restoration as an integral part of measures on nature conservation, in particular, on minimization of the destructive effects of industry on the surrounding landscape is needed to compensate for ecological damage" [11]. The environmental rehabilitation technologies are developed by foreign experts. It is in foreign practice that the concept of past ecological damage is applied, which is still at the stage of scientific study and development in Russia.

4. Ways to solve social and environmental problems

The problem of restoration of contaminated areas has state status in all countries of the world. In Russia, the "Program-Target Approach to the Formation and Implementation of State Environmental Policy" has been implemented since 2003, including at the regional level. At the same time, it should consider not only the possibility of restoration of contaminated areas and disturbed landscapes, but also the improved regulation of the development of single-industry cities and settlements as sources for creating environmental problems. Since natural ecosystems, especially mountain areas, do not follow the administrative division of regions, it is impossible to solve the problems of sustainable development at the city level. The cities of Zlatoust district are considered as a group of regions similar in types of industry, historical and cultural potentials, and related to one mountain ecosystem and a single set of problems.

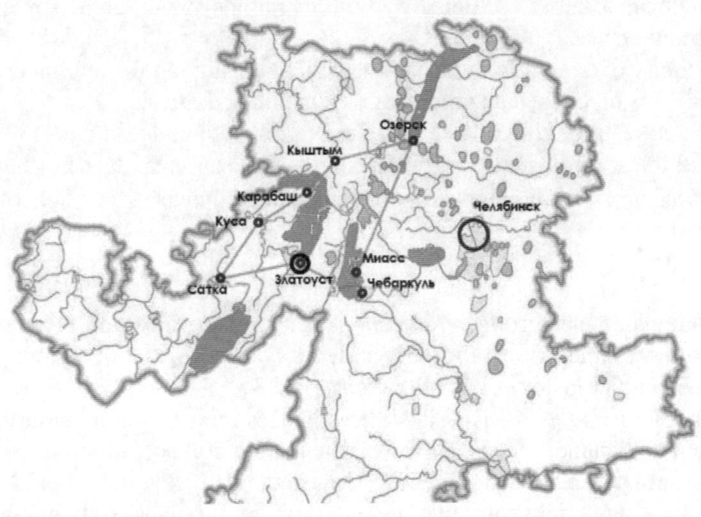

Figure 2. Ratio of the frame of the cities of Zlatoust district and protected areas. Fragment of the map of Chelyabinsk region

Satka, Miass, Kusu, Karabash, Chebarkul, Ozersk and Kyshtym can be considered as a local agglomeration with its center in the city of Zlatoust. Urban development is proposed to be carried out according to cluster type, which allows reducing the load on individual urban and natural territories, and increasing the socio-economic efficiency of the group as a whole [Figure 2]. The development of the district should be ensured not by the increased use of environmental resources, especially in conditions of increased susceptibility and vulnerability of mountain ecosystems, but by the improvement of existing production structures and the development of their cooperation. In accordance with the problems considered and the tasks set for the development of mountain areas of the Southern Urals, the following recommendations for future development can be formulated:

- Preservation of the production complex with a significant modernization of the technological process as the most important factor in improving the ecological state of the environment, which includes:
  system application of energy-saving and environmentally friendly technologies in the production, sewage treatment;
  search and development of biosphere-rehabilitation and environmentally compatible construction methods [12];
  the use of advanced technologies, as shown by the historical development of the territory, plays a huge role in the production efficiency and safety. In early 1970, the Artinsk ironworks located in Zlatoust district transferred to the manufacture of braids by rolling, which is still a unique technological know-how.
- Increase in the share of recreational areas in the urban and mountain natural environment and the regional settlement system:
  reforestation;
  defining of the boundaries of protected areas and creation of buffer recreational zones between protected areas and settlements;
  possibility of reserving territories as links for the green frame;
  restriction of uncontrolled anthropogenic extension towards natural territories [13];
  integrated management of tourist activity [14].
- Strengthening of educational and cultural potential through the creation of a local multi-functional cluster:

creation of local forms of settlement (agglomerations) in order to streamline the existing regional structures;
use of proximity of large urban formations while maintaining the agglomeration autonomy
intracluster division of specializations of production activities;
restoration and restructuring of free industrial territories and buildings.
formation of the security status of former industrial areas as landmarks, their valorization

- Microstructural urban planning to ensure the internal integrity of each of the cities of the agglomeration.

5. Conclusion

Thus, the global problem of the environmental management and the development of mountain areas are particularly relevant to the Southern Urals. This is due to an active role of the industrial and mining function with its three-century history of development.

The modern economic paradigm of Russia still recognizes mining and metallurgical production as the main production of the Southern Urals. However, the imperfection of technologies and the excessive industry have a devastating effect on natural ecosystems and, as a result, on people's health.

The example of Zlatoust district shows that mountain areas are rich not only in mineral resources. In addition, the mixture of functions and the creation of integral local agglomerations contribute to the stable economic and social development of small historical cities and to the comprehensive restoration of disturbed natural areas and elimination of subsequent adverse effects.

References

[1] Godde P M, Price M F and Zimmermann F M 2000 *Tourism and development in mountain regions* (New York: CABI Publishing)
[2] Lubovnyy V Ya 2018 *Single-industry cities of Russia: origin, evolution of development and regulation* (Moscow: Econ-Inform)
[3] Iofa L E 1951 *Cities of the Urals (Economical and geographical. research)* (Moscow: Geografgiz)
[4] Matyushkin D N 1977 *Jasper Belt of the Urals* (Moscow: Art)
[5] Simonov A V 1977 *Kusa* (Chelyabinsk: Bainov P. Artistic Cast Iron. Kusa)
[6] Kozlov L A 1961 *Satka industrial* (Chelyabinsk: Chelyabinsk Publishing House)
[7] Morozov V V 1983 *Miass-city in the golden valley 3rd ed.* (Chelyabinsk: South Ural Publishing House)
[8] Akhmin L M 1968 *Karabash* (Chelyabinsk: South Ural Publishing House)
[9] Dunyashin A 1979 *Chebarkul – Motley Lake of the Urals* (Sverdlovsk) **4** 118–125
[10] Mikulina E M, Blagovidova N G 2013 *Architectural Ecology* (Moscow: Academy Publishing Center)
[11] Pikulko V M 2017 *Renovation of natural systems and elimination of objects of past ecological damage: a monograph* (Moscow: INFRA-M)
[12] Blagovidova N G, Perkova M V and Tribuntseva K M 2015 Features of Design of Ecovillages in Depressed Areas in the City *Res. J. of Applied Sci.* **10(10)**, 608–19
[13] Holling C S 1973 Resistance and stability of ecological systems *Annual Review and Ecology and Systematics* (4) ed C S Holling (US: Annual Review) 1–23
[14] Zhelezov G 2016 Models and strategies for sustainable management of mountain territories in central and southeastern Europe *Sustainable Mountain Regions: Challenges and Perspectives in Southeastern Europe* 385–391 (Dordrecht: Springer)
[15] Hall C M and Lew A A 1998 *Sustainable Tourism, Geographical Perspective* (New York)

IOP Conf. Series: Materials Science and Engineering **663** (2019) 012021 doi:10.1088/1757-899X/663/1/012021

Engineering-geological and hydrogeological features of the dzuarikau-tskhinval gas pipeline installation and environmental aspects of its operation in high seismicity conditions

O G Burdzieva, V B Zaalishvili, Kh O Chotchaev, D A Melkov and B V Dzeranov

Geophysical Institute – the Affiliate of Vladikavkaz Scientific Centre of the Russian Academy of Sciences, 93a, Markova Str., Vladikavkaz, 362002, Russia

E-mail: cgi_ras@mail.ru

Abstract. Despite the strategic importance and social orientation of the project to install the world's highest mountain gas pipeline to South Ossetia, the damage to the environment is great. In contrast to flat terrain, highland landscapes due to large-scale natural (landfalls, avalanches, mudflows, landslides, etc.) or anthropogenic (installation of oil and gas pipelines) negative impacts either not restored at all, due to the progressive influence of factors interventions, or the recovery takes decades. In high-mountain areas, soil destruction is accelerated due to developing sources of landslides and ravines, mudflows, creeps and collapses, active lateral infiltration of groundwater, leading to rapid dehydration of soil and biocenosis degradation. Neglecting the scientific basis for minimizing damage to the environment, the primacy of business over environmental consequences, the lack of fundamental and qualified expertise of projects relating to interference with the natural environment, the legislative vulnerability of ecological systems have become commonplace and have a very steady tendency to take root. The result of consumer attitudes towards ecosystems is estimated by the example of these works by extensive disafforest, deep cutting of extended slope intervals at high mountains and, finally, significant deviations from the original project. At the same time, there are no works on recultivation, which, however, are doomed to failure in the existing orographic and seismological conditions of the Dzuarikau-Tskhinval gas pipeline route. The route of the Dzuarikau-Tskhinval gas pipeline enters the Kassar (valley of the Ardon River) gorge near the Tamiysk resort and then it cuts through both slopes of the Ardon River to Zaramag HPP, and then passes through the picturesque valleys of the Mamikhdon and Zemegendon Rivers, where the gross impact on nature is most clearly observed.

1. Introduction

The best option for installation the Dzuarikau-Tskhinval gas pipeline was considered in the following direction: Kassar gorge (valley of the Ardon river), further along the valleys of the Mamikhdon and Zemegendon rivers, through the Kudarsky pass along the Dzhodzhora River to Kvaisa and then along the highway to the town Tskhinval. This outline was taken as the basis for which the engineering and geological features of the site were to be studied by complex engineering and geological investigations. It was supposed to perform the strategic task of installation a gas pipeline with minimal damage to the environment, so that the environmental indifference of business goals did not replenish the other waste

IOP Conf. Series: Materials Science and Engineering **663** (2019) 012021 doi:10.1088/1757-899X/663/1/012021

of humanity by the disfigured landscape of the route, in which humanity is suffocating according to Niels Bohr [1–7].

Surveys for the preparation of a final design concept were started by the researchers of the Geophysical Institute of the VSC RAS in September 2005 and were aimed at investigation the conditions for the development of hazardous geological processes (lithological, hydrogeological, seismotectonic, anthropogenic) and the extent of their impact on the objects of the proposed construction and elements of their infrastructural planning, classification of soils composing the route, determining their physico-mechanical characteristics and predicting dangerous exogenous processes, environmental justification of project documentation for the construction of the pipeline.

According to the technical specification the soil conditions were investigated to a depth of 2.5 m along linear part of the gas pipeline over natural crossings (rivers, ravines, talus, landslides) and artificial barriers (embankments, dumps) – up to 6 m, and, if necessary, up to 8–10 m (tectonic zones, debris cones). In some areas of the development of adverse processes, detailed engineering and geological conditions were carried out with an increase in the depth of drilling up to 15 m and the formulation of geophysical works using the vertical electrical resistance method.

Based on the survey materials, the soils along the gas pipeline route were classified into engineering-geological elements, taking into account their age, genesis, textural and structural features and nomenclature type, as well as regulatory and design characteristics were calculated. The total length of the route, within which engineering and environmental surveys are carried out, is 162 km. The layout of the gas pipeline route and the position of the geophysical profiles of geolocation sounding are shown in Figure 1.

According to the complexity of the terrain, altitude, geological and mining conditions, the route of the Dzuarikau-Tskhinval gas pipeline has no analogues in the world. The pipeline crosses a height of 3148 m, crosses 30 avalanche sections, covers 75 km of the northern and southern slopes of the Main Caucasus Range with heights exceeding 1500 m. 15 tunnels with a total length of 1848 m were driven in the rocky soils, and 29 bridges were built, the cost of the project was 15 billion rubles. Pipeline with diameter 426 mm and working pressure of 50 atm. has capacity of 252.5 million m^3 per year, which allows satisfying consumption with full gasification of the Republic. Typical mining-geological and landscape conditions of the route are characterized by Figures 2, 3.

2. Materials of field works

The materials of regional engineering and geological work, monitoring of exogenous geological processes in the territory of the RNO-A and specialized work carried out by the North Ossetian Geological Survey Expedition, Geophysical Institute of VSC RAS on the area of the proposed construction of the gas pipeline were used [5–12].

3. Geological research

The main geomorphological, geotectonic potential factors of negative impact on the pipeline include: geomorphology, quaternary formations, tectonics, erosion processes.

3.1 Tectonic features of the area

The tectonic features of the area is the alternation of ridges and intermontane depressions of the all-Caucasian orientation (280–295°), intersected by valleys of the submeridional direction. The main geomorphological units of the Northern Slope of the Greater Caucasus in the strip of the gas pipeline route include: Lesistyi ridge, with absolute elevations up to 1200 m, fully forested and folded with conglomerates, tuff sandstones, gravelites and ashes of Neogene; Pastbishnyi ridge with absolute elevations up to 1800 m in the majority (up to 80 %) is forested and represented by carbonate strata (limestone, marl, dolomite) chalk; Skalistyi ridge with the highest points of 3438 m (mountain Kariukhokh), composed of carbonate sediments (dolomites, limestones, dolomitic limestones) of the upper Jurassic. The northern and partially southern slopes are covered with forest up to 2500 m; Bokovoi ridge, the highest in the system of the Greater Caucasus ridges, with maximum absolute levels of catchment areas [16], affecting the gas pipeline zone, 4404 m (mountain Adaykhokh) and 4427 m (mountain Tepli), composed of terrigenous igneous and metamorphic rocks of the Central uplift zone. Dwarf forest (birch) is fixed at up to 2600 m on the northern slopes. The Glavnyi (Main) ridge separating the North Caucasus and Transcaucasia with absolute elevations reaching 3,938 m (mountain Khalats), composed of Cretaceous terrigenous-carbonate deposits of flysch (limestones, marls, sandstones, shales). The gas pipeline route passes through this ridge at an altitude of 3148 m (Kudarskiy passage). Forest cover reaches the level of 2500 m.

On the southern slope (the territory of the South Ossetia) the main geomorphological units intersected by the gas pipeline route include: Rachinskiy Ridge, which is a second order structure relative to the Glavnyi (Main) ridge, has a northeast orientation, is composed of Cretaceous-age carbonate sediments, reaches absolute marks of 3500 m and it joins the Glavnyi (Main) ridge west of Zekar Pass (at an altitude of 3620 m). The gas pipeline route crosses this ridge in the vicinity of the Ertso Passage at the level of 1820 m; The Suram ridge (Likhskiy), along the northern and southern spurs of which the gas pipeline runs, connects with the Racha ridge in the vicinity of Lake Ertso. The ridge is composed of reefogenic limestones of the Upper Jurassic, porphyrite and tuff sandstones of Bayos. The absolute elevation of the ridge does not exceed 2470 m (mountain Ribisi). The ridge is completely forested to a height of 1800–1900 m.

3.2 Quaternary sediments

Quaternary sediments are represented by alluvial deposits of the floodplain and terraces. Modern alluvium consists of 65–75 % of boulders, pebbles and gravel, 20–25% of sand particles and 5–10 % of dusty and clay fractions. The thickness of modern alluvium of Ardon and Dzhodzhora rivers is 5–25 m, in the estuarine parts of its large tributaries 4–17 m, but on the rapid areas is almost absent. The thickness of the proluvial sediments of the modern – Upper Pleistocene age varies from the first meters to 30 meters. The deluvial deposits have a thickness of 0.3–0.8 meters, less often up to 9 meters (at the foot of the slopes, especially southern) and are represented by gruss-rock debris soils with sandy-clayey aggregate 20–30 %. Colluvial crust deposits on terrigenous rocks contain 77–80 % of detrital material, 10–13 % – of sandy fraction, about 10% of silty-clay. The sediment thickness is up to 10 m, less often 30–40 m. The eluvial formations along the gas pipeline route are fairly widespread and developed on the passage section (Kudarskiy passage) and the crest parts of the Dzhava and Surami ridges. The thickness of the eluvium developed, both in the rocks of the Cretaceous flysch and the terrigenous deposits of Neogene, and in the porphyrites of Bayos reaches 2.5 m in some places. Landslides formations are a type of conventional slope deluvium and differ from it by a larger fractional composition, increased natural moisture, less compaction and stability. The landslide formations can reach several tens of meters in the work area. Landslides for the most part have a seismogenic nature, localizing along zones of powerful tectonic ruptures and by the mechanism of mass movement, with rare exceptions, can be attributed to collapses.

3.3 Tectonic features of the area

The tectonic features of the area include its location at the junction of two megaplates (Scythian and Transcaucasian), which distinguishes it by its geological structure.

The main rupture structures intersected by the pipeline route, which may adversely affect it, are shown in Figure 4.

On the territory of South Ossetia, the Dzhava zone of seismic dislocations is distinguished. The most characteristic elements of this zone, regardless of 1991 events, can be considered young raptures in the Kemult region at the confluence of the Patsa and Keshelta rivers. Given the size of both individual linear forms (up to 1–10 km), and the displacement magnitude of the of the fracture wings (from 1 to 50 m), as well as the length of the system as a whole and its individual elements (3–10 km), we can talk about seismic events with intensity of 9–10 points and higher in this sector of the Greater Caucasus.

3.4 Erosion Processes

The territory is characterized by a high hazard degree for the development of exogenous geological processes. Very complex engineering-geological, hydrometeorological and seismic conditions require the implementation of ubiquitous integrated engineering protection of structures against combinations of interdependent catastrophic and hazardous processes [11,12, 17].

When analyzing the damage from the impact of exogenous geological processes using the example of a 20-year operation of Transcaucasian Highway, it turned out that in the overall balance the erosion processes account for 45 %, mudflows –30 %, avalanches – 10–15 %, rock slide-scree – 8–10 % and landslides – 1–2 %.

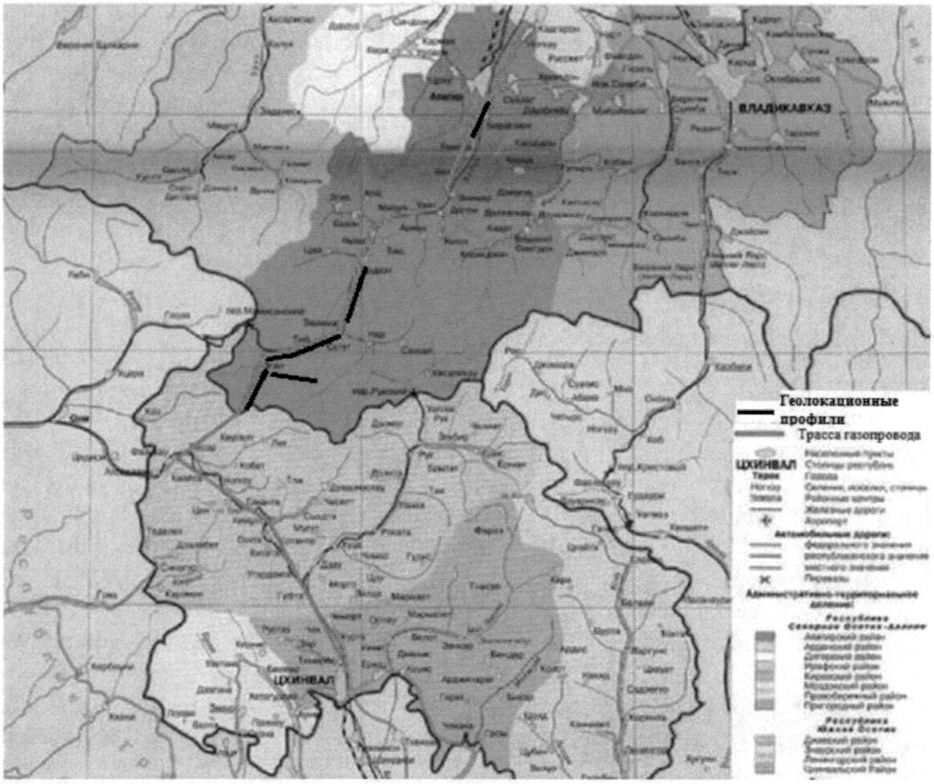

Figure 1. Route of the gas pipeline Dzuarikau-Tskhinval

Erosion processes include rainwash, gully erosion, lateral and bottom erosion of rivers [18]. The intensification of erosion processes on a regional scale is due to the fact that at present the orogen of the Greater Caucasus is captured by a tectonic uplift at speeds from 2 to 14 mm/year.

In high-mountain areas, the length of the erosion network is 2.5–3.4 km/km2, and the depth of local erosion bases is 1000–1500 m. Under these conditions, water flows are characterized by high speed (up to 3 m/sec and more) and perform significant erosion work.

The most intense linear erosion develops in rock slide-scree, deluvial-gravity and landslide deposits, shale and mudstone of the northern and southern intermountain depressions. This is manifested in the formation of numerous ravines and gullets up to 4 m deep and a sharp increase in their growth.

Figure 2. One of the crossings

Figure 3. In the Kassar Gorge

The greatest negative impact on economic facilities, especially roads, has lateral river erosion. Caving in the banks as a result of lateral river erosion is most intensely manifested in the basin of the river Ardon from headwaters to the exit from the mountains to the plain.

The rate of development for lateral erosion according to long-term observations in the mountainous part of Ossetia reaches 1–5 m/day during the flood period, depending on the material composing the watercourse banks.

3.4.1 Mudflows

Mountainous of Ossetia, both South and North, is one of the most mudflow-prone regions of the Caucasus. Highly dissected relief, modern glaciation, high seismicity, huge reserves of loose detrital material, high moisture content, create favorable conditions for the mudflow formation [19, 20]. According to the volume of one-way mudflow, all the basins are divided into four groups:

- very powerful – with a single release of more than 100.000 m^3 of solid material;
- average power – from 10.000 to 100.000 m^3;
- weak – from 1000 to 10.000 m^3;
- micro-mudflows (sloping) – less than 1000 m^3.

The most common and active mudflows are in the high-altitude area in the area of the Bokovoy and Glavnyi (Main) ridges, where about 50% of all mudflow sites are concentrated.

Among the local mudflow sources there are three main types (according to geomorphological features): mudflow cuts, mudflow gulchs, rocky sources.

For Mountainous Ossetia, the accumulation of the necessary amount of detrital material in the rocky mudflow source, sufficient for the formation of the mudflow, requires 8–10 years. This is especially evident in the territories of development of terrigenous sediments.

3.4.2 Rock slide-scree processes

About 90 % of the entire length of the route passes through loose-debris sediments, much of which is the product of rock slide-scree processes.

The reasons for the rock slides in the conditions of sharply dissected relief and intensively stationed rocks, especially in the zone of the Central elevation of the Greater Caucasus, are: powerful geodynamic processes; phenomena of accumulated seismic effects, when the effect of collapse is manifested not from one powerful seismic impact, but from a multitude of weak ones; seismic impact – powerful seismic shocks up to 7–9 points [21], which discharge in mountain conditions occurs in the ridge and near-the-tip parts of the relief, leading to the collapse of large blocks of rock and loose rocks; the action mechanism of lateral repulse cracks, that is especially effective in the high-altitude zone with sharp daily temperature differences. In this case, the main role is played by microcracks, in which the effect of freezing moisture is especially strong.

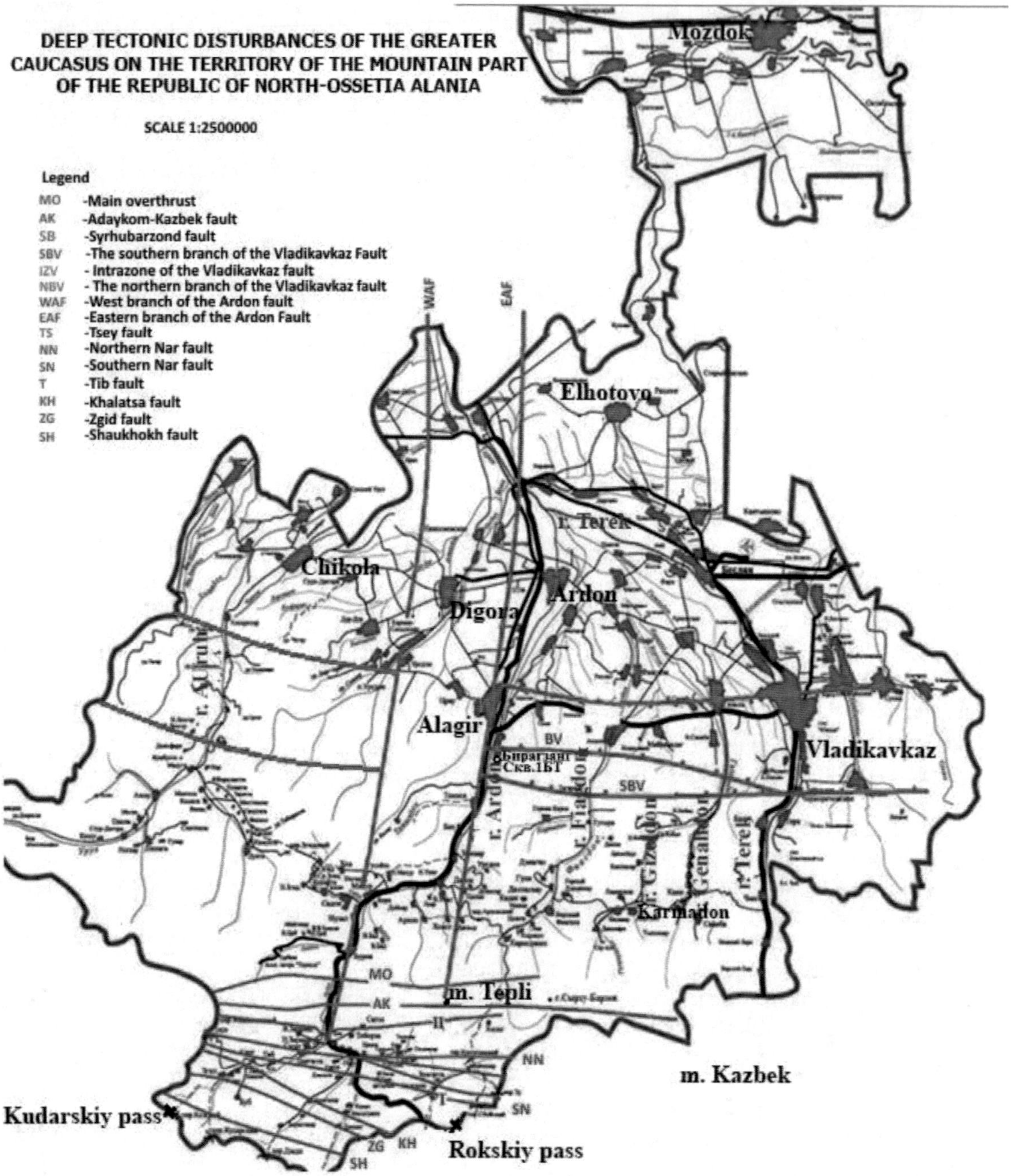

Figure 4. Location of Ardon, Vladikavkaz and Adaykom-Kazbek fault groups according togeophysical data

The example of the effect of geodynamic impact can be a sudden collapse of rocks with a volume of about 5000 m^3 in the zone of the Nuzal thrust (1990) on the left side of the Ardon River in the area of PK 518 + 00. The road was completely blocked over 150 m.

An example of a collapse as a result of short-term powerful seismic effects can be the presence of a huge (up to 1 km^2) plume of large-block (up to 10 m in diameter) accumulations of granite composition on the left side of the Ardon river (region of the Buron tunnel), which was formed as a result of a localized seismic impact about 500 years ago. The ancient road and customs buildings were littered with blocks of granite.

Along the entire pipeline route, any steep-slope (up to 50°–90°) section composed of hard-rocks can be attributed to rock slide-prone, but the most likely areas include the development of powerful tectonic disturbances (Nuzal thrust, the range of tectonic faults of north bourder of Central elevation – big Bad and Tsey faults, the zone of main Thrust, that limits the Central elevation from south, Tib fault zone – the right side of the Mamikhdon River).

Scree is the most common type of rock slide-scree processes, which are particularly intense on the slopes, formed by shale and thin-layer rocks of the terrigenous formation of the lower-middle Jurassic. At the foot of such slopes with a steepness of more than 30°, almost everywhere there are plumes of active fine-grained debris, ungrassed and almost devoid of aggregate. The steepness of scree plumes is 28–35°.

On slopes composed of igneous and metamorphic rocks, scree processes are less developed, scree material is coarser. The steepness of scree plumes is 37-39 °. Large and active debris is developed mainly in the upper reaches of the rivers of the high-mountainous zone.

On scree deposits, passes, about 50 km of the pipeline route. The examples of scree deposits of high mobility can be scree zones of the development of carbonate flysch in the upper reaches of the Mamikhdon River.

3.4.3 Landslides

With a fairly widespread development of landslides on the territory of the proposed construction of the gas pipeline, its route was chosen so that it crosses just 6 landslide areas, only two of which can be classified as active, the rest are stabilized. A detailed description of these objects is given in the picket description of the pipeline route. The absolute majority of these formations is confined to zones of powerful regional faults and has a seismogenic nature, and can be attributed to landslide with a high degree of probability.

3.4.4 Avalanches

The area of works, and especially its high-mountain part, is characterized by a wide occurrence of avalanches. One-time snowfall with a capacity of up to 1m is an ordinary phenomenon here [22].

The most common and large ones are channelled avalanches, which are permanently tied to lower forms of relief (beams, couloirs, canyons), and the smaller ones are snow slides, which descend from slopes that have anthropogenic or natural cutting [23–25].

A special type of avalanche should include avalanches of semi-liquid consistency, which are formed after rain precipitation on the slopes overloaded with snow. Such avalanches go at low speed (about 1–2 m/s) and have low energy and reaching the traffic lane and turning around 90^0, they pass directly along it, without going beyond its limits at a distance of hundreds of meters. Subsequently, the water and snow mass freezes into a single ice body.

A one-time recorded avalanche mass emission in the area of the route reached 300 thousand m^3. Most avalanches belong to the high-speed category (more than 100–150 km/h) and have powerful shock waves that can heave practically any equipment (cars, bulldozers, etc.) into the air. As a rule, the shock wave capture zone exceeds the width of the area affected by snow masses in 2–3 times.

Most of the avalanches overlap the river bed with a transfer of snow mass to the opposite side. The so-called "jumping" avalanches (when the bulk of the avalanche breaks off the stream bed at the site of a sharp increase of the slope (cliff) steepness and falls into the river channel, flying several hundred meters through the air) should be referred to the most dangerous avalanches. In this case, in the affected canyon, the effect of the repeated impact of reflected shock waves is manifested, in which steel supports of power transmission lines, which have no wind resistance, acquire the most bizarre forms.

Snow density of avalanche catchment ranges from 0.06 to 0.4 g/cm^3. In a discharge zone (avalanche cones) it is up to 0.5 g/cm^3. As a rule, up to 70 % of snow cover out of the entire area of the avalanche catchment comes down. Instrumentally measured avalanche velocities show the values of 10 to 60 m/s.

The avalanches of fresh snow at a precipitation intensity of more than 1 mm/hour, a temperature of minus 4 ° and below and a wind speed of 6 m/s and more. The daily critical amount of precipitation is about 10 mm, and for the descent of many large avalanches, it is equal to 20–44 mm/day. In general, in the Greater Caucasus, the distribution of avalanches according to genetic types is as follows: 7 5% of their total number was observed during snowfalls, 8 % – during spring snowmelt, 6 % – during thaws, 2 % – during blizzards and 9 % – during avalanches of sublimation recrystallization.

4. Geophysical study

Geophysical study was performed on individual local sections of the route, where it was necessary to determine the thickness of soils, the nature of the change in the direction of the relief uprising from the project line of the route, the presence of flooded areas in the section of a certain profile that is of interest in determining the location of the route, to differentiate the soils by material and granulometric composition, to assess a landslide activity of a certain soil massif.

The Georadar OKO-2M with the AB-150 MHz antenna unit in the sounding mode [26] (when the transmitting and receiving antenna units are rigidly interconnected) was used as a measuring complex. The scan step was set up at 300mm.

As it is known, soils as dielectrics have different values of dielectric capacity (σ). While transmitting the electromagnetic field at the boundary of two media with different values of σ, a wave reflection occurs. The reflected wave can be strengthened or weakened due to the polarization of the medium, which is superimposed or reduces the intensity of the primary wave. The isolation of such reflected waves, their correlation and interpretation form the basis of the geolocation method [27–29].

The transmitter set by the operator in the time mode sends electromagnetic pulses, during the interval between the pulses, the receiving antenna receives a reflected signal, which is amplified many times due to multiple accumulations (64,128,256 times) of single measurements over the time interval between switching on of the transmitter.

GPR works were carried out at the maximum penetration depth of an electromagnetic wave, determined by the option of 511 points, which means the detailed of registration, i.e. the equipment records 511 points at the maximum possible penetration depth, which characterizes the vertical detail. Thereat, in case of low values of σ, the depth of penetration will be greater and vice versa, but the number of measurements, in either case, will be the same (511). In order to speed up the work, this value can be reduced in 2 and 4 times.

The process of setting up the equipment, establishing the recording parameters and synchronizing with the operator's personal computer at the initial stage of scanning the soils of the route is shown in Figure 5.

The obtained data were processed in the Geoscan-32 program, which includes various methods of filtering, amplifying, smoothing, selection of the field of spectra, taking the vertical coordinate (relief) into account, selection of layers and their coloring and filling with conventional signs (Fig. 6, 7).

Figure 6 shows a geologic-geophysical profile according to geolocation sounding along the profile located on the eastern slope of the Zemegendon River valley 300–350 m away from the thalweg. There are five different layers of soil.

The upper gravelly layer, which is underlain by a thick layer of clay and sandy-loamy sediments, located in the north on a layer of flooded coarse-grained blocky deposits, which wedge out to the south and form a landslide massif, shifted along the surface of the similar coarse-grained rocks, but not watered. The soils are potentially landslide and can shift to the valley of the Zemegendon River.

The profile of Figure 7 is passed transversely to the Zemegendon River valley to the east of its confluence with the River Kozikomdon. Here the composition of soils is similar to the cut in Figure 6, but the aquifer is not on the cutting line, has no continuation into the valley and can flow to the north. At the same time, a layer of loam can serve as slickensides for surface rubble – sandy sediments.

5. Discussion of the research results

IOP Conf. Series: Materials Science and Engineering **663** (2019) 012021 doi:10.1088/1757-899X/663/1/012021

Being based on geolocation scanning of local sections of the route allows singling out four types of Quaternary formations with a total capacity of 10 m. Under the covering rubble – sandy deposits a steeply dipping clay-loamy interlayer with a thickness of up to 2 meters is noted. It can serve as a slickenside with the activation of landslide processes by the anthropogenic or seismogenic impact. The bottom of clay-loamy formations is saturated with water, since the underlying blocky-rubble stratum is saturated with water, and this fact favors the occurrence of landslide processes. The thickness of the Quaternary formations towards the thalweg of the river increases and the multimeter side cutting on the part of the hanging road shelf will cause a gravitational slide of the overlying loose mass.

Figure 5. Members of the Geophysical Institute at work with Georadar

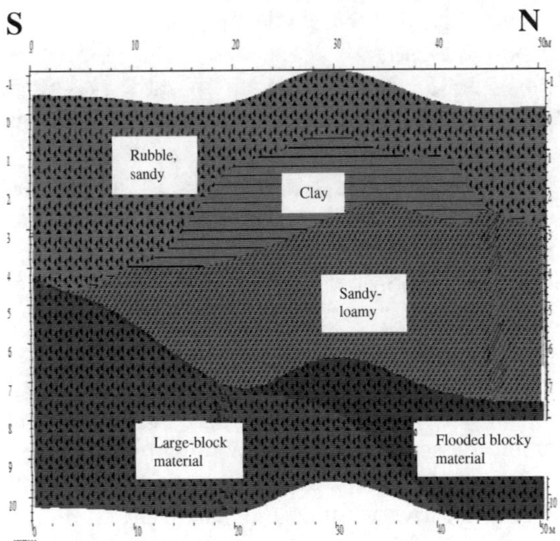

Figure 6. Geologic-geophysical profiles of the route sections on the eastern slope of Zemegendon River 300 m away from the thalweg

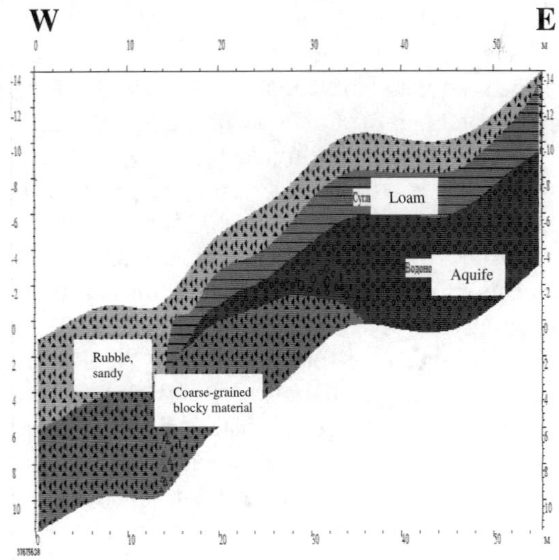

Figure 7. Geologic-geophysical profiles of the route sections of the eastern slope of the gas pipeline route from the mouth of the river Kozikomdon

As a result of the analysis of the spatial variability of particular indicators of the physical and physical-mechanical properties of soils determined by laboratory methods, taking into account the data on the geological structure and lithological features of soils in the area of the projected gas pipeline 20 engineering-geological elements have been identified, the tensile strength of which varies from 18 MPa to 300 MPa. Such a wide range of properties implies the development of special conditions for the successful implementation of the project. In this regard, a very important step is the organization of local environmental monitoring, which involves determining the main components of the environment that need monitoring [30, 31], determination of background values or initial data on the day of the beginning of the facility construction and realization of stationary observations to determine changes in indicators of the environment state. The proposed line of the road shelf cutting, located on the eastern slope of the Zemegendon River valley significantly upward the river, unfortunately, was not taken into account in the construction of the route. The cutting was carried out near the river, which caused intense lateral infiltration of groundwater, removal of loamy cement and activation of landslide processes along the entire cutting line.

The action program of the environmental monitoring group during the construction stage was limited to the following:

- preventing unauthorized use and land pollution outside the boundaries of the branch for storage of materials and discharge of oil products onto the ground [32];
- monitoring the use and reclamation of temporary allotment areas for construction objects.

The actions program of the environmental monitoring group during operation consists in monitoring the emission of pollutants from pumping stations, emergency points and launch sites.

6. Conclusion

The gas pipeline route in the range of PK 231 + 00 – PK 264 + 00; PK 283 + 00 – PK 339 + 00 and PK 560 + 00 – PK 665 + 00 passes through the territory of the North Ossetian State Reserve (beech, oak, linden, pine, yew, alder, aspen).

Line of precipitation which may cause a landslide is equal to 30-50mm/day; daily precipitation up to 70 mm/day occurs once in 10–15 years.

The background seismicity of the route area of the projected gas pipeline is equal to 9 points. This area is characterized by a high hazard level of exogenous geological processes development (mudflows, landslides, rock slide-scree and erosion processes, avalanches, carst).

The total length of landslide and rock slide-scree plots is 30052 m, and avalanche-hazardous is 5815 m. The total length of areas with shallow groundwater occurrence (less than 3.0 m depth) is 9525 m, as a rule, in alluvial and diluvial sediments.

The total length of areas with shallow rock bedding (depth less than 2.0 m) is 23054 m. The gas pipeline route in the range of PK 256 + 00 –PK 1100 + 00 is intersected by 12 powerful tectonic zones (from 100 to 2500 m) of a regional scale, which are still active today. Zones of the newest seismic dislocations were recorded both on the northern and southern slopes of the Glavnyi (Main) ridge, indicating high-intensity seismic shots.

This leads to the formation of the Action Program of the environmental monitoring group at the construction stage, which was to prevent unauthorized use and land pollution outside the boundaries of the branch for storage of materials and discharge of oil products onto the ground and providing monitoring of the use and reclamation of the temporary allotment areas for construction objects.

The actions program of the environmental monitoring group during operation consists in monitoring the emission of pollutants from pumping stations, emergency recovery points and launch sites.

Thus, in order to exclude possible accidents in mountain conditions, it is absolutely necessary to conduct special studies in the development of the project of laying the pipeline route along the so-called "weak" low-strength soils and the creation of a system of instrumental and visual monitoring of the condition of the underlying pipeline array.

IOP Conf. Series: Materials Science and Engineering **663** (2019) 012021 doi:10.1088/1757-899X/663/1/012021

References

[1] Qi S S, Hao F H, Ouyang W and Cheng H G 2012 Characterizing landscape and soil erosion dynamics under pipeline interventions in Southwest China, *18th Biennial Isem Conf. on Ecological Modelling for Global Change and Coupled Human and Natural System. Proc. Environmental Sci.* Z. Yang and B. Chen (eds.) pp 1863–71

[2] Sahley C, Vildoso B, Casaretto C, Taborga P, Ledesma K, Linares-Palominoa R, Mamani G, Dallmeier F and Alonso A 2017 Quantifying impact reduction due to avoidance, minimization and restoration for a natural gas pipeline in the Peruvian Andes *Environmental Impact Assessment Review* **66** 53–65

[3] Lidskog R and Elander I 2012 Sweden and the Baltic Sea pipeline: Between ecology and economy *Marine Policy* **36(2)** 333–8

[4] Zaks T, Burdzieva O and Zaalishvili V 2019 Impact of Noise, Gamma Radiation and other Geophysical Factors on Population Health *Akustika* **32** 206–10

[5] Alborov I, Burdzieva O and Zaalishvili V 2019 Technology for the Maintenance of Acoustic Comfort on the Transcaucasian Highway in the Zone of the Residential Areas Adjusting the Motor Road *Akustika* **32** 211–5

[6] Zaalishvili V, Dzhgamadze A, Gogichev R et al 2018 Changes in the Qualitative Characteristics of Groundwater of the Ossetian Artesian Aquifer *Int. J. of Geomate* **15(51)** 22–30

[7] Burdzieva O G, Zaalishvili V B, Beriev O G et al 2016 Mining Impact on Environment on the North Ossetian Territory *Int. J. of Geomate* **10(19)** 1693–7

[8] Zaalishvili V B, Melkov D A, Dzeranov B V et al 2018 Integrated Instrumental Monitoring of Hazardous Geological Processes Under the Kazbek Volcanic Center *Int. J. of Geomate* **15(47)** 158–63

[9] Shempelev A G, Zaalishvili V B and Kukhmazov S U 2017 Deep structure of the western part of the Central Caucasus from geophysical data *Geotectonics* **51(5)** 479–88

[10] Zaalishvili V B, Nevskaya N I, Nevskii L N et al 2015 Geophysical fields above volcanic edifices in the North Caucasus *J. of Volcanology and Seismology* **9(5)** 333–8

[11] Rogozhin E A, Gorbatikov A V, Zaalishvili V B et al 2015 New data on the deep structure, tectonics, and geodynamics of the Greater Caucasus *Doklady Earth Sci.* **462(1)** 543–5

[12] Gorbatikov A V, Rogozhin E A, Stepanova M Yu et al 2015 The pattern of deep structure and recent tectonics of the Greater Caucasus in the Ossetian sector from the complex geophysical data *Izvestiya Physics of the Solid Earth* **51(1)** 26–37

[13] Zaalishvili V B, Nevskaya N I and Mel'kov D A 2014 Instrumental geophysical monitoring in the territory of northern Caucasus *Izvestiya Physics of the Solid Earth* **50(2)** 263–72

[14] Milyukov V K, Yushkin V D, Kopaev A V, Zaalishvili V B et al 2014 Monitoring Current Vertical Movements of the Northern Caucasus by Absolute and Relative Gravimetry *Measurement Techniques* **56(10)** 1105–10

[15] Akoeva L A and Zalikhanov M Ch 1987 *Natural phenomena in the Caucasus and their intensification under the influence of economic activity* (Moscow: Hydrometeoizdat) Khashirova T Yu, Lamerdonov Z G and Kuznetsov E V 2007 The conceptual model of the protection of mountain and foothill landscapes as a natural and man-made environmental management complex *Melioration and Water Economy* **6** 43–6

[16] Zerbaliev A M and Guseynova M R 2009 Study of anti-erosion technology of gravity irrigation of land *Environmental Management* **2** 43–8

[17] Agibalova V V 1983 *Mudflows in North Ossetia* (Ordzhonikidze: Ir)

[18] Kononova N K and Malneva I V 2003 The occurrence of emergency situations caused by the development of dangerous geological processes in the North Caucasus, and their forecast for the next decade *Problems of forecasting emergency situations: Reports and speeches at the II sci. and pract. conf.* (23 October 2002) (Moscow, Center Antistikhiya Ministry of Emergency Situations of Russia) pp 160–6

IOP Conf. Series: Materials Science and Engineering **663** (2019) 012021 doi:10.1088/1757-899X/663/1/012021

[19] AS USSR, Schmidt Institute of Physics of the Earth 1976 *New catalog of strong earthquakes on the territory of the USSR from ancient times to 1975* (Moscow: Nauka)

[20] Zalikhanov M Ch 1981 *Snow-avalanche regime and prospects for developing the mountains of the Greater Caucasus* (Rostov-on-Don) 376 p

[21] Socratov S A, Seliverstov Yu G, Shnyparkov A L et al 2013 Anthropogenic influence on avalanche and mudflow activity *Ice and snow* **2(122)** 121–8

[22] Zaalishvili V B and Vorobiev V G 2006 Monitoring of glaciers and determination of the boundaries of avalanche danger in mountain areas *Earthquake Engineering. Safety of buildings* **6** 32–6 (Moscow: VNIINTPI)

[23] Oliveros F, Hernandez E and Soto G 2017 Application of geotechnical criteria for the occurrence of earth flows (avalanches) on the right of way of pipeline transportation system of Camisea in the coast zone of Peru *Proc. of the ASME 2017 Int. Pipeline Geotechnical Conf. IPG2017* (25–26 July) (Lima, Peru)

[24] Vladov M L and Starovoitov A V 2001 Interpretation of GPR data *Exploration and protection of subsoil* **3** 11–4

[25] Paramonov N B, Shishkov D L, Zaripov M N and Egorov G A 2016 Processing of geolocation information *Issues of radio electronics* **3** 74–9

[26] Alterman A D and Lushnikov N D 2018 Geolocation capabilities and its features *Theory. Practice. Innovation* **11(35)** 15–9

[27] Kiselev D A 2018 Research methods for processing geolocation data. In the collection: Innovations in science and practice *Collection of articles on the materials of the VII Int. Sci. Pract. Conf* In 5 parts pp 258–61

[28] Angelici F M, Grimod I and Politano E 1999 Mammals of the Eastern Niger Delta (Rivers and Bayelsa States, Nigeria): An environment affected by a gas-pipeline *Folia Zoologica* **48(4)** 249–64

[29] Jones I L, Bull J W, Milner-Gulland E J, Esipov A V and Suttle K B 2014 Quantifying habitat impacts of natural gas infrastructure to facilitate biodiversity offsetting *Ecology and Evolution* **4(1)** 79–90

[30] Baynard C W 2011 The landscape infrastructure footprint of oil development: Venezuela's heavy oil belt *Ecological Indicators* **11(3)** 789–810

IOP Conf. Series: Materials Science and Engineering **663** (2019) 012022 doi:10.1088/1757-899X/663/1/012022

Directional protection of substation equipment

N S Buryanina[1], Y F Korolyuk[1], M L Koryakina[1], E V Lesnykh[2] and K V Suslov[3]

[1]Chukotka branch of North-Eastern Federal University, 3, Studencheskaya str., Anadyr, 689000, Russia
[2]Siberian Transport University, 191, D. Kovalchuk str., Novosibirsk, 630049, Russia
[3]Irkutsk National Research Technical University, 83, Lermontov str., Irkutsk, 664704 Russia

E-mail: bns2005_56@mail.ru

Abstract. The introduction of microprocessor technology in the management of energy objects, including diagnostics, into relay protection and automation, allows combining and unifying relay protection of individual objects into a single system. One of the methods of unified relay protection may be directional protection. It is possible to develop and offer directional protection for generators, transformers, which are currently not used. An obstacle to the use of directional protection of generators and transformers is the presence of aperiodic components in the short-circuit currents, that introduce distortions in determining the angle between the short-circuit currents and voltages. At the "Power Supply" Department of the North-Eastern Federal University named after M.K. Ammosov a method that excludes aperiodic components in the secondary currents of current transformers of power system objects was developed, which made it possible to exclude incorrect actions of directional protections, including when saturating the current transformer magnetic cores. The method is based on processing four instantaneous samples of currents separated by identical sampling intervals.

1. Introduction

The local power system can be represented as the union of individual objects connected by bars, often also by power lines. Local objects can be generators, transformers, connecting them to the line. As a rule, local systems are settlements, industrial and national economic objects. Figure 1 shows an example of a local power system. The system consists of two generators, one of which operates on generator voltage buses, the second one through a transformer on high voltage buses U_2 – a line that combines the system under consideration with a combined one.

For relay protection of the external network with voltage U4 short circuits (SC) 1 and 2 are internal. An external SC for the protection of generators and transformers is a SC at point 1. The directions of the currents relative to the voltage of this point are shown in Figure 1. The disconnection is affected by protection of the transformer, which connects the voltage buses U1 and U2 and the generator-transformer block. With a SC at point 2, the currents I1, I2 and I3 change the direc-tion to the opposite, the generator is turned off by protection. Line W is disconnected at SC at point 1.

2. Research objective

IOP Conf. Series: Materials Science and Engineering **663** (2019) 012022 doi:10.1088/1757-899X/663/1/012022

It is proposed to determine the direction from the place of measurement of the currents supplied to the microprocessors in the SC side by means of the parameters of the direct and reverse sequence. The instantaneous values of the currents of the direct and inverse sequences are proposed to be calculated via phase currents without zero components [1]:

$$i_1(t) = \frac{i_A(t) + \frac{i_B\left(t+\frac{T_C}{4}\right) - i_C\left(t+\frac{T_C}{4}\right)}{\sqrt{3}}}{2} \qquad i_2(t) = \frac{i_A(t) - \frac{i_B\left(t+\frac{T_C}{4}\right) - i_C\left(t+\frac{T_C}{4}\right)}{\sqrt{3}}}{2} \tag{1}$$

where $i_1(t)$, $i_2(t)$ – instantaneous values of direct and reverse sequence currents at time t; $i_A(t)$, $i_B(t)$, $i_C(t)$ – instantaneous values of phase currents A, B and C at time t; T_C – the period of sinusoidal currents expressed in radians.

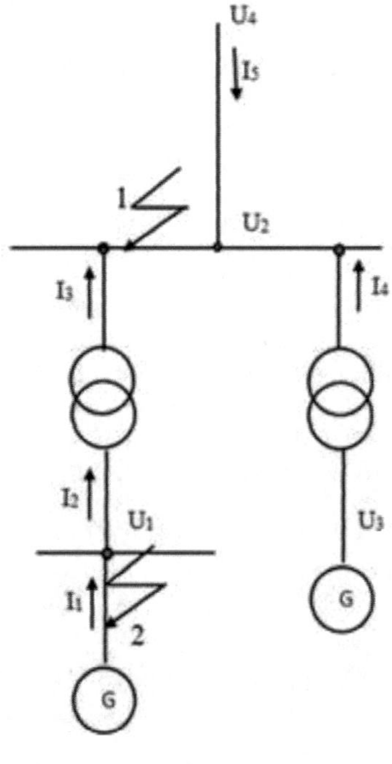

Figure 1. Example of local power system

In rare cases of three-phase short-circuits near the installation sites of directional protection, it is impossible to reliably obtain information about voltages due to their magnitude. The secondary voltages supplied to the analog relay protection devices in this case are either zero or may be distorted. The introduction of digital control of electric power industry facilities allows conducting the analysis of modes of several objects simultaneously. To implement digital directional protection in these cases, it is possible to use the voltage of other elements of the electrical system. For example, with a SC at point 1, when the voltages of all three phases of U_2 are equal to zero, the voltages U_1, U_3 and even U_4 can be used.

This circumstance allows recommending digital directional protection as a universal for lumped objects.

3. Materials and methods

SC currents, as a rule, consist of periodic sinusoidal and aperiodic components [2]:

$$i_{\text{кз}}(t) = I \cdot \sin(\omega t + \psi - \varphi) - I \cdot \sin(\psi - \varphi) \cdot e^{\frac{-t}{T_a}} \tag{2}$$

where ω – angular frequency expressed in radians and equal to $2\pi f$; f – network frequency expressed as 1/s; I – current amplitude; $T_a = L/R$ – the constant of the network which is equal to the ratio of inductance and equivalent electromotive force of the network to the short circuit point; R – active resistance.

The unknowns in equation (2) are determined using four samples of instantaneous currents separated by the same sampling intervals Δt. The authors proposed to calculate the instantaneous current values of currents with a significant decrease in aperiodic components using three samples:

$$i_{calc}(n\Delta t) = \frac{2 \cdot i(n \cdot \Delta t) - i((n-1) \cdot \Delta t - i((n+1)\Delta t)}{4 \cdot sin^2(\omega \cdot \frac{\Delta t}{2})} \tag{3}$$

where n – sample sequence number.

Essentially, if the current contains sinusoidal and aperiodic components, the aperiodic component in the calculated current value decreases and the sinusoidal component is reproduced without distortion. This allows selecting the aperiodic component of the current. By substituting the values of the samples into the following formula, one can calculate aperiodic components at the moment of time $2\Delta t$ and $3\Delta t$. It is possible to eliminate sinusoidal components from the short-circuit current as

$$i_{calc}(2\Delta t) - i(3\Delta t) \text{ and } i_{calc}(\Delta t) - i(2\Delta t) \tag{4}$$

The time constant Ta and the current phase φ are defined as:

$$T_a = \frac{-\Delta t}{\ln(\frac{i_{calc}(2\Delta t) - i(3\Delta t)}{i_{calc}(\Delta t) - i(2\Delta t)})} \tag{5}$$

$$\varphi = arctg(\omega \cdot T_a) \tag{6}$$

The current amplitude is defined as:

$$I = \frac{i(n \cdot \Delta t)}{\sin(\omega \cdot n \cdot \Delta t + \psi - \varphi) - \sin(\psi - \varphi) \cdot e^{-n \cdot \Delta t / T_a}} \tag{7}$$

4. Research

Let us consider the example of a current equal in amplitude in relative units of 2 (two), the initial phase of the current is zero, the network time constant from the equivalent EMF to the SC point is 0.01 s., the sampling interval $\Delta t = 0.000625$ s., φ angle of the resistance of the network from the equivalent EMF to the SC point – one radian.

$$i_{sc}(t) = 2 \cdot \sin(\omega t + \psi - \varphi) - 2 \cdot \sin(\psi - \varphi) \cdot e^{\frac{-t}{0.01}}$$

The dependence of the current on time is shown in Fig. 2

Four samples with four-digit precision:

$$i(\Delta t) = 0.0395, i(2\Delta t) = 0.1533, i(3\Delta t) = 0.3324, i(4\Delta t) = 0.5656.$$

Calculated current values at time points $2\Delta t$ и $3\Delta t$:

$$i_{calc}(2\Delta t) = \frac{2 \cdot (2\Delta t) - i(\Delta t) - i(3\Delta t)}{4 \cdot \sin^2(\omega \cdot \frac{\Delta t}{2})} \qquad i_{calc}(3\Delta t) = \frac{2 \cdot (3\Delta t) - i(2\Delta t) - i(4\Delta t)}{4 \cdot \sin^2(\omega \cdot \frac{\Delta t}{2})}$$

$$i_{calc}(2\Delta t) = -1.69922, i_{calc}(3\Delta t) = -1.40777$$

The time constant T_a and the phase current φ

IOP Conf. Series: Materials Science and Engineering **663** (2019) 012022 doi:10.1088/1757-899X/663/1/012022

$$T_a = \frac{-\Delta t}{\ln\left(\frac{i_{calc}(2\Delta t) - i(3\Delta t)}{i_{calc}(\Delta t) - i(2\Delta t)}\right)} = 0{,}00999\ c. \qquad \varphi = \mathrm{arc}(\mathrm{tg}(\mathrm{Ta}\cdot\omega)) = 1.264\ \mathrm{rad}$$

The amplitude of the current is:

$$I = \frac{0.395}{\sin(\omega \cdot 0.000625 - 1.2624) - \sin(-1.2624) \cdot \mathrm{e}^{-0.000625/0.00999}} = 1.9998\ \mathrm{r.u.}$$

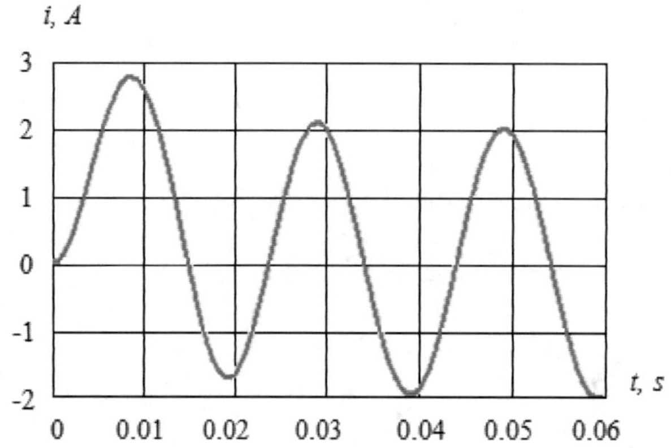

Figure 2. Example of the change in SC current over time.

Errors in the calculation of values that determine the currents are not more than 0.1%.
The change of the aperiodic component in time is described by the equation:

$$i_A(t) = -I \cdot \sin(\psi - \phi) \cdot e^{-t/T_a} \tag{8}$$

The sinusoidal component is defined as:

$$i_{sin}(t) = i(t) - i_A(t) \tag{9}$$

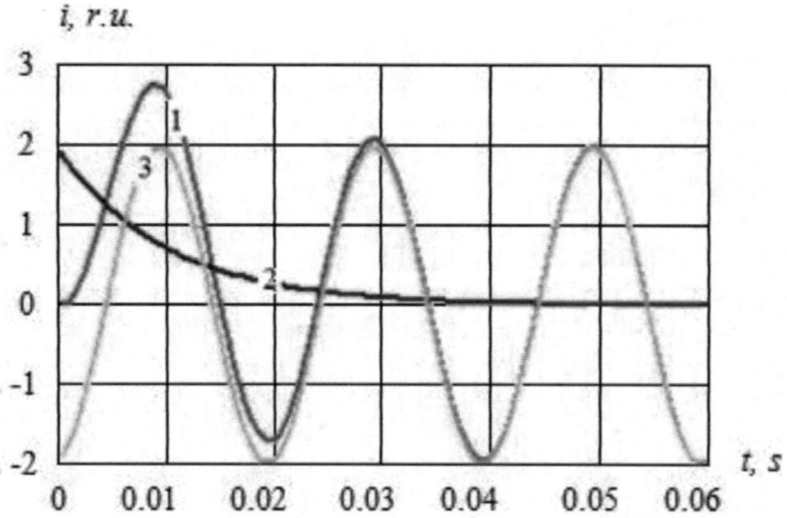

Figure 3. Changes in SC current over time (curve 1), aperiodic component calculated according to (8) (curve 2), sinusoidal component calculated according to (9) (curve 3).

The change in currents over time (1 is the total SC current, 2 is the aperiodic component of the current, 3 is the sinusoidal component of the current) is shown in Fig. 3

Equations (1) involve the calculation of the currents of B and C phases at intervals of time that are separated from the time of measuring the current of A phase by a quarter of the period. I.e., the processing time of the currents is – 5 ms. By this time, the magnetic circuits of current transformers in some cases may become saturated, and the results of the calculation will be incorrect.

An idealized picture of the saturation of the magnetic circuits of current transformers is shown in Figure 4. For half the period, the secondary current in time has two sections: an ideal transformation and a section with an error where part of the secondary current is missing. If the secondary current is decomposed into a Fourier series, then the first harmonic has a frequency greater than the nominal one and is smaller in magnitude than the secondary current reduced to the primary one.

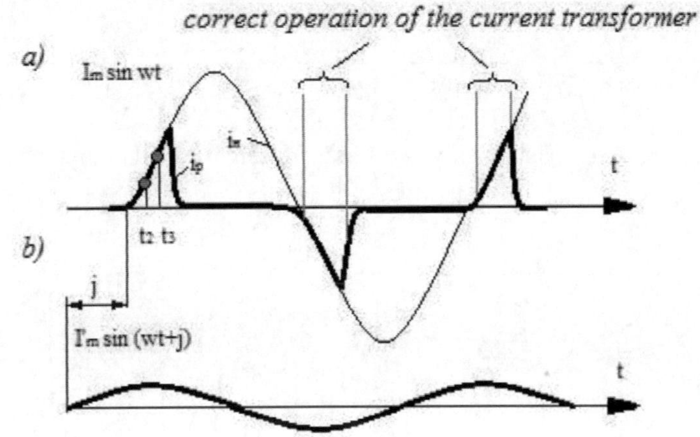

Figure 4. Primary and secondary currents at saturation of magnetic circuits of current transformers.

The method of four samples has three advantages that distinguish it from all currently used ones in digital control:

- uses information about short-circuits in a very short period of time (up to 2.0–2.5 ms), which does not allow tuning away from the saturation of the magnetic conductors of current transformers;
- completely excludes aperiodic components from secondary currents entering the measuring bodies of digital relay protection and automation;
- determines the sinusoidal components of currents on time intervals outside the measurement interval (after 2.0–2.5 ms).

The third advantage is especially valuable when calculating secondary parameters, such as direct and inverse sequence currents, powers, resistances, etc. Four samples are sufficient to implement any protections that respond to the periodic components of the mode parameters. Figure 5 shows the time dependences of the active and reactive powers calculated as the product of the voltage complex and the conjugate current complex. Dependences 5 and 7 – the current is calculated according to (3), dependencies 6 and 8 – the current is calculated according to (3–9). According to the work (3), power calculation errors reach 25–100 % over the time interval 0–0.01 s. When calculating the currents according to the works (3–9), the errors in calculating the powers are zero.

Sampling intervals of 0.5 and 0.625 milliseconds can be obtained as part of a sampling interval equal to a quarter of the current network frequency period. In its turn, the sampling interval in a quarter of the period of the network frequency is obtained by fixing zero-voltage transitions on the active and inductive resistances of the RC – link, as shown in Figure 6 [3].

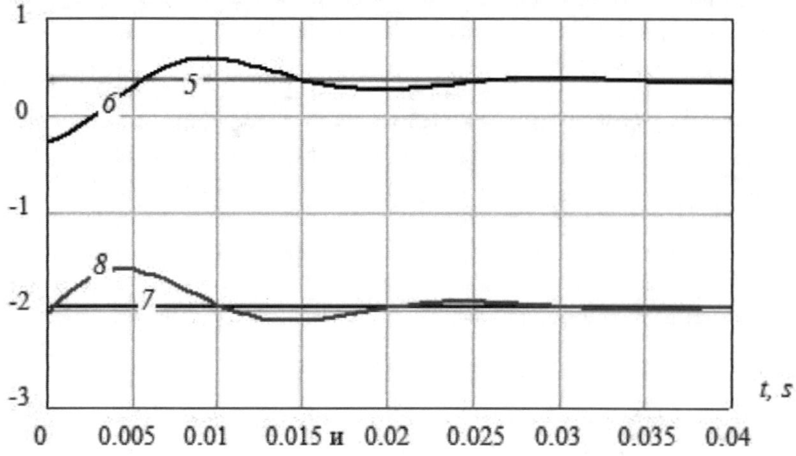

Figure 5. Active and reactive powers in different calculation methods of of SC currents.

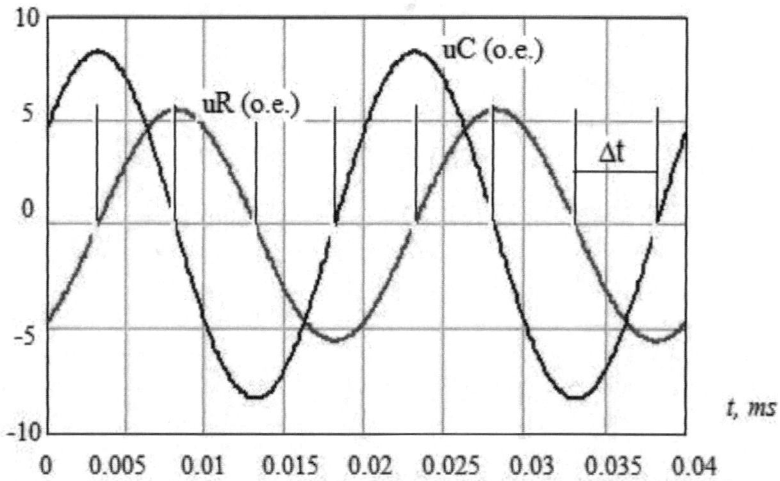

Figure 6. Ensuring sampling intervals equal to a quarter of the period of the network frequency.

The sampling interval equal to a quarter of the period of the network frequency can be used in the development of emergency control algorithms [4].

5. Conclusions

1. The method which makes it possible to isolate the periodic components of short-circuit currents from currents equal to the sum of the periodic and aperiodic components has been developed.

2. The algorithm that implements the method uses a total of four samples, separated by three sampling intervals, which makes it possible not to take into account the possibility of saturation of the magnetic circuits of current transformers, removing information in areas of sufficiently accurate transformation.

3. The method allows calculating the currents of the direct, inverse sequences, and any combinations thereof, resistance, power, used in the relay protection and emergency control automation at any moment of time.

4. Directional protection is proposed for generators and transformers. In order to exclude "dead" zones with close three-phase short circuits, it is proposed to use voltages of adjacent elements of the electrical network that are connected to the protected objects.

5. Transferring the method to reading information for the operation of automation allows increasing the sampling interval to a quarter of the period of industrial frequency.

6. Acknowledgments

The study was carried out with the financial support of the Russian Foundation for Basic Research and the Subject of the Russian Federation – the Republic of Sakha (Yakutia) № 18-48-140 010.

References

[1] Buryanina N S, Korolyuk Y F and Lesnykh E V A 2016 *Method for determining the instantaneous values of the parameters (currents and voltages) of the forward and reverse sequences* Pat. no EN, 265 63 49, MPK H02J 13/00 (2016/01) no 2016137597; Appl. 20.09.2016, publ. 05.06.2018, Bul. no 16

[2] Stroev V A 1996 *Transients of electrical systems in examples of illustrations* (Moscow: Znak Publ) 223 p

[3] Buryanina N S, Korolyuk Y F and Lesnykh E V 2016 *Method of reference of instantaneous values of voltages* Pat. 2 625 172, RU, IPC H02J 13/00 (2006.01) Russian Federation, no 2016105446; Appl.17.02.2016; publ. 12.07.2017, Bul. no 20

[4] Greshnov E B and Korolyuk Y F 1982 About the choice of the sampling interval of the analog parameters entered into the computer for programmable protections *Electronic Simulation* **5** 71–74 (Kiev)

IOP Conf. Series: Materials Science and Engineering **663** (2019) 012023 doi:10.1088/1757-899X/663/1/012023

Directed protection of the generator

N S Buryanina[1], Y F Korolyuk[1], M L Koryakina[1], E V Lesnykh[2] and K V Suslov[3]

[1]Chukotka branch of North-Eastern Federal University, 3, Studencheskaya str., Anadyr, 689000, Russia
[2]Siberian Transport University, 191, D. Kovalchuk str., Novosibirsk, 630049, Russia
[3]Irkutsk National Research Technical University, 83, Lermontov str., Irkutsk, 664704, Russia

E-mail: bns2005_56@mail.ru

Abstract. In addition to periodic components of industrial frequency, the aperiodic ones arise in windings and on the generator output leads under the conditions of shortcuts. These aperiodic components slow down the operation of the main differential protection of a generator up to ten periods of a network frequency with the application of saturated magnetic cores of current transformers as well as zero current relays and transformer differential protection [1]. This slowing down promotes accidents and can even lead to generator's winding and steel burn-out at high-ampere currents. The present paper considers the ways to reduce the impact of aperiodic components on the operation of digital relay protection of generators and generator-transformer blocks and proposes generator's directional digital protection.

1. Introduction

Digital protections gain considerable attention. Predominately, the algorithms of existing analog protections are implemented [2–4]. The difference between digital relay protections and analogue ones is that analogue protections receive operational information continuously while digital ones receive it discretely at equal intervals of time called sampling intervals and are processed by microprocessors. Particular attention is paid to the protections which apply information about changes in currents [5, 6]. This article proposes a method for determining a sinusoidal component of the short circuit current on the basis of four samples of instantaneous values.

Methods of implementing digital directional protection of a generator and a generator-transformer block are considered. An algorithm aimed to exclude aperiodic components from the measured currents is proposed and investigated. The measurement time of four samples does not exceed 2.0 – 2.5 milliseconds with the sampling intervals from 0.5 to 0.625 milliseconds. During this period the magnetic cores of current transformers do not saturate, i.e. relay protection receives undistorted information.

2. Research objective

Winding and steel burnout is the damage to generators stators caused by short circuits. Damage recovery is carried out in the factory by generator dismantling, which often results in the long-term shortage of electricity among consumers of a large territory even in the case of current repairs at the stations. The process of assembling generators after repairs can take several days. Accidents in the power system during this period can lead to a long period of power cutoff of both domestic and industrial consumers

[2]. The amount of a burned-out metal substantially depends on the short circuit duration. Aperiodic components in short-circuit currents slow down relay protection response time up to 0.2 s or more. Therefore, modern relay protection is launched after analog or digital filtering [3 – 5]. In addition, the current distance generators protections fail due to zero voltage [1]. Therefore, the development of relay protection algorithms shortening the faults identification period by several times being triggered when the voltage is zero proves to be rather relevant [5, 6]. Unfortunately, turning analog relay protections into digital ones does not contribute to a significant reduction in the time of its operation. The present paper proposes a digital relay protection algorithm able to recognize a short circuit during 2–2.5 ms.

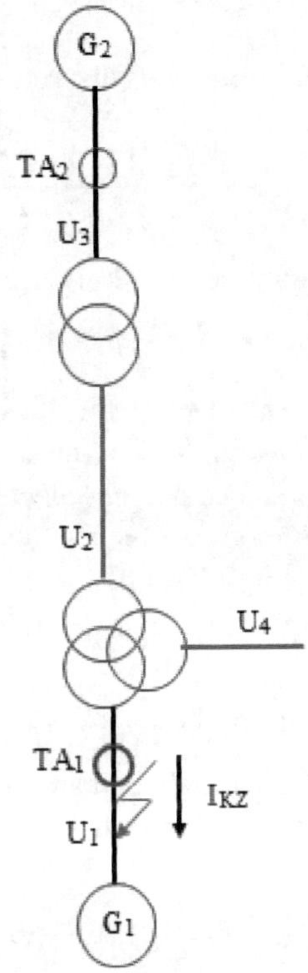

Figure 1 Scheme of the power system's electrical network

3. Materials and methods

Let us consider a part of the electrical network shown in Fig. 1. A three-phase short circuit occurred on the generator buses. The voltage dropped to zero. Remote protection is not able to determine the short circuit current direction and can be triggered at short circuit on the generator buses outside the protection zone. The probability of an external fault is several times higher since the generator current transformers are installed in the generator chamber and the generator buses from the camera to the generator voltage switchgear or the generator-transformer unit transformer are ten times longer than the buses from the generator terminals to its chamber. Additionally, there is a chance of a false shutdown. The situation can be improved by applying voltage U_2 rather than voltage U_1 due to the fact that U_2 has a value equal to the product of a short circuit current by the transformer T_1 resistance reduced to the generator voltage and surpasses the voltage U_1 by 30^0 angularly. This voltage is sufficient for correct short circuit identification. It is possible to use voltages U_3 and U_4.

As it was mentioned above, an aperiodic component, which is always present in the fault current, firstly, slows down relay protection and, secondly, can lead to its false work.

IOP Conf. Series: Materials Science and Engineering **663** (2019) 012023 doi:10.1088/1757-899X/663/1/012023

Let us consider the ways to exclude the aperiodic component within 2.0–2.5 ms. If we fix the values of four measurements of instantaneous currents at equal sampling intervals Δt, then at the time instants 2Δt and 3Δt and a sinusoidal current i(t) = sin(ωt) we obtain the following:

$$i_v(\Delta t) = \frac{2i(2\omega\cdot\Delta t) - i(\omega\cdot\Delta t) - i(3\omega\cdot\Delta t)}{4sin^2\left(\omega\cdot\frac{\Delta t}{2}\right)}$$

$$v(2\Delta t) = \frac{2i(2\omega\cdot\Delta t) - i(\omega\cdot\Delta t) - i(3\omega\cdot\Delta t)}{4sin^2\left(\omega\cdot\frac{\Delta t}{2}\right)} \quad (1)$$

where ω is angular frequency, t is current time, Δt is discretization interval per second.

Generally,

$$i_v(t) = \frac{2i(n\cdot\omega\cdot\Delta t) - i((n-1)\cdot\omega\cdot\Delta t) - i((n+1)\cdot\omega\cdot\Delta t)}{4sin^2\left(\omega\cdot\frac{\Delta t}{2}\right)}$$

Let us consider this method by the example of a short circuit current equal to [7]:

$$i(t) = \frac{E_{eq}}{\sqrt{R_{eq}^2 + X_{eq}^2}}(\sin(\omega\cdot t + \psi - \varphi) - \sin(\psi - \varphi)\cdot e^{\frac{-t}{T_N}}) \quad (2)$$

where t is current time, E_{eq} is equivalent electric force, R_{eq}, X_{eq} is equivalent active and reactive resistances with respect to a short circuit point, T_N is the time constant of the network relative to the short circuit point, ψ is an angle between an equivalent electromotive force and short circuit current, φ is an angle between an equivalent electromotive force and short circuit current.

Let us compute the values of the currents $i_v(t)$ at the time (Δt) and (2Δt) according to (1) and distract i(Δt) and i(3Δt) correspondingly. The time constant is determined as:

$$T_N = \frac{-\Delta t}{ln\left(\frac{i_v(2\Delta t) - i(3\Delta t)}{i_v(t) - i(2\Delta t)}\right)} \quad (3)$$

Angle φ is determined via the network constant and angular frequency:

$$\varphi = arctan(T_N\cdot\omega) \quad (4)$$

The aperiodic component is as follows:

$$i_A(t) = \left[U_{eq}/\sqrt{R_{eq}^2 + X_{eq}^2}\right][-\sin(\psi - \varphi)]\cdot e^{\frac{-t}{T_N}} \quad (5)$$

The sinusoidal component is determined as follows:

$$i_{sin}(t) = i(t) - i_A(t) \quad (6)$$

while a total current of short circuit is determined as follows:

$$i(t) = \frac{U_{eq}}{\sqrt{R_{eq}^2 + X_{eq}^2}}(\sin(\omega t + \psi - \varphi) - \sin(\psi - \varphi)\cdot e^{\frac{-t}{T_N}}) \quad (7)$$

The sinusoidal component and its derivative can be determined through the following equations at any time:

$$i_{sin}(t) = \frac{U_{eq}}{\sqrt{R_{eq}^2 + X_{eq}^2}}(\sin(\omega\cdot t + \psi - \varphi)) \qquad i'_{sin}(t) = \frac{U_{eq}}{\sqrt{R_{eq}^2 + X_{eq}^2}}(\cos(\omega\cdot t + \psi - \varphi)) \quad (8)$$

The amplitude value of the short circuit current is equal to its argument:

$$I = \sqrt{i_{sin}^2(t) + i'_{sin}(t)^2} \qquad \varphi = arctg(i'_{sin}(t)/i'_{sin}(t)) \quad (9)$$

Active and reactive powers are determined as follows:

$$P = \frac{U\cdot I}{2}\cdot\cos(\varphi), \qquad Q = (U\cdot I)/2\cdot\sin(\varphi) \quad (10)$$

The complex power is calculated through the product of the voltage and the conjugate current complex.

IOP Conf. Series: Materials Science and Engineering **663** (2019) 012023 doi:10.1088/1757-899X/663/1/012023

$$S(t) = \frac{(u_{sin}(t) + u'_{sin}(t) \cdot j)}{2} \cdot \frac{(i_{sin}(t) - i'_{sin}(t) \cdot j)}{2} \quad (11)$$

4. Research

Let us consider an example of a short circuit on the generator tires with the following parameters: 120 MW power, rated voltage equal to 10.5 kм, cos φ is 0.85, x_d`` is 0.189, x2 is 0.23. The time constant of the stator windings is 0.03 s. Three-winding transformer 250/220 parameters are as follows: power is 250 MVA, rated voltage of the higher winding is 242 kv, medium voltage is 37.5 kv, low voltage is 10.5 kv, U_k h-av is 11%, high-voltage U_k h-l is 24%, Uk av-l is 36%, ΔP_{kz} is 0.6 MW, ΔP_{xx} is 0.207 MW, high voltage windings are grounded. Line parameters are the following: L is 100 km, R is 0.33 ohm/km, X is 0.36 ohm/km.

The amplitude of the periodic component is 23.7 kA and the maximum aperiodic component is equal to the periodic one. Four measurements since the occurrence of a short circuit are as follows: sampling interval of 0.625 milliseconds: i (Δt) = 0.4547 кA, i (2Δt) = 1.7891 кA, i (3Δt) = 3.9333 kA, i (4Δt) = 6.7866 кA.

According to (3), the time constant with an accuracy of 4 characters is T_N = 0.0297 s. By putting T_N into (4) we get φ = 1.464 radian or 83.88^0. The effective value of the sinusoidal current is 23.7 кA. According to (5), i_A (t) = 23.7 · [–sin (ψ-φ)] · $e^{\frac{-t}{T_N}}$ кA. If i (Δt) = 0.45 кA, i (2Δt) = 1.78 кA, i (3Δt) = 3.93 кA, i (4Δt) = 6.78 кA, T_N = 0.03235 s, we get φ = 1.473 radian or 84.4^0, the effective value of the sinusoidal current is 23.69 кA. A further decrease in the measurement accuracy practically does not affect the calculation of the current amplitude. If the currents are measured with two decimal places after comma, the calculated amplitude of the sinusoidal component decreases to 23.695 кA. The periodic component is obtained by applying (6): i_{sin} = i (t) – i A (t) = 23.7 (sin (ωt – φ)). Power is determined according to (11). The calculated currents and powers are given in Fig. 2 and 3.

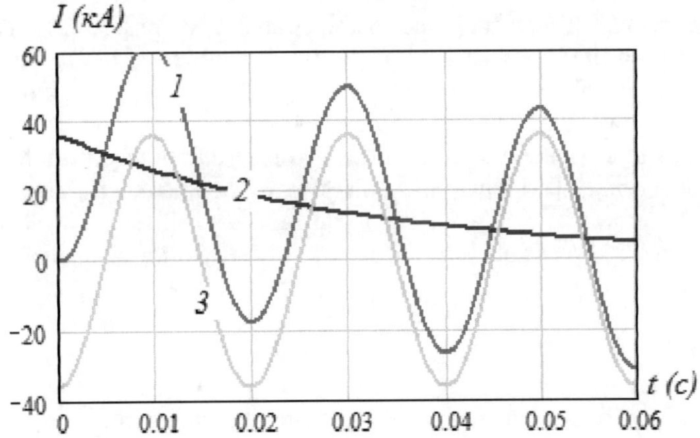

Figure 2. Short circuit current (1), aperiodic component (2)

The instantaneous values of the currents of the forward and reverse sequences can be defined as follows:

$$i_1(t) = [i_A(\omega \cdot t) + (i_B\left(\omega \cdot t + \frac{\pi}{2}\right) - i_C\left(\omega \cdot t + \frac{\pi}{2}\right))/\sqrt{3}]/2 \quad (12)$$

$$i_2(t) = [i_A(\omega \cdot t) - (i_B\left(\omega \cdot t + \frac{\pi}{2}\right) - i_C\left(\omega \cdot t + \frac{\pi}{2}\right))/\sqrt{3}]/2 \quad (13)$$

The outcomes are the same when using the following equations:

$$i_1(t) = ([i_A(\omega \cdot t) + i_B\left(\omega \cdot t + \frac{2\pi}{3}\right)]/3) + ([i_C\left(\omega \cdot t - \frac{2\pi}{3}\right)]/3) \quad (14)$$

$$i_2(t) = ([i_A(\omega \cdot t) + i_C\left(\omega \cdot t + \frac{2\pi}{3}\right)]/3) + ([i_B(\omega \cdot t - \frac{2\pi}{3})]/3) \quad (15)$$

Equations (12) and (13) have fewer calculations.

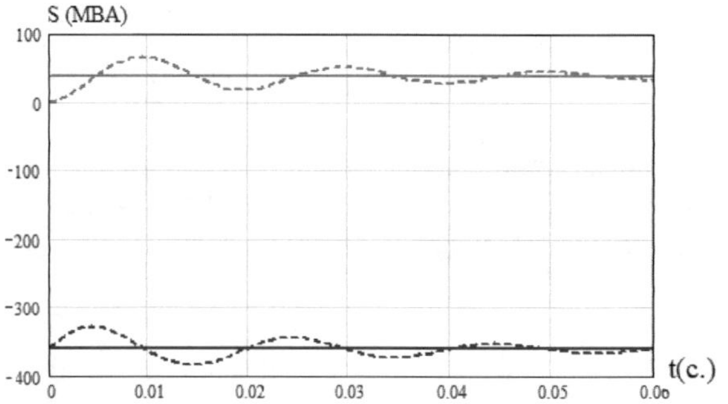

Figure 3. Design power. The dashed line is the product of voltage by the current in accordance with (1) and (6), solid line is the product of voltage by the conjugated current complex calculated in accordance with (11)

The advantage of the proposed method is the opportunity to apply only four measurements of short-circuit currents. In this case, the instantaneous values of the aperiodic and sinusoidal periodic components, the amplitude values and arguments of the short-circuit currents, the active and reactive powers in the elements of the electric network for at least three or four periods of the frequency of the electric network are calculated. The proposed method enables to create high-speed protection with a short circuit identification time for a quarter of the frequency period of the electrical network. It is possible to create algorithms for all protections using currents and algorithms for determining the location of damage on power lines. The aperiodic components in the fault currents have a significant effect on the accuracy of determining fault locations by modern fault localization systems. The error of current measurements during 2.0–5.0 ms can reach 30 % of the line length. Saturation of the magnetic cores of current transformers as well as zero current relay and differential relay, which slow down protection at least up to 0.2 s, does not affect the accuracy of separating sinusoidal components from the short circuit currents.

5. Conclusions

The method of selecting periodic components of short circuit currents from currents equal to the sum of the periodic and aperiodic components has been developed.

The algorithm based on this method uses only four samples separated by three sampling intervals, which eliminates the possibility of saturating the magnetic cores of current transformers by using information in the areas of sufficiently accurate transformation.

The method enables to calculate the currents of the forward and reverse sequences as well as any combination of these sequences, the resistance and power used in relay protection and emergency control at any time by measuring only four short-circuit currents in the following range: 2.0–2.5 ms.

Directional protection for generators is being proposed. It is proposed to use the voltages of adjacent elements of the electric network connected to the protected objects with the purpose to eliminate dead zones with closed three-phase short circuits.

6. Acknowledgments
The study was carried out with the financial support of the Russian Foundation for Basic Research and the Subject of the Russian Federation – the Republic of Sakha (Yakutia) № 18-48-140 010.

References
[1] Fedoseev A M and Fedoseev M A 1992 *Relay protection of electric power systems: Textbook for universities* 2nd ed, *Revised and enlarged* (Moscow: Energoatomizdat) 528 p
[2] *IEC, Electromechanical elementary relays* 2006 Part 7: Test and measurements procedures, IEC 61810–7
[3] Gurevich V 2006 *Electric Relays: Principles and Applications* (Boca Raton: CRC Press)
[4] Bojan Gergič and Darko Hercog 2019 *Design and implementation of a measurement system for high-speed testing of electromechanical relays Measurement* vol 135 pp 112–21
[5] Saleh K A, Zeineldin H, Al-Hinai A and El-Saadany E F 2015 Optimal coordination of directional overcurrent relays using a new time-current-voltage characteristic IEEE Trans *Power Delivery* **30(2)** 537–44
[6] Enríquez A C and Martínez E V 2006 *Enhanced time overcurrent coordination Electric Power Systems Res.* 76(6–7) April 457–65
[7] *Electromagnetic transients in electric power systems* 2012 Course of lectures (Nizhnevartovsk: Publishing House of the NGU) 199 p

IOP Conf. Series: Materials Science and Engineering **663** (2019) 012024 doi:10.1088/1757-899X/663/1/012024

Methodology of geochemical search for hydrocarbon deposits in lithological complexes of North-Yarudeysky rampart in Western Siberia

S V Dubrova[1] and V German[2]

[1]Russian State Pedagogical University named after Herzen, Building 5, 48, Moika River Embankment, 199034, St. Petersburg, Russia
[2]Petro Geochemical Service GmbH, Nunburg, Philippstraße 46, 90431, Germany

E-mail: dubrova.stanislav@gmail.com

Abstract. The technology of passive adsorption of hydrocarbons in soil-ground gases (more than 80 hydrocarbon compounds from ethane C_2 to C_{20} phytane) with the use of special sorbent modules is a new and very effective method for oil and gas fields' exploration. This technique being relatively inexpensive and easy to use for fieldwork is applied in almost all oil and gas regions of the world. In recent decades it has also been effectively used in Russia (Volga-Ural, Timan-Pechora, West Siberian and other oil and gas basins). The article presents the outcomes of geochemical explorations at the Shuginsky site of the Yarudeyskaya license area of the Nadym district in Western Siberia.

1. Introduction

Intensive exploitation of hydrocarbons resulting in the rapid depletion of deposits with easily recoverable reserves requires a new method of industrial deposits search and exploration, namely, deposits delineation, and grading the areas with previously established oil and gas potential. Until recently such problems were solved by applying a complex of geophysical methods, i.e. recording perturbances of natural electric and magnetic fields. However, nowadays using only geophysical methods does not meet the reliability requirements for the fossil fuels deposits prognosticating [1, 2].

More reliable industrial outcomes were obtained by geophysical and gas-geochemical methods integration. Initially gas-geochemical studies were carried out by drilling shallow wells (5–6 m deep) within a certain network with the subsequent soil sampling and extraction for being exposed to hydrocarbon analysis. It turned out later that higher quality testing results could be obtained by applying special sorbent modules in the holes with a diameter of up to 1.5 cm at the depth of 50 cm. In this case, no drilling is required and laying sorbent modules is implemented manually, which greatly reduces the cost of field work.

The methodology is based on a well-known Pearson's theory of the so-called hydrocarbon earth depth breathing. It is the phenomenon of diffused penetration of hydrocarbons into the space above deposits in the form of microfiltration, which is confirmed by being used in numerous oil and gas fields. The main goal of geochemical studies is to forecast deep deposits of gas, oil and condensate basing on the interpretation of soil gas sampling data, which are the basis for deciphering ascending gas emanations.

IOP Conf. Series: Materials Science and Engineering **663** (2019) 012024 doi:10.1088/1757-899X/663/1/012024

2. Materials and methods

Ground gas samples are taken from a relatively shallow depth by special sorbent modules introduced into the bore holes at the depth of about 50 cm and a diameter of up to 1.5 cm for a period of 17 to 21 days depending on the season of work. The proposed system of passive adsorption of the gas fraction significantly differs from the previous one, which used 5–6-meter wells for sampling deep gases. It dramatically reduces the cost of fieldwork. All samples of ground gas are analyzed using thermal desorption with the further gas chromatographic separation and mass spectrometric detection of a wide profile of organic compounds (including the thermogenic ones) in the range of C_2–C_{20}, which enables to make conclusions concerning not only the intensity of hydrocarbon breathing but also the qualitative composition. A distinguishing characteristic of the technique is a reliable and detailed deep gases diagnosis enabling to distinguish background hydrocarbons and to categorize the gases according to the genesis, specifically, plant, microbiological, surface contamination, source rock, etc.

3. Results and discussions

Let us consider the results of applying this methodology. These results were obtained after conducting geochemical studies at the Shuginsky local uplift in the Yamalo-Nenets Autonomous Okrug in 2011. Additionally, the results were processed in 2019. Geochemical studies based on passive ground gas adsorption technology were first carried out at the Yarudeysky uplift in winter 2005–2006 along seismic profiles with a sampling step of 500 and 1000 m. One profile (transect) was passed through the Yarudeyskoye – South Shuginsky – Shuginsky uplift and detected positive geochemical anomalies interpreted as predicted hydrocarbon deposits. According to the outcomes of geochemical studies conducted within the identified anomalies, the prospects of the structure aimed to be used for drilling a prospecting well were confirmed. In 2007, a well P-5 was drilled in the Yarudeyskaya structure with regards to the results of geochemical work. An industrial inflow of gas condensate and gas with a flow rate of more than 100 thousand m^3 per a day in this well was obtained when testing the interval from 3030 to 3050 m from the reservoir UN_{12}. In winter 2007–2008, geochemical studies of the Shuginsky structure over a network of 1000x1000 m with condensation of up to 500 m in the areas of the proposed wells drilling was also carried out. The set of hydrocarbon compounds in the samples at the Shuginsky uplift under study correlates with the set of similar compounds in the productive gas condensate well Yarudeiskaya 5 used as a reference.

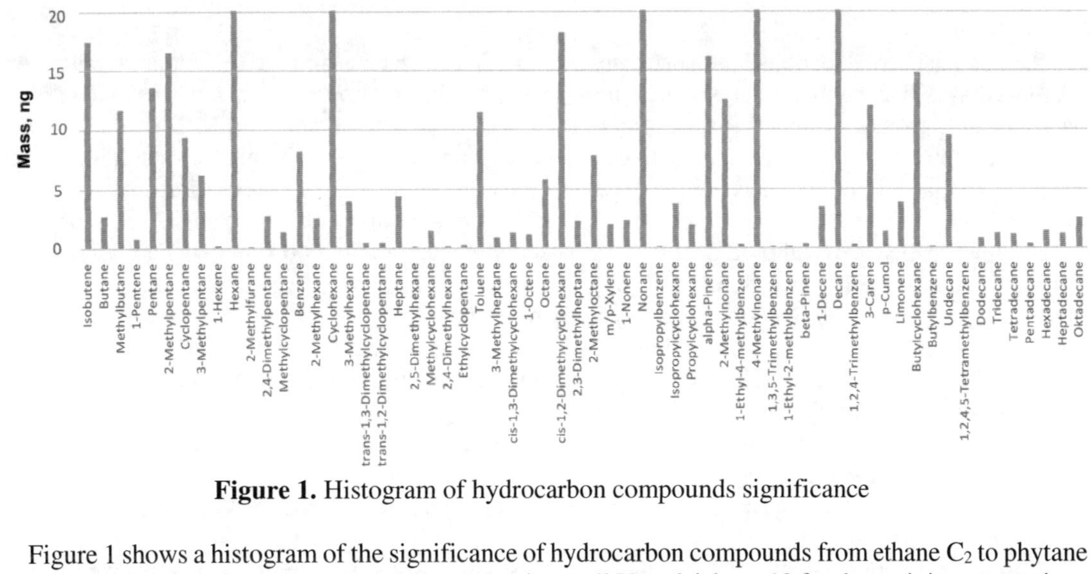

Figure 1. Histogram of hydrocarbon compounds significance

Figure 1 shows a histogram of the significance of hydrocarbon compounds from ethane C_2 to phytane C_{20} according to samples taken around the production well Yarudeiskaya 10 for determining emanations over hydrocarbon deposits. The higher the bar of the histogram, the more significant this compound is for identifying the oil-like qualitative characteristics of emanation. This is a robust model because it is independent of one or more compounds. Such a complex characteristic of hydrocarbon models remains stable in most hydrocarbon deposits.

A mathematical projection of each sample on the axis dividing the centroid of the classes of finite members is constructed for reliable interpretation of geochemical data (oil/gas or dry class). At the same time, the dry class centroid has a zero probability of membership in the oil/gas interval while the oil/gas class centroid has a 100 % probability in its interval (Fig. 2).

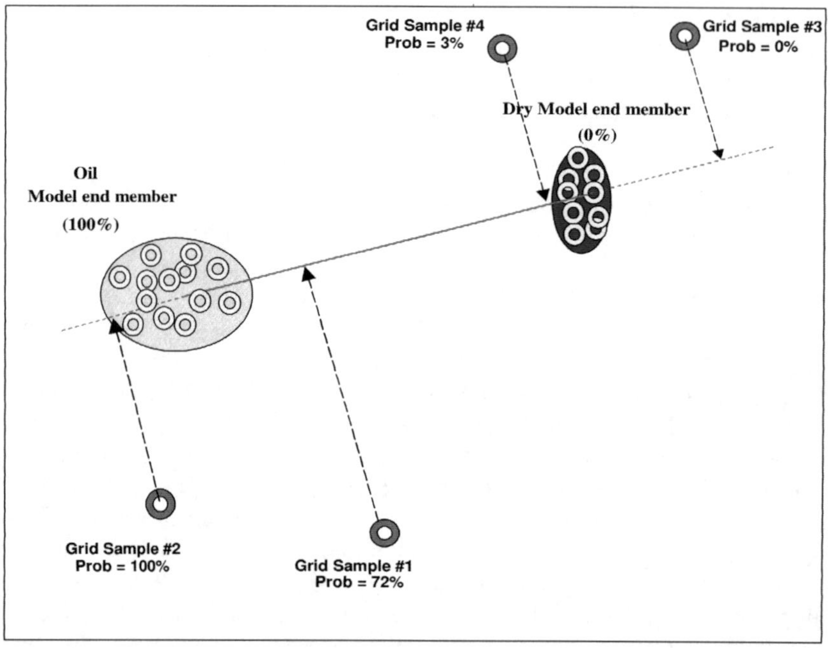

Figure 2. Schematic illustration of separating centroids of model classes by the discriminant analysis method. The probability value is assigned to each sample by mathematical projection (dashed lines) on the axis of maximum separation between the final members of the model (thick line)

Several maps of the object prospectivity are issued for each standard (or for a standard well Yarudeiskaya 10 and cluster analysis as in the case under consideration) in case when several reference wells are used to determine the prospectivity of a site with regards to oil and gas [3]. The results of work within the Shuginsky site are presented on geochemical probability maps (Figure 3). According to these maps, potential oil and gas deposits in the study area are elongated and submeridional. The results obtained may indicate the oil and gas potential of local structural traps within the Shuginsko-Yarudeysky small rampart. Geochemical profile (transect), passed in winter 2005–2006 from the Yarudeysky uplift through the Shuginsky and South Shuginsky uplifts confirms the local nature of the potential hydrocarbon deposits distribution separately in each structure.

The reasons for this focal distribution of geochemical anomalies have not been fully figured out. Perhaps the main role is played by the presence of small closed local uplifts with a changing lithological composition of reservoir rocks along the lateral.

Nowadays it is possible to compare all materials and make a considered conclusion about the oil and gas potential of the Yarudeysky licensed area. The linear geochemical anomalies elongated in the submeridional direction indicate the significant role of the faults of both conducting and localizing structures.

Unfortunately, the insignificant area of research prevents from estimating the localizing role of the Shuginsko-Yarudeysky small rampart as a whole.

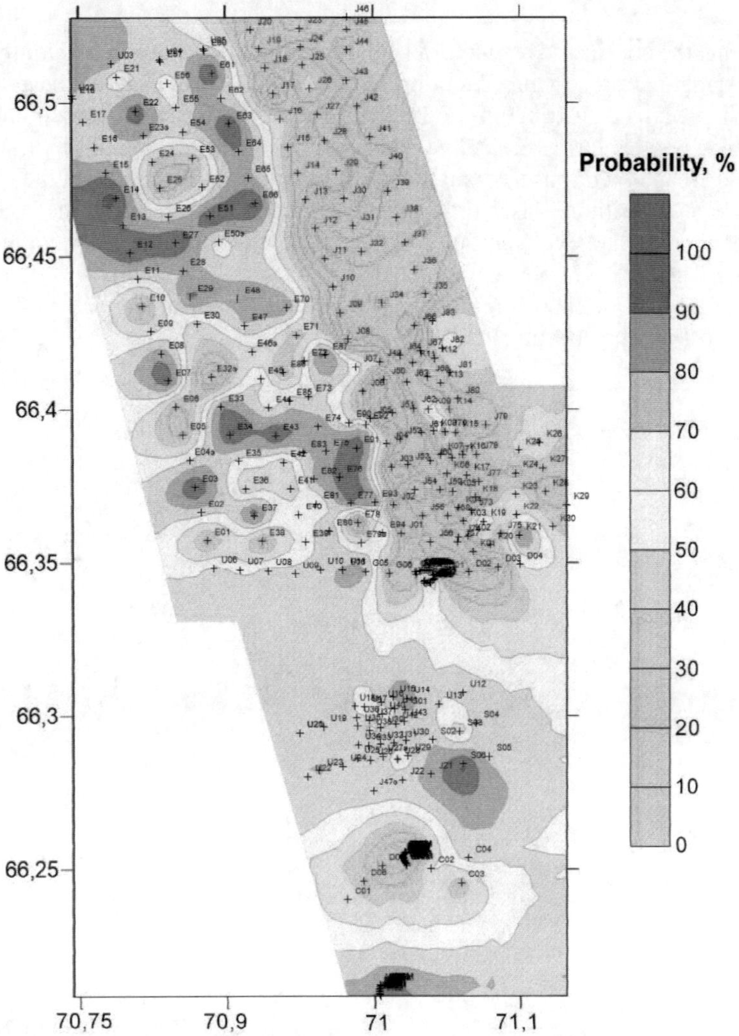

Figure 3. Probability distribution of product availability by type 10 well in the Yarudeysky licensed area (interpretation of 2019 results)

The results of drilling exploratory wells have appeared to be essential for assessing the oil and gas potential of the Yarudeysky licensed area. In total, six exploratory wells (Yarudeyskaya 1, 2, 3, 4, and 81, 82) had been drilled in the past years before the geochemical work started and only one of them received a small inflow. Industrial flow rates were obtained following the application of the hydrocarbons passive adsorption technology in 2006 and 2008 and in all four wells (well Yarudeyskaya 5 at the Yarudeisky uplift and the wells Yarudeyskaya 7, 8 and 10 at the Shuginsky uplift) drilled before 2011. Indisputably, a rather costly drilling of deep wells in the area proved to be more effective taking into account the obtained geochemical results.

Geochemical exploration work based on the proposed methodology was also carried out at other sites in the West Siberian oil and gas basin. In addition to the Yarudeysky and Shuginsky sections considered in the article, similar work was carried out on a large number of other objects. The oil content of the structures with the revealed positive geochemical anomalies was confirmed by drilling pioneer and exploration wells. In addition, hydraulic fracturing was planned. Geochemical studies carried out at, say, the East-Yetypurovskaya structure proved to be dead-end, which led to cancelling the planned well drilling.

4. Conclusion
The article states that the considered technology of passive adsorbing soil-ground air hydrocarbons with their subsequent statistical processing (resulting in geochemical mapping of the probabilities of the forecast hydrocarbon deposits distribution in the studied area) by means of sorbent modules is a direct method for searching and exploring oil and gas fields with a fairly high reliability (up to 90 %). It was used in all outlined areas of work as a means for sorting seismic structures with the objective to select the targets for exploratory drilling. Based on experience of other countries, this method can also be used at the stages of exploration and exploitation of hydrocarbon deposits for selecting the optimal location of production wells. The cost of implementing the described method per unit of work in Western Siberia is less than 25 % of the cost of 2D seismic work. The minimum cost of a geochemical project does not exceed 5 % of the cost of a prospecting well.

5. References
[1] Khan S D and Jacobson S 2008 *Geol Soc Am Bull.* **120(1/2)** 95–105
[2] Schumacher D 2000 *The leading edge* **19(3)** 258–61
[3] Suslick S B and Schiozer D J 2004 *J. Pet Sci Eng* **442(1–2)** 1–9

IOP Conf. Series: Materials Science and Engineering **663** (2019) 012025 doi:10.1088/1757-899X/663/1/012025

Technological fluids on biopolymer basis for repair wells

E V Egorova[1], Yu S Minchenko[1] and I I Bosikov[2]

[1]Astrakhan State Technical University, 16, Tatishcheva St., Astrakhan, 414056, Russia
[2]North-Caucasian Institute of mining and metallurgy (State Technological University), 44, Nikolaeva St., Vladikavkaz, 362021, Russia

Abstract. The article describes the commercial experience of using various compositions of process fluids during repair and insulation works and substantiates the practical value of solutions on a biopolymer basis. The results of laboratory studies are presented and the compositions of process fluids on a biopolymer basis for well repair are proposed. The preservation of the natural permeability of the reservoir in the near-wellbore zone is one of the main requirements for quality well repair. However, as the field practice shows, the overhaul is almost universally carried out with the use of process fluids, the parameters of which are chosen without taking into account the specific geological conditions, the lithologic and physical properties of the reservoir and the physicochemical characteristics of its fluids.

1. Introduction

The cost of biopolymer systems is higher than that of traditional process fluids con-tainting synthetic polymers, but their use is economically beneficial, since it minimizes the negative impact on the productive layer, and the development time of wells de-creases after carrying out repair work.

Repair and restoration work (RRW) are accompanied by killing and temporary blocking of the well bottom zone. One of the main reasons for the loss of productivity of wells in fields and underground gas storage facilities (UGS) with a large capacity of highly permeable reservoirs is an unreasonable approach to the use of process fluids. RRW wells without taking into account their physicochemical characteristics and geological and physical features of reservoirs, leads to intensive absorption of working fluids, loss of circulation, deterioration of filtration-capacitive properties (FCP) of productive formations, a sharp decline in well productivity. Therefore, the development of compositions of process fluids for plugging with temporary blocking of the bottom hole zone during well repair to prevent contamination of the bottom hole formation zone (BFZ) is an actual problem for many fields and UGS.

2. Relevance of the topic

The analysis of works devoted to the study of improving the performance of gas wells in various geological conditions, allows summarizing the theoretical and practical research data, as well as identifying the requirements for the gel-forming compound for temporary blocking of the BFZ: manufacturability (the content of the minimum number of components, simplicity download); homogeneity; inertness to the rocks composing the reservoir; do not reduce the permeability of the BFZ; possess gas retention capacity and exclude fluid migration pathways through the BTC; have high values of adhesion strength with enclosing surfaces; complete removal from the well by a pressure drop that does not exceed the critical depression on the formation (exclude the use of chemical processing of the PPP).

3. Essence of the issue

The analysis of some known compositions of process fluids for the construction and repair of wells is presented below.

The authors of the fastening technology [1] propose an annular space (BTC) of the production casing above the reservoir filled with a viscous fluid trap with a density that provides hydrostatic pressure above the reservoir by 5–10 %. A composition based on monoethanolamine, polyacrylamide, aluminum sulphate, drilling mud and anticorrosive inhibitor can be used as a viscous hydraulic seal. The proposed water seal has a reduced strength, which is associated with the ingredient composition, in particular the type of cross-linking system. The use of a trivalent metal salt as a gelling agent provides a weak three-dimensional structure of a cross-linked polymer. In addition, the water seal has insufficient plastic strength, over time, syneresis can occur – the separation of water incorporated into the cross-linked polymer. As a result, gas migration may occur from the reservoir to the BTC. Also, this composition has a reduced adhesion, in connection with which it is not possible to use it effectively.

One of the new ways to create sealing screens in MCD warning technologies is the use of a hydraulic seal [2–5]. The hydraulic lock is understood to mean a working fluid with the required technological properties, which forms an impermeable screen in the well BTC, preventing gas from entering the cement stone and preserving the reservoir proper-ties of the formation.

The insulating material of the water seal in the work [4] includes used drilling mud, poly-acrylamide, ferrochrome lignosulfonate, carboxymethylcellulose, and a corrosion inhibitor. Additionally, the solution may contain bentonite, chalk or gypsum. This compound has a fast gelation, resulting in a difficult-pumped system, the use of which may be limited due to the inability to pump the compound to the depth of the reservoir. The increase in viscosity can also be facilitated by the presence of solid filler over the optimum with given amounts of the polymers used.

In the work [6], a three-phase foam is used, which is unstable and changes its properties in time. Upon contact with the reservoir water, the liquid phase of the foam dissolves in it, and the foam collapses, which leads to imbalance "well-reservoir". Foam has a low gas breakthrough gradient, which is why it is pumped into the well to the mouth.

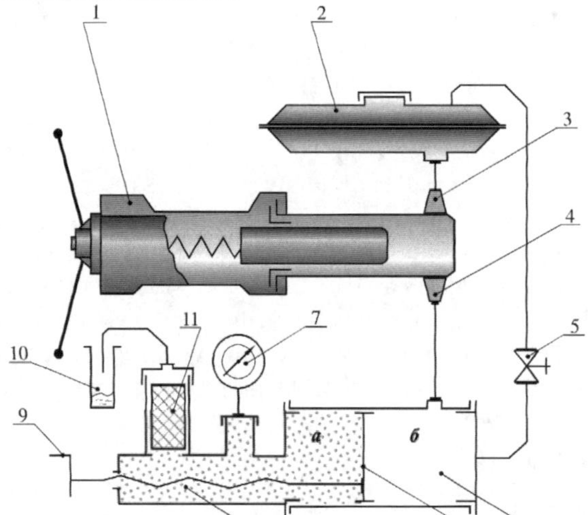

Figure 1. Installation diagram for the study of the blocking ability of process fluids: 1 – high pressure hydraulic press; 2 – capacity for oil; 3 – receiving valve; 4 – discharge valve; 5 – valve for pressure relief; 6 – hydraulic cylinder; 7 – manometer; 8 – filtration chamber; 9 – piston return screw; 10 – measuring capacity; 11 – permeable medium model; 12 – separation piston

Table 1. Effect of known blocking compounds on the permeability of artificial sandstone

IOP Conf. Series: Materials Science and Engineering **663** (2019) 012025 doi:10.1088/1757-899X/663/1/012025

№ exp.	Blocking fluid	№ of sample	t water bath,°C	Repressio, MPa	Breakpressure , MPa	K_o, mcm^2	K_b, mcm^2	β, %
1	Water	11a	20	0	-	0.28	0.14	51
		4a	20	0	-	0.75	0.42	55
		6a	20	0	-	0.79	0.60	76
		9a	20	0	-	0.96	0.89	93
2	Clay solution Weighted with barite	1	50	3.5	0.20	0.10	0.09	75
		2	50	3.5	0.21	1.02	0.77	88
3	Clay solution thickened with CMC	1a	50	3.5	1.10	0.25	0.23	91
		2a	50	3.5	1.20	0.15	0.14	94
4	Blocking fluid + chalk	3	50	3.5	0.06	0.14	0.09	71
		4	50	3.5	0.06	0.75	0.26	35
		5	50	3.5	0.06	0.87	0.51	58
5	Blocking fluid + + filler	6	50	3.5	0.08	0.79	0.35	44
		7	50	3.5	0.20	0.14	0.11	75
6	Hydrocarbon-based blocking fluid	8	50	3.5	0.03	0.15	0.09	64
		9	50	3.5	0.07	0.96	0.41	43
7	Gel-forming cement slurry	10	50	3.5	-	0.16	0.11	71
		11	50	3.5	-	0.34	0.21	62
8	Gelling slurry for temporary isolation	12	50	3.5	2.70	0.69	0.47	68
		13	50	3.5	3.20	0.13	0.11	86
9	Gelling composition	14	50	3.5	-	0.95	0.71	75
		15	50	3.5	-	0.16	0.15	92
10	Gelling ompound	16	50	3.5	1.80	0.14	0.09	63
		17	50	3.5	1.60	0.15	0.08	55
11	C HS-1600	18	50	3.5	0.15	0.30	0.24	80

Figure 1 shows the experimental setup that simulates the process of clogging of the reservoir during well repair.

In order to select the composition for the temporary blocking of the formation that meets the above requirements, as well as to quantify the plugging capacity of the gas reservoir rocks, laboratory studies were conducted. During the experiments, the clogging ability of known compositions of technological systems was determined. Artificial cement-sand cores with different permeabilities, with a special technology, were used as permeable samples. To exclude chemical clogging, the core material did not contain a swelling substance (clay). Drilling solutions (weighted and thickened), blocking liquids, gel-forming compounds, salt solutions and technical water (table 1) were used as well-known compounds of technological systems.

Residual permeability of the core was determined by the method of [7].

Indicator of the degree of clogging ability of the studied liquids on the core adopted the coefficient of recovery of permeability β, determined by the formula:

$$\beta = (K_b/K_o)\cdot 100\ \%, \qquad (1)$$

where K_b – the coefficient of permeability of the core after its interaction with blocking fluid (after decolmatisation); K_o is the coefficient of initial permeability of the core.

From table 1 it follows that even water for gas reservoirs can be a mechanical plugging. From experience No. 2, it can be seen that with a low initial permeability of the core and exposure to mud, the coefficient β is very high and decreases as K_o increases. The formed mud cake has a restraining effect on the penetration of the mud filtrate into the formation. This is explained by the fact that with an increase in K_o, the surface tension of the filtrate and water significantly decreases, which facilitates the process of its displacement by gas. However, the situation changes under the influence of highly viscous systems on the core, especially those containing a solid phase [7].

From the experimental data it can be seen that a high degree of clogging ability ensures presence of inert-blocking solutions, which are essentially chemically neutral substances contained in a reservoir and contains a solid phase, which is similar to bentonite when exposed to solutions of thixotropic properties.

Minimizing the damage to reservoir productivity in terms of reducing its natural permeability, which is especially important when operating wells under Abnormal Seam Pressure conditions, allows the use of biopolymers at the microbial level as the basis of working fluids for well repair. The practical value of biopolymers is determined primarily by their ability to sharply change the rheological properties of aqueous systems in low concentrations: increase viscosity, form gels, serves as stabilizers for suspensions and emulsions. These properties have attracted the attention of oil and gas producers, and biopolymers in the last two decades have been tested and applied in the practice of exploration and production drilling, changes in the bottom hole formation capacity, improved efficiency of oil displacement processes, and in recent years also during work over [7].

Biopolymers of the “xanthan” type produced in the West are obtained in the form of a dry powder, which considerably reduces the cost of transportation of the product, since the solution contains 98-99% water in the finished form. At the same time, the “dry product” causes a number of problems: the repeated increase in the cost of production, the need to use toxic substances in the process of isolation and drying to ensure term of containment (expiration), environmental pollution, the need for special equipment for preparing a solution from powder.

For the temporary blocking of the reservoir, we propose a viscoelastic compound (VUR) based on SupraXan biopolymer [8]. The main chain of SupraXan is composed of D-glucose units linked by 1,4 b-glycosidic bonds, and in the branches of the core is a trisaccharide consisting of β-D-mannose, β-D-glucuronic acid and α-D-mannose. Glucuronic acid residues and acid pyruvic groups impart an anionic character of [5-8] to xanthan molecules. SupraXan molecules in aqueous solutions are prone to self-association and with an increase in the ionic strength of the solution or the concentration of the polysaccharide a gel is formed. The cross-linking reagent in the composition of the solution is TEAT-1.

According to the results of laboratory studies, it is most effective to apply the gel-forming compound without fillers under conditions of low permeability of the reservoir (≤ 0.1 mcm^2). In order to expand the range of application of the composition for various mining and geological conditions, introduced fillers differ in chemical nature and degree of dispersion. In order to more clearly present the change in the coefficient of core permeability recovery (β) and the change in core plugging factor (γ) from the

IOP Conf. Series: Materials Science and Engineering **663** (2019) 012025 doi:10.1088/1757-899X/663/1/012025

content of clogging filler, graphs were prepared for the gel-forming compound with fiberglass (figure. 2, 3).

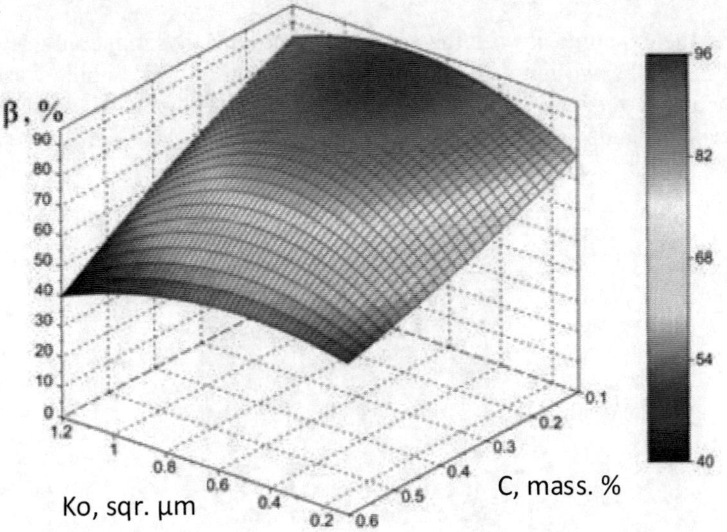

Figure 2. Change of core permeability recovery coefficient (β) from the content of clogging filler (C).

The usage of the proposed composition is optimal for cores with a permeability of 0.5–0.9 mcm^2 at a fiber concentration in the gel of 0.25–0.40 % of total mass. The thin inert fiber under the influence of pressure reliably blocks the "cracks" of the test core sample. As part of the solution, it is located perpendicular to the core crack and forms a solid shield on the sample surface due to damped filtration. Due to the minimal penetration of the gel into the core (1.0–3.0 mm), the sample was released at a pressure three times lower than the repression pressure.

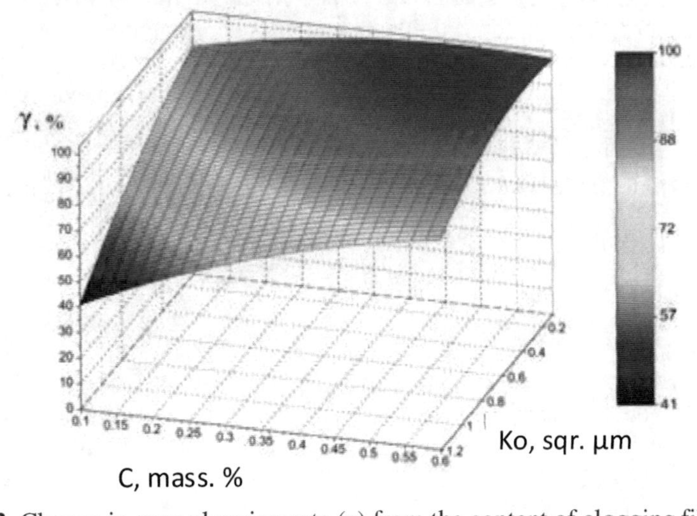

Figure 3. Change in core plugging rate (γ) from the content of clogging filler (C).

Temporarily isolating the core with a permeability of 1 mcm^2 is effectively a gel system with cord fiber. For the core permeability from 0.2 to 0.54 mcm^2, the best performance was obtained when processing the cross-linked gel system with basalt fiber filler.

In real conditions, the composition for temporary blocking will interact with the rock. Therefore, additional studies of the adhesive properties of the composition on artificial rock samples were carried out.

In the course of the experiments with the compositions under investigation, gas was returned from the reciprocating gas to the syringe with air supply of 1–2 cm^3. An air bubble was formed inside the composition, which was retained within the system without rising to the surface (Figure 4). This indicates a sufficient strength of the gel structure during its interaction with the rock, which allows concluding that the adhesive and gas-blocking ability is high.

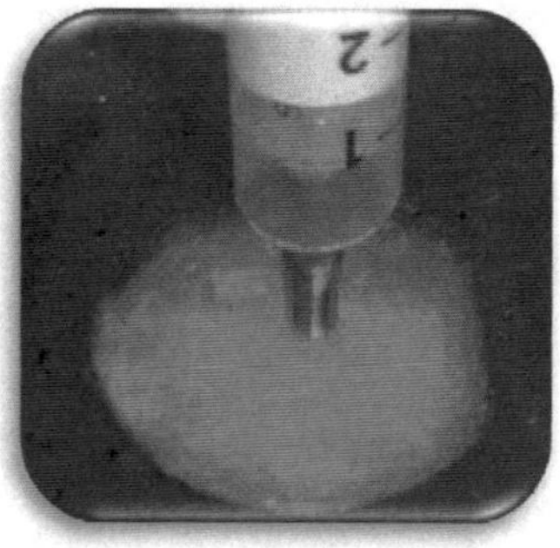

Figure 4. Test of holding capacity of the gel-forming composition

We also offer compositions of process fluids based on biopolymer (Saraksan, SupraXan), which can be used for flushing sand plugs, sidetracking and well completion. Biopolymer liquids have reduced filtration properties and pronounced pseudoplastic, and the content of antifreeze in the compositions allows the use of these liquids at negative temperatures in winter. The main physico-chemical properties and structural and rheological parameters of process fluids are presented in table 2.

Biopolymer fluids have improved structural and rheological properties that ensure their increased retention and transport capacity. The latter are characterized by the nonlinearity coefficient -*n*. The smaller the n, the more the solution exhibits pseudoplastic properties. This means that viscosity decreases with increasing relative shear rates and, conversely, viscosity increases with decreasing relative shear rates. Reducing the constant ***n*** allows you to improve the removal of rock and cleaning of the well due to the alignment (flattening) of the velocity profile of fluid flow in the annular space.

Table 2. Main properties of process fluids based on biopolymer.

Indicator	Value
Biopolymer liquid withoutantifreeze	
Densityρ, kg/m^3	600 ÷ 842
Filtration Φ, cm^3/30 min	2,0 ÷ 4,0
Plastic Viscosity η, mPa·s	13 ÷ 28
Dynamic shear stress τ_o, dPa	144 ÷ 373
Static shear stress for 1 and 10 min $\theta_{1/10}$,dPa	29 ÷ 124 / 34 ÷ 168
Nonlinearity coefficient*n*	0,30 ÷ 0,40
Freezing point, ° C	– 3 ÷ – 5
Biopolymer liquid with antifreeze	
Densityρ, kg/m^3	630 ÷ 990
Filtration, м	2,0 ÷ 4,8
Plastic Viscosity η, mPa·s	16 ÷ 31
Dynamic shear stress τ_o, dPa	129 ÷ 254
Static shear stress for 1 and 10 min $\theta_{1/10}$,dPa	34 ÷ 91 / 34 ÷ 115
Nonlinearity coefficient, *n*	0,37 ÷ 0,47
Freezingpoint, ° C	– 12 ÷ – 39

Compared with the water-soluble synthetic polymers traditionally used in gas production, in particular, PAA, biopolymers have a number of significant advantages, including such that allow their use in very harsh environments where the use of synthetic polymers is ineffective.

The cost of biopolymer systems is higher than conventional process fluids containing synthetic polymers. Despite the high cost of biopolymers, their use is beneficial, since the negative impact of biopolymer fluids on the reservoir is minimized, and the time spent on developing wells decreases after repairs are performed using them. By virtue of these advantages, these systems are effective, and they have no alternative under certain conditions.

The use of biopolymers, which are products of the life of microorganisms, in process fluids for wells with high temperatures (over 100 ° C) are not found in the field practice of oil and gas producing enterprises, since the effectiveness of microbial biopolymers is generally limited to temperatures of 60–80 ° C. In this regard, we can conclude about the feasibility of using biopolymers in the compositions of process fluids with weighting additives for the overhaul of wells in the conditions of ASP (abnormal seam pressure) with compliance of the specified temperature limits of usage

For the use of biopolymers of microbial origin in the overhaul of wells with ASP at higher temperatures, they need to be adjusted during the production process or the development of multicomponent biopolymer fluids characterized by increased heat resistance.

4. Conclusion

1. The analysis of modern process fluids used for repair and restoration work was carried out, the results of which substantiated the effectiveness of the use of solutions on a biopolymer basis.
2. The requirements for process fluids for temporary blocking of the formation were determined.
3. The proposed composition for the temporary blocking of the reservoir in the form of a polymer-containing gel with clogging fillers. It has been tested for gas holding capacity and contact strength (adhesion) with rock and metal. With the help of experimental installations, the coefficient of permeability recovery and the blocking rate of the core were determined.
4. The proposed composition of process fluids based on biopolymer (Saraksan, SupraXan), which can be used for flushing sand plugs, sidetracking and well development. Biopolymer liquids have reduced filtration properties and pronounced pseudoplastic, and the content of antifreeze in the compositions allows using these liquids at negative temperatures in winter.

Acknowledgments

The research was carried out with the financial support of the Russian Foundation for Basic Research and the Ministry of Education and Science of the Republic of South Ossetia in the framework of the scientific project No. 19-511-07001.

References

[1] Klyuev R V and Bosikov I I 2016 *2nd Int. Conf. on Industrial Engineering, Applications and Manufacturing (ICIEAM)* vol 53 pp 1–5
[2] Klyuev R V, Bosikov I I, Gavrina O A and Revazov V Ch 2018 *MATEC Web of Conf.* vol 226 pp 1–5
[3] Klyuev R V, Bosikov I I, Gavrina O A, Revazov V Ch and Madaeva M Z *Proc. of the VIII Sci. and Techn. Conf. "Contemporary Issues of Geology, Geophysics and Geo-ecology of the North Caucasus" (CIGGG 2018), Advances in Engineering Res.* vol 182 169–74
[4] Mustafayev S D and Gasymova S A 2019 *Oil industry* **2(1144)** 58–60
[5] Gaydarov M M-R, Khubbatov A A, Gaydarov A M, Norov A D, Khrabrov D V and Potapova I A 2019 *Oil industry* **2(1144)** 33–8
[6] Nedozhdiy A A, Zavyalov A S and Vasileva V S 2019 *Oil industry* **2(1144)** 39–43
[7] Rustamov I F and Cherepanov A V 2017 *Oil industry* **2 50-53**
[8] Shakhverdiev A Kh 2017 *Oil industry* **2 58–63**

IOP Conf. Series: Materials Science and Engineering **663** (2019) 012026 doi:10.1088/1757-899X/663/1/012026

High-tech drilling of wells in offshore fields using import-substituting technologies

A G Alekseev[1] and E V Egorova[2]

[1]LUKOIL-Nizhnevolzhskneft LLC, Admiralteyskaya st., Astrakhan, 4140241, Russia
[2]Astrakhan State Technical University, 16, Tatishcheva st., Astrakhan, 414025, Russia

E-mail: egorova_ev@list.ru

Abstract: The article considers the relevance of increasing the efficiency of development of offshore shelf deposits in the North Caspian, using the example of the oil and gas condensate field named after Y. Korchagin. The development of offshore fields is associated with the use of the latest expensive technologies aimed at highly efficient development of facilities with hard-to-recover hydrocarbon reserves. The involvement of field operators in the development of expensive imported technologies leads to a significant increase in the cost of oil produced. In this regard, there is a need for testing and introducing similar domestic technologies into the offshore field development process to optimize the hydrocarbon production process.

1. Introduction

Currently, as we study new promising areas in terms of oil and gas potential, the complexity of developing deposits located in hard-to-reach areas of the earth's surface and associated with hard-to-recover reserves is increasing.

Such objects include the field them. Yu. Korchagin, located in the Russian part of the Caspian Sea.

Effective development of facilities with hard-to-recover oil reserves on the shelf is a daunting task. It can be solved only by using the latest development technologies, for example, such as drilling extended horizontal wells (horizontal wells), branched horizontal wells (BHW), in combination with downhole equipment that allows you to control and regulate fluid flow from different intervals of the reservoir.

When drawing up a technological development scheme in order to increase the efficiency of the development of the oil rim of the Neocomian deposits of the field named after Yu. Korchagin proposes a development system by drilling a long-range horizontal well. Production wells with horizontal shafts up to 5000 m long are evenly distributed over the reservoir area, near the oil-water contact (WOC), parallel to its surface [1]. The length of the wells is determined by the size of the field, their trajectory is less dependent on the characteristics of the geological structure of the formation. Due to the large length of the well, a high density of wells is achieved and oil displacement by gas occurs more evenly.

After a gas breakthrough, the wells are transferred to periodic work or their flow rate is regulated (reduced). Thus, the development of the reservoir is carried out while limiting the gas factor to a value that allows for normal well operation. Breakthrough gas is pumped back into the gas cap to maintain reservoir pressure in it at the initial level, while produced water is transferred to the marginal zone through water-absorbing wells.

However, the use of the described field development system is hindered by the fact that the wells that were drilled had a high complexity of drilling, completion and development, due to the complex profile of the well. The horizontal drift was more than double the depth of the wells vertically.

2. Factors that improve the quality of well drilling

To ensure high-quality well drilling in these conditions, the most important factors are:

1. Conducting high-tech geosteering, providing a high percentage of the reservoir in the horizontal part of the trunk.
2. Consideration of all risks and negative factors of well drilling based on high-quality geomechanical modeling.

In preparation for geosteering and compiling a geomechanical model, factors were noted that increased the degree of uncertainty of the calculations. These factors are primarily associated with a lack of input data [2]. These include:

- High structural uncertainty (especially the northern slope of the structure).
- Weak coverage of the study area with well data.
- Lack of full inversion / interpretation of seismic data reliable impedance and density cubes.
- Incomplete interpretation of acoustic log data for geomechanics.
- Probable differences in the petrophysical interpretation of data by various methods (for example, clay content in exploratory and production wells).
- Lack of knowledge of the Volga layer, namely the lack of results of mechanical core testing, the lack of direct measurements of the minimum horizontal stress.

As a result of the finite element calculation for each cell, Schlumberger specialists determined all the components of the stress tensor acting on it. In general, the direction of stresses is maintained over the entire area and is ~ 150 ° for the maximum horizontal stress (~ 60 ° for the minimum). Near the faults, a slight rotation of the vertical stress occurs, while the directions of the horizontal stress practically do not change. Thus, the calculation of the 3D stress state allows one to consider the effect of discontinuous disturbances on the magnitude and stress orientation. In the future, taking into account the influence of the development on reservoir pressures can assess the risks of reactivation of displacements along the fault planes and predict the complications associated with this, as well as plan procedures to avoid such problems or to reduce their impact [3].

3. The results of the use of import-substituting technologies

At present, the LUKOIL-Nizhnevolzhskneft's geological service independently conducts geo-navigation of wells using the Russian software PetroInfoComplex.

This software in comparison with both imported and Russian counterparts has several significant advantages in geological support of drilling, namely it allows [4]:

1. Dynamic use of seismic data (2D / 3D) with automated rebuilding of a deep-speed model considering geo-navigation data (Fig. 1). At the same time, deep constructions can be verified using depth anisotropic migration before summation (PSDM) implemented in the PetroInfoComplex ProSystem software.
2. 3D geological modeling, which is one of the layers in the geosteering window (Fig. 2).
3. 1D and 3D geomechanical modeling, including considering seismic data. Risk assessment of well drilling considering the stress-strain state of the medium, calculation and change of the well trajectory considering the above factors.
4. The use of real-time attribute seismic analysis according to the data of acoustic impedance, Hilbert transforms, prediction of fault zones and fractures, etc. Moreover, all attributes are calculated directly in the software during geo-navigation.
5. Opportunities for recording and using other data: 4D seismic exploration (if available), electrical exploration and other geophysical methods for optimizing well drilling and several other software capabilities.

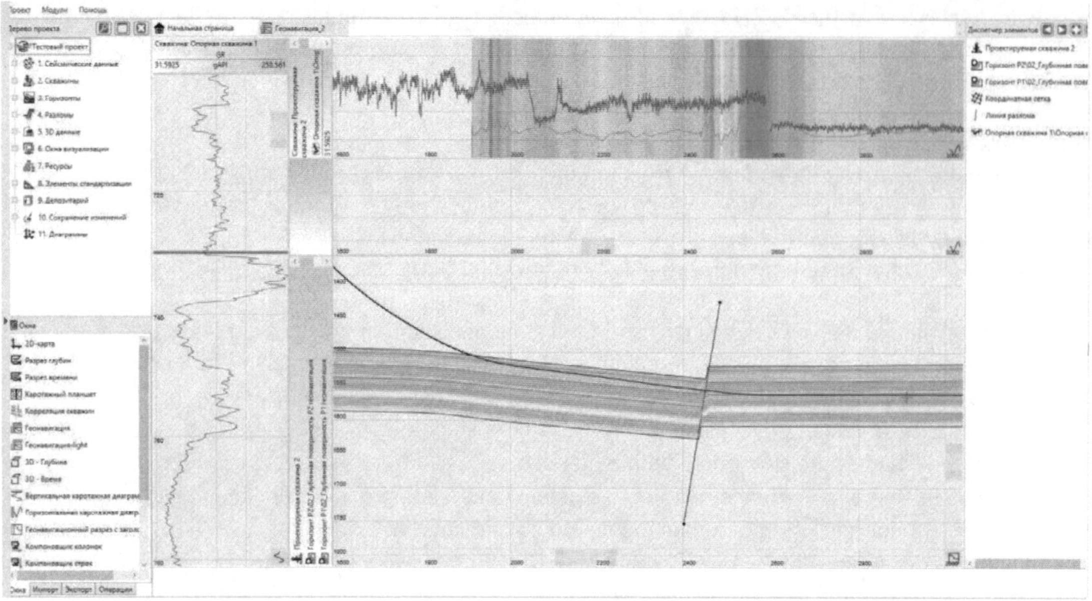

Figure 1. Geosteering in the PetroInfoComplex software along a section with a fault

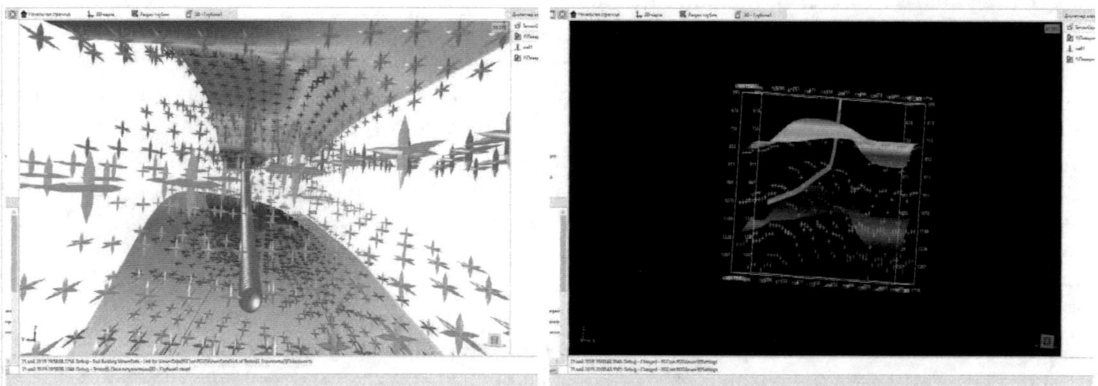

Figure 2. 3D geomechanical modeling in PetroInfoComplex software

4. Conclusions

Thus, the specialists of the geological service of LUKOIL-Nizhnevolzhskneft LLC are successfully implementing high-tech drilling of wells in offshore fields using import-substituting technologies.

The results show a serious development of domestic technologies that can compete with the technologies of the world's leading oil and gas companies. One of the main factors for the further growth of import substitution during the development of offshore fields is the education and training of highly qualified specialists who can develop and implement this area of the oil and gas sector.

To this end, LUKOIL-Nizhnevolzhskneft LLC actively interacts with Astrakhan universities. Cooperation is actively developing with the Astrakhan State Technical University (ASTU), which graduates specialists in the field of applied geology and oil and gas business. Leading specialists of LUKOIL-Nizhnevolzhskneft LLC are involved in the educational process – they teach basic disciplines, conduct seminars, and participate in conferences. Such cooperation of production workers with the educational institution allows us to transfer tremendous experience and knowledge of high technologies to new generations of oil workers.

IOP Conf. Series: Materials Science and Engineering **663** (2019) 012026 doi:10.1088/1757-899X/663/1/012026

Considering the experience gained in the field of geological support for offshore drilling and field development at a high technological level using import-substituting technologies, it is planned to create a basis for training students and graduate students on the basis of ASTU, as well as conduct special advanced training courses for industry specialists. The unique experience of developing offshore fields, which will be passed on to new generations of specialists in the oil and gas industry, will give impetus to the development of this direction for all regions where the development of deposits in offshore and other difficult natural conditions for the development of hydrocarbon deposits is relevant.

Reference

[1] Mandrik I E, Minlikaev V Z, Somov V F, Delia S V, Perminova N N and Krist T M 2008 Improving the technology of development and development of the field to them *Oil industry* **8** 52–4

[2] Golenkin M, Shtun A, Khaliullov I, Sadykov L and Zakharov I 2014 Integration of methods for predicting and identifying faults and fracture zones in order to reduce risks when crossing wells with extra-long horizontal sections (6000 m) Korchagin Field *Rus. Technical Oil and Gas Conf. and SPE Exhibition for Exploration and Production* (Moscow)

[3] Bogdanov S, Delia S, Lacin D, Akhmetov M, Gagaev Yu and Udodov A 2012 Drilling wells with large world-class waste in the northern part of the Caspian Sea *Rus. Oil and Gas Techn. Conf. SPE*

[4] Valisevich A, Shtun S, Zvyagin V, Golenkin M, Sadykov L, Ocheretyany A, Khakov A and Golubtsov R 2015 Evolution in the application of geomechanics at the field named after Yuri Korchagin, North Caspian Sea *Caspian Oil and Gas Techn. Conf. SPE*

Application of a smart active adaptive system on the basis of renewable energy sources for electricity supply to an agricultural facility in highland regions of the Caucasus

A R Elbazurov, M I Malsagov and G R Titova

Moscow Power Engineering Institute (National Research University), 14, Krasnokazarmennaya Street, Moscow, 111250, Russia

E-mail: essame@yandex.ru

Abstract. This paper presents the collated results of the research conducted by the authors and made in the electric system of an agricultural holding. The article is devoted to the development of a smart electric grid which provides energy efficient electricity supply to responsible consumers of an agricultural facility. The research shows the main components of the smart electric system, their operation mode, their role in enhancing energy conservation and efficiency. The authors prove the efficiency of smart energy systems and renewable energy sources in electricity supply to an agricultural facility in the Chechen Republic.

1. Introduction

The technologies and principles for the distribution of generation have been getting more and more important for the last 20 years due to the increasing demand for electricity conservation and efficiency along with reduction of electricity component in the cost of industrial production. Nowadays the systems of distributed generation are the most efficient addition the centralized electric grid in case it is not reliable enough to supply power to the most responsible consumers. Taking into account abundant sources of wind energy in highland regions of the Chechen Republic, which lack centralized power networks and have no reliable electricity supply, using local sources of energy is highly recommended.

All over the world the distribution of electric generation aims at the decentralization of electric power systems along with using local energy sources and enhancing efficiency of electric supply to industrial, agricultural and housing facilities by approaching of the source to the consumer in case of lack or remote location of the centralized energy grid [1].

Taking into account the fact that the power system of the Chechen republic is undergoing reconstruction there are a number of problems which must be solved. They are the scarce of generation facilities, problems with reliability requirements fulfilment due to the deterioration of the significant part of the equipment and lack of processes automation. About one third of substations 35, 110KW, have been put into operation on the basis of a single-phase voltage transformer with no possibility of power supply redundancy. This condition of the electric power system along with the rise of energy consumption cannot provide reliability and quality of power supplies for the consumers.

According to the authors the problems mentioned above, in particular, the scarce of generation facilities and unreliability of electricity supply can be solved by creation of an autonomous micro-energy system which operates a smart electricity system with an active adaptive network on the basis of renewable sources of energy such as the wind and the sun for electricity supplies to an agricultural holding.

2. Smart electric power system with an active adaptive network

Nowadays in all the developed countries of the world much attention is paid to electric power systems which use the newest equipment and technologies providing reliability and cost-effectiveness of electric systems.

Undoubtedly it is an acute task for the Russian electric power industry. The electric grid as a structure connecting the generation and the consumers, plays the key role in the renovation of electric power system on the basis of new principles. Due to the new technologies used in electric grids based on equipment features adjustment to the operating mode, active interaction between the generation and the consumers we can create an efficiently operating system. The integral part of this system are modern information diagnostic systems, control automation systems with all their components participating in production, transmission, distribution and consumption of electric power.

What is meant here is the creation of a smart electric power system with an active adaptive network (SEPS AAN) which is the most advanced system based on multiagent control of its operation and development. The purpose of this system is the provision of the efficient consumption of all types of resources (natural, human, social, productive) through the flexible interaction of its components (all generation types, electric grids and consumers) on the basis of modern technologies and a single smart control system [2].

The differences of this system from the traditional one are the following [3]:

- a number of active components which allow altering technological network settings;
- a big number of detectors measuring the current operating settings necessary to evaluate the network condition in normal, pre-fault, emergency and post-fault operating mode;
- the system of data collection, transmission and processing and the programs of adaptive control which can immediately influence the active network components and the electric devises of the consumers real-time;
- the presence of the necessary control elements and devices which can real-time immediately alter the topological network settings and also influence the related power assets (generation and consumption);
- a controlling system providing interaction between the network and the power generating facilities and allowing to react instantly to any changes in the operating mode of the electric power system;
- the ability of immediate evaluation of the current situation and forecast of the future situations in the whole power system and its parts and also influence to the power assets and equipment to prevent or contain electricity supply in case of emergency and post-fault recovering;
- a high operation speed of the controlling system and data exchange, assuring cycle control of the system's condition, its parts and components with different timing cycles at different control levels.

The priority of the system's factors and conditions as well as the system's reliability and cost-effectiveness as a whole are the basis of an active adaptive network development and power system control.

Before developing a power layout with an active adaptive system on the basis of renewable sources of energy for an agricultural facility, we will prove the advantages of using renewable energy sources in the Chechen Republic

3. Climatic and energy characteristics of the sun in the Chechen republic

The location of the Chechen Republic between 42 and 46 degrees north latitude provides intensive inflow of solar radiation. Solar energy resources, expressed in terms of radiation balance, in plain and foothill areas consist 50 to 55 kcal/cm^2 per year. The duration of insolation is about 330 days per year. Days without sun are rare – 34 to 40 days in plain and foothill areas and 10–12 days in mountainous areas. The largest number of dull days is 61 in the plain areas. As a whole during a year the cloud cover reduces the inflow of direct radiation for 20 to 25 % of the potential.

The total radiation is the sum of direct and scattered radiation on a horizontal surface. The total radiation in the Chechen Republic peaks from May to July. The total radiation intensity varies between 280 and 300 MJ/m^2 in the foothill areas. In the mountainous areas it ranges from 360 to 400MJ/m^2 [4].

The NASA database provides the average monthly size of the total solar radiation falling on a horizontal surface on the ground surface for each month for a period of 25 years from July 1993 to June 2018, Table 1. Each month the average value is the arithmetic mean of 3 hour values for that month. The values are given for the village of Shelkovskaya in the Chechen Republic with an area of 167.21 km^2.

Table 1. Long-term (25 years) monthly average values of insolation falling on a horizontal surface (kWh/m^2 per day)

Month	I	II	III	IV	V	VI	VII	VIII	IX	X	XI	XII	Year
E^y_g	1.61	2.36	3.30	4.51	5.67	6.07	5.92	5.05	3.94	2.69	1.67	1.30	3.67

The advisability of combined power supplies can be shown using the example of the power supplies to a cattle-breeding complex for 800 head with the total peak load of 257,6 kWh and an annual power consumption of 875,84 MWh.

The average value of solar energy gross potential per year is calculated on the following formula:

$$E_g^y = \sum_{k=1}^{d} E_{g\,k}^y \,/\, d \tag{1}$$

d – number of estimated months, E^y_g – inflow of solar radiation on a horizontal ground for k – a month.

The solar energy gross potential in the Chechen Republic is estimated at 3,67 kWh/m^2 per day:

$$E_g^y = (1.61 + 2.36 + 3.3 + 4.51 + 5.67 + 6.07 + 5.92 + \\ +5.05 + 3.94 + 2.69 + 1.67 + 1.3)/12 = 3.67 \text{ kW} \cdot \text{h/m}^2 \cdot \text{d} \tag{2}$$

Then the value of an annual solar energy gross potential is:

$$E_g^y = 3.67 \cdot 365 = 1340 \text{ kW} \cdot \text{h}/\left(\text{m}^2 \cdot \text{year}\right) \tag{3}$$

The solar energy gross potential in the village of Shelkovskaya taking into account its land area is:

$$E_g^y = E_g^y \cdot S \Rightarrow E_g^y = 1340 \cdot 167.21 \cdot 10^6 = 224.061 \cdot 10^9 \text{ kW} \cdot \text{h/year}. \tag{4}$$

These calculations show that the village of Shelkovskaya has a significant solar energy potential which can be used as a source of electricity for a cattle-breeding complex power supplies.

4. Climatic and energy characteristics of the wind in the Chechen republic

According to the wind stream gross potential the Chechen Republic has a medium level of wind energy.

For efficient utilization of wind energy it is necessary to have full information about the wind as a natural phenomenon and a source of energy. The environmental characteristics needed to estimate wind energy gross potential are [5]: wind speed temporal variations; an average wind speed V; wind speed recurrence t (V), %; wind speed duration P(V), %; correction factors which take into account changes in the wind on the territory due to heterogeneity of the underlying surface; average vertical wind speed profile; top wind speed $F^e_{max.}$; specific power N_s and specific energy E_s of the wind, wind energy resources of the region.

According to the NASA database the multi-year average wind speed in the village of Shelkovskaya is 4.67 m/s at the altitude of 10 m and 5.91 m/s at the altitude of 50 m, table 2.

Table 2. Long-term (10 years) monthly and multi-year average wind speed at the altitude of 10 and 50 meters over a landscape like an airport

Month	I	II	III	IV	V	VI	VII	VIII	IX	X	XI	XII	Year
$\overline{V},\overline{V}_0$ (10m)	5.11	5.14	4.79	4.59	4.43	4.12	4.26	4.41	4.51	4.61	4.93	5.17	4.67
$\overline{V},\overline{V}_0$ (50m)	6.47	6.50	6.06	5.81	5.61	5.22	5.39	5.58	5.72	5.84	6.24	6.55	5.91

According to the multi-year average wind speed values at the altitude of the weathervane at the wind speed V>5 m/s there are good conditions for wind energy utilization, V<4 m/s utilization of wind energy isn't recommended, 4<V<5 m/s a business case for wind energy utilization is needed [6].

Along with the multi-year average wind speed the specific power of the wind flow is also an original characteristic of the wind intensity total level in the estimated region. It allows indicating the prospects of wind resources utilization: with specific power less than 100 (N_s<100 W/m^2) it is not recommended, more than 400 (N_s >400 W/m^2) there are good conditions, 100 < Ns < 400 W/m^2 – a business case is needed.

Let us calculate the specific power and the specific energy of the wind flow at the altitude of 50 m to estimate the prospects of wind energy utilization.

In the calculation of the specific power by different means will take into account that the air density does not change and equates ρ_0=1,226 kg/m^3.

1) Ns is growing according to the multi-year wind speed value V_0:

$$N_s = \frac{1}{2}\rho V_0^3 = 126{,}54 \text{ W/m}^2 \tag{5}$$

2) Calculation of Ns according to a range of speed observations Vi (a range of 3 hour observation):

$$E_S = \sum_{i=1}^{8760} N_{Si} = \frac{1}{2}\rho \sum_{i=1}^{8760} V_i^3 \Rightarrow N_S = E_S/(8760/3) = \sum_{i=1}^{2920} N_{Si}/2920 =$$
$$= (\frac{1}{2}\rho \sum_{i=1}^{2920} V_i^3)/2920 = 126{,}92 \text{ W/m}^2 \tag{6}$$

3) Calculation of Ns according to multi-year recurrence t(V):

$$E_S = \frac{1}{2}\rho T \cdot V_j^{gr3} t(\Delta V_j^{gr})\,; \quad N_S = \frac{1}{2}\rho V^3. \tag{7}$$

The results of the calculation are shown in Table 3.

Summing the specific energy values E_{sj} in the full range of the observed speeds we calculate the multi-year annual specific energy E_s and the specific power value N_s of the wind flow:

$$E_{Sj} = \sum E_{Sj}; \quad N_{Sj} = \frac{E_S}{T} = 230 \text{ W/m}^2 \tag{8}$$

Table 3. Results of the specific power and specific energy calculation

V^{gr}_j	1	4,5	8,5	12,5	16,5	22
$t(\Delta V)^{gr}_j$	10	52	31	7	0	0
N_{sj}	0,61	55,86	376,46	1197,27	2753,67	6527,22
$Э_{sj}$	0,54	254,45	1022,31	734,16	0,00	0,00

According to these calculations the power values range from 126 to 230 W/m^2. For the following calculations we will take the lowest value N=126,54 W/m^2.

The power generated by a wind plant Fuhrländer FL100 with a rated power of 100 kW is calculated as follows:

$$N_{WPP}(\upsilon_{hb}) = N'_{g.p.}(\upsilon_{hb})\eta_{ww}\eta_r\eta_{gen} = 0{,}125\rho\pi D_{ww}^2 \upsilon_{hb}(t)^3 \eta_{ww}\eta_r\eta_{gen} \tag{9}$$

were
$$N^{'}_{g.p.}(\upsilon_{h\delta}(t)) = 0,5\rho F_{ww}\upsilon_{h\delta}(t)^3$$
$$F_{ww} = 0,25\pi D^2_{ww} \quad (10)$$

Electricity generation by a wind plant in this region according to recurrence of the wind speed for period T is calculated as follows (kWh):

$$E_{WPP}(T) = \sum_{j=1}^{N^{gr}} \left[N_{WPP}(V^{-h\delta}) t(\Delta V_j^{gr}) \right] T \quad (11)$$

The calculation of electricity generation by a wind plant fuhrländer fl100 shows the result 294443 kWh.

The gross potential of the wind flow in the village of Shelkovskaya:

$$E_g = E_g \cdot \frac{S}{20} = N_g \cdot T \cdot \frac{S}{20} = 126,54 \cdot 8760 \cdot \frac{167,21 \cdot 10^3}{20} = 9267,534\ \text{MW} \cdot \text{h/km}^2 \cdot \text{year} \quad (12)$$

The annual electricity consumption of a cattle-breeding complex for 800 head:

$$W_y = P_{\max} T = 257,6 \cdot 3400 = 875840\ \text{kW} \cdot \text{h} = 875,84\ \text{MW} \cdot \text{h} \quad (13)$$

where T = 3400 h. is the number of hours at the peak load [7].

These calculations show that the village of Shelkovskaya has a significant wind energy potential which can be used as a source of electricity for a cattle-breeding complex power supplies.

5. Power supply scheme based on smart electric power system with an active adaptive network

A power layout of an agricultural facility with an active adaptive network based on the combined sources of energy which was developed on the basis of the calculations and the relevant electric installations (Figure 1).

This power layout works as follows. In a normal operating mode all three divisions of the facility get electricity from a wind power plant (WPP) and energy accumulates in a storage battery (SB). In case the weather is bad for the WPP the detectors of the wind velocity send a signal to the control element, "the brain" processes this data and reacting to the active components switches off the WPP. Using the special detectors the system checks the level of insolation for solar power redundancy. These actions come instantly. If the weather does not suit the solar power plant the controlling system switches on the diesel power plant (DPP). The DPP is switched on without delay because there are diesel dynamic uninterruptible power supply sources installed (DDUPSS). After some period of time when the sun comes out if there are no wind the relevant detectors monitoring the condition and the work schedule of the DDUPSS send a signal to the control element. In its turn the controlling system transfers the 3rd category load to the supply from the SB controlling the voltage of the SB. Simultaneously the system switches the 1rd and 2rd category load in the solar power plant (SPP) which can get electric power directly from the SPP or the SB where the power has been accumulated from the SPP. All the commands mentioned above are executed through the smart active adaptive controlling system by means of the related detectors and active components.

The efficiency factors during the construction of SEPS AAN: demand management and loss reduction with the full elimination of commercial loss will reduce electricity consumption at least at 5%; active adaptive regulation of consumption mode will reduce the growth of the peak electrical demand reducing the required power input of power plants; significant capacity growth of power transmission lines and cross sections reducing the necessity of investing money in power lines construction; electricity supply reliability growth at the expense of preventive and adaptive management SEPS AAN and its elements will multiply reduce the damage from the system accidents and power supply interruptions.

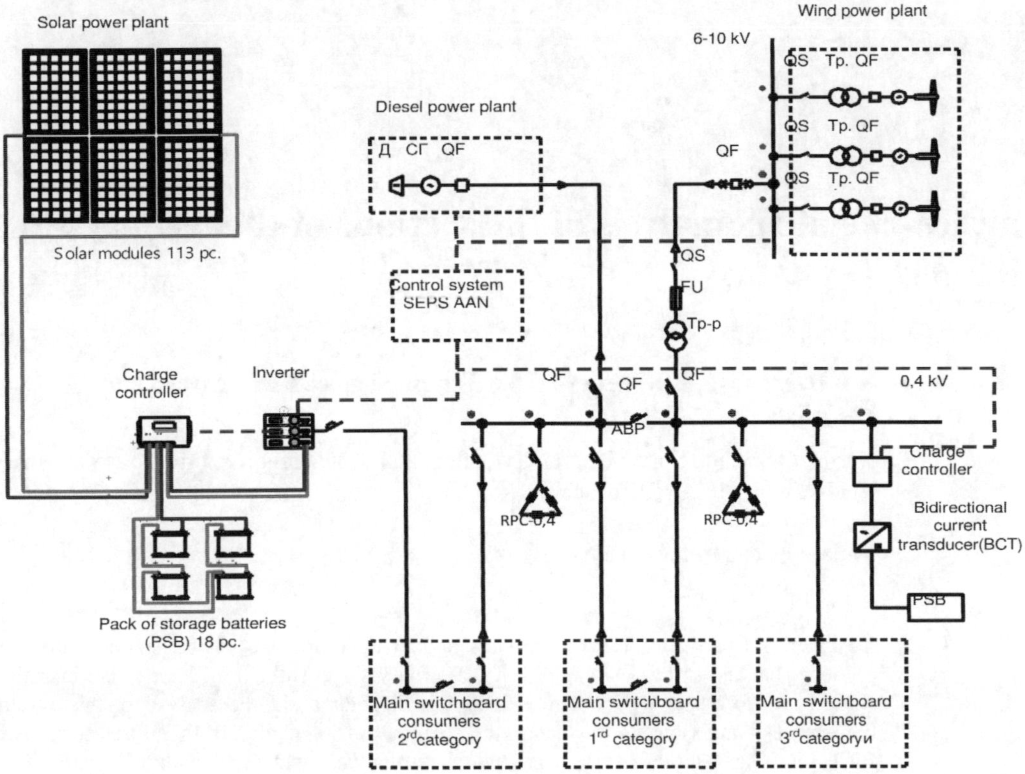

Figure 1. Electricity supply to a cattle breeding facility from a wind, solar and diesel power plants with a smart active adaptive controlling system

6. Conclusion

Therefore by the means of gradual refuse of the centralized power supply, development of distributing generation systems and integrated introduction of alternative and renewable energy sources (wind and solar energy) we should create a flexible cross-functional and efficient smart system of complex monitoring and control of energy sources consumption and quality of electric power, which plays the key role in energy efficiency and conservation.

References

[1] Martinez-Velasco J A and Martin-Arnedo J 2007 *Distributed Generation Impact on Voltage Sags in Distribution Networks* (Barselona) pp 9–10
[2] Berdnikov R N, Bushuev V V and Vasiliev S N 2012 *The concept of an intelligent power system with an active-adaptive network* (Moscow: Federal grid company of the Unified electric Power system) 238 p
[3] Dorofeev V V and Makarov A A 2009 Active-adaptive network – new quality of the UES of Russia *Energyexpert* **4** 29–34
[4] Kerimov I A, Debiev R A-M and Magomadov H I 2007 Resources of solar and wind energy of the Chechen Republic *Engineering Herald of the don* **1** p 11
[5] Deryugina G V, Malinin N K and Pugachev R V 2012 *The main characteristics of the wind. Wind resources and methods of their calculation* (Moscow: MPEI publishing house) 257 p
[6] Bezrukikh P P, Arbuzov Yu D and Borisov G A 2002 *Resources and Renewable Energy Efficiency in Russia* (St. Petersburg: Science) 314 p
[7] Budzuko I A 2000 *Electricity supply for agriculture* (Moscow: Kolos) 536 p

IOP Conf. Series: Materials Science and Engineering **663** (2019) 012028 doi:10.1088/1757-899X/663/1/012028

Physico-chemical analysis of the sections of the ternary $BiCl_3$-LiCl-AgCl systems

K B Dzeranova, L K Yesieva, I Ya Ter-Barsegova, F A Agayeva and I M Bigayeva

North Ossetian State University after K.L. Khetagurov, 44-46, Vatutina Str., Vladikavkaz, 362025, Russia

E-mail: esieva.mila@yandex.ru

Abstract. Recently, the use of halogens and their compounds in industry and the national economy has increased. The halides of the ternary systems Ag – Bi – Li of are insufficiently studied. So, to create semiconductor materials on their basis, the analysis of this system has become very important. In this paper, the character of interaction of the binary sections in the $BiCl_3$-LiCl-AgCl ternary system is studied through a complex of methods of physical and chemical analysis, precisely by differential thermal and X-ray phase analyses. The analysis of the obtained data allowed to establish the eutectic character of interaction in the LiCI-$AgBiCI_4$system. It was confirmed that the addition of $AgBiCI_4$ lowers significantly the melting point of the second component. Also, it was defined as the eutectic character of interaction with limited solubility of the components in the $AgBiCI_4$-$LiBiCI_4$ and $AgBiCI_4$-$LiBi_4CI_{13}$ systems.

1. Introduction

The development of physical and chemical bases of halogenation is explained by the increasing role of halogens in various manufacturing processes of chemical raw materials to obtain pure and superpure materials. Halides of various metals appear to be non-conventional materials in this field.

2. Problem statement

The aim of this paper is to analyse binary sections of LiCl-$AgBiCl_4$, $LiBiCl_4$-$AgBiCl_4$, $LiBi_4Cl_{13}$-$AgBiCl_4$ in the $BiCl_3$-LiCl-AgCl ternary system using the complex methods of physical-chemical analysis: differential thermal and X-ray phase analysis [1, 2].

3. Scientific novelty

A quasi-binary section of LiCl-$AgBiCl_4$ and polythermal sections of $LiBiCl_4$-$AgBiCl_4$, $LiBi_4Cl_{13}$-$AgBiCl_4$ in the $BiCl_3$-LiCl-AgCl ternary system are studied firstly.

According to the literature data [3, 4, 5] the LiCl – AgCl binary system is characterized by a number of solid solutions; the $BiCl_3$-AgCl system is characterized by the formation of the $AgBiCI_4$ chemical compound at a ratio of 1:1; the $BiCl_3$ – LiCl system is characterized by the formation of chemical compounds at a ratio of 1: 1 and 1: 4, i.e. the $LiBiCL_4$ and $LiBi_4CI_{13}$ compounds.

4. Materials and methods

To analyse the section of the LiCI-$AgBiCl_4$ compound the alloys of chemical compounds or solid solutions were taken.

Pre-weighed samples (3.0000 g) were thoroughly mixed and homogenized to bring into equilibrium at temperatures below the melting temperatures of the respective compositions and time regimes. The mixture of chlorides were vacuumed and sealed in the molybdenum vessels of Stepanov (5 cm^3) and annealed to provide the equilibrium state. The vessels were then cooled to room temperature in a furnace.

Differential thermal analysis was carried out on a thermal analyzer (DTA-850), the heating rate of the samples was 1–25 deg/min. The standard was calcined aluminum oxide, HC model number. To stabilize the temperature, the required time was 10 minutes, after which the study began.

X-ray phase analysis was performed on XRD-7000 Maxima diffractometer manufactured by Shimadzu with vertical goniometer by a continuous flow method. The obtained diffractograms were identified and compared with the database of X-ray card index of PDF-2 2012.

The measurement process was completely controlled by the computer. Using the software, the measurement results were subjected to various profile processing (smoothing, background subtraction), peak parameters processing (peak search, systematic error correction, internal /external standard method).

The accuracy in determining the reflection angles $\Delta\ \theta$ did not exceed 0.02 °C. The hygroscopic samples were protected from air contact by X-ray amorphous film. The original components were identified using X-ray card index ASTM, interplanar distances were found in the tables of Hiller J.L.

5. Discussion of results

The alloys from the area of formation of solid solutions or chemical compounds of the $AgBiCl_4$ – LiCI system are of practical interest as solid electrolytes [6-11]. According to the results of DTA a polythermal section was constructed. The melting temperature of the primary components are respectively equal to 100°C ($AgBiCl_4$) and 610 °C (LiCI). The following phases are found on the diagram of the analysed system: L, three double phases: L+ LiCI, L+ $AgBiCl_4$, LiCI + $AgBiCl_4$, two triple L +LiCI + AgCl, L+ AgCl + $AgBiCl_4$ phases. The analysis showed that the addition of LiCl increases the melting point of the alloys, while the addition of $AgBiCl_4$ lowers the melting point of the alloys; and this corresponds to the 1st Gibbs-Konovalov law. Interplanar distances are gradually increased for alloys, hence the solid phase of the analysed system is relatively rich in LiCl.

The XRPA results confirm the presence of primary crystallization fields.

In the analysed $AgBiCl_4$ – $LiBiCl_4$ system a polythermal section is firstly built, in which there are no solid-phase transformations at the temperature below the solidus level.

The system has three single-phase areas: L, α – and β-solid solutions based on the initial components, there are three double areas: L + α, L + β, α + β and the area of L + α + β non-variant equilibrium. The solidus curve was calculated theoretically by the Schroeder – Le-Chatelier equation, and the experimentally obtained data are in good agreement with them. The analysis of XRPD data confirmed the absence of solid-phase transformations, which is a consequence of a monotonic change in the interplanar distance.

in the LiCI-$AgBiCl_4$, $LiBiCl_4$-$AgBiCl_4$ sections of the LiCI-AgCI-$BiCl_3$ system the analysis of peak intensities showed that the transition from 100 % of the $AgBiCl_4$content to 100% of LiCl content in the alloys of the intensity and interplanar distances change monotonically, indicating the preservation of the structure of the samples (tabl. 1–2).

IOP Conf. Series: Materials Science and Engineering **663** (2019) 012028 doi:10.1088/1757-899X/663/1/012028

Table1 The results of X-ray phase analysis of the LiCI – $AgBiCI_4$ section of the alloys

The contents of $AgBiCI_4$, mol %									
100		75		60		30		0	
d, nm	I	d, nm	I	d, nm	I	d, nm	I	d, nm	I
0.713	100	0.725	93	0.738	92	0.725	100	0.714	4
0.641	9	0.689	33	0.670	3	0.660	8	0.691	4
0.568	5	0.651	30	0.641	4	0.530	12	0.653	5
0.367	76	0.619	28	0.615	4	0.479	10	0.598	13
0.342	50	0.583	48	0.598	4	0.452	12	0.467	9
0.296	43	0.527	15	0.471	4	0.432	8	0.427	9
0.276	88	0.504	15	0.455	3	0.406	12	0.423	8
0.268	74	0.488	18	0.419	2	0.398	10	0.407	5
0.244	57	0.446	18	0.400	4	0.366	46	0.394	6
0.219	17	0.419	48	0.366	30	0.343	39	0.380	15
0.194	40	0.395	35	0.353	3	0.318	14	0.359	5
0.184	26	0.375	18	0.350	3	0.309	12	0.350	5
0.175	5	0.360	23	0.332	6	0.277	39	0.345	5
0.171	8	0.340	23	0.318	36	0.274	19	0.319	4
0.168	14	0.320	40	0.305	8	0.268	80	0.299	10
0.166	40	0.316	25	0.293	4	0.255	12	0.284	4
0.160	40	0.306	15	0.285	4	0.240	17	0.274	4
0.157	22	0.289	23	0.275	100	0.220	14	0.271	100
0.152	17	0.285	25	0.266	13	0.195	23	0.259	11
0.146	12	0.278	100	0.250	6	0.194	25	0.255	8
0.137	14	0.269	40	0.244	15	0.187	19	0.254	9
		0.253	23	0.233	3	0.182	35	0.226	8
		0.239	25	0.227	2	0.176	8	0.221	9
		0.233	13	0.226	3	0.166	42	0.220	10
		0.228	13	0.219	4	0.160	12	0.211	8
		0.216	18	0.216	4	0.157	15	0.207	6
		0.196	53	0.198	4	0.156	19	0.201	4
		0.191	10	0.195	23	0.152	15	0.200	4
		0.168	20	0.192	3	0.146	15	0.196	5
		0.160	18	0.184	4	0.144	4	0.193	5
		0.145	15	0.183	7	0.140	10	0.192	10
		0.140	15	0.181	5	0.138	12	0.184	5
		0.139	18	0.167	10	0.137	6	0.181	4
		0.137	12	0.166	8	0.136	8	0.180	6
		0.135	12	0.160	5	0.134	8	0.176	8
		0.129	18	0.161	4	0.129	8	0.172	8
		0.124	15	0.157	12	0.126	19	0.170	4

IOP Conf. Series: Materials Science and Engineering **663** (2019) 012028 doi:10.1088/1757-899X/663/1/012028

Table 2. The results of X-ray phase analysis of the $LiBiCl_4 – AgBiCl_4$ section of the alloys

The contents of $AgBiCl_4$, mol %									
100		80		50		20		0	
d, nm	I	d, nm	I	d, nm	I	d, nm	I	d, nm	I
0.569	5	0.661	7	0.660	9	0.660	17	0.542	15
0.367	76	0.640	10	0.640	9	0.640	8	0.511	15
0.342	50	0.610	10	0.619	6	0.619	7	0.461	100
0.297	43	0.550	10	0.567	9	0.550	12	0.410	5
0.276	88	0.535	10	0.550	9	0.530	7	0.369	40
0.268	74	0.519	10	0.530	15	0.520	7	0.334	20
0.244	57	0.480	12	0.520	9	0.519	8	0.297	5
0.219	17	0.475	7	0.519	6	0.485	7	0.294	10
0.194	40	0.450	17	0.484	9	0.469	34	0,257	10
0.184	26	0.428	12	0.466	12	0.450	82	0.229	20
0.175	5	0.415	17	0.445	9	0.415	8	0.197	15
0.171	8	0.390	17	0.428	9	0.370	8	0.185	15
0.168	14	0.380	7	0.415	9	0.360	24	0.170	5
0.166	40	0.370	7	0.388	15	0.340	33	0.153	5
0.160	40	0.360	10	0.370	9	0.319	82	0.151	5
0.157	22	0.340	12	0.360	9	0.300	17		
0.152	17	0.320	60	0.345	9	0.295	17		
0.146	12	0.289	10	0.330	9	0.289	8		
0.137	14	0.285	7	0.300	100	0.278	100		
		0.278	100	0.276	97	0.260	24		
		0.260	12	0.220	15	0.240	17		
		0.220	17	0.190	51	0.209	8		
		0.190	62	0.165	12	0.190	82		
		0.179	17	0.160	18	0.185	17		
		0.165	17	0.158	12	0.165	24		
		0.160	18	0.150	15	0.155	7		
		0.140	17			0, 140	17		

The melting temperatures of the primary components are respectively equal to 100°C ($AgBiCl_4$) and 210 °C ($LiBi_4Cl_4$).

The analysis of the results of XRPA for the samples section of $LiBi_4Cl_{13}$-$AgBiCl_4$ confirmed that the intensity of the reflexes corresponding to the solid solution based on $AgBiCl_4$ decreases with decreasing the relative content of this phase (tabl. 3).

Table 3. The results of X-ray phase analysis of the alloys of the $LiBi_4CI_{13}$ – $AgBiCL_4$ section

The contents of $AgBiCL_4$, mol %									
100		90		50		20		0	
d, nm	I	d, nm	I	d, nm	I	d, nm	I	d, nm	I
0.730	100	0.738	100	0.738	100	0.738	100	0.734	80
0.367	76	0.659	15	0.638	12	0.658	24	0.555	20
0.342	50	0.51	25	0.625	15	0.634	12	0.514	10
0.316	48	0.641	18	0.603	12	0.608	12	0.490	100
0.297	43	0.615	25	0.590	76	0.598	24	0.467	80
0.276	98	0.580	12	0.550	45	0.550	6	0.428	12
0.268	81	0.576	33	0.503	18	0.525	5	0.380	20
0.266	74	0.528	24	0.450	18	0.490	14	0.370	20
0.244	57	0.515	13	0.441	15	0.467	7	0.336	30
0.219	17	0.462	25	0.434	36	0.443	12	0,317	20
0.194	40	0.436	52	0.423	12	0.423	10	0.260	26
0.184	26	0.445	20	0.374	15	0.370	6	0.244	20
0.175	5	0.400	38	0.342	48	0.350	50	0.235	25
0.171	7	0.343	46	0.318	15	0.342	35	0.229	40
0.168	14	0.319	21	0.279	30	0.318	27	0.215	90
0.166	40	0.379	42	0.271	88	0.279	7	0.210	80
0.160	40	0.249	38	0.263	9	0.271	21	0.197	10
0.157	16	0.226	21	0.249	12	0.256	32	0.187	10
0.152	22	0.225	18	0.242	25	0.249	8	0.185	15
0.146	17	0.220	13	0.236	19	0.236	5	0.167	8
0.137	12	0.197	17	0.215	21	0.220	4	0.162	2
		0.167	17	0.210	12	0.197	5	0.158	8
		0.145	17	0.197	18	0.169	10		
		0.130	25	0.185	12	0.166	6		

6. Conclusion

The $AgBiCL_4$ – $LiBi_4CI_{13}$ system was analysed firstly, and a polythermal section was built. The melting temperatures of the initial components are respectively equal to 100°C ($AgBiCL_4$) and 224 °C ($LiBi_4CL_4$). The system has three single-phase areas: L, α – and β-solid solutions based on the initial components, two L + β, L + α double areas and a L + α + β triple area. The analysis of XRPA results showed that the intensity of reflexes corresponding to $AgBiCL_4$-based on the solid solution decreases with decreasing relative content of this phase.

In the analysed system it was made the projection of the liquidus surface of the $BiCl_3$-LiCl-AgCl ternary system. It consists of four fields of primary crystallization phases: $BiCI_3$, LiCI, AgCI, $LiBi_4CI_{13}$ and solid solutions based on $AgBiCL_4$ and $LiBiCL_4$.

The XRPA results confirm the presence of primary crystallization fields.

Table 4 presents the non-variant equilibria.

Table 4. Non-variant equilibria

The character feature of the point	t °C	Composition, mole.%			The equilibrium phases
		BiCI	LiCl	$AgCl_3$	
E_1	120.0	13.4	23.3	63.3	L↔ $LiBi_4CI_{13}$ + $AgBiCL_4$ + $BiCI_3$
E_2	100.0	66.6	6.2	27.2	L ↔$AgBiCL_4$ + AgCl + $LiBiCL_4$
E_3	125.0	25.0	41.7	33.3	L ↔$LiBi_4CI_{13}$ +$AgBiCL_4$ + AgCI

IOP Conf. Series: Materials Science and Engineering **663** (2019) 012028 doi:10.1088/1757-899X/663/1/012028

On the basis of experimentally obtained data, the lines of joint crystallization were drawn and the coordinates of non-variant points were determined.

Thus, in the present paper, the phase equilibria in the $BiCl_3$-LiCl-AgCl ternary system were analysed utilizing the methods of differential thermal and X–ray phase analysis. The paper establishes that in the LiCI-$AgBiCL_4$system the eutectic nature of the interaction is observed. The addition of $AgBiCI_4$ significantly lowers the melting point of the second component. The eutectic character of interaction with limited solubility of components in the $AgBiCI_4$-$LiBiCl_4$ and $AgBiCI_4$-$LiBi_4CI_{13}$ systems is determined.

References

[1] Trunin A S 1997 *Complex methodology of the multicomponent systems* (Samara: Samara State Technical University)

[2] Trunin A S and Meshalkin A V 2010 *Methodical manual for maintaining of the unit of differential thermal analysis DTA-500* (Samara: Samara State Polytechnic University)

[3] Dzeranova K B and Trunin A S 2005 *Phase diagrams of systems of Bi (III) halides and halides of the elements of I and II groups of the Periodic Table of D I Mendeleev* (Vladikavkaz)

[4] Kodzasova S A and Dzeranova K B 2016 The $BiCl_3$-AgCl system. In the collection: *Actual problems of chemistry, biology and biotechnology. The Materials of X All-Russian sci. conf.* (Vladikavkaz: North Ossetian State University Publishing house) pp 222–3

[5] Dzeranova K B and Trunin A S 2007 Thermodynamic study of the $BiCL_3$-MCL (M = LI, NA, K, CS) systems in the ideal associated solution model *Inorganic Materials* **43(9)** 1015–7

[6] Dzeranova K B 2007 Phase relations in the BiI_3-ZnI_2 system *Inorganic Materials* **43(10)** 1142–4

[7] Dzeranova K B 2004 Physico-chemical study of the quasi-binary section of $LiZnBiI_6$-ZnI_2 of the BiI_3-LiI-ZnI_2 ternary system *The Proceedings of higher educational institutions. The North Caucasus. Ser. Natural sci.* **51** 35–7

[8] Dzeranova K B 2004 Thermal analysis of binary system of Li and Zn iodides. *The Proceedings higher educational institutions. The North Caucasus. Ser. Natural Sci.* **54** 60–3

[9] Dzeranova K B 2004 The prediction of phase formation in the M(I)G-BIG_3systems *Izvestia of Rus. State Pedagogical University named after A I Herzen* **4(8)** 94–9

[10] Kaloev N I, Kubalova L M, Dzeranova K B, Toporovskaya E A and Paramazova S E 1998 Thermodynamic investigation of the binary BiI_3-RbI, BiI_3-CsI, and BiI_3-TiI systems in the approximation of the ideal associated solutions *Rus. J. of Inorganic Chemistry* **43(1)** 125–7

[11] Dzeranova K B 2007 Interaction in the BiI_3-ZnI_2 system *Inorganic materials* **43(10)** 1270–2

[12] Dzeranova K B 2004 The analysis of physico-chemical interaction in the melted materials of Bi iodides with iodide of Zn. *SPb Izvestiya RSPU after A I Herzen* 23

IOP Conf. Series: Materials Science and Engineering **663** (2019) 012029 doi:10.1088/1757-899X/663/1/012029

Analysis of the use of downhole pumping equipment (DPE) at the Tevlinsko-Russkinskoye field

M D Gadzhiev

Tyumen Industrial University, 38, Volodarsky St., Tyumen, 625000, Russia

E-mail: gadzhievmd@tyuiu.ru

Abstract. The problem of operating ECP in the zone of low production rates is very relevant for a number of leading oil companies, since most of the wells are operated by such pumps. The adopted values of the boundaries of the working part of the characteristics of the pump are indicators: 0.75–1.25. According to preliminary estimates, 31 % of the ECP at the Tevlinsko-Russkinskoye field are operating outside the operating range, which leads to a decrease in the operating efficiency of downhole pumping equipment (DPE). The average well production rate for oil was 8.6 tons per day, for liquids – 89.5 tons per day, for water content – 90 %. This article presents an analysis of the ECP operation in the Tevlinsko-Russkinskoye field. The work describes the complications of the DPE when it is used beyond the right and left borders of the working area. On the basis of pilot field works, the possibility of using alternative DPE has been described. At the end of the article highlighted the main problems relating to the operation of wells and equipment.

1. Introduction

The choice of a pumping unit for the operation of marginal wells is not always as obvious as it would seem. It is considered, for example, that the use of bottom hole pump on this fund is more preferable, since they have higher (compared to electric centrifugal pump) efficiency and are characterized by relatively low power consumption at low feed rates.

Unfortunately, in the last 2 decades, subsoil users in Western Siberia in the operation of oil wells with flow rates from 10 to 30 m^3/day. and the decision fully gave preference to installations of electric centrifugal pumps (ECP). In this connection, with the high water cut of these wells, the low efficiency of the ECP of low productivity and the low cost of oil on the world market, these wells become unprofitable. [1]

2. A brief analysis of the mining well stock

As of January 1, 2017, the production fund of the Tevlinsko-Russkinskoye field amounted to 1,663 wells, of which: ECP – 1636 (98 %), submersible pump installation (SPI) – 14 (1 %), while separate production (WSP) – 13 wells (1 %).

In the operating fund there are 1564 wells, in the idle – 27 (1.6 %), in inactivity – 97 (5.8 %).

By the end of 2016, the operating fund of the Tevlinsko-Russkinskoye field amounted to 1,560 wells. In comparison with the beginning of the year, the number of wells equipped with ESP systems increased from 1540 to 1545, in turn, the number of wells operated by the SPI decreased (from 6 to 2).

Technological indicators at the end of the year: the flow rate of oil and liquids decreased (by 0.57 and 2.97 tons/day, respectively), the water content of the production increased (by 0.97 %).

3. Operation of wells using ECP

Almost all oil production in the Tevlinsko-Russkinskoe field is provided by electric centrifugal pump – 4,717 thousand tons (99 % of the total oil produced in 2016).

Figure 1 presents an analysis of the work of the DPE of the existing stock of wells operated by ESPs, which revealed:

- 69 % or 1064 wells are operated within performance limits;
- 12 % or 192 wells behind the right working area;
- 19 % or 293 wells behind the left work area.

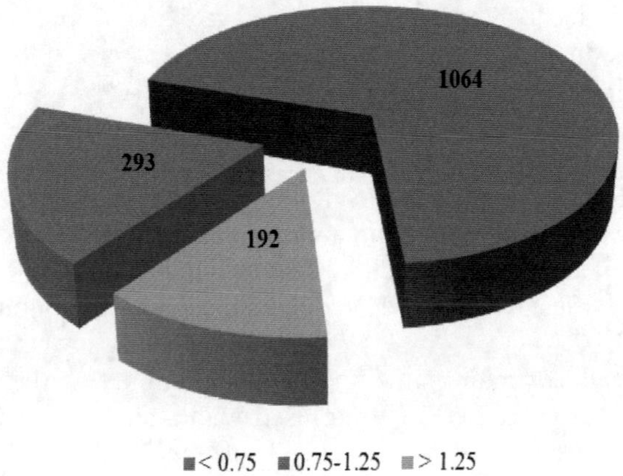

Figure 1. Distribution of existing production wells (number) along the boundaries of the ECP working area

3.1 Operation of the pump beyond the right edge of the working area

The pump operation beyond the right edge of the working area occurs with reduced efficiency, as well as with large wear of the upper textolite washers of the impeller. This, in turn, sooner or later leads to their abrasion and friction of the metal on the metal, accelerates the wear of working bodies and increases the likelihood of their jamming and breaking the shaft.

3.2 Operation of the pump beyond the left edge of the workspace

The left boundary of the working area is determined by the flow, below which the head begins to drop. The service life of the lower support bearing is significantly reduced by the increasing axial force acting on the impeller. Cooling conditions deteriorate as the pumping rate decreases. Heating of the pumped liquid in an electric motor can reach significant (critical) values. The likelihood of premature failure of the ESP increases due to the lack of necessary cooling of the installation units and, as a result, a decrease in cable insulation (R-0).

Excessive heating contributes to the formation of scaling, which also causes a further reduction in supply and increases the likelihood of seizure of the working bodies. Operation beyond the left boundary may lead to unstable operation, stopping and restarting equipment, thereby increasing the load on the insulation of the SEM and cable, increased consumption of hydraulic oil, the risk of contamination by mechanical impurities, etc.

In order to avoid premature failures of deep hole pumps, in cases of reducing the dynamic level, it is recommended to carry out measures to limit fluid withdrawal (fitting wells, reducing engine speed), reducing the size of the pump, changing the mode of operation (from ECP to bottom hole pump), or lowering the ECP of a larger size with the short periodical work mode.

IOP Conf. Series: Materials Science and Engineering **663** (2019) 012029 doi:10.1088/1757-899X/663/1/012029

4. Analysis of the work of the periodic wells

As of January 1, 2017, 233 production wells (15 % of the total operating ECP fund) of the Tevlinsko-Russkinskoye field are operated in a periodic mode of operation.

The considered wells are operated mainly by domestic-made ECPs (229 wells), imported ones – TDK (4 wells), with a nominal capacity of 15 to 160 m^3/day.

One of the reasons for the withdrawal of DPE from continuous operation were:

- low coefficient of productivity (on average 0.23 m^3/day.atm);
- reduction of reservoir pressure (on average by 28 % of the initial);

4.1 Methods of operation of the periodic wells

Automatic reclosing mode (automatically reclosing). The automatic reclosure mode can lead to a large amplitude of bottomhole pressure, a decrease in the volume of production in comparison with continuous operation. This is due to an increase in the average dynamic level of the reservoir fluid above the pump intake, a decrease in depression to the reservoir and, as a result, a reduction in the flow of reservoir fluid into the well.

109 wells are in operation in the field in the AR mode. On average, in one cycle, the operation time was 1.1 hours (65 minutes), the accumulation was 2.5 hours (148 minutes). At the same time, the operational indicators are as follows: dynamic level – 1991 m (the depth of the DPE descent is 2347 m), liquid flow rate – 17 m^3/day, oil – 4.6 t/day, with a water content of 59 %.

Short re-work mode. The optimal selection of operating times and well shutdowns in the accumulation, does not significantly reduce the dynamic level, contributing to the preservation of continuous fluid withdrawal.

In this mode, on average, the operating time is 0.3 hours (20 minutes), the accumulation is 0.6 hours (36 minutes), 124 wells are in operation. ECP are used with a capacity from 25 to 125 m^3/day. The average parameters of the wells: the dynamic level of 1993 m, the flow rate of the liquid 20 m^3/day, oil 4.8 tons/day, the water-cut rate was 68 %.

5. Application of alternative DPE in the fild

According to generally accepted criteria for the selection of the method of operation of production wells, the most optimal, for objects with low productivity Tevlinsko-Russkinskoe oil field, is the use of centrifugal pumps [2–4]. Alternative mining methods are difficult:

5.1 Use of electric screw pumps

Pilot field tests of screw pumps complete with a control station manufactured by Netzsch Pumps Rus in 2015 were carried out at two wells in the Chamber of Commerce and Industry of Povneftegaz. As a result, in the first well, the UEVN did not fulfill the warranty period (24 hours), in the second one it was 5 days (the well did not enter the mode). Both failures were due to pump seizure due to elastomer swelling. In other words, the cage swelling is caused by an incorrect selection of the type of elastomer (in particular, the content of aromatic hydrocarbons in borehole fluids is not taken into account).

Screw pumps, in wells, work in harsh conditions under the influence of many adverse factors: aromatic compounds, high temperature and pressure, abrasive mechanical impurities, gas (including hydrogen sulfide and carbon dioxide). In this connection, increased requirements are imposed on the materials of the clips.

An increase in the operating time of screw pumps can be achieved through the use of holders made of elastomers, the composition of which is selected on the basis of specific well conditions, in particular, the physicochemical properties of the formation fluid.

In 2017, the OPI of high-pressure electric pumps with a modified elastomer composition was resumed at the fields of LLC LUKOIL-Western Siberia, for oil production from low-yield wells, with a large depth of descent (more than 2000 m).

In the case of positive results and appropriate decisions on further implementation, it is recommended to consider the use of this DPE in the low-yield fund of the Tevlinsko-Russkinskoye field.

5.2 Installation of jet and hydraulic piston pumps
The lack of sufficient operating experience and its own facilities for maintenance and repair, prevents the introduction of tandem installations ECP-JP (jet pumps with submersible drive).

Thus, for wells with a low productivity factor, a significant dynamic level and a small inflow of production from the reservoir, the method of periodic operation is most preferable.

6. Causes of premature repairs of ECP
In January 2017, 221 premature repairs were carried out on the well stock operated by the ECP.

The main reasons for repairs:
- salt deposition of DPE (PCT, ECP) 9 %;
- ECP operation in difficult conditions of 5 %;
- corrosion of DPE 3 %;
- clogging of ECP with mechanical impurities of 2 %, table 1.

Table 1. The reasons for all premature repairs of the ESP for 2016

Reason for repair	Number of repairs	
	pc.	%
GTM	143	64,7
The fault of service organizations	29	13,1
Salt Deposition	20	9,0
ECP operation in difficult conditions	10	4,5
Corrosion of DPE	7	3,2
ECP clogging	5	2,3
Crash	6	2,7
Uncontrolled operation of ECP	1	0,5
Total	221	100

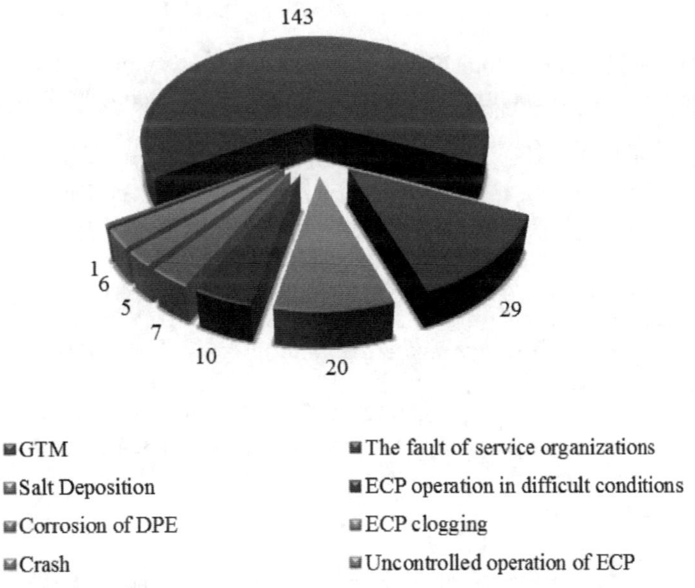

Figure 2. The distribution of existing production wells ESP for reasons of repair

7. Conclusion
Based on the analysis of technological modes of operation of wells and the analysis of equipment failures given above, the main problems relating to the operation of wells and equipment were highlighted.

7.1 Operation of ECP outside the operating range
According to preliminary estimates, 31 % of the ECP work beyond the boundaries of the working area: 19 % beyond the left border, 12 % of the ECP – beyond the right border. These wells are characterized by a decrease in the operating efficiency of downhole pumping equipment and an increase in the risk of premature failure of DPE. In subsequent repairs, it is necessary to select the DPE that ensures a given flow rate of the reservoir fluid with the ECP indicators closest to the optimal flow, head, and power, based on the operating characteristics of the unit size. For newly commissioned wells, as well as for wells, after conducted geological and technical measures, including hydraulic fracturing, it is recommended that the ESP unit be lowered with a valve engine (for subsequent control of the extraction of produced fluid).

In order to avoid premature failures of DPE, in cases of reducing the dynamic level, it is recommended to carry out measures to limit fluid withdrawal (fitting wells, reducing engine speed), reducing the size of the pump, changing the mode of operation (from ECP to SPI), or lowering the ECP of a larger size with the SPW mode (short periodical work).

7.2 Periodic well stock
At the mining ESP fund, 233 wells (15 % of the operating ECP fund) are operated periodically.

According to the results of the analysis, for some of the wells it is possible to switch to a more favorable one – a permanent mode of operation, by selecting an ECP corresponding to the current pressure-flow rate characteristics for WM.

7.3 Application of alternative DPE (jet, piston pumps)
Due to the lack of operating experience, its own maintenance and repair facilities is not recommended.

The ability to use the installation of electro-screw pumps – only the results of their repeated, field test trials, planned in 2017.

References
[1] Korabelnikov M I and Dzhunisbekov M Sh 2016 Analysis and ways to increase the efficiency of mechanized oil production from marginal wells in crisis conditions *Bull. of SUSU Ser. Energy* **16(1)** 75–9
[2] Mishchenko I T, Bravicheva T B and Ermolaev A I 2005 *The choice of the method of operation of oil fields with hard-to-recover reserves* (Moscow: Oil and Gas) 448 p
[3] Drozdov A N 2008 *Technology and technology of oil production by submersible pumps in difficult conditions: a tutorial* (Moscow: MAKS Press) pp 286–309
[4] Glavtyumenneftegaz 1979 *Guidelines for the selection of mechanized methods of operating oil wells. Enterprise standard* STP-59-00-006-78
[5] Lyapkov P D 1987 *Selection of the installation of a submersible centrifugal pump borehole* (Moscow: MING) 71 p
[6] Ivanovsky V N, Darishchev V I, Sabirov A A, Kashtanov V S and Beijing S S 2002 *Well pumping installations for oil production* (Moscow: Oil and Gas) 824 p
[7] Yang Y, Zhang H, Tian G, Xu Y, Wang C and Gao J 2019 Performance Analysis of a Multistage Centrifugal Pump Used in an Organic Rankine Cycle (ORC) System under Various Condensation Conditions *J. of Thermal Sci.* **4** 28
[8] Yaremenko O V 1976 *Tests of pumps: reference manual* (Moscow: Mechanical engineering)
[9] Yurchuk A M and Istomin A 1979 *Calculations in oil production, textbook for technical schools* 3rd, etc (Moscow: Nedra) 271 p
[10] Persianyantsev M N 2000 *Oil production in inflated conditions* (Moscow: Publisher Nedra-

IOP Conf. Series: Materials Science and Engineering **663** (2019) 012029 doi:10.1088/1757-899X/663/1/012029

Busines Center) 653 p ISBN 5-8365-0052-5
[11] Ibragimov N G, Khafizov A R and Shaydakov V V 2003 Complications in oil production (Ufa: LLC Publishing House of Scientific and Technical Literature) 302 p ISBN 5-94920-023-3
[12] Zhu J, Zhang J, Cao G, Zhao Q, Peng J, Zhu H and Zhang H-Q 2019 Modeling flow pattern transitions in electrical submersible pump under gassy flow conditions *J. of Petroleum Sci. and Engineer.* 471–84
[13] Gizatullin F, Khakimyanov M I and Shafikov I N 2019 Energy efficiency of well electric submersible pumps for oil production *IOP Conf. Ser. Mater. Sci. and Engineer.*

Hydropower potential implementation as an important stage of the environmental and economic frames formation in the mountain valleys

R O Kalov[1] and A B Vagapov[2]

[1]Kabardino-Balkarian State Agrarian University, 1v, Lenin Ave., Nalchik, Russia
[2]Chechen State University, 32, A. Shipipov str., Grozny, Russia

E-mail: calov.r@yandex.ru

Abstract. The hydrographic network within the mountain-piedmont zone is characterized. The necessity of the revaluation of the economically feasible rivers' energy potential is grounded. The main hydrological elements of river basins applied to the research topic have been identified. A landscape assessment of the mountain valleys specificity considering the creating hydropower facilities issue has been carried out. The most preferred for the development watercourses are designated and described, considering statistical materials on their water flow dynamics. The method of compensation for deficient statistical materials on the water flow dynamics is represented. The features of the high-mountain zone hydropower potential are highlighted. A comparative analysis of the energy reserves of the main mountain rivers is given. The role of ice-free feeding minor rivers in a hydroelectric project is determined. The method of determining the water flow security is described, the corresponding coefficients are determined. 16 topographically convenient mountain-valley extensions were identified as potential cisterns for reservoirs. Dam and lake methods for regulating effluent are evaluated on the acceptability. An algorithm for determining the mountain streams' channel energy potential of is given. A comparative analysis of the specific power of the rivers in relation to their different sections has been carried out. The most important limiting factors for the hydropower projects implementation have been identified. A working hypothesis for the use of environmentally friendly hydro energy has been formulated. In the first approximation, a scheme for selecting favorable landscapes for the hydropower facilities construction has been drawn up, and the corresponding map chart has been made. Criteria for selecting the power and type of power plants adequate to the natural base are indicated. The forecast of hydropower development is formulated in conjunction with the expected levels for electricity demand in the future.

1. Problem Statement

Kabardino-Balkaria mountain zone is one of the republican disadvantaged regions in terms of the hydropower competitive advantages realization. Meanwhile, by the beginning of the Second World War, every large mountain village had its own hydroelectric power plants (HPP): 20 plants of different capacities functioned in the mountains; another 22 objects were under installation. The production of own electricity was reasonably estimated as a guarantor for the sustainable provision of population with it, a real factor in reducing the cost of production and services [1].

Given the current trend of growth in the business projects cost structure, the electricity cost and inadequate tariff burden on low-income mountain dwellers, the feasibility of involving the unsold hydropower potential of the zone has only increased. In addition, there is always the growing relevance of using renewable energy on the global agenda.

IOP Conf. Series: Materials Science and Engineering **663** (2019) 012030 doi:10.1088/1757-899X/663/1/012030

Therefore, the hydropower resources reassessment under the transformation of the rivers nourishment nature and their flow regime on the background of global climate change; hydropower devices placement, proportional to the watercourses natural potential, remains strategic for the modernization of the economic and social infrastructure in the mountains. The designated priority harmoniously fits into the project proposed by the authors for constructing balanced ecological-economic zones in the main mountain valleys of the Republic.

2. Research Questions

Potential carriers of the hydropower resources of the Kabardino-Balkaria mountain areas are the basins of the Malka, Baksan, Cherek Rivers.

3. Purpose of the Study

Assessment of the current conditions for the involvement in the economy of the mountain rivers' hydropower potential, the identification of optimal areas for the use watercourses, the choice of sites for the priority hydroelectric power plants placement.

4. Research Methods

On the expeditionary study basis, a research was conducted on the Lochskii canyon in order to regulate the river Malka flow. Based on archival statistical materials of past years, a working hypothesis was formulated for the use of mountain rivers' hydropower. The comparative-geographical method was used in assessing the hydro-morphological parameters of watercourses. The mathematical method was used to identify the average long-term flow of rivers and their potential energy capacity.

5. Findings

Physical and geographical features of the study area led to the formation of three river basins' types: high mountain, main mountain and foothill [2]. They differ in the nature of the nourishment, the structure of the basin and the longitudinal profile, as well as the potential power of rivers' different sections.

Depending on the nature of nourishment, in the following river basins elements are highlighted:

1. The nourishing high-mountain fan of river tributaries of the 1st, 2nd, 3rd and higher orders (area of glacier flow), whose water content has increased by about 10% in recent decades (Table 1);
2. The nourishing system of the foothill rivers with mostly snow and rain feed;
3. Separate sources of groundwater;
4. The main mountain valleys that collect the waters of all influx types and divert them outside the mountain areas.

In accordance with the designated hydrological elements of the basins, zonal demarcation was carried out relating to the hydropower resources distribution:

I – the zone of the mountain rivers' main valleys (Terek, Malka, Baksan, Urukh, Cherek, Chegem);

II – zone of high mountain runoff (Donguz-Orun, Iusengi, Adyl-su, Irik, Adyr-su, etc.);

III – zone of minor rivers of ice-free nourishing (Nalchik, Kich-Malka, Shalushka, etc.). The hydropower resources of each zone have specific features.

When analyzing the potential of hydropower resources zones, achieve materials were not enough to characterize the intensity of the increase in river flow along the length. Therefore, the assessment of potential channel capacity, in some cases, is carried out by known water flow rates and a schematic typical longitudinal profile, in others, by a known longitudinal profile and water flow rates, adopted by analogy with adjacent basins. In isolated cases, interpolation and extrapolation were hampered by a sharp differentiation of high-altitude microclimates and the lack of reliable data on the level of correlation between river runoff dynamics and global climate change, therefore approximate generalizations of their energy characteristics were made.

Table 1. Mountain-foothill areas hydrographic network characteristics [3]

Main River	1st order influxes	2nd order influxes	3d order influxes	Basin area, km²	River length, km	Fall of the river, m	Average slope, %	Average annual water consumption, m²/s
Malka	—	—	—	9819	201	2879	14.3	112.4
	Baksan	—	—	6822	161	2154	13.4	102.2
	—	Cherek	—	3031	119	1840	15.5	52.3
	—	—	Cherek-Khulamskii	640	46	1358	29.5	13.9
	—	Chegem	—	923	102	1975	19.4	13.9
	—	Gundelen	—	579	56	2223	40.0	5.2
Urukh	—	—	—	1256	251	2601	10.4	27.4

To assess potential hydropower resources, the water flows workload associated with gravity has been identified. The method of continuous channel accounting was based on the rivers breakdown into sections with approximately equal slopes [4]. The work of the watercourse is calculated by the difference in water levels at the beginning and end of the selected part of the river using the formula:

$$N = g \sum_{H_y}^{H_u} \frac{Q_1 + Q_2}{2} \Delta H, \tag{1}$$

where g – gravitational acceleration, m/s; Q_1 and Q_2 – water consumption in m³/s at the beginning and at the end of the section, having a fall ΔH; H_s (H_u) and H_e(H_y) – absolute heights of the source and entry in meters; N – water flow.

The analysis allows conducting a hydropower assessment in the context of the 3 designated zones. The high mountain runoff zone covers the hydropower resources of high-mountain small rivers, mainly with glacier nourishing, which are main mountain rivers' influxes. The zone can also include small rivers that have lost their glacial nourishing, but which geomorphological features indicate their postglacial sources, which formed the typical longitudinal profile. Due to the alpine snow nourishing, the seasonal distribution of the latter is similar to glacial rivers. The peculiarity of the high-mountain rivers' hydropower resources is a high specific power with small runoff volumes and high pressures at their steep (downfall) source sections [5].

Valleys are included in the main mountain rivers space, crossing both the zone of high-mountainous areas and the underlying watercourses that do not have ice nourishing. The main part of hydropower resources is concentrated in the middle course of the zone, since runoff from both mountainous areas and ice-free nourishing rivers [6] rolls down into the mountain rivers' valleys. This rivers' group has a flatter parabolic longitudinal profile and maximum power in the middle course. Downstream, the power density decreases in line with the decrease in the channel slope, despite the increasing gross flow.

The secondary rivers zone of glacial-free nourishing is partly mountainous and the entire foothill area. Foothill rivers are energetically less significant, with a steep parabolic profile and diminishing downstream power. Maximum power within the zone is in the upper or in their middle parts.

Due to the high variability of water consumption, the full involvement of the indicated capacities is possible only under the condition of seasonal, even better – year-round regulation. Unregulated stock cannot be used highly efficiently [7]. The part of the water resources, which are suitable for development without flow regulation, is determined by the authors according to the high-availability water consumption. At the same time, the provision ratio is conditional and identified as the ratio of the water flow of this provision to the value of the average long-term water flow. The rates of transition from average annual water consumption to annual water consumption (100 %), nine-month (75 %) and six-month (50 %) provision are shown in Table 2.

Table 2. Average annual water provision of the main KBR rivers

Name of the river	Average annual water consumption, m^3/s	100 % provision		75 % provision		50 % provision		Potential annual average power (MW)
		water flow, m^3/s	provision coefficient	water flow, m^3/s	provision coefficient	water flow, m^3/s	provision coefficient	
Malka	13.5	5.0	0.37	7.5	0.55	10.8	0.80	196
Baksan	33.9	9.0	0.26	12.0	0.35	20.0	0.59	394
Cherek	40.5	9.0	0.22	11.0	0.27	19.0	0.47	564
Chegem	12.6	3.0	0.24	3.6	0.28	6.3	0.50	174
Terek	147.2	57.0	0.39	70.0	0.48	111.7	0.76	168
Urukh	24.2	3.9	0.16	5.7	0.24	13.0	0.54	72

Table 2 suggests that the water consumption of year-round provision for most watercourses varies within 16–26 % of the average annual water flow and only for rivers Malka and Terek – close to 40 %. If we start from the minimum and installed capacity of hydroelectric power plants as 1: 2, the estimated water consumption for rivers with glacier nourishing is close to nine months provision (75 %), and the flow is used only by 25–30 %. For mountain rivers with non-glacial nourishing under the same conditions, water consumption is approximately six months provision (50 %), and runoff is used at 50–60 %. Consequently, the flow of rivers with glacier nourishing is used less efficiently.

Thus, the vast majority of hydropower resources in the mountains must be regulated [8]. The morphology of mountain valleys, in general, is favorable for the construction of regulatory objects. In relation to them, 16 topographically convenient mine-valley extensions were previously identified, they are: 2 – on the Malka River, 5 – on the Baksan River, 2 – on the Cherek River, 1 – on the Cherek-Khulamskii River, 2 – on the Chegem River, 4 – on the small influxes of the Baksan (Danguz-Orun, Iussengi, Syltran-Su, Gundelen).

In the mountains, there are alternative (lake) regulation. The most preferable object for lake regulation with respect to the energy and economic parameters of Baksan is the Syltrankel limnosisystem located in the Kyrtyk River at an altitude of 1700m north-west from Upper Baksan Village. It is more expedient to regulate the Cherek River by the Giybaskel Lake, which is localized southwest of the Upper Balkaria village [2]. These reservoirs also have the most favorable starting natural characteristics for erecting hydro-accumulating stations on them; therefore they can become top-priority energy facilities.

There are no functionally significant lakes in the Malka River basin, therefore here we can only consider the weir way of influence on the flow dynamics. A more preferable variant of its regulation is the creation of a reservoir on the Lochansky canyon basis. It has more acceptable topographical and geological conditions. With a dam height of 105–110 m, the useful volume of the reservoir will be about 80 million m^3, which is sufficient to regulate the entire main river flow.

An important lever for increasing the river flow utilization efficiency is maximizing the energy involvement of the entire river fall [9]. In the process of studying this property, the hydropower channel potential was estimated by dividing the water courses into sections. The boundaries of the sites were selected on a topographic map at the fractures points of the rivers longitudinal profiles and in the entry of the most significant influxes (the yield of at least 5 % of the main river flow was taken as the criterion of significance).

Proceeding from the above, the Cherek and Baksan rivers have the largest specific power, the Malka –the smallest. In them, the energy capacities decrease in the entry areas and have maxima in the upper or middle third of their length. The hydropower potential of the entry of small high-mountain rivers with preserved glacier ourishinf is favorable: it is quite high, which is due to their extremely steep fall, caused by over-deepening of the erosion base under glacial landscapes [10].

At the same time, the energy values of some mountain rivers are limited by certain circumstances. End moraine soils and fluvial-glacial deposits are the limiting factor for the implementation of hydropower projects in the upper reaches of the Chegem River, since they are not a reliable basis for dams. The upper parts of Adyr-Su do not have applied significance, which means they except from the general river fall.

The involvement of high-mountain rivers in their average flow is problematic: debris flows and mountain landslides often clutter the valleys, creating gradations of longitudinal profiles and significant elevation differences with increased specific power of the watercourse in short sections [11]. Geological conditions may also become blocking development factors: karst landscapes of valleys confined to limestone and shale zones (Tiubele Kamyk Bylymskoe landtype associations on the Baksan River; Cherek gorges) are not a reliable basis for the construction of dams.

Thus, the distribution of hydropower resources reserves in the mountains in the first approximation can be estimated as follows: for high-altitude watercourses – 18 %, main mountain rivers – 76 %, foothill – 6 % of potential. Taking into account the cumulative properties of river systems, a schematic map has been compiled as one of the options for locating hydraulic systems and their combinations in river basins (Figure 1).

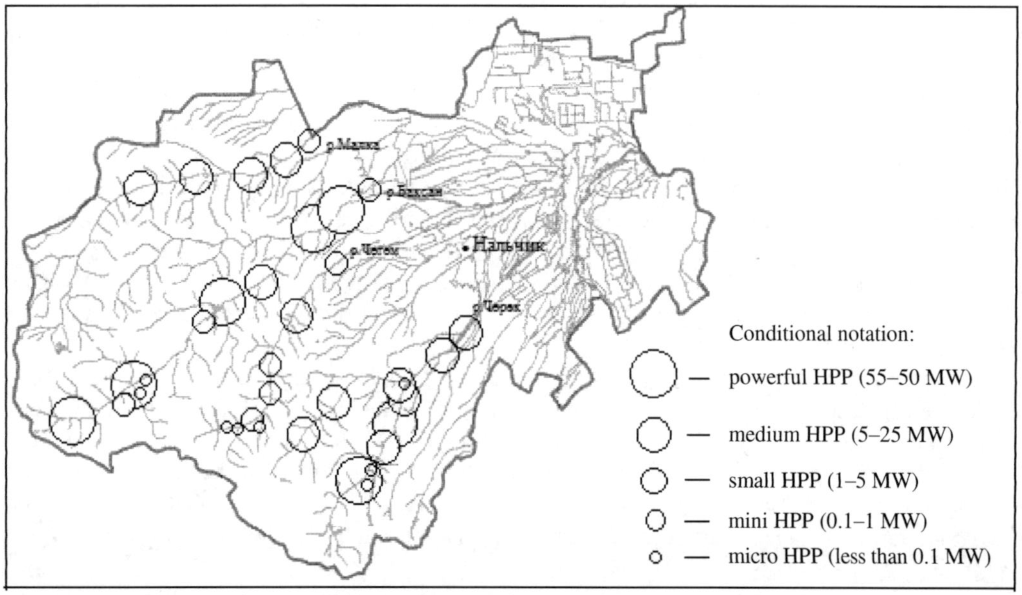

Figure 1. Approximate scheme of involvement of the mountain-piedmont zone hydropower potential

The choice of power plants' power and type is important. It should be determined considering the nature of the valleys occupancy, the intensity of economic activity, the formalized priorities of the region's development for the long term [12]. 6 climbing camps and one hotel are already functioning in the small Adyl-Su basin. In the foreseeable future, we predict a further expansion of the recreational economy, the beginning of industrial production and drinking water bottling, scientific institutions revitalization, rescue structures consolidation, and therefore the construction of mini hydroelectric facilities will be in demand. Low volume of construction and installation works; acceptability of small-sized power plants spatially combined with decentralized power consumers; insignificant specific capital investments level the limiting non-critical physical-geographical factors.

The further involvement of Baksan hydropower resources should take into account federal projects for the Elbrus ski infrastructure modernization, for the restitution of mining environmental management on the Tyrnyauzskogo field basis with a wider range of commercial products, limestone building materials production [9].

Due to the particularly high specific power of water in the middle reaches of the main mountain rivers, relatively large (more than 25 MW) water installations should be selected for them. In their lower course, the power density decreases; the valleys slopes, the morphology and geology of the valleys do not allow concentrating high water heads. Covered with a thick layer of sediment, disintegrating in the river branches are unfavorable for the construction of hydropower facilities. In addition, the density of power grids is quite high there, so consumers can choose the best connection option for themselves [10].

The experience of past construction in the republic shows that hydropower does not make major negative changes in the natural environment. The construction of hydropower plants will allow attracting a part of the unemployed both in the sphere of construction and its maintenance, and the subsequent operation of hydroelectric power plants. The industry will give a new impetus to the economic development of the mountain zone and, importantly, it will remain a permanent factor in motivating its economy development.

6. Conclusion

1. Landscape-climatic features, hydrographic parameters of mountain streams and their hydropower potential can be the basis of a profitable, annually renewable hydraulic energy source.

2. The main carriers of the Kabardino-Balkaria mountain zone hydropower resources are the basins of the rivers Malka, Baksan, Cherek.

3. High annual variability of water flow necessitates regulation of mountain river flow.

4. Flow regulation through the construction of large valley reservoirs is limited by a sharp differentiation of physiographic conditions, an increase in environmental risk, a large amount of construction and installation work, and their significant capital intensity.

5. The most expedient way to mountain rivers is the construction of low-capacity pumping stations on the natural mountain lakes base, which can simultaneously perform regulatory and energy-producing functions.

6. The most preferable for the construction of pumped storage stations are the karst lake Giibashkel (to regulate the flow of the Cherek River), the lake Syltrankel (to regulate the flow of the Baksan River). It is more expedient to transform the Malki River runoff by erecting a dam with the maximum use of favorable topographic and geological properties of the Lokhranskii canyon.

7. The distribution of hydropower resources in the mountains is as follows: for high-altitude watercourses –18 %, main mountain rivers – 76 %, foothill – 6 % of potential.

8. The Cherek and Baksan rivers have the largest specific power, the Malka River has the smallest. In them, the energy capacities decrease in the entry areas and have maxima in the upper or middle third of their length.

9. Phased involvement of the hydro potential cost-effective part in the power system, including autonomous supply, will reduce the cost of current and future business projects and the electricity costs for mountain people.

10. In case of the optimistic scenario for the renewable energy development, Kabardino-Balkaria will have the prospect of electric power cooperation with the regions of the South of Russia and the Transcaucasian states depending on the balance situation of the resource in the power systems.

7. Acknowledgments

The study was carried out with the financial support of the Russian Foundation for Basic Research in the framework of the research project No. 19-010-00882.

References

[1] Kalov R O 1999 *Economic geography of the Kabardino-Balkarian Republic* (Nalchik)

[2] Kalov R O 2018 On the issue of environmental risks optimization in the mountain-valley environmental management conditions *Bull. of the Int. Academy of Ecology and Life Safety* **4** 56–60

[3] State Water Cadastre 1998 *Annual data on the regime and resources of surface waters of the RSFSR land* (Obninsk)

[4] Altunin V S and Dmitruk V I 1987 Determination of river flow with a lack of hydrometric observations *Hydrotechnical and land improvement* **2** 20–4
[5] Malik L K 2000 Problems of development of small rivers hydropower potential in Russia: environmental and socio-economic aspects *Fundamental problems of water and water resources at the turn of the third millennium* 627–30 (Tomsk)
[6] Milkov F N 1981 River basin as a para-dynamic landscape system and environmental management issues *Geography and natural resources* **4** 11–8
[7] Kalinin V G and Piankov S V 2002 The use of hydrographic characteristics of rivers and their basins in hydrological calculations *Meteorology and hydrology* **11** 75–80
[8] Siniugin V Iu 2008 *Pumped-storage power plants in modern electric power industry* (Moscow: ENAS)
[9] Kalov R O 2016 On the issue of unrealized hydropower potential of mountain-valley landscapes of the KBD Ne*ws of KBSTs RAS* **6(74)** 28–33
[10] Kalov R O, Kiloev D.D and Elmurzaev R S 2018 *Ecological and economic design of environmental management within river basins Ecology of urban areas* **4** 94–8
[11] Antipov A N and Fedorov V N 2000 *Landscape and hydrological organization of the territory* (Novosibirsk: SB RAS)
[12] Korytnii L M 2001 *Basin concept in environmental management* (Irkutsk: SB RAS)

IOP Conf. Series: Materials Science and Engineering **663** (2019) 012031 doi:10.1088/1757-899X/663/1/012031

Activities and technical solutions for ecological management of the city Vladikavkaz

D V Kasharin[1], T N Meisner (Kovaleva)[1], M Yu Kosichenko[1] and N A Kovalev[2]

[1]Platov South-Russian State Polytechnic University, Novocherkassk, 346428, Russia
[2]Moscow State Technological University STANKIN, Moscow, 127055, Russia

E-mail: designtatyana@yandex.ru

Abstract. The article is devoted to the activities ensuring city of Vladikavkaz ecohumanisation by engineering protection and development of recreational areas. The analysis of natural and climatic conditions and social aspects of the city deveopment. Technical solutions and their justification on the example of the right-bank territory of Vladikavkaz, as well as the development of Park areas are proposed.

1. Introduction

The development of the city of Vladikavkaz in the Republic of North Ossetia (Alania) is an example of a rapid and successful transition from the status of a fortress to the status of a Capital. Vladikavkaz was built on the site of a swamped territory of the Terek river catchment area of and transformed into a historical center, which became the cultural heritage of the Republic. It has an important political, social and industrial purpose. The intensive development of modern buildings makes significant changes in the appearance of the city and makes it necessary to control the processes associated with changes in the urban situation, prevention of degradation of the natural landscape and distortion of the pristine appearance of historical and cultural monuments within the boundaries of the historical center of Vladikavkaz [1]. Since the history of each individual landscape is unique, the dynamics of its development require planning to change it. Successful planning requires analysis of both rapid changes and gradual evolution, which contributes to a better description, prediction and appointment of activities in landscape planning [2].

Landscape transformation is one of the conceptual elements of ecohumanistic strategy in modern city [3]. Vladikavkaz has already passed the technogenic stage of development and now it is necessary to pay attention to environmental safety and humanizing development of recreational areas, not forgetting the engineering protection of the landscape. Vladikavkaz according to modern research is one of the largest cities with accumulated seismic activity [4]. Therefore, when designing recreational areas, it is especially important to give preference to earthquake-resistant structures of shore protection construction. According to the zoning plans of the city of Vladikavkaz (industrial, residential and recreational areas (fig. 1)) the most problematic is the landscape of the residential zone from the right bank of the Terek river, as well as the currently empty landslide site from the left bank, which has a significant slope. These areas are subject to erosion and landslide processes, which leads to negative environmental and humanizing consequences [5]. It requires planning of landscaping in two stages. At the first stage, it is important to implement measures for engineering protection of the river bank, ensuring the least impact on the historical landscape and ecology of the territory. The use of new soil-reinforced structures of shore protection structures, including those built on weak biogenic soils, allow to restore and improve the landscape, creating conditions for its further transformation and use. The second stage includes eco-humanizing measures for the improvement of the territory by expanding the park areas, which give s a lot of information and recreational areas, combined with the device of shore protection structures (fig. 1).

Figure 1. Map of Vladikavkaz with the designation of the territory for planning a recreational zone: a) the planned area of reconstruction; b) the increased fragment of the territory recommended for planning of a recreational zone; c) 1 – erosion and landslide site, a zone of reconstruction of a landslide site

2. Measures for engineering protection of residential area of Vladikavkaz from erosion and landslide processes on the right Bank of the Terek river

2.1. Prospects of application of new technical solutions

One of the priority measures is to carry out works on building of shore protection constrction that ensure the stability of the coastal zone together with the national economic objects located on its territory. Also, during these activities, the preservation and expansion of existing beaches, parks and other recreational areas is ensured. Existing rigid structures of shore protection structures do not provide sufficient reliability and their destruction occurs when the base subsidence [6]. In the absence of high-quality local building material and subsidence soils, the most effective is the use of soil-reinforced structures. Theoretical and experimental studies conducted by foreign and domestic scientists: Schlosser, Long, Hausman, Mogilkov, Hristo, L. N. Timofeeva, G. M. Kaganov, I. M. Evdokimova, etc., show the prospects and feasibility of soil reinforcement with the use of local building materials [6-8]. To justify the use of structures of soil-reinforced structures, we consider the design provisions for their stability.

The deformation of the reinforcement equal to the lateral deformation of the soil is determined by the following relationship [5, 9]:

$$\varepsilon_{\text{а}} = \Delta_{\text{а}} = \sigma_0 A_{\text{гр}} / E_{\text{а}} A_{\text{а}}, \tag{1}$$

где $E_{\text{а}}$ – where is the modulus of elasticity of the reinforcement material, MPa.

The tensile stress in reinforcing strip, in accordance with the theory of Rankine, at depth z can be represented as follows (Fig. 2):

$$F_{\text{а}, z} = K_{\text{а}} \gamma z S_y S_z. \tag{2}$$

That is, the maximum tensile force perceived by the lower reinforcing layer, according to formula 2, is defined as:

$$F_{\text{а, max}} = K_{\text{а}} \gamma H S_y S_z. \tag{3}$$

The safety factor of the reinforcing strip at break is determined taking into account the formula (3) and is equal to:

$$K_S^N = R_p b_a \delta / K_{\text{а}} \gamma H S_y S_z. \tag{4}$$

We accept coefficients that take into account the reinforcement area, as well as the decrease in the strength of the reinforcing material during the operation of Kc for composite polymer materials. Then formula 9, taking into account formula 5, is reduced to the form:

$$K_S^N = \mu R_p K_c / K_a \gamma H = \mu R_p K_c \Big/ \mathrm{tg}^2\left(\frac{\pi}{4} - \frac{\varphi}{2}\right)\gamma H . \quad (5)$$

The friction force holding the reinforcement in the soil mass is determined by the Rekin's formula:

$$F_f = 2 f \gamma z b_a \cdot l_a , \quad (6)$$

where f – the coefficient of friction between the ground and the material reinforcing strip in the area of the compacted backfill; – the length of reinforcing strip taken not less than 1,5 m.

The factor of safety of the reinforcing strip on pulling out, according to formulas 2 and 6, is defined on the following dependence:

$$K_S^C = F_f / F_{a,z} = 2 f b_a l_a / K_a A_{\text{тр}} . \quad (7)$$

Primed and reinforced structures from polymer composite materials include four main structural elements: coffer the front surface (front wall) with the use of rubber-córdoba, geotextile and other materials; anchor elements of the same material, including fued of soil shell; reinforcing strip of composite materials (straight, double, oblique); soil, interacting with the anchoring devices, flexible connections, anchoring and drainage systems. The proposed new technical solutions of soil-reinforced structures provide the lowest weight of the structure due to the use of a flexible front wall of geosynthetic material (fig. 2 a, b) [10-12]. When calculating the strength and stability of these structures, two types of limit state are considered: internal strength and overall stability of the entire structure.

a b c

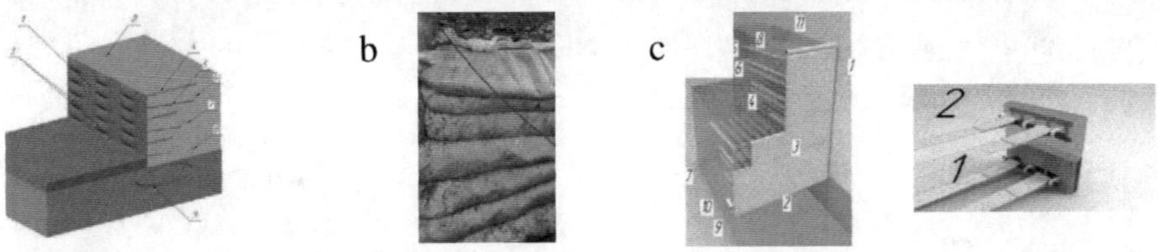

Figure 2. Engineering solutions of ground-reinforced shore protection structures: a – method of creating a ground reinforced supporting structure and a device for its implementation (Pat. RF 2352713): 1 – front wall; 2 – bulk soil; 3 – flexible connections; 4 tape shells; 5 – corrugated and flat armor; 6 – drain shell; 7 – special holes; 8 – drainage system; 9 –anker blocks;b – ground-reinforced shore protection structure on the foothill section of the river Gizeldon; c – Engineering solutions of ground-reinforced shore protection structure with a front wall made from a multi-layer composite material: 1 – a front wall; 2 face elements made from composite materials resistant to wearing; 3, 4 –horizontal and vertical sections of the front elements 2; 5, 6 – inclined lower and upper reinforced tapes respectively; 7 – ground massif (for example, technogenic soil); 8, 9 – ground-filled multi-shell lower and upper anchors; 10-drainage; 11 – textile material with seeds

The latter is determined taking into account the total tensile force in the valve at the design level, determined by the following formula [6, 8]:

$$N_0 = N_1 + N_2 = \int_0^{l_{a1}} 2b_1 \cdot m_1 \cdot f_1 \cdot \gamma \cdot z \cdot dl + \int_0^{l_{a2}} 2b_2 \cdot m_2 \cdot f_2 \cdot \gamma \cdot z \cdot dl = \quad (8)$$
$$= 2\gamma \cdot h \cdot (bm_1 \cdot f_1 \cdot l_{a1} + bm_2 \cdot f_2 \cdot l_{a2}),$$

Where N_0 is the total tensile force in all layers of reinforcement, corresponding to the surface of collapse, kN/m; N_1, N_2 – the tensile stress in reinforcing strip with a different tilt kN/m; , b_1, b_2 – width straight and sloping reinforcing strip with a different tilt, m; m_1 and m_2 – are the coefficients of the slope of reinforcing strip z – height of the considered layer of soil over reinforcing strip, m; f_1, f_2 – a coefficient of friction of the soil on the valves; l_{a1}, l_{a2} - the length of armament, m; γ – weight of unit volume of the mound, kN/m3.

To assess the overall stability of reinforced soil, we will use one of the most common methods, which is based on the round-cylindrical method.

To assess the overall stability the safety factor is determined as the ratio of the moments of resistance forces and shear moments:

$$K_{зап} = M_R / M_F \quad . \tag{9}$$

For structures made of reinforced soil, the moment of resistance is determined by the following formula (Fig. 4) [4, 5]:

$$M_R = \sum_{i=1}^{k} F_{T,i} \cdot r + \sum_{i=1}^{k} F_{C,i} \cdot r = \sum_{j=1}^{n} \left(F_{а,j} \cdot \cos\beta_j + F_{а,j} \cdot \sin\beta_j \operatorname{tg}\varphi_i \right) \cdot r1_j \ , \tag{10}$$

where $F_{T,i}$, $F_{C,i}$ – respectively, the conditional forces of friction and resistance in the I-section compartment is directed along the tangent to the slip surface, kH; $F_{а,j}$ – is the resistance of the reinforcing element defined equal to the calculated stress of reinforcement at break kH; $F_{а,j} = A_а \cdot R_{а,j}$; $R_{а,j}$ - the strength of the j- reinforcement layer in the gap; r – the radius of the curve slip, m; β_i – angle between the radius of the curve of the slide and the vertical; φ_i – the angle of internal friction of the i–section; $r1_j$ - the distance from the center of moments; $r1_j = d_j / \cos\beta_j$; d_j - the shoulder lines of action $F_{а,j}$ about the center of moments (fig. 4).

The moment of shear forces will be equal to:

$$M_F = r \sum_{i=1}^{k} G_i \sin\beta_i \ , \tag{11}$$

where G_i -the weight of the i-section into which the potentially unstable array is divided; α_i - the angle of inclination of the sliding platform within the i-section.

The assurance factor can be represented taking into account the dependencies (14-16) in the following form:

$$K_{зап} = r \sum_{i=1}^{k} \left(\sigma_{ni} \operatorname{tg}\varphi_i + c_i \right) \frac{l_i \cdot S_y}{\cos\alpha_i} + R_а \cdot A_а \cdot \sum_{j=1}^{n} d_j \cdot \left(1 + \operatorname{tg}\beta_j \cdot \operatorname{tg}\alpha_j \right) \Big/ r \sum_{i=1}^{k} G_i \sin\beta_i \ , \tag{12}$$

where σ_{ni} – is the effective normal stress at the elementary shear site, which can be defined as; $\sigma_{ni} = G_i \cos\alpha_i / l_i$; c_i – the specific clutch of the i-section [4].

The value of the calculated value of the allowable tensile stress for the layer σд appointed by the results of special tests to assess the long-term strength. For the preliminary calculations, the value of σд allowed to take a proportion of short-term strength of reinforcing strip under uniaxial tension $R_а$: for woven materials, rigid meshes of polyamide, polyester raw materials $\sigma_д = 0,6R_а / \delta$, polypropylene raw materials $\sigma_д = 0,3R_а / \delta$ – non-heat-treated or heat-treated needle-punched reinforcing strip of polyamide, polyester raw materials сырья $\sigma_д = 0,25R_а / \delta$, polypropylene $\sigma_д = 0,1R_а / \delta$.

Selection of the number of layers of reinforcement is performed according to the formula:

$$n = 0,53 K_{зап.тр} \sum G_i \left(\cos\alpha_i - \sqrt{\cos^2\alpha_i + 4\sin^2\alpha_i} \right) - \sum \sigma_{pi} l_i S_y \Big/ \sigma_д A_а \ , \tag{13}$$

где – требуемый коэффициент запаса устойчивости откоса.

В случае предварительного испытания материала армоленты, ее длина определяется по следующей формуле:
where $K_{зап.тр}$ – required slope stability factor.
In the case of preliminary tests of the reinforcing strip material, its length is determined by the following formula:

$$l_{3,i} = 0,5R_{а} / \sum \rho_i \cdot h_i \cdot \text{tg}\,\varphi' + c' , \quad (14)$$

where ρ_i, h_i – the density and thickness of the soil layers located above the top of the layers.
It was found that in the range of the smallest reinforcing strip length for sand minimum is 3 m, sand and clay – 3.5 m, and for loess and loam soils – the maximum length of not more than 4.2 m [9-12].
The safety factor of the formulas must be greater than or equal to the stability factor:

$$K_{зап} \geq \gamma_n \gamma_{fc} / \gamma_c , \quad (15)$$

where, γ_n, γ_{lc} – respectively, the reliability coefficients of the responsibility of the structure and the combination of loads; γ_c – coefficient of working conditions.

The calculated values of the stability coefficient with the appropriate combination of loads should not exceed the value of $\gamma_n \gamma_{lc} / \gamma_c$ more than 10 %, unless this is due to the characteristics of the structure.

Since the considered structures belong to class IV, take the coefficients of resistance to SP 58.13330.2012 $\gamma_n = 1,10$; $\gamma_{lc} = 1,0$; $\gamma_c = 0,95$. Then $\gamma_n \gamma_{lc} / \gamma_c = 1,16$ and the condition of the General stability of slopes of a ground construction will have the following form:
$1,28 > K_{зап} \geq 1,16$.

However, for greater reliability (according to the results of research by prof. Kaganov G. M.), the condition of stability of the soil-reinforced slope is proposed based on the following condition [6]:

$$K_{зап} \geq 2,25 . \quad (16)$$

The calculation of the front wall of the shore protection structure is performed in accordance with the theory of instantaneous soft shells, taking into account the deformation of the material.

To determine the parameters soil-reinforce structures developed algorithm to create the program of calculation of parameters for soil-reinforce coast protection works on a computer, which can be used to ensure reliable operation of bank protection structures in the considered residential area [12].

The use of these structures, which have a minimal negative impact on the surrounding landscape and the ecological state of the environment, contributes to the implementation of the first stage of landscaping of the coastline of the Terek river, passing through the Central part of the residential zone of Vladikavkaz.

3. Humanizing measures for the improvement of the city

Due to technical measures (strengthening of the river banks) we have partially solved the problem of environmental impact. After that, we proceed to the second stage of planning – humanizing protection measures. In the process of urbanization and socio-economic development of society, along with the solution of environmental and functional problems, there were prerequisites for solving the problem of preserving cultural heritage and traditions [13-16]. A solution to the above task involves planning along the fortified banks of the river recreational areas, contributing to: 1) expand the use of bank protection structures; 2) the problem of inaccessibility of the territories; 3) humane impact these structures have on the person; 4) the restoration and preservation of the cultural heritage of the region.

When planning recreational areas, it is important to take into account both the seismic characteristics of the terrain and the features of the Terek river, which has heavy alluvial regimes that bring significant changes to the landscape during floods. Due to alluvial regimes, new sleeves are formed spontaneously and chaotically, which adversely affects the landscape and urban infrastructure [17-19]. To stabilize the negative impact on the landscape and create favorable environmental structures, it is proposed to use the properties of regulatory constructions (spurs) [20] and shore protection constructions, which delay sediments of solid runoff and eventually form new landscape forms created by nature and controlled by man.

In these geographical conditions, it is appropriate to design Park recreational areas with a developed network of terraces, pedestrian bridges and recreation areas with different design schemes. Analogous to the creation of parkland on wetlands can serve as Park Yanweizhou in Jinhua city, Zhejiang province of China [14]. The Park is located at the mouth of three rivers in a swampy area, which was a prerequisite for the design of flooded bridges for pedestrians, which during the flood are closed for walking.

The creation of recreational areas with the additional use of local plants and the inclusion of regional artifacts in the landscape contributes to the protection and preservation of the ecosystem, as well as the preservation and development of architectural and cultural traditions of the region [17].

It is also necessary to use unfavorable areas of the left bank of the Terek river, including a long dangerous landslide along the sleeve, as a recreational zone, provided that retaining walls are created, on the terraces of which sections of different-level Park zones and a network of pedestrian and Bicycle paths can be organized. Modern designers do not note the aesthetics of concrete as a building material for the facial finishing of building structures and to improve the stability of the infrastructure, they offer technologies based on biological methods, combining the use of concrete and biomaterial for shifting surfaces [21]. An alternative to bicycle paths near the shore can serve as plants located on shore protection structures and retaining walls, strengthening the soil cover and creating an aesthetic landscape.

To achieve sustainable development of Vladikavkaz it is necessary to ensure the interaction of the entire spectrum of factors: recreational architecture, planning organization, economic, environmental, humanization and cultural indicators, supported by the support of government agencies and the population [22].

4. Conclusion

The vector of ecohumanistic approach in the development of the city is aimed at achieving sustainability of the city. Considered in our study, a separate indicator strategy ecohumanistic [18]– eco-friendly construction technologies, materials and activities in landscape planning, we have combined and established synergies between the technical and humanities areas. It helps to closer horizons for achieving the global objectives of sustainable urban development, recognized at the international summit of the UN in 2016 a guide for all mankind. Using new design of soil and reinforcing bank protection structures on low nutrient soils, the algorithm and the methodology of calculations for determining their parameters, while expanding the functionality of facilities and humane influence of structures on people, we reduce the negative impact on the landscape and contribute to its sustainable development taking into account the positive evolutionary patterns and ecological safety.

References

[1] Sze S M 1969 *Physics of Semiconductor Devices* (New York: Wiley–Interscience)

[2] Dorman L I 1975 *Variations of Galactic Cosmic Rays* (Moscow: Moscow State University Press) p 103

[3] Caplar R and Kulisic P 1973 *Proc. Int. Conf. on Nuclear Physics (Munich)* vol 1 (Amsterdam: North-Holland/American Elsevier) p 517

[4] Szytula A and Leciejewicz J 1989 *Handbook on the Physics and Chemistry of Rare Earths* vol 12, ed K A Gschneidner Jr and L Erwin (Amsterdam: Elsevier) p 133

[5] Kuhn T 1998 Density matrix theory of coherent ultrafast dynamics *Theory of Transport Properties of Semiconductor Nanostructures (Electronic Materials* vol 4*)* ed E Schöll (London: Chapman and Hall) chapter 6 pp 173–214

[6] Tsallagov S F and Tezieva M D 2014 Vladikavkaz: the past and the present Materials of SCSMI (STU) **21** 57–67

[7] Marcucci D J 2000 *Landscape and Urban Planning* **49** 67–81 Retrieved from: https://doi.org/10.1016/S0169- 2046(00)00054-2

[8] Kovaleva T N 2014 Basic conceptual approaches to the study of urban space *Bull. of Krasnodar University of the Ministry of Internal Affairs of Russia* **23** 97–9

[9] Milukov V K, Mironov A P, Steblov G M et al 2017 Satellite geodetic monitoring of the Vladikavkaz active fault zone *Izv. Solid Earth* **53** 598 Retrieved from: https://doi.org/10.1134/S1069351317040061

[10] Dmitrak Yu V, Vaskov I M, Cherkashin V I , Yusupov A R and Tamaeva M P high energy land slides, their possible impact on the reservoirs in the mountain valleys of the east caucasus *Sustainable Development of Mountain Territories* **10(2)** 199–214

[11] Kasharin D V 2018 Justification for the construction of the beds of ground-reinforced shore protection structures for small watercourses *Soil mechanics in geology and foundation engineering: mater. of the Int. Scientific and Engineer. Conf.* (Novocherkassk: Platov South-Rus. State Polytechnic Un. (NPI)) pp 663–71

[12] Kasharin D V 2010 Optimization of the selection and justification of the parameters of mobile structures of engineering protection made of composite materials in the catchment area *News of Vedeneyev RSRIG* 50–60

[13] Kaganov G M, Evdokimova I M and Shevchenko K I 2004 *Hydraulic structures made of reinforced soil* (Moscow: Research Institute – Nature) p 607

[14] Kasharina T P, Kasharin D V et al 2007 *Method for building a ground reinforced retaining structure and device for its implementation*: Pat. 2327845 of the Russian Federation: MPK E02B3/06, announced 31.07.2007; publ. 04/20/2009, Bul. no 11

[15] Kasharina T P and Kasharin D V 2005 *Worksheet for the ground-reinforced structures of protective hydraulic structures* (Rostov-on-Don: Yuzhvodproekt) p 35

[16] Kasharin D V et al 2008 *Guidelines for the installation of ground-reinforced foundations of engineering structures for highways* (NTS FSO "Management" Rostovmeliovodkhoz no 25)

[17] *Calculation of ground reinforced foundation: Certificate of state registration of computer programs* no 2010616390.

[18] Guzmán P C, Pereira Roders A R and Colenbrander B J F 2017 Measuring links between cultural heritage management and sustainable urban development *An overview of global monitoring tools* **60** 192–201 Retrieved from: https://doi.org/10.1016/j.cities.2016.09.005

[19] XX2 century, 2014–2018 XXII CENTURY. Discoveries, expectations, threats. *Science, engineering, technology, medicine, futurology, social trends. News, publications* Retrieved from: https://22century.ru/popular-science-publications/arch-2015

[20] Kovaleva T N 2014 *Modernization of Urban Space in the Context of Solving Current Environmental Problems Humanitarian and Socio-Economic Sci.* pp 49–52

[21] Kovaleva T N 2016 *Ecohumanization of a modern city* (Moscow: Rusains) p 152

[22] Alekseevskii N I and Sidorchuk A Yu 2017 Morphology and dynamics of active bed forms in the Terek River channel *Water Resources* **44** 226–36 DOI: 10.1134 / S0097807817010031

[23] Vorobev A E, Ortsukhaeva Z Sh and Alferova N N 2015 The formation of channel deformations in the lower river Terek *Peoples' Friendship University of Russia* Retrieved from:: http://vestnik.osu.ru/2015_10/82.pdf

[24] Kosolapov A E, Kosolapova N A, Matveeva L G and Chernova O A 2018 Rational use of water resources in the economy of the North Caucasus mountain areas *Sustainable development of mountain areas* **10(1(35))** 48–62

[25] Koronovskiy N V Fluidogenic Breccias in the Darial Gorge of the Terek River (Central Caucasus Mountains) *Moscow University Geology Bull.* **72(6)** 399–404 DOI: 10.3103/S0145875217060059

[26] Tatczewski R 2015 Formation of Sustainable Infrastructure Using Microbial Methods and Humanization of Man-made Environment vol 3, pp 1704–11 DOI: 10.1016/j.promfg.2015.07.991

[27] Smoliar N, Bredun V and Toronchenko O 2018 Urban green zones planning concept and ecological functionality *Int. J. of Engineer. and Technol.* **7** 522–7 DOI: 10.14419/ijet.v7i3.2.14582

IOP Conf. Series: Materials Science and Engineering **663** (2019) 012032 doi:10.1088/1757-899X/663/1/012032

Analysis of non-sinusoidal voltage at metallurgical enterprises

R V Klyuev[1], O A Fomenko[2], O A Gavrina[1], A A Sokolov[1], O A Sokolova[1], Dzeranov B V[1], I D Morgoev[1] and S G Zaseev[3]

[1]North-Caucasian Institute of mining and metallurgy (State Technological University), 44, Nikolaeva St., Vladikavkaz, 362021, Russia
[2]Southern Federal University, 105/42 Bolshaya Sadovaya Str., Rostov-on-Don, 344006, Russia
[3]Gorsky State Agrarian University, 37, Kirova St., Vladikavkaz, 362040, Russia

Abstract. The article presents the results of the study of non-sinusoidal voltage in the power supply system of non-ferrous metallurgy enterprises. The non-sinus voltage coefficients are calculated for the operation of valve inverters and welding machines. The range analysis of the noah caste of higher harmonic components by voltage was carried out and it was established that the values of the correlation coefficients between the non-sinusoidal coefficients of voltage and the fraction of noah and pointer castes of electricity consumers for the production of hard alloys are equal to: 0.56 and 0.37, which makes it possible to limit the range of priority tasks for calculating the required filter-compensating devices. It is necessary to conduct further studies of higher harmonics in the system of power supply of industrial enterprises and the development of methods for calculating and optimal distributing filter-compensating devices in enterprises.

1. Introduction

At modern industrial enterprises, non-linear loads have become widespread. These include, first of all, various types of valve inverters, mainly thyristor, electric arc and electric resistance welding plants, electric arc steelmaking and ore heat-treating furnaces, gas-discharge lamps, power magnetic amplifiers and transformers. These loads consume current from the network, the curve of which turns out to be non-sinusoidal, and in many cases non-periodic; as a result, non-linear distortions of the mains voltage curve or, in other words, non-sinusoidal modes occur.

Non-sinusoidal modes adversely affect the operation of power electrical equipment, systems of relay protection, automation, telemechanics and communications. The economic damages arising from the impact of higher harmonics are mainly due to the deterioration of energy performance, reduced reliability of electrical networks and reduced service life of electrical equipment. In some cases, there is deterioration in quality and a decrease in the number of manufactured products. Therefore, the progressive introduction of a valve drive and electrical technology has led to the importance and relevance of solving the problem of higher harmonics in electrical networks.

The main range of issues that make up the content of this problem comes down to the following: assessment of the electromagnetic compatibility of sources of higher harmonics and other loads, i.e. the effect of harmonics on electrical installations, and the resulting economic damage; quantification of higher current harmonics generated by various non-linear loads, and prognosis of the values of higher current and voltage harmonics in electrical networks; decreasing levels of higher harmonics [1–9].

2. Characteristics of non-sinusoidality contributor in the power supply system of an industrial enterprise

One of the largest non-ferrous metallurgy enterprises in RNO-Alania is Pobedit OJSC. The main products of the plant are hard alloys, tungsten and molybdenum billets. The most energy-intensive is the power building, in which the consumers of the noah caste of the rank *H*-distribution are concentrated: electrolyzers FB-500 M; $n_1 = 6$; gas blowers VK-50; $n_2 = 2$; calcining furnaces; $n_3 = 5$; compressors 202 GP 20/2; $n_4 = 3$. Let us evaluate the level of higher harmonics for the transformer-converter unit of the electrolysis units of the power building of Pobedit OJSC. A transformer TDNPV-12500/10-U1 and converter unit VAKV2-12500/450 are installed at the substation of the power building, the parameters of which are $S_t = 6800$ kVA, $U_2 = 375$ V, $u_k = 11\%$, $U_d = 450$ V, $I_d = 2{\cdot}6250 = 12500$ A, $\eta = 97\%$, cosϕ = 0.93. The converter is made with a push-pull circuit of the sequential type, equivalent to the twelve-phase rectifying mode $m = 12$, as a result of which the current and voltage curves contain the 11, 13, 23, 25th harmonics.

The main contributer of the non-sinusoidal distortion of the voltage in the power supply system (PSS) of the enterprise are the semiconductor silicon converters of the type VAKV2-12500-450, designed to rectify the current in electrolyzers of the type FV-500M, in which the distilled water is decomposed into hydrogen and oxygen. Hydrogen is the main reducing agent in the production of hard alloys in resistance furnaces and welding machines. The rated power of the electrolyzers is 2993 kW. The replacement scheme of the power supply plant is shown in Figure 1.

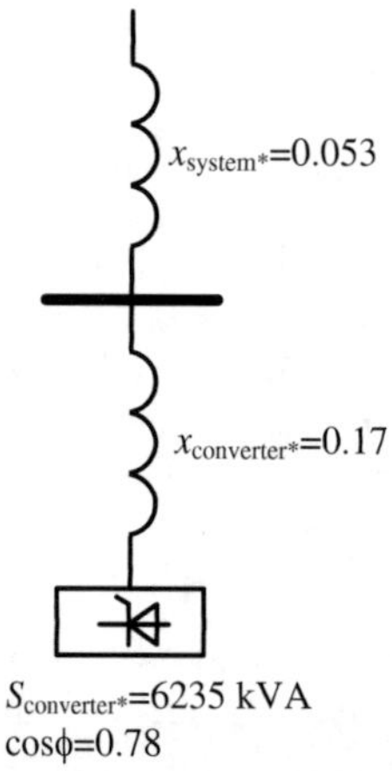

Figure 1. Substitution scheme of the enterprise power supply system

Figure 1 shows the calculated values of system resistances ($x_{system*}$) and converter ($x_{converter*}$) in relative units, power consumed by the converter ($S_{converter}$) and its power factor ($\cos\phi$). The effective values of the 11th, 13th, 23rd and 25th harmonics and the sinusoidal distortion coefficient of the voltage curve k_U are determined:

$$U_n = \lambda_\gamma \frac{m}{6}\frac{\Delta U_1}{\sqrt{2}}, \tag{1}$$

$$k_U = x_{system*}\sqrt{\frac{3}{\pi}\cdot\frac{\sin\phi}{x_{system*}+x_{converter*}} - \frac{9}{\pi^2}}, \tag{2}$$

where λ_γ is the coefficient determined depending on the angle of commutation of the converter γ=0,17 for the *n*-th harmonic; ΔU_1 is the depth of the main commutation distortion, $\Delta U_1 = 1.25$ kV.

The calculated values of the parameters are: $k_U = 4.84$ %; $U_{11} = 0.25$ kV; $U_{13} = 0.23$ kV; $U_{23} = 0.12$ kV; $U_{25} = 0.14$ kV.

3. Results of studies of non-sinusoidal voltage on the buses of the switchgear of the enterprise with valve converters

As a result of the research, the values k_U and $k_{U(n)}$ for interfacial voltages at different load currents (I_{load}, A) of a single rectifier unit [10-12] were determined.

The amplitude-frequency spectrum of harmonic oscillations of phase-to-phase voltages of normative normal $k_{U(n)normal}$ and maximum permissible values $k_{U(n)normal} = 1.5 \cdot k_{U(n)limit}$ are shown in Figure 2. The total values k_U and $k_{U(n)}$ of electrolyzers receiving power from 4-6 rectifier units, may exceed the maximum permissible values.

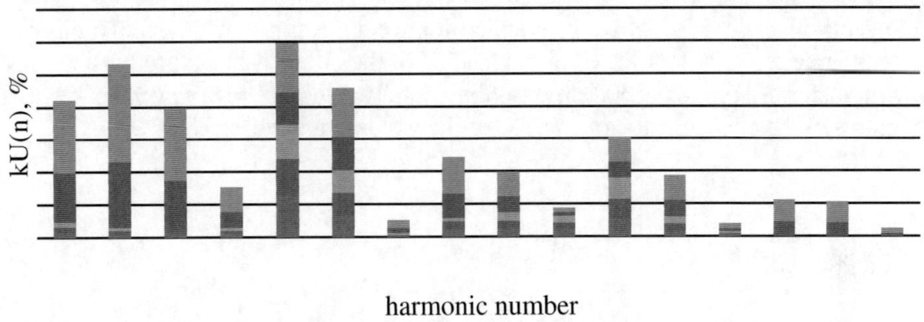

Figure 2. Amplitude-frequency spectrum of harmonic oscillations

It is established that the values k_U are respectively:

$$k_{UA\text{-}B} = 2.7\ \%;\ k_{UB\text{-}C} = 2.6\ \%;\ k_{UC\text{-}A} = 2.4\ \%.$$

Let us check the non-sinusoidal coefficient using the calculation formula, taking into account the three-phase short-circuit current on the buses of the switchgear with voltage 110 kV. According to the power system, the value of the three-phase short-circuit current (s.c.) on the substation buses (PS) RP 110 kV is $I_{\kappa 1}^{(3)} = 10.3$ kA. The non-sinusoidal ratio of electrical networks with valve converters is calculated by the formula:

$$k_U = \frac{S_{nominal.converter}}{S_{short\ circuit}} \cdot \sqrt{\frac{0.955 \cdot \sin\phi}{\frac{S_{converter}}{S_{short\ circuit}} + x_{converter}} - 0.91}\,, \tag{3}$$

where $x_{converter}$ – is the inductive resistance of the converter unit. The formulas feature the parameters of the power supply system of JSC Pobedit, the supply transformer and the valve converter. The calculated value of the non-sinusoidal coefficient amounted $k_U = 6.7$ %, i.e. exceeds the permissible requirements of GOST $k_U = 5$ %, and therefore in the PSS it is necessary to install filter-compensating devices (FCD), providing compensation for the deficit of reactive power of the main frequency, filtering higher harmonics, and balancing the network voltage. FCD is placed in the node connecting the nonlinear load. The total reactive power generated by the filters is selected from the condition of the balance of reactive power in the load node.

In the course of performing the calculations, it was obtained that the power of the filter-compensating device generated to compensate for the reactive load of consumers is 3.2 MVAr. The higher harmonic currents of the valve converter are: $I_{11} = 50.4$ A, $I_{13} = 43$ A.

According to the calculated filter power values and harmonic currents, filters F11-10-1600 and F13-10-1600, $I_{н} = 80$ A were selected. The total power of the filters $Q_{FCD} = 3200$ kVAr. To check the results of the calculations, experimental measurements of k_U were made on RU 6 kV buses. The measurements were carried out using the PKK-57 complex control device.

According to the results of experimental measurements, the value of the non-sinusoidal coefficient on 6 kV switchgear buses is $k_U = 7{,}1$ %, which agrees quite well with the calculated value. The spectrum

IOP Conf. Series: Materials Science and Engineering **663** (2019) 012032 doi:10.1088/1757-899X/663/1/012032

of voltage and current harmonics contains only odd harmonics and the most pronounced of them are the 11th, 13th, 23rd, 25th harmonics.

4. Results of the studies of non-sinusoidal voltage at the plant with welding machines

One of the most important technological conversions, in which the production of tungsten billets is carried out, is the shop of refractory metals No. 1. The main equipment of the workshop is resistance furnaces (rotating (VKP), 13 pipe (TP), muffle) and welding machines. Using the devices AR5, PKK-57, Energotester PKE, the non-sinusoidality of voltage and current arising during operation of welding machines was studied [10–14].

Welding machines TSEP-223A are intended for welding tungsten billets. To ensure the required welding temperature of 2200-2400 °C, direct heating of the billets is used through an OSU 100/0.5 type step-down transformer or OSU 200/0.5, providing output values U_2=18 V, I_2= 4.7 kA. Figure 3 shows the histograms of the changes in $k_{U12(n)}$, $k_{U23(n)}$ of the welding machines TSEP-223A.

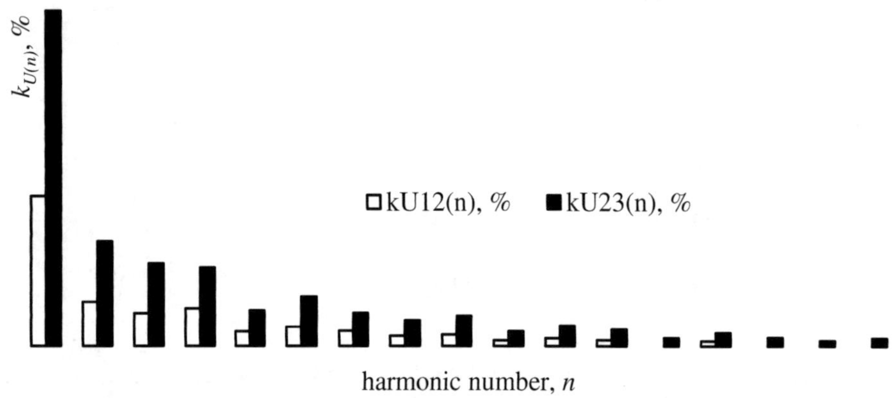

Figure 3. Histograms of change $k_{U(n)}$

From Figure 3 it can be seen that during the operation of welding machines, harmonics arise with voltage n = (3–35), while the value $k_{U(n)}$ varies from 0–67,58 %, the most influenced are 3, 5, 7, 9, 11 and 13 harmonics.

The values of voltage non-sinusoidal coefficients (k_U), are determined, which for welding machines are: k_{U12}=33.82 %, k_{U23}=76.59 %. The value of kU in the PSS of welding machines significantly exceeds the normally acceptable values of $k_{U\ normal}$ = 8 % at U = 0.38 kV.

5. Results of studies of non-sinusoidal voltage at the enterprise with welding machines

It is necessary to tighten the requirements for HH voltage to the required k_{Uallow} =2%. Therefore, already now at the enterprises of non-ferrous metallurgy it is necessary to conduct additional studies of the HH and the development of measures to reduce non-sinusoidal voltage. Based on the analysis of HH for all castes of consumers of carbide production, the rank distribution of HH is constructed, shown in Figure 4.

From figure 4 it is clear that HH are ranked as follows: HH with n = 13, 11, 5 – noah caste; HH with n = 17, 3, 7, 19 – pointer caste and HH with n = 9, 23, 15, 25 – locust caste.

Figure 5 shows the dependences of k_U and W on the rank of consumers ($k_U = f$(Rank), $W = f$(Rank)) for consumers of noah (rank 1–6) and pointer (rank 7–11) distribution castes.

The correlation coefficients between the total value of the nth harmonic component of the voltage ($k_{U(n)}$, %), the voltage non-sinusoidality coefficient (k_U, %) and the share of the noah and pointer castes of electricity consumers (W, %) are 0.56 and 0.37 respectively.

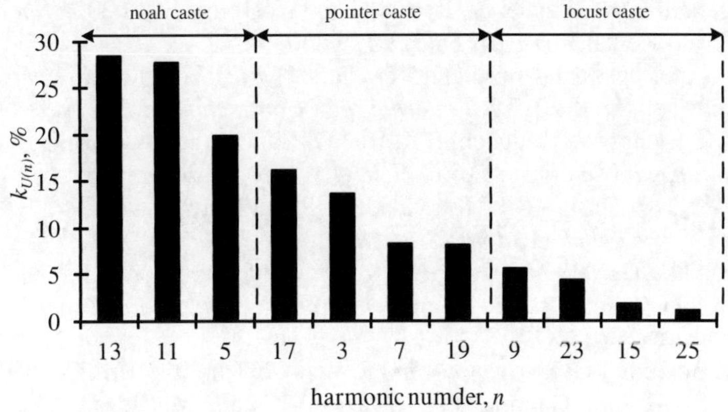

Figure 4. Rank distribution of HH across all castes of carbide consumers

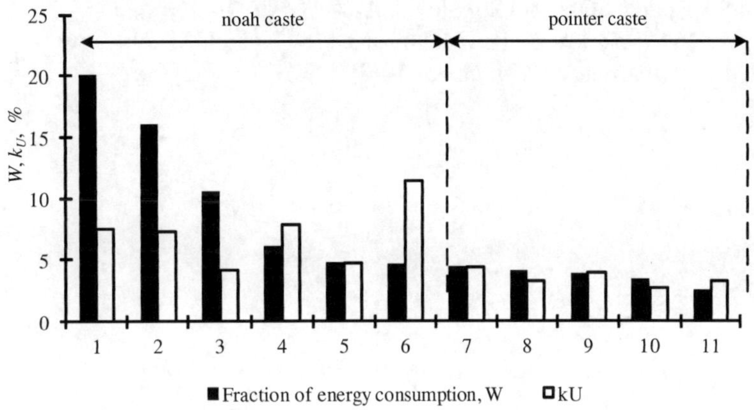

Figure 5. Dependencies $W=f(Rank)$, $k_U=f(Rank)$

6. Conclusion

The method for the experimental study of the amplitude-frequency characteristics of the HH voltage in the 6 kV PSS Pobedit JSC has been developed. The analysis showed that the values of $k_{U(n)}$ and k_U of individual non-linear consumers exceed the values allowed by GOST.

The range analysis of the voltage of the noah caste of HH by voltage (n = 11, 13, 5) was carried out and it was established that the values of the correlation coefficients between $k_{U(n)}$, k_U and the share of noah and pointer castes of electric power consumers (W, %) for carbide production are: 0.56 and 0.37, which allows us to limit the range of priority tasks for calculating the required FCD.

It is necessary to conduct further studies of HH in PSS of industrial enterprises and the development of methods for calculating and optimally distributing FCD in PSS of 6 kV of non-ferrous metallurgy enterprises.

7. Acknowledgments

The results of the work are reflected in the Grant project of the President of the Russian Federation for the support of young scientists, MK-1324.2007.8, on the topic "Research and development of mathematical models of electricity quality indicators at non-ferrous metallurgy enterprises".

References

[1] Ahmad B A, ElSheikh H H and Fadoun A *2015 International Conference on Industrial Engineering and Operations Management (IEOM)* 1–8

[2] Patrao C, Delgado J, Almeida de A T and Fonseca P 2011 *11th International Conference on Electrical Power Quality and Utilisation* 1–6

[3] Norouzi H, Abedi S, Jamalzadeh R, Rad M G and Hosseinian S H 2014 *Energy* **68** 140–7
[4] Bhonsle D C and Kelkar R B 2016 *Energy* **115** 830–9
[5] Hernandez J C, Ortega M J, Cruz J and Vera D 2011 *Electric Power Systems Res.* **81** 1247–57
[6] Gujar N S and Talati S K 2017 *7th International Conference on Power Systems (ICPS)* 111–5
[7] Ricciardi T R, Freitas W, Taniguchi F K, Hax G, Moya R and Archilli G B 2018 *18th Int. Conf. on Harmonics and Quality of Power (ICHQP)* 1–6
[8] Khramshin V R, Khramshin R R, Karandaev A S and Medvedev V N, 2015 *Int. Siberian Conf. on Control and Communications (SIBCON)* 1–6
[9] Morsi W G and El-Hawary M E 2011 *Electric Power Systems Res.* **81** 1381–7
[10] Klyuev R V and Bosikov I I 2016 *2nd Int. Conf. on Industrial Engineer., Applications and Manufacturing (ICIEAM)* **53** 1–5
[11] Klyuev R V, Bosikov I I, Gavrina O A and Revazov V Ch 2018 *MATEC Web of Conf.* **226** 1–5
[12] Klyuev R V, Bosikov I I, Gavrina O A, Revazov V Ch and Madaeva M Z 2018 *Proc. of the VIII Sci. and Technol. Conf. "Contemporary Issues of Geology, Geophysics and Geo-ecology of the North Caucasus" (CIGGG 2018)*, *Advances in Engineer. Res.* vol 182 169–74
[13] Petrov Yu S, Gabaraev O Z and Sokolov A A 2015 *Gornyi Zhurnal* **8** 25–7
[14] Galachieva S V, Sokolov A A, Makhosheva S A and Sokolova O A 2018 *Sustainable Development of Mountain Territories* **10(3)** 329–35

IOP Conf. Series: Materials Science and Engineering **663** (2019) 012033 doi:10.1088/1757-899X/663/1/012033

Ensuring the consumer reliability based on retrospective analysis

R V Klyuev[1], O A Fomenko[2], O A Gavrina[1], A A Sokolov[1], O A Sokolova[1], M T Plieva[1], A A Kabisov[1] and E Yu Ikoeva[3]

[1]North-Caucasian Institute of mining and metallurgy (State Technological University), 44, Nikolaeva St., Vladikavkaz, 362021, Russia
[2]Southern Federal University, 105/42, Bolshaya Sadovaya Str., Rostov-on-Don, 344006, Russia
[3]Gorsky State Agrarian University, 37, Kirova St., Vladikavkaz, 362040, Russia

Abstract. The article discusses the research methods of indicators of structural and functional reliability for electricity consumers based on a retrospective analysis, which makes it possible to evaluate the causes of under-supply of electricity to consumers and to develop measures to eliminate these causes. The analysis of the reliability of overhead power lines of the power system. The analysis of the causes of failures for 2012–2018 showed that about half of the causes of failures are not established at all, other causes of failures are approximate. The exceptions are the causes of failures due to a malfunction of the relay protection and automation devices, the proportion of which in the total number of failures does not exceed 2–3 %. In order to obtain an objective assessment of the influence of climatic factors on the reliability of the electric power system, a special technique has been developed. This technique allows quantifying the average values of the correlation coefficients between the failure rates of overhead power lines and climatic factors.

1. Introduction

The reliability of power supply is understood as the reliable supply of consumers with electricity of the required quality with possible economic performance indicators. The interruption of power supply is possible in case of failure of a certain group of elements (complete failure), however such cases are rare, more often there is a decrease in the transmitted power due to power shortage, as well as overloading of the remaining elements in operation (partial failure).

Under the conditions of operation of electric power systems (EPS), the reliability of power supply is quantitatively evaluated by the parameters of circuit and operating reliability. Under the reliability of the circuit the probability of providing electricity to consumers of EPS is meant. In particular, during the energy analysis at the enterprises of non-ferrous metallurgy of the Republic of North Ossetia-Alania: Electrozinc OJSC (lead-zinc production), Pobedit OJSC (hard-alloy production), reliability indicators were investigated.

2. Key indicators structural reliability

For a quantitative assessment of structural reliability, single and complex indicators are used. The life (t_{lifei}) and restoration time ($t_{rest.timei}$) are taken as single indicators of the reliability of recoverable elements of EPS. The failure means the termination of an element's performance (full or partial). The operational condition of the equipment between the two failures determines the life. The complex indicator represents a generalized assessment of the reliability of an element and is characterized using the availability rate:

$$k_{av.rate} = \frac{t_{life_i}}{t_{life_i} + t_{rest.time_i}}, \tag{1}$$

The failures of elements are considered as rare events, therefore, for practical calculations, average reliability indices of the same type of elements are used. These indices are calculated on the basis of statistical data for a certain period of time:

$$\bar{t} = \frac{1}{N}\sum_{i=1}^{N} t_{life_i}, \tag{2}$$

$$\bar{t}_{rest.time} = \frac{1}{N}\sum_{i=1}^{N} t_{rest.time_i}. \tag{3}$$

The failure flow of EPS elements is the simplest Poisson flow, thus the probability of failures can be quantified:

$$q = \lambda \cdot e^{-\lambda t}. \tag{4}$$

where $\lambda = 1/t$ – is the rate of the failure flow.

The probability of equipment failure-free operation can be quantified by the availability rate:

$$p = 1 - q \approx k_{av.rate}. \tag{5}$$

Investigating the change in reliability indicators by years, it is possible to estimate the expected values for the future, i.e. to perform a retrospective analysis of the elements reliability. This provides the basis for the development and adjustment of the timing of the withdrawal of equipment in scheduled preventive maintenance or conducting preventive tests.

The failure of an element or its withdrawal for repair may break the power supply circuit of consumers, but this is mainly determined by the power supply circuit and the level of redundancy. In case of failure of one element, the power is supplied through the other, and the probability of failure of both elements (active resistance, R and inductance, L) is equal to the product of the failure probability of each element. The situation is similar when the element is brought in for repair, but in this case the situation of the second element failure and, accordingly, the formation of a power supply interruption are possible. In this case, the probability of interruption of the power supply of a specific consumer will be determined by the sum of the probability of successive chains of elements of the electrical network:

$$Q = \sum_{i=1}^{k_1} q_i + \sum_{i=1}^{k_2} (q_k \cdot q_l)_i + \sum_{i=1}^{k_1} q_{plan.out_i} + \sum_{i=1}^{k_2} \left(k_{h.f._1} \cdot q_k \cdot q_{plan.out1} + k_{h.f._2} \cdot q_{plan.out1} \cdot q_l\right)_i. \tag{6}$$

where q_i is probability of accidental failure of elements;

$q_k q_l$ is the probability of a crash at the same time of two elements;

$q_{plan.out i}$ is the probability of planned outage.

$q_k q_{pll}$ is the probability of failure of the element k at the time of repair of the element l;

$k_{h.f.1}$, $k_{h.f.2}$ is the coefficient of hardware failure of an element of type k at the time of repair of an element of type l, if they are of the same type, then $k_{n1} = k_{n2}$.

The retrospective analysis of the structural reliability of the power supply system allows developing measures not only to improve the reliability of the elements of the electrical network, but also to change the power supply circuit.

Assessing the probability of interruption of power supply, it is necessary to take into account that the failure of elements usually does not lead to a complete interruption of the power supply to consumers, but the resulting overload of the remaining elements forces to limit the transmission of electricity, which is equivalent to a partial interruption of the power supply. Therefore, the reliability of power supply is additionally evaluated by functional or operating reliability using the capacity factor of the elements:

$$k_{cap.factor} = \frac{P \cdot \tau}{P_{inst.power} \cdot 8760}. \quad (7)$$

where P is the actual loading of elements;

P_{inst} is the installed power of the element or the maximum transmitted power;

τ is the number of hours of elements use.

The combined use of indicators of structural (6) and functional (7) reliability makes it possible to more fully assess the reliability of electricity supply to consumers for the development of measures to improve it. In this case, for the purposes of forecasting, it is necessary to combine them functionally with one generalized indicator, for example, the coefficient of providing consumers with electricity service:

$$k_{el.serv} = \frac{W_{ins.energy} - W_{act.cons}}{W_{ins.energy}}. \quad (8)$$

where $W_{inc.energy}$ is increment of energy determined by the declared power for the year;

$W_{act.cons.}$ is actual electricity consumption.

When forecasting the reliability of power supply to consumers, a mathematical model is compiled based on the multiple regression equation:

$$P = b_0 + b_1 \cdot Q + b_2 \cdot k_{cap.factor}. \quad (9)$$

where: b_0, b_1, b_2 are regression coefficients.

Based on the statistics of failures and operating modes, the average indicators of structural and functional reliability (6-8) are determined and the regression coefficients are calculated using the least squares method. Based on the compiled model (9), projected indicators for the future are calculated. The evaluation of the projected reliability indicators allows assessing the degree of change in the reliability of power supply.

In order to improve reliability, the measures are usually taken to upgrade the reliability of components (preventive tests, repairs, replacement of obsolete equipment). In order to improve the circuit reliability, a redundancy circuit (including parallel elements) and modal (replacement with a line of a higher bandwidth) are provided. Similarly, the transformers are replaced with more power, and also provide the inclusion of additional power sources. Planned measures are evaluated quantitatively based on changes in the parameters of the circuit and operation reliability, i.e. the effectiveness of these measures is evaluated.

Conducting a retrospective analysis of the reliability of power supply allows assessing the causes of under-supply of electricity to consumers and developing measures to eliminate these causes [1–7].

3. Analysis of failures of overhead power lines in electric power systems

The main task of the electric power systems (EPS) is the reliable supply of consumers with electrical energy. In turn, the reliability of EPS can be assessed by two main components: structural and functional reliability. The first is determined by both the reliability of individual elements and the circuit of their connection. The second is determined by the operation mode of the system.

One of the largest EPS of the North Caucasus is the North Ossetian branch of IDGC of the North Caucasus.

The analysis of the reliability of EPS elements showed that the least reliable element of the system is power lines, which account for more than 70% of all failures. As a system-forming distribution network, JSC Sevkavkazenergo mainly considers 110 kV overhead lines (VL-110). At present 81 Vl-110 kV overhead lines with a total length of 1009 km are in operation. These air lines operate in difficult, widely varying conditions, and are exposed to a number of external factors, the main of which are: changes in wind speed, icy rime deposits, daily and seasonal changes in air temperature, chemical elements and their compounds (for example, salt , alkali, acid).

Using the database of element failures of the North Ossetian branch of Sevkavkazenergo, the curves of the distribution of the frequency of failures over the years were constructed. Figure 1 shows the regression models of failures of 110 kV overhead lines.

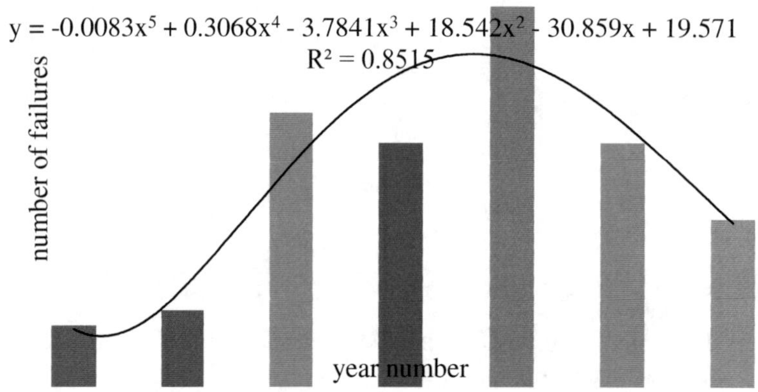

Figure 1. Regression model of failures for VL-110 kV overhead lines

The main indicators of the structural reliability of the electrical network are indicators of the reliability of the elements of the EPS under study. Currently, there are 81 110 kV overhead lines and 33 substations in operation, in which 61 transformers are installed.

The distribution of the frequency of failures among them is unequal, so about 20 % of the lines were disconnected on average 2–3 times a year for various reasons and for different durations, including the response time of the automation. Over 30% of overhead lines for the entire study period 2012–2018 shut off no more than 2–3 times. The disconnection of transformers during this period, on average 2–3 times a year. Therefore, the focus on improving the reliability of EPS elements should be given to overhead power lines.

To analyze the causes of failures of 110 kV overhead lines, typical causes are selected from the database and new tables are formed for them, where elements that failed due to similar features are entered. As it is known that the causes of element failures can be divided into two types: non-objective and objective. Objective reasons include the effects of climatic conditions (thunder, wind, flood, ice, etc.). Non-objective reasons include a violation of the operating mode (personnel errors, failure of relay protection and automation) and a culture of operation and maintenance (violation of preventive test schedules for the destruction of elements by unauthorized persons, etc.). A special position is occupied by unidentified causes of failures, according to the records in the dispatch service logs.

The distribution of causes of failures for 2012–2018 found that about half of the causes of failures was not defined. This indicates a lack of control over the maintenance of operational logs and the responsibility of personnel to account for the causes of failures. The main share of failures accounted for urban electrical networks. Thus they account for about 40 % of failures. According to the number of failures, some lines fail 5-6 times a year for a period from a few minutes to 2–3 hours. The distribution due to failures after unspecified reasons is due to outages, unsatisfactory operation of automation devices (APV, AVR, MTZ, etc.). Under climatic conditions, the share of failures for urban electrical networks is insignificant, but for mountainous and rural areas it acquires a stronger impact.

The distribution by district is also unequal. In comparison with urban electrical networks, the failure rate for other areas is much less and the proportion of unidentified causes is smaller [8–12].

The main share of failures, more than 40%, is accounted for by urban electrical networks; the emergency downtime of transmission lines varies from a few minutes to 2–3 hours. The distribution of the main climatic factors by months for zone I is shown in Figure 2.

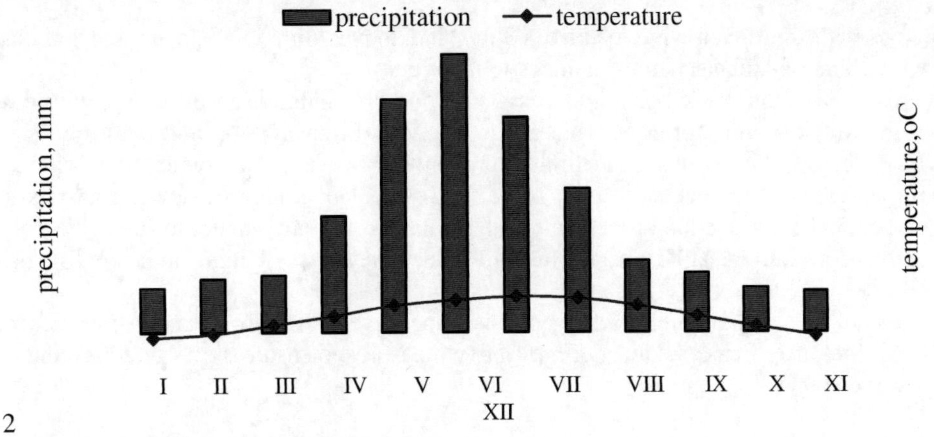

Figure 2. Average monthly precipitation and air temperature

The analysis of the causes of failures for 2012–2018 showed that about half of the causes of failures are not established at all, other causes of failures are approximate. The exceptions are the causes of failures due to a malfunction of the relay protection and automation devices, the proportion of which in the total number of failures does not exceed 2–3 %.

In order to obtain an objective assessment of the influence of climatic factors on the reliability of 110 kV overhead lines, the special technique has been developed. This technique allows quantifying the average values of the correlation coefficients between the failure flows of 110 kV overhead lines and climatic factors.

For the correlation analysis, the samples were made for the main climatic factors based on the data presented in the reports of the Hydro-meteorological Center of the RNO-Alania for 2012–2018. The main climatic factors, the probability of which influence on the reliability of 110 kV overhead lines is obvious, include: air temperature from 0 to –5 °C (ice deposition condition on overhead lines), wind speed (from 3 m/s and more), cloudiness (in points from 8 to 10 and above), snow rain, thunderstorm, hail - all in conventional units, as the relative frequency of occurrence per unit of time. The distribution of climatic factors by months of the year is given in Table 1.

Table 1. Distribution of climatic factors by months

Climatic factors	Frequency of factor recurrence by months, times/month											
	1	2	3	4	5	6	7	8	9	10	11	12
temperature	4.75	4.25	2	0.25	-	-	-	-	-	0.255	0.5	6.25
wind	-	0.25	-	1	0.5	0.5	1.5	0.25	0.5	0.25	0.75	1
Cloudiness	3.25	1.5	4.5	4.75	2.25	1.25	2.5	4	2.5	3	2.5	4.75
Snow	1	1	1.75	0.75	-	-	-	-	-	-	1.25	4.25
Rain	-	-	0.75	4.5	4.75	5.5	3.75	2.25	4.5	1.25	1	-
Thunderstorm	-	-	-	0.75	3.25	3.5	3	1.25	1.5	-	-	-
Hail	-	-	-	0.25	1.25	1.25	1.25	-	0.25	-	-	-

The average values of paired correlation coefficients are shown in Table 2

Table 2. Assessment of the significance of selected climatic factors

Weather factors	Tempe rature	Wind	Cloudiness	Snow	Rain	Thunder storm	Hail
Average correlation values	0.674	0.58	0.68	0.767	0.714	0.75	0.436

4. Conclusion
Based on the conducted research, the following recommendations can be made in order to improve the reliability of EPS.

1. In areas with significant wind load from 3 m/s and above, vibration dampers must be installed to prevent wire vibration and mechanical damage to the wires.
2. In areas with intensive icing, it is necessary to organize monitoring of the icing process on the overhead lines, review and shorten the time for inspection of transmission lines and improve network utilization, develop and introduce ice melting to prevent the destruction or swinging of wires.
3. In urban areas, it is necessary to increase the level of operation of power transmission lines, strengthen preventive inspection and cleaning of insulators that are subject to intensive pollution by industrial emissions into the atmosphere. This will reduce the flow of failures in heavy fog, drizzle and snow.
4. In areas of intense lightning discharges, it is imperative to install lightning protection cables on the overhead lines, to increase the level of preventive measures to ensure the required ground resistance value of overhead lines.

5. Acknowledgments
The results of the work are reflected in the Grant project of the President of the Russian Federation for the support of young scientists, MK-1324.2007.8, on the topic "Research and development of mathematical models of electricity quality indicators at non-ferrous metallurgy enterprises".

References
[1] Doletskaya L I, Kavchenko V P and Solopov R V 2015 *Sci.* **6** 1–13
[2] Timashova L V, Nikiforov E P and Nazarova I A 2014 *Energy of a single network* **5** 6–14
[3] Tso G K, Liu F and Liu Ke 2013 *Proc. Computer Sci.* **17** 752–8
[4] Santana P H and Bajay S V 2016 *Energy reports* **2** 62–6
[5] Franc B, Filipovic-Grcic B and Milardic V 2016 *Electric Power Systems Research* **138** 78-84
[6] Oseni M O and Pollitt M G 2015 *Energy Economics* **52** 277–86
[7] Pezzini P, Games-Bellmunt O and Sudria-Andrey A 2011 *Reviews of renewable and sustainable energy* **15** 2028–41
[8] Klyuev R V, Yegorova E V, Bosikov I I and Tsidaev B S 2018 *Sustainable Development of Mountain Territories*, **10(3)** 392–403
[9] Klyuev R V, Bosikov I I, Gavrina O A and Revazov V Ch 2018 *MATEC Web of Conferences* **226** 1–5
[10] Klyuev R V, Bosikov I I, Gavrina O A, Revazov V Ch and Madaeva M Z 2018 *Proceedings of the VIII Science and Technology Conference "Contemporary Issues of Geology, Geophysics and Geo-ecology of the North Caucasus" (CIGGG 2018)*, *Advances in Engineering Res.* **182** 169–74
[11] Petrov Yu S, Gabaraev O Z and Sokolov A A 2015 *Gornyi Zhurnal* **8** 25–7
[12] Galachieva S V, Sokolov A An, Makhosheva S A and Sokolova O A 2018 *Sustainable Development of Mountain Territories* **10(3)** 329–35

IOP Conf. Series: Materials Science and Engineering **663** (2019) 012034 doi:10.1088/1757-899X/663/1/012034

Technical support of the power lines design - as a linear structure in difficult mountain conditions

L I Kortiev[1], R V Klyuev[1], R P Kulumbegov[1], A L Kortiev[1], I I Bosikov[1], O A Gavrina[1], and Madaeva M Z[2]

[1]North-Caucasian Institute of mining and metallurgy (State Technological University), 44, Nikolaeva St., Vladikavkaz, 362021, Russia
[2]Grozny State Oil Technical University named after academician MD Millionshtchikov, 100, Isaeva ave., Grozny, 364051, Russia

Abstract. The article highlights the process necessary at the present stage of organizing the scientific and technical support of projects for the construction of overhead power lines (OPL) in difficult mountain conditions. The development of mountain areas is associated with overcoming various technical, technological and organizational difficulties, sometimes reducing the expected economic, environmental and social benefits. In order to obtain the maximum effect from the activities carried out during the construction of power lines, the article substantiates the method of scientific and technical support - an expert opinion of a scientist on the safe operation of power lines, which will provide economic and social benefits to society.

1. Introduction

The development of a mountainous area, as the practice of the development of society shows, begins with the construction of a hiking or pack pass and then its improvement to the possibility of transporting materials and structures for the construction of essential facilities for human life. The creation of social and living conditions of a person includes pulling of power lines, water, gas lines, etc.

In the middle of the last century, when the question of connecting the North Caucasus and Transcaucasian energy systems arose, this became possible thanks to the then existing Georgian Military Road.

The location of the power lines along this historic highway did not represent any special interference. Especially in the most pass area in the area of the Cross Pass, where the elevated plateau and relief for the design and construction of high-voltage transmission lines made it possible for the multivariate choice of the direction of the line. In fact, the only problem was the removal of power lines from avalanche-prone areas. Then design organizations, both the North Caucasian and Georgian energy builders and, accordingly, the designers, coped with this problem quite successfully and two powerful North Caucasian and Transcaucasian power systems were looped back. It should be noted that in addressing the issue a particularly difficult area was the Darial Gorge section, where, along with the avalanche hazard, Russian specialists met with conditions for pulling power lines on stone-landslide, landslide and mud-prone areas, so that by the complexity of relief, geological and other conditions, an example of construction Power lines through the Cross Pass in Caucasus is instructive for science and practice.

2. Relevance of the topic.

The stability of the power lines depends on the conditions of its operation. In stable operation of power lines under flat operating conditions, there are no special external influencing factors. Another question is if it is pulled in mountain conditions, where the stability of the operation of power lines is associated with geological, relief and other crushing processes from different slope (surface-exogenous) phenomena. In such conditions, the operation of power transmission lines is a difficult task, requiring tremendous efforts, since the steady state of the society and its social and living conditions are connected with its steady work. Monitoring the sustainability of power lines, the timely elimination of weak "points", preventing outages from winter and natural external factors are important, especially the preservation of social stability [1-9].

On the example of operation of a single power transmission line to South Ossetia, its operators annually encounter cases of avalanche crashes, since power lines in difficult areas are not monitored and there is no scientific and technical support (expertise of unsuccessful technical solutions by scientists or a group of scientists).

3. Essence of the question.

Upon completion of the construction of the Crossing Highway through the Main Caucasus Range along the Roki Pass - Transkam, the question arose of looping up the South portal of the Roki Tunnel and the Dzhava substation with the Sevkavkazenergo electrical networks. For designers it was not worth a lot of energy to find the most advantageous variant of connecting these two points [10-13].

In the course of the construction of the highway, the builders built a “service tractor road” for driving the road equipment from the Northern section to the Southern section through the Magsky Pass, at an absolute height of 2880m. This process of transportation of road equipment was previously carried out by railway through Beslan - Baku - Tbilisi - Tskhinval with a loss of time for almost two weeks. With the laying of the specified tractor road, for example, a bulldozer started working in another construction site after five hours. A clear benefit was achieved [14-16].

In cramped mountain conditions, when any open linear structure - road (iron or automobile) or power transmission lines are under the watchful eye of avalanche mass, mudflow, or other natural phenomena, their protection becomes a difficult issue (Figure 1).

Figure 1. Power transmission tower designed in a dangerous place.

Earlier, when the importance of the tractor road was settled, the economic, tourist, pasture, operational and other values were called the factors of its necessity, but no one had thought of using it for pulling power lines. This appointment gave her energy builders. The raising and transfer of high-voltage power lines through the Main Caucasian ridge was carried out in the past century for the second time, but this time its absolute height rose higher than 400m. Naturally, the operating conditions of power lines at such an absolute height presented particular difficulties in the relief and climatic conditions.

The invention itself with respect to the main construction of power transmission towers reached a fantastic development in the practical sense of performance (Figure 2).

Figure 2. The original fantasy in the construction of power lines.

This is what concerns the plain conditions.

It was mentioned above that in the middle of the last century, designers and builders of high-voltage power lines showed samples of bold engineering solutions for the transfer of wires and the original placement of the power lines themselves on the slopes with the difference in vertical and horizontal lines not seen before at the absolute height of their placement to 2400 m.

During the operation period of power lines in winter conditions, there are often breaks in the spans of wires due to avalanche emissions in areas of power lines that are poorly protected from avalanches. Analysis of the research issue showed that the power lines under consideration was designed and built in summer conditions free from avalanches. Operating conditions in such complex avalanche-hazardous conditions were not sufficiently examined and investigated. The results of the annual winter operation show that 2-3 times the disconnections due to breaks in the winter and avalanche conditions for the operation of power lines, resulting in such an important in economic, strategic, political, social and other values of the object in an unstable state. To all the complex process of operating power lines through the Magsky Pass, it is enriched with the imagination of engineering. In difficult river gorges on the slopes, a method of anchoring of transverse stretch marks was undertaken, on which instead of power transmission towers wires are fixed according to the following scheme (Figure 3) [17-18].

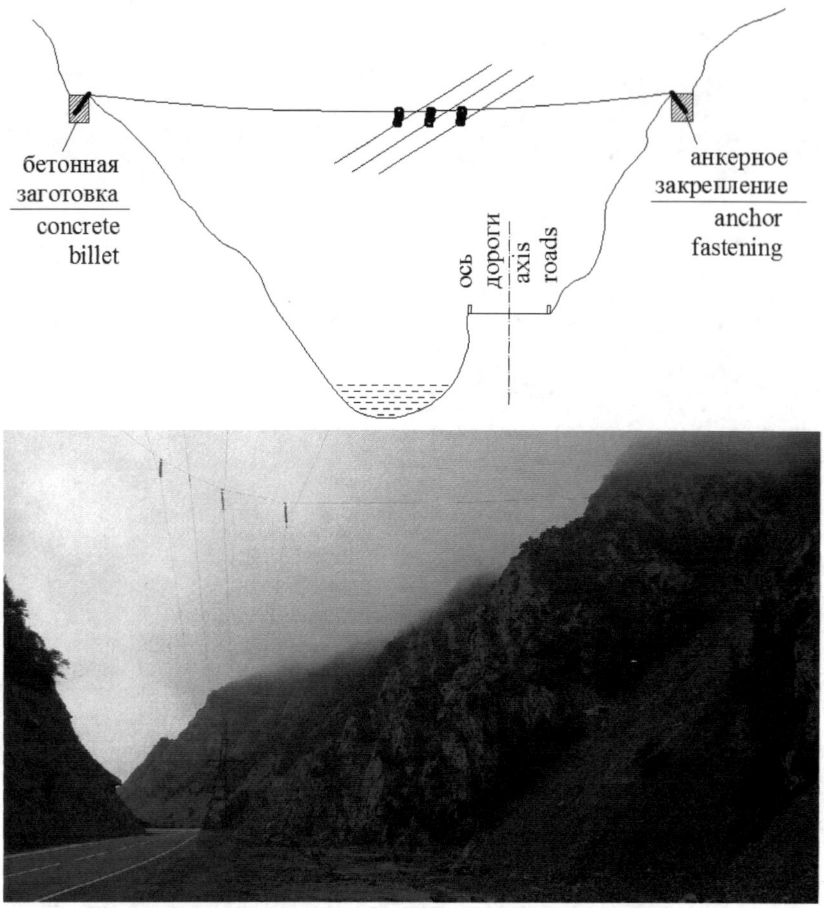

Figure 3. Anchorage sites in difficult mountain conditions.

Here, on the Magsky Pass, for the first time energy builders and designers had to lift power lines to the height of the Magsky Pass of 2880 m.

The considered method is widely used on the Kosar site of the power lines in the Ardon gorge. In order to ensure the safe and uninterrupted operation of power lines through the Magsky Pass, this method should be introduced in particularly complex mountain areas, including the Dzhomagskiy emergency-hazardous area.

Thus, the analysis gives grounds to conclude that the scientific and technical support of the project or expert opinion on the safety of power lines should be carried out at the stage before the construction of power lines, and if it was built earlier (using the example of the Dzhomagsky site of the Magsky Line), then such scientific and technical support should be implemented immediately in order to ensure its safe operation.

4. Conclusion

Trouble-free and uninterrupted operation of power lines in mountain conditions depends on the creation of safe conditions in the area of its pulling and also on natural and climatic conditions.

Operational monitoring should be carried out continuously, especially in the winter period.

The implementation and organization of scientific and technical support of the design and estimate documentation should be carried out from research and development centers whose performers have the appropriate qualifications.

5. Acknowledgments
The research was carried out with the financial support of the Russian Foundation for Basic Research and the Ministry of Education and Science of the Republic of South Ossetia in the framework of the scientific project No. 19-511-07001

References
[1] Buryanina N S, Korolyuk Yu F, Maleeva E I and Lesnykh E V 2018 *Sustainable Development of Mountain Territories* **T.10 3** 404-410
[2] Pospelov P I and Kortiev A L 2018 *Sustainable Development of Mountain Territories* **T.10 4** 624-630
[3] Tomaszewski M, Ruszczak B, Michalski P and Zator S 2019 *International Journal of Critical Infrastructure Protection* **25** 139-151
[4] Zaalishvili V B, Rekvava P A and Melkov D A 2018 *Sustainable Development of Mountain Territories* **T.10 4** 509-517
[5] Mueller C 2019 *Energy Policy* **130** 341-346
[6] Wang J, Xiong X, Hu J and Lu X 2019 *International Journal of Electrical Power & Energy Systems* **110** 232-245
[7] Crespo A, Sola A, Moreu P and Gomez J 2018 *IFAC-PapersOnLine* **11** 211-216
[8] Bamigbola O, Ali M and Oke M 2014 *Applied Mathematics and Computation* **241** 214-221
[9] Lehtomaki M, Kukko A, Matikainen L and Hyyppa J 2019 *Automation in Construction* **105** 1-16
[10] Bernardino J, Bevanger K, Barrientos R and Dwyer J 2018 *Biological Conservation* **222** 1-13
[11] Song B and Li X 2014 *Neurocomputing* **129** 350-361
[12] Tempesta T, Vecchiato D and Girardi P 2014 *Landscape and Urban Planning* **126** 53-64
[13] Brinkerink M, Gallachoir B and Deane P 2019 *Renewable and Sustainable Energy Reviews* **107** 274-287
[14] Madhja A, Nikoletseas S and Voudouris A 2019 *Computer Networks* **152** 87-97
[15] Klyuev R V and Bosikov I I 2016 *2nd International Conference on Industrial Engineering, Applications and Manufacturing (ICIEAM)* **53** 1-5
[16] Klyuev R V, Bosikov I I, Gavrina O A and Revazov V Ch 2018 *MATEC Web of Conferences* **226** 1-5
[17] Klyuev R V, Bosikov I I, Gavrina O A, Revazov V Ch and Madaeva M Z, *Proceedings of the VIII Science and Technology Conference "Contemporary Issues of Geology, Geophysics and Geo-ecology of the North Caucasus" (CIGGG 2018), Advances in Engineering Research,* **V.182** 169-174
[18] Joalland O, Pereau J and Rambonilaza T 2019 *Energy Economics* **80** 75-85

IOP Conf. Series: Materials Science and Engineering **663** (2019) 012035 doi:10.1088/1757-899X/663/1/012035

Influence of underground mining of Taseevskoye field on landscape and open-pit remining

M V Kostromin and Yu T Popova

Transbaikal State University, 30, Aleksandro-Zavodskaya St., Chita, 672039, Russia

E-mail: kostrmv@yandex.ru

Abstract. The Baley ore body unites Taseevskoye and Baleyskoye gold mines. The internal structure of Taseevskoye field is defined by the system of longitudinal and cross deformations splitting the Baleyskaya depression into a series of large blocks. One of the main features of open-pit remining of the First ore zone of Taseevskoye field is that mining operations will continuously happen in a zone dangerous in terms of caves. Caving systems in underground mining lead to displacement of some mineral resources to underlying horizons. Hence, the main provisions are defined to study regularities of earth surface displacement within a field. Depending on the correlation of parameters the earth surface is not deformed or deformed as failures, cracks or smooth movements. The development of depression in the direction and speed increases towards the second mining. Thus, within the contours of the designed pit the rock massif is located in the zone of deformations with developed displacement under the influence of underground mining operations. Contours and sizes of dangerous zones are defined for open-pit mining operations in the zone of displacement.

1. Introduction

The Baley ore body (area – 40 km^2) located in Zabaykalsky Krai 350 km southeast of Chita includes Taseevskoye and Baleyskoye gold fields.

Until 1964 the Baley ore field was developed by open-pit and underground methods. Now the Baley mine is temporarily abandoned.

The Taseevskoye field has been developed since July, 1948. The field was found in 1941 and during the period from 1948 to 1994 16.3 million tons of ore, from which 6.4 million ounces of gold at gold content of 12.2 g/t, were extracted here. This gold was extracted from the epithermal system of cracked gold-bearing quartz veins located on a circular site with the diameter of 1000 meters.

The field was mainly developed via the underground method, which was concentrated on three main reef systems. A wide mineralized zone with lower gold content located within argillic alteration envelope surrounding fissure veins was left undeveloped [1].

The ores of Taseevskoye field are processed at Taseevskaya factory. The Taseevskoye field represents a large ore site of gold-quartz mineralization located in the central part of the central block of the Baley fault trough composed at the highest levels of Jurassic and cretaceous tuffogenic and sandy-conglomerate deposits and at the bottom it is composed of folded Proterozoic rocks [2].

The field resembles a magnificent long deposit, a stretched mineralization band located northeast (300 m wide and up to 2 km long). About 50 vein-shaped bodies grouped in seven ore zones and having abrupt steep pitches in the northwest are discovered within its limits. According to Protodyakonov's scale, quartz ores belong to the 14th category of hardness, their volume weight makes 2.4 t/m^3, fragmentation index –1.0, humidity – 1.87 % [3].

The Taseevskoye field is presented by several almost parallel ore zones were currently the first, second, third, fourth and intermediate ore zones are in operation. The lower boundary of A+B+C reserves in place is the horizon of 516 m. The internal structure of Taseevskoye field is defined by the system of longitudinal and cross deformations splitting the Baleyskaya depression into a series of large blocks. In turn, each of them is split by more or less dense network of deformations thus explaining the generally high level of tectonic break of the area. Thus, the fractural tectonics within the considered site of the field is diverse starting from a standard karst and ending with large disjunctive dislocation like faults and shifts stretching in various directions [4].

2. Opening of a field and the system of development

The Taseevskoye gold field was opened by vertical shafts. The northeast flank of Taseevskoye field is opened from a surface with a shaft to the horizon of 300 m, and a "blind" mine shaft 4 from the horizon of 216 m to the horizon of 466 m. Ore bodies in all horizons are opened with main crosscuts with the section of 8.1-9.0 m^2. The mines are fixed as follows: wooden support with spacing or as a longwall with incomplete frames, arch concrete, rock bolting. All types of supports are used here depending on the stability of rocks [5].

A variety of mining-and-geological conditions predetermined the use of several development systems: with ore shrinkage used for development of steep veins with a capacity of 0.8–3.0 m by solid host rocks; overhead stoping with spacer propping – for vein capacity of 0.4–0.8 m of firm walls and 0.4–3.0 m of soft walls with angles of incidence of more than 50°; sub-level caving – for development of large steeply dipping ore bodies with the capacity of more than 3.0 m, in rocks with tectonic faults [6].

Losses within all development systems reach 20–25 %, ore dilution – up to 46.3 %.

Due to huge mineral losses the development of the so-called First ore zone via the open-pit method is planned in the ceiling and rock pillars.

The backfill was developed only in places of underground structures. Rock and low-grade ore filling was used as a stowage material. Underground development resulted in deformations of the massif of the First ore zone. Failures, some of which reach considerable sizes, were formed on the surface. Their volumes change from 600 m^3 to 12000 m^3 [8].

3. Observations over displacement

Observations over surface and rock displacement within a field began in 1956-57 and were carried out by mining surveyor of the Taseevo mine; in 1960-62 these works were performed by Soyuzmarkshtrest, from 1960 to 1971 the Soyuzmarkshtrest expedition studied rock displacement and defined key parameters of this process. From 1976 to 1981 the observations were made by Chita Polytechnic Institute [8]. Within a design contour of the Taseevskoye pit three displacement zones are defined [9]:

- zone of smooth displacement;
- fractured zone;
- caving zone.

The basic provisions to study regularities of earth surface displacement in Taseevskoye field are the following:

1. Nature and parameter of rock displacement are defined by form and sizes of ore bodies, applied development systems, correlation of sizes of the developed space.
2. Full development of earth surface displacement and minimum angles of displacement are obtained when the sizes of the developed space on the strike and falling of ore bodies are equal or make more than the depth of development.
3. Zones of dangerous displacements and cracks are limited to angles of 50°–70°.

4. The zone of funnels and failures is the most dangerous site of a shift trough. Funnels are formed as a result of the by-pass of loose rocks or stowage material from top to bottom horizons and during development of giant ore bodies [10].

5. Collapses, as a result of which funnels are formed, are stretched dome-shaped perpendicular to interbedding. Depending on the capacity of superincumbent rock and rock front the angle of rapture fluctuates from 75° to 85°.

In terms of the influence of development of some ore bodies on nature and parameters of displacement four main groups may be identified:

1. Ore bodies, which development causes failures, funnels and zones of strong vertical shifts of rocks on the earth surface.
2. Ore bodies, which development causes periodic development of dangerous zones of displacement of the earth surface and formation of funnels over top horizons.
3. Ore bodies, which development is followed by smooth subsidence of a surface with critical sizes of deformation and without formation of failures and large cracks.
4. Ore bodies, which development does not cause dangerous displacement of the earth surface.

The first group includes ore bodies of quite big and very big capacity (up to 20 m and more) developed without filling.

Ore bodies of big capacity developed using stowage material and ore bodies of average capacity (up to 6 m) with the depth of formation of the upper boundary making less than 100 m developed without filling belong to the second group.

The third group is composed of ore bodies of average capacity with the depth of the upper boundary making more than 100 m from the surface developed without filling.

Ore bodies of small (2 m) and average capacity with the depth of the upper boundary from the surface of more than 300 m developed with filling belong to the fourth group.

At present, there are active earth movements in the First ore zone due to undermining of the massif with underground mining operations, which is confirmed by surface subsidence and cavings. Through the development of top horizons (up to 166 m) the removal of failures belongs to the top part of veins. With inclusion of lower horizons there is a shift of newly formed failures towards the hanging layer followed by rock advancing. The Taseevskoye field is presented by rocks of average stability, which besides uncontrolled caving phenomenon, are characterized by their ability to maintain keep dangerous voids under roof arches. Currently, the only reliable method to define voids within a field is to drill exploratory wells [11].

4. Study and analysis conducted by Transbaikal State University

The study conducted by Transbaikal State University show that the earth surface of the First ore zone has no sites with set displacement process. The development of subsidence by direction and speed increases towards second mining, i.e. in northwest direction.

According to surveying, there are about 300 thousand m^3 of subsurface voids in the First ore zone.

One of the main features of open-pit remining of the First ore zone of Taseevskoye field is that mining operations will continuously happen in a zone dangerous in terms of caves. Caving systems in underground mining lead to displacement of some mineral resources to underlying horizons [12].

Being conformed with the situation created in the First ore zone it shall be noted that

1. Open-pit remining with the application of standard technological schemes is connected with stabilization of displacement of the rock massif. This process can be quite long for the following reasons:

a) in the center of a pit there are 7 blocks with concrete pillars on the 216 m horizon, which can temporarily constrain the displacement process, and during blasting operations – cause sudden failure since the development of lower horizons was conducted without fitting;

b) metasomatites lying under deposits can temporarily constrain the removal of failures to the surface.

Besides, stabilization of rock displacement will require immediate and long preservation of underground works, which will affect planned production at Taseevskoye mine.

2. Transition of underground works to the system with fitting is almost excluded since there is no access to developments, the massif is broken, developments are deformed and filled up. Delivery of fittings to some voids will require penetration of many large diameter wells. All this is technically difficult and obviously not favorable in terms of economic efficiency.

Correct assessment and accounting of displacement degree of a rock massif represent the greatest complexity. Under the influence of underground mining operations on the surface of the First ore zone the displacement trough, within which funnels and failures reaching considerable sizes are systematically formed, was created. Underground mining operations on horizons of 166 m, 216 m cause the repeated addition of loose stowage material from top to underlying horizons thus forming additional voids in the massif [13].

The stated conditions and features of mining operations define nature and parameters of displacement of the earth surface and the entire developed massif. Complex mining-and-geological conditions, unevenness of development of steeply deeping ore bodies (falling, stretching and across the strike, selective development of a vein), discrepancy and lack of information on mining operations of the last years and a big variety of factors create serious difficulties in studying and forecasting the nature of rock displacement and to a certain extent affect the solution of open-pit remining of top horizons [14].

The study conducted by Transbaikal State University revealed many sites of profile lines where surface inclinations considerably exceed admissible values. The qualitative assessment of this phenomenon required the definition of the maximum speed of subsidence in sites with average maximum subsidence of reference points and the calculation of relative inclinations. According to the instruction, a critical inclination is the inclination of 4 mm/m. Table 1 shows the results on local sites of profile lines.

Table 1. Maximum inclinations and incremental slopes of the earth surface

No. lines	No. reference	Incremental slopes, mm/m		Inclinations
		2015.01–2016.01	2016.01–2016.02	2009–2016 mm/m
I-I	32-33	0.79	1.25	14.04
II-II	27-28	0.33	0.26	9.90
III-III	13-14	2.40	0.78	10.42
V-V	13-14	1.81	0.67	13.86
VI-VI	13-14	0.93	0.13	15.45

The analysis of these results shows the concentration of zones of intensive displacement of the earth surface on two areas characterized by the greatest connection with the highest inclinations of the earth surface. All failures and funnels are concentrated on these areas. One of such areas is in the region of crossing profile lines 4-4, V-V with lines 3-3, IV-IV, III-III. The second area of intensive displacement is in the region of crossing of lines 4-4 with lines I-I and II-II.

The area maintaining stable condition of the earth surface without funnels and failures is located between areas with lowered deformations and inclinations. This feature in the nature of displacement of the earth surface can be explained by high quality and safety of a reference point during underground operations on the 216 m horizon. It is fair to assume that there will be no narrowing of areas of active deformations of the earth surface with the formation of a uniform zone of failures. This is also confirmed by failures 5, 6 and 7, which were formed in areas of active deformations of the earth surface.

The analysis of deformation of the earth surface shows that displacement has not finished yet. Subsidence of the earth surface increases and it is followed by sudden failures and funnels. The stability of the earth surface is defined by the system of development applied during underground works and depends on mining-and-geological and physicomechanical factors: strength properties of rocks, capacity of an ore body (Figure 1).

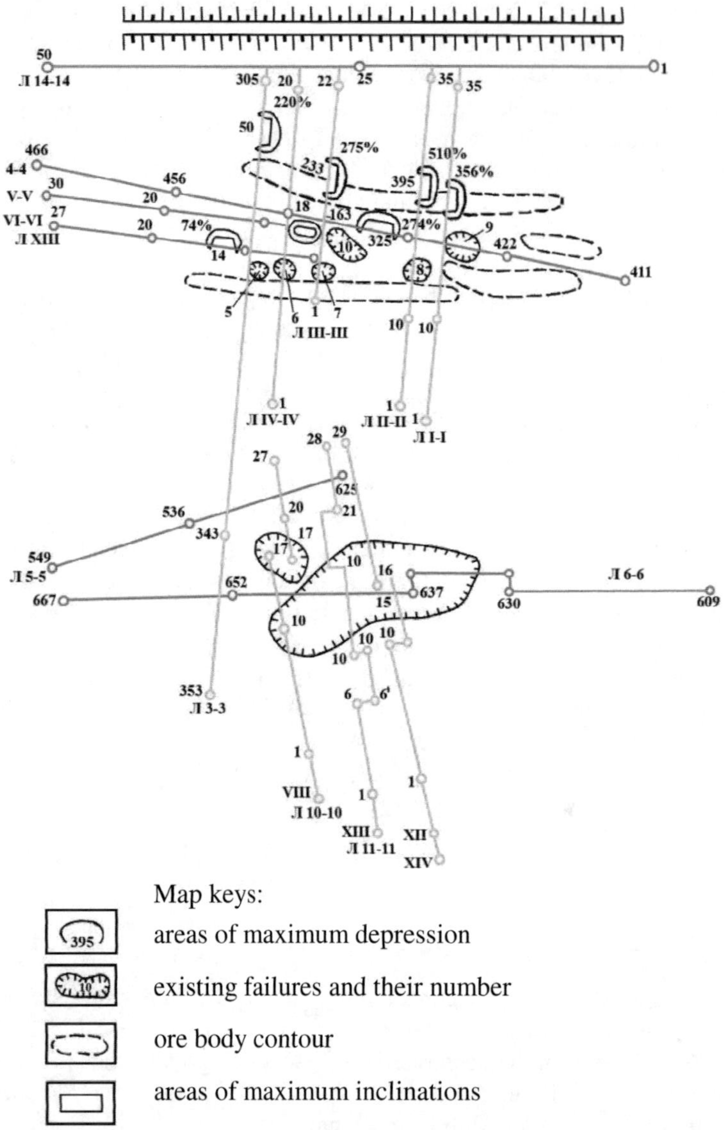

Figure 1. Surface plan of Taseevskoye field in settlement isolines (cavings)

Depending on the correlation of the above parameters the earth surface is not deformed or deformed in the form of failures, cracks or smooth displacement. With the increase of the development depth and at relatively unchanged sizes of the developed space by falling and stretching the deformation of the earth surface decreases due to their smoother distribution in a displacement trough. The speed of caving development on average makes 2.5 m/month. At such speed the voids on the horizons of 166 and 216 m shall be removed. However, the comparison of volumes of voids and failures shows that this did not happen. Such delay in removal of voids is explained by heterogeneity of the massif with faster displacement of the cave roof along weak rocks and with delay in places where rocks have high mechanical strength. Since some underground voids are not removed, we can expect the formation of new failures on a surface of the First ore zone.

IOP Conf. Series: Materials Science and Engineering **663** (2019) 012035 doi:10.1088/1757-899X/663/1/012035

5. Conclusion

Thus, within the contours of the designed pit the massif is located in the zone of deformations with developed displacement under the influence of underground mining operations [15].

Materials of observations and analytical study of the nature of displacement of the earth surface and the massif allow concluding the following:

1. The process of displacement of the surface of the First ore zone continues intensively.
2. There is the increase of the displacement trough by falling of an ore body and in the northeast direction.
3. Within a displacement trough there are two zones of increased deformations and inclinations. Between these zones there is an area of the earth surface without failures and funnels, and it is possible to assume that the narrowing of active deformation zones with the area of smooth depression and formation of a uniform zone of failures is unlikely to happen.

Contours and sizes of dangerous zones are defined for open-pit mining operations in the zone of displacement.

References

[1] Yurgenson G A 2011 Shallow gold and silver deposits, condition of formation and mineralogical and geochemical technology of their deep search and assessment *Scientific notes of Transbaikal State University* **1(36)** 136–45

[2] Spiridonov A M, Zorina L D, Letunov S P and Prokofiev V Yu 2010 Fluorite mode of ore formation of the Baleyskoye gold and magmatic system (East Transbaikalia) *Geology and geophysics* **51(10)** 1413–22

[3] Sekisov A G and Manzyrev D V 2009 Mineralogical-geochemical and nanostructural features of gold-bearing ores of the Zabaykalsky Krai fields *Bulletin of Chita State University* **2** 25

[4] Yurgenson G A 2003 *Typomorphism and ore formations* (Novosibirsk: Nauka)

[5] Rylnikova M, Kalmykov V and Ivachov N 2003 Combined Mining of copper pyrite deposits *Russian Mining* **4** 21–4

[6] Kaplunov D P, Kalmykov V N and Rylnikova M V 2003 *Combined geotechnology* (Moscow: Ore and metals)

[7] Tsarkov V A and Dobroskokin V V 2000 Current trends of gold extraction from radical ores *Mining J.* **5** 81–2

[8] Goris J M 1996 Cable Bolt Support Technology in North America. Ground Control Mining. Cable Bolt Supports *The 15th Int. Conf. on Mountain Pressure* (Golden, Colorado, United States) pp 135–7

[9] Shpilko S I 2005 Choice of polymer materials for anchor rods *Engineering and technology of open-pit and underground mining* **331** 116–20

[10] Dzhigrin A V, Burchatsky V M and Shpilko S I 2007 Results of the study of physical-mechanical properties of composite materials used for the production of polymer anchors *Coal* **11** 97–101

[11] Eremenko A, Eremenko V and Viktorov S 2009 Technological Problems ofth1.on Ore Mining in Siberia *Proc. of the 6 Int. Conf. on Physical Problems of Rock Destruction* ed Jian Zhao (Beijing, China) pp 270–3

[12] Zvyagilsky E L 2000 Calculation of depression parameters of a vertical pit mouth of the recovered mine *Geotechnical mechanics* **18** 26–31

[13] Weakly L F 1982 Ore moving logistics for room and pillar mines in the vilvr nym-trend *Mining Engineering* **4** 403–4

[14] Dawson L R 1995 Developing Australia's First Block Cave Operation at Northparkes Mines Endeavour *Australasian Institute of Mining and Metallurgy* **26** 155–64

[15] Richard H C 1984 Long rock bolt support at the new broken hill consolidated limited *Pras Austral* **251** 21

Technical and economic assessment of existing and projected enterprises for the development of placer mines

M V Kostromin[1] and T Y Panina[2]

[1]Transbaikal State University, Chita, Russia
[2]Irkutsk State Agrarian University named after A.A. Ezhevsky, Irkutsk, Russia

E-mail: kostrmv@yandex.ru, panirais@mail.ru

Abstract. The development of placer mines is a significant source of replenishment of Russian reserves with precious stones and precious metals. Currently, the development of placers is carried out in the vast territory of the Far East, Transbaikal, Eastern and Western Siberia, Altai, Ural. The efficiency of mining operations is significantly influenced by geological, mine engineering, climatic factors, the physical and mechanical properties of rocks, the physical state of placers (frozen, thawed), location (distance from settlements and communication routes), the percentage of clay material, etc. Based on these factors, the technical and economic indicators of the same-type mining machines (for example, drag diggers) often differ significantly (sometimes several times). In this regard, the method of complex technical and economic assessment of operating and projected mining enterprises for the development of placer mines has been developed. The method includes the evaluation of the mines according to the following technical and economic indicators. 1. Indicators characterizing operating conditions. 2. Indicators characterizing the potential of an enterprise. 3. Indicators characterizing the enterprise work results. Mathematical processing of all groups of factors in natural units of measurement is not possible due to the difference in units of measurement and a large amount of variation. Therefore, it is necessary to bring all the indicators in the appropriate form. These requirements coincide with the system of indexing (coding) of the source data, in compliance with the necessary principles. The article provides an example of coding. Further, the classification of the enterprises using the Euclidean metric is carried out. The proposed criteria are the basis for a comprehensive assessment of mines, as well as for obtaining linear regression techniques, a simple regulatory function that establishes the reciprocity between the calculated values of the economic results from the conditions and capabilities of the enterprises. The deviation of the calculated value from the actual value allows to judge how effectively (or not effectively) an enterprise operates under the appropriate conditions and opportunities. The article provides examples of determining the results of the activities of particular Transbaikal enterprises by this method.

Keywords: development of placers, assessment of the mines, groups of factors, Euclidean metric, criteria, regulatory function, deviation indices.

1. Introduction

One of the priorities of the mining industry is the extraction of precious stones and precious metals, a significant part of which is extracted with less labor inputs and at the lowest cost from placer mines. Development of placers is carried out in the vast territory of the Far East, Yakutia, Transbaikal, Eastern Siberia, Altai, Western Siberia and Ural.

A feature of currently developed mine fields is that most of them (80–90 %) are difficult to develop and the sands included in them are difficult to reach because of their location in geographic and climatic zones with severe climate, where perennial and deep seasonal permafrost is almost universally distributed (with the exception of the southern regions of Western Siberia, Altai and Ural). About 30–40 % of placers are characterized by a high content of clay and cemented material. In addition to deep-seated Lensky placers, today the similar mines have been explored in the Far East, Yakutia, Transbailkal and other areas, which for the most part have a relatively small capacity of the productive formation and a powerful thickness of dirt rocks (up to 50–70 m and more). All of it greatly complicates the development of placers, is the reason of low productivity, loss of valuable components, reduction of technical and economic performance in general. Most placer mines can be worked out by economical and highly productive dredge, as well as hydromechanized (bulldo-grain-scraper, excavator) methods.

Today the main gold mining areas have a sufficient amount of reserves of explored placer mines, which for a long period of time will retain its importance as a supplier of relatively cheap precious metal that does not require large capital expenses for extraction and enrichment, compared with the ore one. About 70–75 % of these mines (equal to the volume of rock mass) are suitable for the development of the dredge method, which is, as a rule, the most effective of the currently known methods for the development of placers. The prime cost of the metals production from the gold-bearing placers by the dredge method is 1.3–2.5 times lower than the competing hydromechanical method. Modern dredge fleet consists mainly of 80, 150, 250, 380-liter dredges.

The distribution of dredges (for a total scoops capacity) of the total amount is characterized by approximately the following: 50–80 L. (7,5 %); 150–180 L. (9,5 %); 210–250 L. (62 %); 380 h (21 %). The largest dredging fleet in terms of the number of dredges and the total capacity of buckets (over 20 %) develops the placers of Amur region. The dredging fleet of Transbaikal and the Far East for 76 % consists of 250-liter dredges. The duration of the washing season varies widely from 80–90 to 290 days, with the average duration of the dredge work of 180–190 days per year. It depends on many reasons: climatic conditions of the area; availability of sufficient amount of water, especially in spring and autumn periods; the degree of preparedness of landfills and the dredges for winter works; duration and quality of capital repairs, qualification of the dredge crews. Dredges usually start working in April or May finish working at the end of October or in November. Some dredges, when successfully solving issues of peat overburden, thawing and protection of landfills, water supply of the cut and repair works, start working at the beginning or in the middle of March up to the middle or the end December. Capital (winter) repairs pf dredges lasts from 1 to 4 months, begins in January-February and ends usually in April or May, sometimes in March or June. Seasonal performance of a single 250-gallon dredge is 200–500 thousand m^3 thousand with the projected 750–800 m^3 at the same time at landfills where overburden works, thawing and frost protection were carried out qualitatively, under the similar conditions, the dredges can wash up to 1.3–1.6 million m^3 (dredges of Amur region, the Yenisei pool). [14]

As a result, the technical and economic performance of many of the same-type dredges at work, due to the above stated reasons, differ significantly (sometimes several times). Hydromechanized development of placers faces similar challenges. When projecting the development of placer mines for the same reasons, it is difficult to choose a development method, a mining technology, overburden and mining techniques and an enrichment method.

2. Study Relevance

Resulting from the above stated problems, the developing of a methodology for a comprehensive technical and economic assessment of existing and projected enterprises for the development of placer mines was set as the main task of the study. In addition, the possibility of applying a comprehensive assessment of mines was taken into account:

1. When setting the starting prices of mines at auctions. Currently, the starting price is determined on the basis of gold reserves and prevailing prices for it. This does not take into account But the factors that complicate the development of the field (such as distance from the station base, the presence of permafrost, peat capacity and a number of other factors) are not taken into account.

IOP Conf. Series: Materials Science and Engineering **663** (2019) 012036 doi:10.1088/1757-899X/663/1/012036

2. In determining the investment attractiveness of mines. A comprehensive assessment will reveal all the negative and positive factors of the development of a particular field in the search for a potential investor.

3. For facilitating the procedure of determining the size of the subsoil fee. Currently, the fixed intra-industry mining rent in the form of a tax on the extraction of minerals does not fully reflect the conditions of exploitation of mines.

4. To establish tax benefits for mining enterprises, especially when paying the land tax and other taxes.

5. When determining the performance of a mining enterprise in terms of its place among the enterprises of the industry, within the industry or within the region, from the national economic point of view, based on a comparison with the performance of other mining enterprises and primarily with the accumulated best practices.

Previously, the works of various authors were studied. Consideration of theoretical issues of subsoil value appraisal, in solving current problems of subsoil use management (determination of one-time starting payments during auctions (tenders), ranking of mines by degree of attractiveness), was based on the works of: B. I. Benevolsky [6], S. A. Emelyanova [12], A. Y. Katz [13], G. N. Kornev [15], N. K. Nikitina [22], E. A. Podturkina [24] and others. When considering the tax issues in the mining industry, the works of the following scientists were used: V. A. Kurushkin [16], G. S. Mirzekhanov [20], R. K. Sadykov [25], A. E. Cherepovitsyn [26], Y. I. Yukhimov [28] and others. Consideration of the issues of economic efficiency of mining enterprises was based on the fundamental scientific works of Russian scientists: M. I. Agoshkov [1], Arens V. Zh. [2], A. S. Astakhov [3,4], Y. P. Belov [5], J. K. Galiev [7,8,9], S. N. Goncharenko [10], F. D. Larichkin [17], V. K. Maximov [18, 19], Y. V. Mossakovsky [21], I. M. Shchadov [27], M. A. Yastrebinsky [29] and others.

For all the importance of research by these scientists, it should be noted that insufficient attention is paid to a comprehensive assessment of mineral mines, in particular, placer gold mines, which would take into account the constantly changing technical and economic conditions for their development and thereby allow to establish the level of efficiency of mining enterprises.

3. Statement of the Problem

The stated aspects of the lack of the problem development led to the setting of goals and objectives of the study on the development of a methodological approach to solving problems in the field of a comprehensive technical and economic assessment of placer gold mines and determining on this basis the effectiveness of their development.

The main objectives of the study are the following:

- to assess the current state of the theoretical foundations of a comprehensive technical and economic assessment of mineral mines;
- to clarify the specifics of open mining works and the nature of technical and economic indicators of mining enterprises;
- to determine the possibility of a comprehensive technical and economic assessment of the development of placer gold mines;
- to substantiate the mechanism of comparison of the initial quantitative and qualitative information of the developed placer gold mines;
- to develop and test the methods of complex technical and economic assessment of placer gold mines in the Transbaikal territory;
- to evaluate the effectiveness of the implementation of the methodology.

4. Theoretical part

The methodology provides for the evaluation of placer gold mines in the following groups of technical and economic indicators:

IOP Conf. Series: Materials Science and Engineering **663** (2019) 012036 doi:10.1088/1757-899X/663/1/012036

1. Indicators characterizing the operating conditions that include: 1) the distance to the nearest railway station (L r. st.), km; 2) the category of the road from the mines to the train station (CR); 3) the amount of sand mineral mines (Vsnd), thousand of m3; 4) volume of overburden (V overb.), thousand of m3; 5) the average power of peat (PP), m; 6) length of the placer (Lp), m; 7) the width of the placer (Wp), m; 8) coefficient of overburden (C overb.), m3/m3 9) the average power of the sand formation (PSF), m; 10) gold reserves (RAu), kg; 11) content (C), mg/m3; 12) damage by permafrost (Pf), %.

The most important factor characterizing the conditions is the price of minerals. Therefore, it is recommended to exclude it from the study because the impact of prices on the economic results of enterprises economic activity will be overwhelming.

2. Indicators characterizing the potential of an enterprise: 1) the capacity of the site for the extraction and washing of sand in one season (Qc), thousand/m^3; 2) performance of the site for stripping peat in one season (Qst.p.), thousand m^3; 3) specific capital investments (Kud), rub/m^3; 4) floating assets (f.a.), rub; 5) coefficient of equipment deterioration (C eq.d.); 6) coefficient of equipment use (C eq.u.); 7) reserve ratio (r.r.).

3. Results indicators: 1) the cost of producing one gram of gold (P.c.), rub/gram; 2) a blank disc-kotirovalnyj income (NPV), RUB; 3) net present value (NPV), rub.

It should be noted that the the belonging of some indicators to this or another group is not certain. And and the selection of information does not imply the receipt of the only correct grouping. Another successful combination of indicators in each of these groups is not excluded.

The use of all groups of factors for mathematical processing in natural units is impossible because of the difference in units of measurement, a significant scope of variation. For example, the distance from the station base is measured in km, mineral reserves in thousand tons, the production capacity of an enterprise in thousand m3 / year, so the indicators are not comparable. This makes it necessary to bring all the parameters into a comparable form, so it is proposed to use the indexing system (coding) of the source data. The following principles must be observed here:

1. Indices (codes) must correspond to the intervals of statistical grouping carried out for indicators in natural units. Only in this case, the indices will reflect the true characteristics of an enterprise, as well as be used for the subsequent calculation of functional dependencies.

2. The need to comply with the direction of the indices involves coding in such a way that it reflects the qualitative characteristics of the studied parameter.

The principle underlying it assumes the assignment of code 1 to the enterprises in better conditions, with the deterioration of conditions, a code of 2, 3, 4, 5, etc. is assigned, and vice versa. For example, the distance from the station base varies from 1 to 500 km, the enterprises are grouped first with the division of the range into groups, then each group is assigned its own index. An example of indexing is shown in the table. 1 [23].

Table 1. The distance to the substation base in an (indexed) encoded form (developed by the author)

Indicator Name	Distance, km		Assignable Index
	"from"	"to"	(code)
Distance to the Nearest Railway Station (L r. s.)	0	50	1
	50	100	2
	100	150	3
	150	200	4
	200	250	5
	250	300	6
	300	and more 300	7

Classification of the enterprises is carried out using the Euclidean metrics [11].

$$d(xi, xj) = \left[\sum_{k=1}^{h}(x_{ki} - x_{kj})^2\right]^{\frac{1}{2}}, \qquad (1)$$

where x_{ki} and x_{kj} – the quantitative values of an attribute for the *i* and *j* enterprises.

It should be noted that The Euclidean distance retains denotative meaning only when all parameters fluctuate in relatively equal ranges. To use the selected metric, it is necessary to normalize the quantitative values of all the features according to formulas 2 and 3.

$$X_{HK} = \frac{X_K - X_{OK}}{S_K} \tag{2}$$

$$S_K = \left[\sum_{i=1}^{m} \frac{(X_{ki} - X_{ok})^2}{m-1}\right]^{\frac{1}{2}} \tag{3}$$

where X_{HK} – the normalized value; X_K – value of an attribute; X_{OK} – the sectoral value; S_K – the mean square deviation of an attribute; m – the number of compared mines.

The proposed criteria are the basis for a comprehensive assessment of mines, as well as for obtaining linear regression techniques, a simple regulatory function that establishes the reciprocity between the calculated value of economic results from the conditions and capabilities of an enterprise.

$$F = f(Ry_i; Rp_i). \tag{4}$$

Deviation (°1*i*) of the calculated value (*F*) from the actual one (Rp_i) allows to judge how effectively (not effectively) an enterprise works under the corresponding conditions and opportunities. The deviation is calculated by the following formula:

$$(\Delta_{1i} = F - Rp_i) \tag{5}$$

In the case of a deviation with a positive value, it can be judged that an enterprise is not working efficiently, does not use internal reserves and favorable operating conditions, in addition, the more the absolute value of the deviation is (Δ_{1i}), the lower performance rates an enterprise gets.

A negative value of Δ_{1i} indicates that an enterprise makes full use of its own capabilities and favorable operating conditions.

The deviation of the average intra-industry value from the actual value is calculated by the formula:

$$\Delta_{2i} = R_{p.cp} - Rp_i \tag{6}$$

where Δ_{2i} – a deviation of the intra-industry value from the actual; $R_{p.cp}$ – a comprehensive assessment of intra-industry average value of economic indicators (for a particular region).

The calculated value of the deviation index of the results of the *i* enterprise works (K_i) shows the deviation in fractions of the units of work results of the average intra-industry value:

$$\mathrm{K_i} = \frac{\mathrm{R_{P.CP}}}{\mathrm{R_{Pi}}}. \tag{7}$$

If the value of the deviation index is greater than one, it can be judged from the fact that an enterprise operates with indicators below the intra-industry ones, and vice versa.

5. Practical Significance, Implementation Results

A comprehensive technical and economic assessment of the efficiency results of the mining enterprises was carried out for 30 placer gold mines of the Transbaikal territory.

For example, during the developing of a placer of the Undra river K_i = 0,6, which means that the enterprise operates with the rates that are 40 % higher than the intra-industry rates, and during the developing of a placer of the Zurgun (K_i = 1,28), the enterprise works with technical and economic indicators that are 28 % lower than the intra-industry rates. To simplify the analysis of the enterprise, it is advisable to rank for Δ_{1i}, Δ_{2i}, and rank the Ki calculation results in a table.

The developed method was tested: with a comprehensive technical and economic assessment of the placer gold mines Klyuch Taliy, in order to find a potential investor for the development of the mines; upon receipt of a bank loan when working out the placer of the river Shunduya; when proving the profitability, it was put into the category of the perspective ones, and the subsoil user received a license for its development. The calculations made it possible to justify the expediency of investing in the development of the Klyuch Taliy field. The deviation of the calculated values from the actual ones is $\Delta_{1i} = -1{,}002$, the deviation of the medium regional value from the actual ones $\Delta_{2i} = -1{,}220$ and the estimated index value of the deviation of the enterprise results is $K_i = 0{,}64$. Thus, it is possible to speak about good opportunities of the enterprise, favorable operating conditions and potentially high technical and economic results that are 22 % higher than the average in the region. At the cost of 215 rubles/gram and the 2008 price for gold of 310 rubles/gram, the net present value of the enterprise will be 4.8 million rubles. As a result, LLC "Aginskaya gold and industrial company" settled the receiving of the investment to $ 38 million in JSC "Sberbank".

The estimated indicators for mining and geological company LLC "Kazakovskoye" that develops a Shunduya placer: the deviation $\Delta_{1i} = -0{,}656$, the deviation $\Delta 2_i = -0{,}398$ and the calculated index value of the deviation $K_i = 0{,}81$, it says that the enterprise has good reserves, but the calculated indicators are slightly worse than the industry average rates. At a cost equal to 274 rub/gram, NPV is 3.6 million rubles. On the basis of the calculations performed by the proposed technique JSC "Sberbank" has implemented lending LLC "Kazakovskoye" in the amount of 5 million.

Also, for the Nelinda placer gold mines that are still not developed, owned by the subsoil user of LLC "Raduga", the deviation of the calculated value from the actual $\Delta_{1i} = -0.835$, the deviation of the regional average from the actual $\Delta_{2i} = -1.037$, they indicate that its mining enterprise that develops it will have good reserves, and the results of its work will be 3.7 % higher than the regional average. With a prime cost of 974.3 rub/gram and a gold price in 2015 of 1700 rubles/gram, the net present value will be 36114 thousand rubles, a payback period will be 1.1 year, capital investments will amount to 52 341 thousand rubles, internal rate of return will be 37 %, index of return – 1.1 %. The negative factors are: the location of the mines is in the North of the Transbaikal territory, the lack of transport communications and infrastructure, a large capacity of permafrost. The starting price of the mines of placer gold "Nelinda" was 576 000 rubles. A comprehensive technical and economic assessment has put the mines in a number of promising and highly profitable ones. As a result, the subsoil user found it possible to raise the price of the mines to 12 038 000 rubles.

6. Conclusion

The introduction of the methodology in a number of placer mines of the Transbaikal territory confirmed its practical significance. The study has shown that it can be used to assess both existing and projected enterprises and to identify their investment attractiveness. The method of technical and economic assessment with some changes and additions to the initial data can be used in the development of other mines (ore, coal, etc.).

References

[1] Agoshkov M I 1984 *Development of Ideas and Practices of Integrated Development of Mineral Resources* (Moscow: IPKON RAN, SSSR)

[2] Arens V Zh 2003 *The Fundamentals of the Methodology of Mining Sci.* (Moscow: Gornaya kn.) 223 p

[3] Astakhov A S 1981 *Economic Evaluation of Mineral Reserves* (Moscow: Nedra)

[4] Astakhov A S 1991 *Economics of Exploration, Production and Processing of Minerals* (Moscow: Nedra)

[5] Belov Y P and Makarkin Y N 2005 The Rental Mechanism of Differentiation of Taxation in Subsoil Use *Mineral Resources of Russia. Economics and Management* **1** 53–6

[6] Benevolskiy B, Blinova Y, Golenev V and Rybin V 2008 About the Methodology for Determining the Starting Sizes of One-time Payments for the Right to Use Subsoil *Mineral Resources of Russia. Economics and Management* **2** 52–6
[7] Galiyev Zh and Kuleshina Y 2014 Comparative Financial and Economic Analysis of the Activities of Mining Enterprises *Mountain News and Analytical Bull.* **9** 234–8
[8] Galiyev Zh and Galiyeva N 2015 Economic and Theoretical Aspects of the Effective Functioning of Mining Enterprises *Mountain News and Analysis Bull.* **8** 260–7
[9] Galiyev Zh K, Galiyeva N V and Tolmachev A G 2011 Economic Rationale for the Development of Placer Gold Mining Enterprises Based on the Formation of a Project Financing System. *Mountain News and Analytical Bull*. 2–5 and 47–54
[10] Goncharenko S N 1998 Evaluation of the Possibility of Increasing the Economic Efficiency of the Enterprise *Mountain News and Analytical Bull.* **1**
[11] Gusarov V M 2003 *Statistics* (Moscow: Yuniti–Dana) 463 p
[12] Yemelyanov S and Polyakov R 2004 Regression Method of Valuation of Gold Ore Mines *Ores and Metals* **3** 34–44
[13] Kats A, Mikhaylov A and Raykhlin B 2003 Market Method for the Rapid Assessment of Deposits, the Reserves of which are Taken to the State Balance *Ores and Metals* **2** 30–9
[14] Kostromin M V, Yurgenson G A and Pozlutko S G 2007 *Problems of Dredged Development of Continental Placers* (Novosibirsk: Nauka) 180 p
[15] Kornev G 2008 Economic Analysis of an Enterprise Using a Statistical Simulation Model *Economic Analysis: Theory and Practice* **8** 51–4
[16] Kurushkin V 2007 Gaps and Problems of the Draft Bill on "Subsoil" *Mountain J.* **6** 4–7
[17] Larichkin F, Ibrokhim A, Glushchenko Yu, Perein V and Aliyeva T 2011 Methodology for Assessing the Environmental and Economic Efficiency of the Integrated Use of Mineral Raw Materials *Mountain J.* **2** 21–4
[18] Maksimov V and Panina T 2008 Study of the Influence of Natural Factors on the Investment Efficiency of the Development of Placer Gold Deposits *VIII All-Russian. Scientific practical. conf. « Kulagin Readings»* (Chita: ChitGU. CH.I.) p 161–3
[19] Maksimov V and Panina T 2007 Prerequisites for a Comprehensive Assessment of Placer Gold Mines. *VII All-Russian. scientific-practical conf. "Kulagin Readings"* (Chita: ChitGU. CH.I.) p 73–5
[20] Mirzekhanov G S 2003 Problems of Taxation of Enterprises of the Gold Mining Industry *Mountain J.. Economics and Management* **11** 3–5
[21] Mossakovskiy Ya V 1994 *Evaluation of the Economic Efficiency of the Introduction of New Technology in Mining Enterprises* (Moscow: MGGU)
[22] Nikitina N and Muravyeva O 2003 One-time Payments for Licensing of Subsoil Use Objects *Mineral Resources of Russia. Economics and Management* **5–6** 22–8
[23] Panina T Y 2014 The Use of Integrated Assessment in Determining the Economic Efficiency of the Development of Placer Mines *Training Notes of the Transbaikal State University* **3(56)** 84–9
[24] Podturkin Y, Kotkin V, Yemelyanov S and Malukhin G 2006 Accounting for Uncertainty and Risk in the Valuation of Mines and Establishing the Size of a One-time Starting Payment for the Right to Use Subsoil *Mineral Resources of Russia. Economics and Management* **2** 62–5
[25] Sadykov R 2005 To the Question of Introducing a Differentiated Tax on Mining *Mineral Resources of Russia. Economics and Management* **1** 64–5
[26] Cherepovitsyn A Y, Larichkin F D and Smirnova N V 2013 Problems of Taxation in the Mining Complex and Ways to Overcome them *Proceedings of the Karelian Scientific Center of the Russian Academy of Sci.* **5** 122–8
[27] Shchadov M I 2002 General Patterns of Development of the Business of Coal Companies *Mountain News and Analytical Bull.* **5** 88–100

[28] Yukhimov Y I and Yukhimova Y Y 2008 Improving the Legislation on Taxation of Mining of Solid Minerals *Mountain J. Economy* **2** 7–11
[29] Yastrebinskiy M A and Gitis L Kh 1999 *Valuation of the Effectiveness of Investments in Mining Enterprises, Taking into Account the Time Factor and Discounting Costs* 2nd ed (Moscow: MGGU)

IOP Conf. Series: Materials Science and Engineering **663** (2019) 012037 doi:10.1088/1757-899X/663/1/012037

Dust content in the air: A case study of the Afanasyevsky open pit mine (Russia)

S A Lemanova

JSC Polymetal Management, 22, Prospekt Narodnogo Opolcheniya, St. Petersburg, 198218, Russia

E-mail: sv3tlem@ya.ru

Abstract. According to experts, the world consumption of mineral raw materials and the rocks extraction have reached about 12 and 100 billion tons per year, respectively. Overall Russian coal production accounts for 259.9 million tons (according to the Ministry of Energy of the Russian Federation and the 'Rosinformugol' agency), metallic ores are estimated at 230 million tons, and building materials reach 1.4 billion tons. The overall growth in mineral production and capacity of ore producers cause an increase in the negative environmental impact of mining extraction and processing. It should be taken into account that due to the depletion of reserves of rich mineral raw materials, the volume of mined rock mass and the amount of processing waste of mineral raw materials increase. It results in the formation of mine dumps, consisting rocks, ore beneficiation wastes, ashes, slags, and sludge. Research data describe critical environmental situation within the mining enterprises, and especially mine dumps, as a result of massive dust emissions. In addition, there is a tendency to further environmental degradation, due to the increase in production capacity without complying with effective measures to reduce dust emissions. Considering the fact that fine particles are priority pollutants in the extraction and processing of mineral resources at mining enterprises, it is necessary to develop new environmentally efficient and economically viable methods of dust suppression.

1. Introduction

Anthropogenic sources of aerosols emitted during the extraction and processing of minerals are one of the largest and specific pollutants of almost all components of the biosphere, a high concentration of which is harmful to human health, forests, agriculture, etc. Therefore, from the waste dumps of mines and quarries, more than 8 million tons of dust per year are blown into the atmosphere, including about 1 million tons per year in Russia [1, 5, 10]. As a result, the dustiness of the air increases tenfold compared with the established norms [4, 22, 29].

The greatest environmental hazards are mining enterprises located in close proximity to settlements, as well as to agricultural and forestry areas [2, 3, 8, 13, 24]. The Afanasyevsky open pit mine is an example of such an enterprise. The mining company develops the Afanasyevsky carbonate field, located in the Voskresensky district of the Moscow Oblast, on the right bank of the Moskva river. The enterprise directly borders agricultural territories of farms and Achkasovo village. In addition, the city of Voskresensk is located within 4 kilometers of the mine dump. The company was commissioned in 1964. The overburden rocks are represented by loams, variegated marls, sometimes intersected by fine-grained dense limestones. Bedding rocks are marls and a layer of dolomitized rocks with an average thickness of 3.5 m and limestone with a thickness of about 7 m. Marl and limestone are used as raw materials for the cement production, dolomitic rocks are processed for crushed stone and lime flour. The total mineral reserves amount to 300 million tons. There are four developed bench within the deposit: overburden rocks, marl, dolomite, and limestone benches. Stripping operations are carried out with walking and bucket-wheel excavators. The total overburden volume is more than 3 million m^3 per year. The annual raw materials output for cement production is more than 4 million tons. Dolomite mining for the production of crushed stone and flour amounts to more than 6 million tons. The development of the Afanasyevsky field is accompanied by aerosol emissions at all stages, leading to air pollution in the nearest settlements, pollution of farm areas with falling dust, which leads to the reduction in yields and products quality. In this regard, the purpose of the research was to assess the state of the components of the natural environment in the impact zone of the Afanasyevsky open pit mine.

2. Methods

In the course of research conducted in 2015–2018, a number of procedures for assessing the state of atmospheric air were carried out. This made it possible to accurately assess the environmental situation in the impact zone of the Afanasyevsky open pit mine.

Analysis of the dust content in atmospheric air in the impact zone of the mine dump was conducted throughout the year, in periods without precipitation under various meteorological conditions. Atmospheric air monitoring points were located directly on the dump territory, on the border of the sanitary protection zone of the production facility and on the residential area of the city in accordance with the average long-term prevailing wind directions [9].

Monitoring of atmospheric air was carried out in accordance with *GOST 17.2.3.01-86 Nature protection. Atmosphere. Air quality control regulations for populated areas* to analyze the dust content in the atmospheric air of the impact zone of the Afanasyevsky open pit mine [15].

Monitoring sites are located taking into account the natural and technical security of the mine dump (the main directions of the average annual wind rose, significant topographic uplifts and depressions, forested areas, and buildings), which makes it possible to accurately identify the patterns of the formation of the atmospheric pollution halo.

Measurement of fine particle concentration in the ground-level air was carried out according to a specially developed program. This program is aimed at identifying areas with the maximum surface concentration of inorganic dust with SiO_2 content from 70 to 20 % under various meteorological conditions. Strict periodicity of measurements was not taken into account. Measurements were carried out using DustTrak TSI-8533 aerosol monitor. This size-segregated mass fraction measurement technique is superior to either a basic photometer or optical particle counter (OPC). It delivers the mass concentration of a photometer and the size resolution of an OPC. Thus, this aerosol monitor provides mass concentration of fine particles with dimensions of –1, –2.5, –4 and –10 µm. Measurement of suspended substances is carried out in a wide concentrations range (0.01–150 mg/m^3), and the relative measurement error does not exceed 20 %. To determine the airflow velocity, atmospheric pressure, temperature, and relative air humidity, the MES-200 meteo-analyzer was used.

During the whole time of field observations, the wind speed changed from 1–2 m/s to 14 m/s (with gusts up to 20 m/s), and the frequency of its directions, on average, corresponded to the wind rose for this region. Such a wide range of changes in the intensity and direction of the wind load and a large number of measurements allowed obtaining reliable data on the dustiness of the ground-level air impact zone of the mine dump under various meteorological conditions. This allowed assertion with a high degree of accuracy on the substantial technogenic load of the Afanasyevsky open pit mine on the atmospheric air of the region.

Over the entire period of field observations of the ground-level air, the air temperature varied from –10 to +30 °C, air humidity did not exceed 65 %, and the atmospheric pressure corresponded to normal values.

3. Results

The assessment and forecast of the state of atmospheric air in the territory of the Afanasyevsky open pit mine were carried out on the basis of the conducted monitoring observations, taking into account the influence of meteorological parameters on the dust distribution in the environment.

The impact of mine dumps dusty surfaces on the environment is closely related to the climatic characteristics of the dumps area [20, 28]. The climate of the region predetermines the intensity and degree of atmospheric pollution, the distance and propagation direction of dust particles depending on the prevailing wind directions. Fig. 1 shows the annual curves of several meteorological factors of the study area, based on the recent 30-years-data. Of particular importance is the consideration of the natural and climatic conditions of mine dumps in identifying agrometeorological phenomena dangerous for the agrarian business (Table 1).

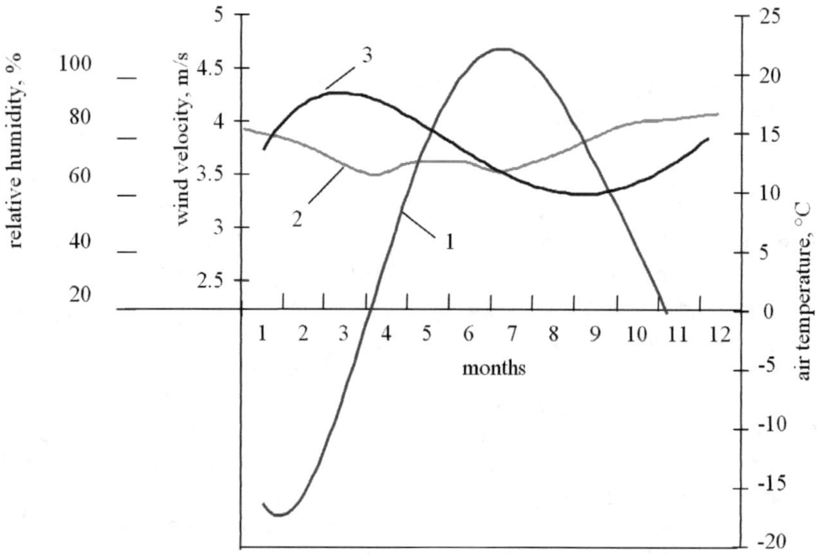

Figure 1. Annual temperature and relative humidity dynamics: 1 – average daily temperature; 2 – relative humidity; 3 – wind speed

Data analysis allows conclusion that temperature, speed, and relative humidity of the air favor dust blowing from the dumps [16]. The monitoring studies showed that fugitive dust emissions (mg/(m^2·s)) parameter make it possible to predict dust dispersion and determine the concentration at any distance (Fig. 2), based on the graphs (Fig. 3, 4) plotted for dust with a particle size of 0–100 μm.

The graphs show that when the wind speed is in the range of 0–5 m/s, the fugitive dust emissions with a particle size of 0–100 μm is insignificant and varies linearly. With an increase in speed above 5 m/s, the fugitive dust emission increases along a parabola.

Thus, all other things being equal, the least air dustiness at any distance from the dump should be expected at wind speeds of up to 5 m/s. This is confirmed by the results of special measurements of the fugitive dust emissions from the dumps surface.

Table 1. The principal agroclimatic indicators of the Afanasyevsky mine impact zone

Indicators	Period of the year	Characteristics
1. Climate type: boreal		
2. Sum of active temperatures, °C	V-IV	2365
3. The temperature of the warmest month, °C	VI-VII	18.3
4. The temperature of the coldest month, °C	XII-II	-10.8
5. Precipitation, mm	I-XII	550
6. Moisture balance, mm	IV-IX	117
6. Probability of dry years, %	year	8.8
7. Season duration		
spring (5–15°C)	beginning	17.04
	end	05.06
	days	49
summer (15–20°C)	beginning	6.06
	end	27.08
	days	83
autumn (5–15°C)	beginning	28.08
	end	11.10
	days	45
winter (0–15°C)	beginning	3.11
	end	2.04
	days	152
10. The main vegetation period (10–15°C)	beginning	3.05
	end	18.09
	days	138
11. Frost-free period	beginning	12.05
	end	20.09
	days	130
12. Biological productivity, points		60–70

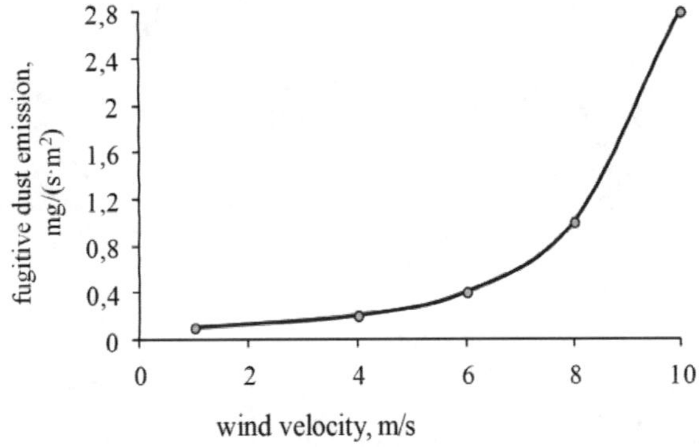

Figure 2. Dependence of fugitive dust emissions on airflow velocity

Studies have shown that air pollution varies with the seasons. That why at the Afanasyevsky mine there is an increase in the concentration of dust in the atmosphere, intensively blown off from the dried rocks from April to September (Fig. 6). In addition, on the contrary, from October to March there is a decrease in the dust concentration in the air due to the high humidity of the rocks.

Dust blowing from the dumps surface depends on the humidity of rocks and dust [17]. This dependence is shown in the graph (Fig. 5), curve 2. Analyzing the meteorological factors, it can be concluded that the dumps are one of the key sources of air pollution. Assessment of emission and transfer of dust can be performed on the basis of known techniques [11, 18, 25]. The amount of dust from open areas (dumps) can be determined by the formula (1):

$$\Pi = S \cdot W_c \cdot \gamma \tag{1}$$

where Π – the amount of dust, kg/s; W_c – fugitive dust emissions (taken into account the wind speed on the dump surface, kg/(m^2·s); S – open surface area, m^2; γ – coefficient of rock mass change (approximately taken equal to 0.1).

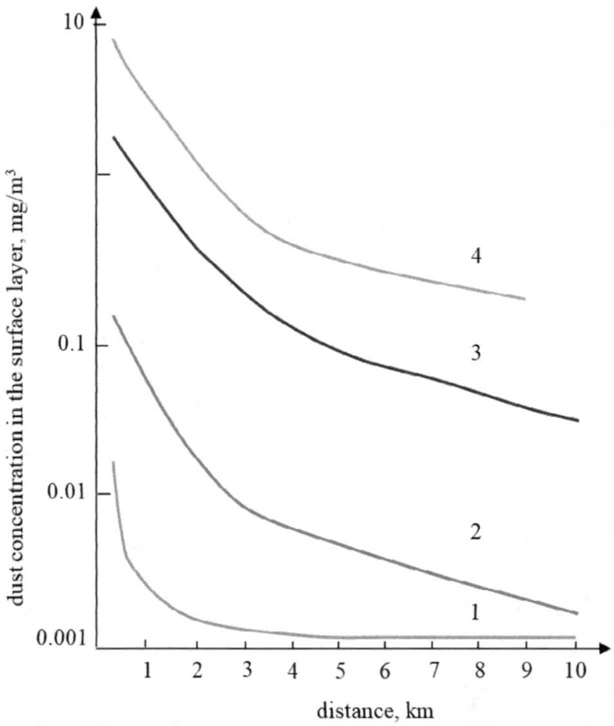

Figure 3. Dispersion in the surface layer of dust blown off from mine dump surface, depending on the wind speed: 1 – 3.5 m/s; 2 – 5.3 m/s; 3 – 6.7 m/s; 4 – 7.8 m/s

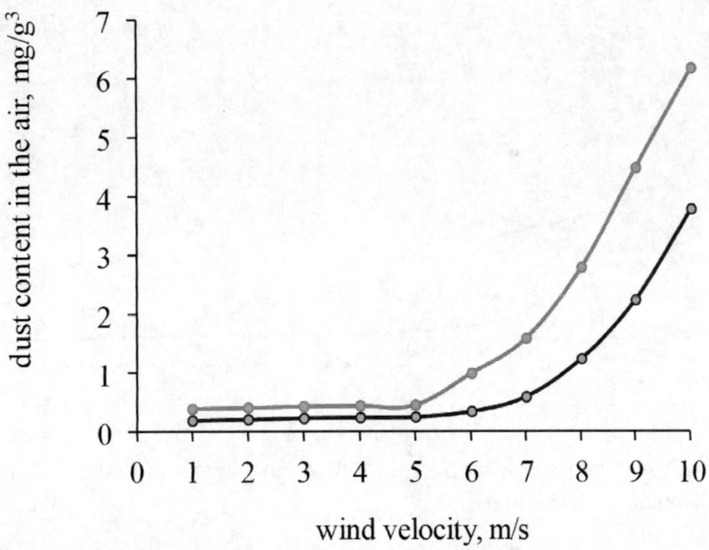

Figure 4. The dependence of the dust content of the air from the air velocity: 1 – dry dust, 2 – dust moistened with water

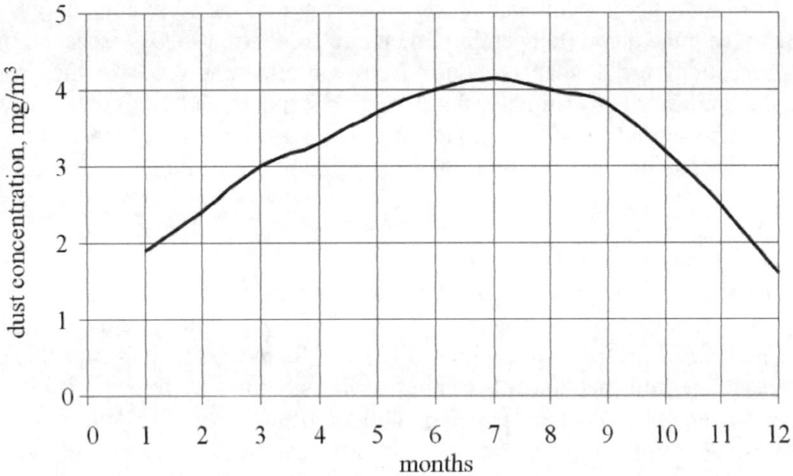

Figure 5. Changing dust concentration in the atmospheric air near the Afanasyevsky open pit mine

The calculated values of changes in the relative dust content in the air and the Δ_{Π} relative mass of precipitated dust Δ_M depending on the distance to the source are shown in Fig. 7.

$$\Delta_{\Pi} = (\Pi / \Pi_0)100, \% \quad (2)$$

where Π is the actual dust content in the air (mg/m^3); Π_0 is the initial dust content at the formation site (mg/m^3).

$$\Delta_M = (M / M_0)100, \% \quad (3)$$

where M and M_0 are the deposited dust masses at the calculated point and at the formation site, respectively, t/(ha·year).

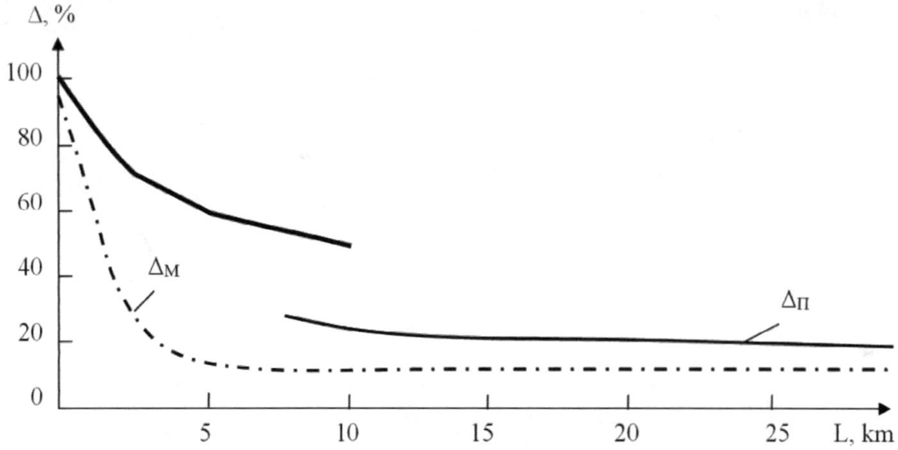

Figure 6.The change in the relative dust content in the atmosphere and $\Delta_{П}$ fallout on the surface Δ_{M} depending on the distance to the source

Changes in the content of harmful components in the atmospheric air over the seasons are explained [12, 14, 26], mainly by the direction and intensity of the heat and mass transfer processes between the mine dump and atmosphere [7, 27]. Heat and mass transfer is determined by the thermodynamic parameters of the atmosphere (temperature, relative humidity, moisture content, air velocity) and the marginal part of the mine dump (temperature, moisture content) [19, 23]. A change in the temperature or moisture content of the atmosphere during daily oscillations leads to the potential difference (gradient) at the boundary between main dump and atmosphere and the increase of the heat or mass (gas) flow, with vapor-liquid phase transitions during the evaporation or condensation [6, 21, 30]. These phenomena change the intensity of fine particle emission from dumps and dust suppression with condensation.

4. Conclusion

Monitoring in the impact zone of the Afanasyevsky open pit mine has revealed the presence of atmochemical pollution halo with a length of up to 25 km^2. Dust fallout on the surrounding landscapes led to the formation of lithochemical pollution halo of 10 km^2, and, as a result, the contamination of agricultural areas. The influence of dust forming surfaces on the environment is closely related to the climatic characteristics of the dumps area. The climate of the region predetermines the intensity and degree of atmospheric pollution, the distance and propagation direction of dust particles. There is a seasonal dynamics of changes in the dust content in the atmosphere, which is explained by the direction and intensity of the heat and mass transfer processes between the mine dump and atmosphere. This study has shown the need to reduce the impact of the Afanasyevsky mine dumps on the environment by reclamation and dedusting system installation in such areas.

References

[1] Abu-Allaban M and Abu-Qdais H 2011 Impact assessment of ambient air quality by cement industry: A case study in Jordan *Aerosol and Air Quality Res.* **11(7)** 802–10

[2] Alekseenko V A, Pashkevich M A and Alekseenko A V 2017 Metallisation and environmental management of mining site soils *J. of Geochemical Exploration* **174** 121–7

[3] Alekseev V I, Marin Y B, Polyakova E V and Machevariani M M 2014 Evolution of zircons from postorogenic intrusive series with Li-F granites, Russian Far East *Geology of ore deposits* **7(56)** 513–30

[4] Alfaro Degan G, Liiello D and Pinzari M 2015 Field evaluation of PM10 detectors in a quarry environment *Int. J. of Sustainable Development and Planning* **10(3)** 361–72

[5] Berry P and Pistocchi A 2003 A multicriterial geographical approach for the environmental

impact assessment of open-pit quarries *Int. J. of Surface Mining Reclamation and Environment* **17(4)** 213–26
[6] Bezel' V S and Zhuikova T V 2007 Chemical pollution: Transfer of chemical elements to the aboveground phytomass of herbaceous plants *Russ. J. of Ecology* **38(4)** 238–46
[7] Bushuev Y Y, Leontev V I and Machevariani M M 2018 Geochemical Features of Au-Te Epithermal Ores of the Samolazovskoye Deposit (Central Aldan Ore District, Yakutia) *Engineer. Materials* **769** 207–12
[8] Danilov A S, Smirnov Y D and Pashkevich M A 2015 Use of biological adhesive for effective dust suression in mining operations *J. of Ecological Engineer.* **16(5)** 9–14
[9] Danilov A, Smirnov Y and Ivanov A 2016 The new method of dust control on tailing dumps (Exemplified by Kovdor mine) *Int. Multidisciplinary Sci. GeoConf. Surveying Geology and Mining Ecology Management SGEM* 743–50
[10] Drebenstedt C and Alekseenko A V 2019 Environmental impact of abandoned mine wastes on an urban area in NW Caucasus *Innovation-Based Development of the Mineral Res. Sector: Challenges and Prospects – 11th Conf. of the Russian-German Raw Mater.* pp 223–9
[11] Ivanov A V and Smirnov Y D 2016 The impact of anthropogenic alluvial arrays on areas settlements depending on the particle size distribution of stored tailings *J. of Ecological Engineer.* **17(2)** 59–63
[12] Ivanov A V, Smirnov Y D and Petrov G I 2018 Investigation of waste properties of subway construction as a potential component of soil layer *J. of Ecological Engineer.* **19(5)** 59–69
[13] Pashkevich M A 2017 Classification and Environmental Impact of Mine Dumps *Assessment Restoration and Reclamation of Mining Influenced Soils* 1–32
[14] Pashkevich M A and Petrova T A 2016 Technology of artificial deposit preservation at the Kursk magnetic anomaly *J. of Ecological Engineer.* **17(4)** 11–6
[15] Pashkevich M A and Petrova T A 2017a Reclamation by Containment: Polyethylene-Based Solidification *Assessment Restoration and Reclamation of Mining Influenced Soils* 235–53
[16] Pashkevich M A and Petrova T A 2017b Technogenic impact of sulphide-containing wastes produced by ore mining and processing at the Ozernoe deposit: Investigation and forecast *J. of Ecological Engineer.* **18(6)** 127–33
[17] Pashkevich M A and Petrova T A 2019 Recyclability of ore beneficiation wastes at the Lomonosov Deposit *J. of Ecological Engineering* **20(2)** 27–33
[18] Pashkevich M A, Alekseenko A V and Vlasova E V 2015 Biogeochemical and geobotanical assessment of marine ecosystems conditions (Novorossiysk city) *Water and Ecology* **(3)** 67–80
[19] Pitz C, Mahy G, Harzé M, Uyttenbroeck R and Monty A 2019 Comparison of mining spoils to determine the best substrate for rehabilitating limestone quarries by favoring native grassland species over invasive plants *Ecological Engineering* **127** 510–8
[20] De Pontes J C, De Lima V L A and Da Silva V P 2016 Environmental impacts of rock blasting with use of explosives in granite quarry of Caico-RN *Geociencias* **35(2)** 267–76
[21] Puzanov A V, Baboshkina S V and Gorbachev I V 2012 Characteristics of heavy metal migration in the natural-anthropogenic anomalies of the North-Western Altai *Geochemistry Int.* **50(4)** 358–66
[22] Sairanen M, Rinne M and Selonen O 2018 A review of dust emission dispersions in rock aggregate and natural stone quarries *Int. J. of Mining Reclamation and Environment* **32(3)** 196–220
[23] Silvester S A, Lowndes I S and Hargreaves D M 2009 A computational study of particulate emissions from an open pit quarry under neutral atmospheric conditions *Atmospheric Environment* **43(40)** 6415–24
[24] Smirnov Y D and Ivanov A V 2018 Investigation of dust transfer processes during loading and unloading operations using software simulation *J. of Ecological Engineer.* **19(4)** 29–33
[25] Strizhenok A and Korelskiy D 2015 Assessment of the anthropogenic impact in the area of tailings

IOP Conf. Series: Materials Science and Engineering **663** (2019) 012037 doi:10.1088/1757-899X/663/1/012037

storage of the apatite-nepheline ores *Pollution Res.* **34(4)** 809–11

[26] Strizhenok A and Korelskiy D 2016 Assessment of the state of soil-vegetation complexes exposed to powder-gas emissions of nonferrous metallurgy enterprises *J. of Ecological Engineer.* **17(4)** 25–9

[27] Syso A I, Kolpashchikov L A, Ermolov Yu V, Cherevko A S and Siromlya T I 2014 Elemental chemical composition of soils and plants in Western Taimyr *Contemporary Problems of Ecology* **7(6)** 636–42

[28] Tejado J J, Guillén J and Baeza A 2016 Assessment of occupational exposure in a granite quarry and processing factory *J. of Radiological Protection* **36(3)** 641–52

[29] Ulm K, Gerein P, Eigenthaler J, Schmidt S and Ehnes H 2004 Exposure and mortality in workers with silicosis in the stone and quarry industry *Arbeitsmedizin Sozialmedizin Umweltmedizin* **39(8)** 436–42

[30] Zhuikova T V, Meling E V, Kaigorodova S Yu, Bezel' V S and Gordeeva V A 2015 Specific features of soils and herbaceous plant communities in industrially polluted areas of the Middle Urals *Rus. J. of Ecology* **46(3)** 213–21

IOP Conf. Series: Materials Science and Engineering **663** (2019) 012038 doi:10.1088/1757-899X/663/1/012038

Study on air anion concentration distribution in Beidaihe, China

L Jiajie[1], L Qingshan[2,3], Q Shuangjin[2], W Zhiguo[2], L Chaoping[2] and W Yang[2]

[1]Metastable Materials Science and Technology State Key Laboratory, Yanshan university, 428, st. Hebei, Qinhuangdao, 066004, China
[2]Institute of health and natural nanotechnology, 28, Shuguxiangyuan, av. Shugu, Qinhuangdao, 066000, China

E-mail: qsli@ysu.edu.cn

Abstract. In this paper, the air negative ion concentration distribution map in Beidaihe district was summarized and completed through several field visits to some representative areas in Beidaihe district or collection of local air detection data over the years. Combined with geomorphic environment, climatic conditions and other factors, this paper analyzes the regularity and characteristics of the concentration distribution of air negative ions in beidaihe, and proposes to build "the Great Wall of Air negative ions".

1. Introduction: The discovery and development of negative ions

Air anion is a natural state, in the air environment with a negative charge of gas molecules or molecular groups. J. Thomson first expressed the characteristics of ions with formulas and established the models of positive ions and negative ions. Eiseer and Geieel then demonstrated the existence of ions, positively and negatively charged particles that are slightly larger than the size of the molecule (Figure 1). In the late 19th century, Philip Lionad [1] (German physicist) was the first to academically demonstrate the effect of negative ions on the human body. He suggested that negative ions in the natural environment were beneficial to human health, and pointed out that the highest concentration of negative ions was around the valley falls. By 1902, Aschkinass and Caspari had confirmed the biological significance of the existence of air negative ions. In the early 20th century, Russian scholars first published the theory of using negative ions to treat diseases. In 1932, the United States RCA hamerson invented the world's first medical air anion generator. Since half a century ago, the research on air anion has experienced a long period of development and application in Europe, the United States and Japan.

2. The generation of air negative ions and its biological significance to human body

2.1 The mechanism of negative ions in nature [2]

There are several mechanisms for the generation of negative ions in nature:

Ray and tip discharge. In the natural world, positive and negative ions of air are generated by ultraviolet rays, cosmic rays and other various rays, which are constantly generated and disappeared to maintain a certain dynamic equilibrium state. At the same time, due to the electrostatic field formed by the ground for the atmospheric ionosphere, the ground is the negative pole, the air negative ions are repelled by the ground, and the positive ions of the air are attracted by the ground. Therefore, under normal circumstances, the ground positive ions are more than negative ions, the positive and negative ion concentration ratios are often greater than one, and the positive and negative ion concentrations are each about 400–700/cm^3.

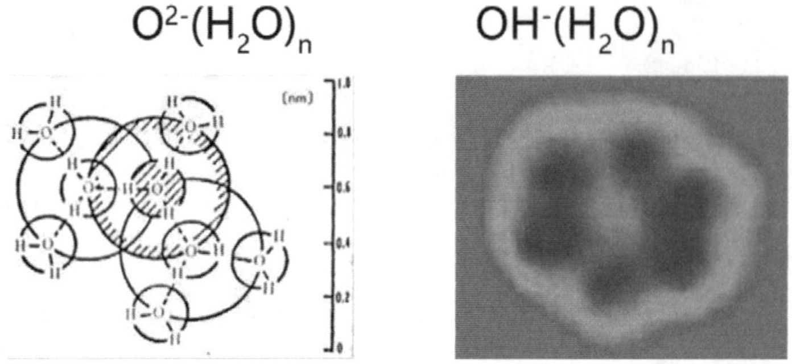

Figure 1. Structure model of air

Figure 2. Brief history of anion development in modern times

Plant photosynthesis. The photosynthesis of plants is the process of converting the absorbed solar energy into carbohydrates and oxygen by using carbon dioxide in the air and water in the soil, and releasing electrons, ie, negative ions, during this conversion.

Physical effects (tidal action, waterfall effects). The formation process of ions is that much particulate matter is rubbed by air, and the electrons in the atomic structure are forced to transfer to form charged ions. The water flow collides and rubs against the liquid itself and the air during the falling or irregular flow, and the droplets generate gaseous negative ions due to the shearing force, so that an atmosphere rich in water and negative ions is formed around the waterfall. This mechanism, also known as the Lenard effect, can be expressed by the following equation:

- $H_2O \rightarrow H_2O^+ + e^-$ (1)
- $H_2O + e^- \rightarrow H^- + OH$ (2)
- $H^- + H_2O \rightarrow OH^- + H_2$ (3)
- $OH^- + nH_2O \rightarrow OH^-(H_2O)_n$ (4)

It can be summarized as: $(H_2O)_n \rightarrow (H_3O^+)(H_2O)_{n-m} + OH^-(H_2O)\ m_{-2}\ (n>>m)$.

IOP Conf. Series: Materials Science and Engineering **663** (2019) 012038 doi:10.1088/1757-899X/663/1/012038

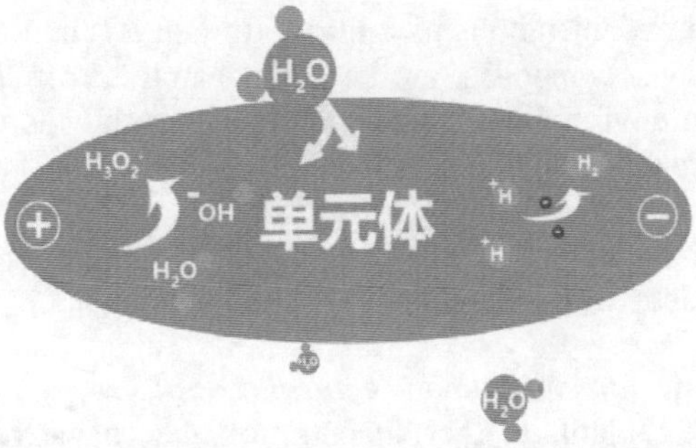

Figure 3. Principle of microelectrolysis of natural minerals

Micro-electrolysis of natural rocks and minerals. At present, there are reported materials for the release of negative ions from natural minerals: Tourmaline, Opal, Hexacyclic stone, Philosophical stone, Ancient sea rock, and clear water stone. Their mechanism of action is: due to the polarity of the mineral unit, microelectrolysis reaction with water molecules in the air under the action of the micro electric field. As shown in Figure 3 below.

2.2 Several methods for artificially generating negative ions

The "corona discharge type" which breaks through the air insulation by applying a large voltage to the humidifier;

"Ray line" of negative ions generated by the radioactivity of radioactive minerals;

Like a waterfall by the energy of the wave impact of negative ions generated by the "wave type";

"High-energy" metastable polymer composites prepared from natural ores. The theory of "ionized air" [3] was proposed and formed. On the one hand, air negative ions were generated by artificial methods under artificial conditions. On the other hand, the utilization of air negative ions has entered a new stage by applying the physiological effects of naturally formed air negative ions under natural ecological conditions. As far as the above types of artificial anions are concerned, "corona discharge" is more and more widely used in household appliances. The negative ions of "wave type" come from nature, but their release efficiency is low. The method of preparing air negative ions through micro-electrolysis of natural ore technology has the characteristics of high efficiency and safety, and more and more attention has been paid to it.

2.3 Biological significance of air negative ions on human body

Anion is a natural reducing agent with biological safety. The content of anion in air is also an important reference index for air environment in various areas. As an "air vitamin", negative ions have many functions and have good effects on the human body and even all living things. Their presence will help the human body to return to the natural balance state. The molecular formula for air anions is $O_2^-(H_2O)_n$, $OH^-(H_2O)_n$, $CO_4^-(H_2O)_n$, and the atom is made up of positively charged nuclei surrounded by electrons. If the air contains too many positive ions or too few negative oxygen ions (<800/cm^3), the air enters the body and stimulates the production of serotonin by platelets in the blood, which is then transported through the circulatory system to the tissues of the body. Serotonin prevents the absorption of oxygen (typical symptoms; Fatigue, dizziness, migraines, inattention, depression, and shortness of breath can be offset by the entry of 1,200 cm^3 of air into the body. When negative ions enter the body through the mouth, nose or directly through the skin, they cause the breakdown of 5-hydroxytryptophan into a non-toxic by-product, 5-hydroxyindoleacetic acid, which is excreted in the urine to eliminate the harm. When negative ion enters human body, still can cause a series of benign reaction. The most obvious: (1) Activate cell vitality, maximize the play of the physiological function of the organs, repair the damaged body, etc.; (2) Negative oxygen ion on the human body has the regulation of central nervous system activities, improve the function of coronary blood flow, but also promote the bronchial cilia movement;

(3) Prevention and treatment, improve respiratory diseases; (4) Regulate hormone secretion and cell potential, promote endocrine and metabolism, improve immune function and resistance to disease. Anion still has eliminate nervous, composed, awake effect, can raise office efficiency. Be in for a long time in relatively good anion environment, to stabilize the person's mood, restore physical strength, prevent and treat certain disease, wait like spirit depression, hypertension, asthma, coronary heart disease, very beneficial.

3. Influencing factors and detection methods of air negative ions

3.1 Influencing factors of air negative ions under natural conditions

From the analysis of geographical latitude, the concentration of air negative ions on the ground depends on geographical factors (waterfalls, beaches, canyons, rural fields, suburban wilderness, towns, etc). Due to geographical differences, as long as there are similar geographical factors, they all present regular distribution [4]. Air anion concentration refers to the number of anions in the air per unit volume, which is expressed in units/cm^3. The number of air negative ions in waterfalls and beaches is usually kept at more than 10000/cm^3, canyons and streams are usually kept at more than 1000/cm^3, suburbs, villages and forests are kept at more than 2000/cm^3, and the content of air negative ions in waterfall mouth is often higher, the highest is close to 40000/cm^3. For the urban environment, the trees and green spaces are significantly less than the suburbs and villages. The roads in the urban areas are mainly cement and asphalt roads, which block the ionization source from the soil. Therefore, the urban air negative ion concentration is much lower than that in the suburbs and rural areas. Under normal circumstances, the air anion concentration is negatively correlated with the fine particulate matter PM2.5 and the concentrations of SO_2, NO_2, and CO, and positively correlated with the ozone concentration. The number of negative ions in urban streets is 100–200/cm^3, while the average number of negative ions in typical residential rooms in big cities is only 40–50/cm^3.

From the perspective of time dimension, the annual change of air negative ions is also significantly different. Due to the large amount of data needed to study the annual change of air anion concentration, systematic data in this respect is relatively scarce. According to the research of some scholars, generally speaking, the concentration of negative ions in urban air is the highest in summer and the lowest in winter, followed by spring and autumn.

3.2 Detection method of air negative ion concentration[5]

The institute of Boyan material design and research in Yanshan university national university science park, Qicai technology of Yanshan university, Polymer innovation of science and technology research institute and Qinhuangdao, keep natural nanotechnology research based on years of release anion material research, the first in February 2008 to Beidaihe environmental protection bureau to air ion concentration observation of district, and published the Beidaihe district air index observation data. This instrument is a DLY 6A2 type air ion measuring instrument (Figure 4) and a WST-08 type atmospheric negative (oxygen) ion measuring instrument (Figure 5).

The air ion meter uses a capacitive air ion collector to collect the charge carried by the air ions. The ion concentration can be converted by measuring the current generated by these charges and sampling air flow. Each air ion is thought to carry only one charge. In the measurement process, the positive and negative air ion with the sampling flow into the collector, under the influence of polarization electric field, ions through the gas when the average speed of (that is, the migration velocity) not only have close relationship with ionic diameter (ion is smaller, the faster), and is proportional to the polarization voltage of electric field intensity, this proportion is called ion mobility. By changing the mobility, different ion concentrations can be measured.

According to professor Li Qingshan of Yanshan university, the seven levels of air negative ion concentration are related to human health [6]. To determine air quality and air freshness, as shown in table 1.

Figure 4. DLY-6A2 type air ion measuring instrument

Figure 5. WST-08 atmospheric negative (oxygen) ion detector

Table 1. Relationship between seven levels of negative ion concentration and human health

Negative ion concentration (a/cm^3)	level	Air freshness	Relationship with human body
More than 10000	Top	Ultra pure and fresh	Recovery and healing effect
5000 ~ 10000	Super	Special and pure and fresh	Prevention of disease
2000 ~ 5000	Level 1	Very pure and fresh	Prevention of disease
1500 ~ 2000	Level 2	Pure and fresh	Improve immunity
500 ~ 1000	Level 4	Not too bad	Health and disease margins
≤500	Level 5	Not pure and fresh	Cause certain diseases

4. Study on air anion concentration distribution in Beidaihe[6]

Professor Li Qingshan's research group from 2008 to 2018, field inspection of several representative areas of Beidaihe in Qinhuangdao, integrate air negative ion concentrations under various conditions to exclude data from extreme individual cases, take the average stable value as follows (Table 2), and draw the "Beidaihe air anion concentration distribution map" (Figure. 6).

Table 2. Relationship between seven levels of negative ion concentration and human health

Location	Negative ion concentration mean ($pieces/cm^3$)	Remarks
Olympic Park	100-500	There are 5000 waterfalls/cm in the landscape[3]
City Center	160-500	The downtown waterfall can reach $10,000/cm^3$
Geziwo	1100-3000	
Lianfeng Mountain Park	1100-10000	The peak reached $15,000/cm^3$
Biluota Park	2100-3000	
Laohushi Park	2000-3000	
Jifaguanguangyuan Park	1000-3000	
Bird watching wetland, lianfeng river estuary	2000-5000	The highest value of $36500/cm^3$ was measured on September 16,2008

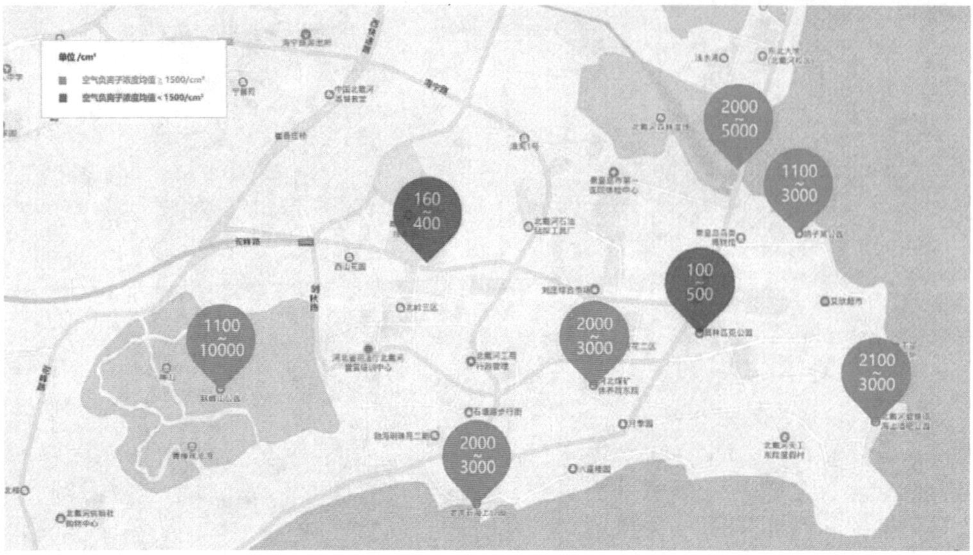

Figure 6. Air negative ion concentration distribution in Beidaihe

Through observational data analysis, preliminary conclusions can be drawn: Beidaihe is located in the northeastern part of Hebei Province, China, backed by Lianfeng Mountain (130 meters above sea level), and the lower-lying beaches go straight into the middle of the northern shore of the Bohai Bay. Therefore, the annual warm and humid coastal monsoon circulation continues. Bringing fresh and humid air to Beidaihe District, the annual average humidity can reach about 65 %. The air anion concentration in the coastal area is mostly maintained at 2000/cm^3 or above. It is located in the "Bird Watching Mecca Wetland" on the Beidaihe Forest Wetland Waterfront. When the weather is fine, the air anion concentration can reach 5000/cm^3, so the waterfront and the suburbs The landscape area is an enriched area of air anions. Under normal circumstances, the air in the urban area is not affected by the production and life, and the concentration of negative ions in the air is not high. In particular, the harmful gases such as nitric oxide generated by the transportation system will rapidly react with the negative ions in the air, so the air negative ion concentration during the peak traffic period is Minimum value [7]. In general, the average value of air negative ion concentration in Beidaihe is more than 2000/cm^3, the highest value is over 10000 (the peak of Lianfeng Mountain), and the minimum value is 100/cm^3, which compares the air negative ion concentration distribution in the country (Figure 7 Air anion concentration distribution map of China), Beidaihe has exceeded the national average level, and is also one of the regions with the best air quality performance in the offshore cities of the North.

According to Beidaihe test of air negative ions distribution and forest air negative ions distribution characteristics such as research and analysis, relying on the unique ecological environment of the local conditions, with the concept of "negative ions build a healthy Great Wall", use negative ion material or artificial anion generation model to strengthen and transform the city, the construction of the densely populated areas, such as using anion asphalt paving [9–10], using anion anion health paint coating and painting the walls, laid floor, equipped with air anion anion humidifier, etc., at the same time, in the scenic spot environment relatively weak anion local construction some landscape fountain, using geographical advantage to build artificial waterfalls, etc., to improve the concentration of negative ions in the air, to meet the needs of the health of natural air, make visitors get relax in tourism and health care effect, At the same time, it can also label and negativeize the "Air anion enrichment area", further strengthen the Beidaihe tourism resources gravitational radiation ability, create first-class tourist livable cities.

Figure 7. Air anion concentration distribution map of China

References

[1] Li L, Du Q, Liu T and Liang S 2017 Research progress in air anion *Modern Agriculture* **12** 30–1

[2] Xiong L, Han Sh, Tang H, Wang M and Huang Yu 2011 Research Progress and Influencing Factors of Air Negative Ions *China Population, Resources and Environment* **21** 184–7

[3] Zhang N 2018 Research progress on the relationship between negative oxygen ion distribution and meteorological factors in China *Chinese Meteorological Society 35th Annual Meeting of Chinese Meteorological Society S11 Urban Meteorology and Environment – The 7th Urban Meteorological Forum [C]. China Meteorology Learn: Chinese Meteorological Society* (Beijing: China Meteorology Learn "Chinese Meteorological Society") p 5

[4] Jianxin C, Baobao Zh and Youjie Zh 2017 Distribution characteristics of air negative ions in coastal and forest environments and their relationship with environmental factors *J. of ecological environment* **26(08)** 1375–83

[5] Xing G, Li Y and Lin J 2008 Determination of Air Negative Ion Concentration and Negative Ion Evaluation Criteria in Beidaihe *Chinese J. of Analysis Laboratory* **38(01)** 112–8

[6] Qing-shan L I, Jun L I U, You-bo D I, Xiang H, Jia L I U and Bai-feng L I 2019 Research Progress in Air Anion Generation Method and Its Detection Technology *J. of Environmental Management College of China* **18(04)** 1–3

[7] Qing-shan L, Jing Zh, De-hai Zh et al 2012 Measuing Air Anion Concentration and Evaluating Tourism Gravity of Beidaihe *Proc. of China Functional Materials Technology and Industry Forum* (Kunming: Yunnan University)

[8] Shan Zh 2013 The Analysis of Gravity on Tourism Resource of Shijiazhuang with Anion, Hebei *Inform. Engineering Res. Institute, USA. Proc. of 2013 Int. Conf. on Agricultural and Natural Res. Engineering (ICANRE 2013). Inform. Engineering Res. Institute* vol 5 (USA: Intelligent Information Technology Application Society) p 7

[9] Xiongwei Zh 2007 *Study on air negative ion level, dust retention and noise reduction in three forests in Lianfengshan Park* (Baoding: Hebei Agricultural University)

[10] Qingshan L I 2005 Research and application of Yanda Qicai negative ion additive *Proc. of the 5th Symp. on Functional Textiles and Nanotechnology Applications* (Beijing: China Textile Science Research Institute) p 5

Geo-ecological problems of artificial reservoirs of Kalmykia (case study of the Tsagan-Nur reservoir)

N Ts Lidzhieva[1,4], S S Ulanova[2], S V Ubushaeva[1] and Zh V Ovadykova[3]

[1]Kalmyk State University named after BB Gorodovikov, 11, Pushkin st., Elista, 358009, Russia
[2]Institute of Complex Studies of Arid Territories, 111, Khomutnikov st., Elista, 358005, Russia
[3]Ukhta State Technical University, 13, Pervomayskaya st., Ukhta, 169300, Russia

E-mail: for-lidjieva@yandex.ru

Abstract. The article presents results of six-year research on the Tsagan-Nur reservoir located on the Caspian lowland. Quantitative values of several indicators were obtained: hydrological (level and area of the reservoir), hydro-chemical (mineralization, chemical elements), environmental (ecotone water-land system). The methodological basis is the ecotone concept according to which there are blocks (areas) experiencing effects of the water body. Analysis of space information showed a significant reduction in the water surface area from 55.4 km^2 (1978) to 6 km^2 in 2018. Mineralization of the reservoir water increased 7 times compared to the average long-term data (10-12 g/l) and amounted to 70.24 g/l in 2018. Drying and deterioration of the water quality increase the saline area and decrease biological diversity of the flora and fauna, make it impossible to use waters of the reservoir for economic purposes.

1. Introduction

Intensification of the anthropogenic impact on natural ecosystems, especially on aquatic ecosystems, contributes to the development of a number of geoecological problems that are extremely acute for arid and sub-arid regions. The most important one for Kalmykia are as follows: salinization of surface waters making them unsuitable for use; pollution of natural waters by toxic elements, waterlogging of lands making it impossible to use them for economic needs, reduction of the water surface, a decrease in the level of groundwater, drying of lakes, formation of takyrs and new saline areas. An important feature of geo-ecological problems is their slow development, the cumulative nature of destructive phenomena and latent manifestation.

Many geo-ecological problems remain unresolved due to the lack of field data. Identification of features of natural systems on the banks of reservoirs in the conditions of arid Kalmykia seems very relevant; it can establish mechanisms and patterns of their development, assess the degree of development and possible hazards.

2. Materials and methods

In order to assess the state of water bodies, since 2001 long-term monitoring has been carried out. Its main task is to analyze quantitative data. Monitoring helps obtain reusable quantitative values of several indicators during the year: hydrological (level and area of the reservoir), hydrochemical (mineralization, chemical composition), environmental (ecotone water-land system). Hydrological indicators (water surface area) were determined using multi-time satellite imagery materials. Generalized requirements for remote sensing information used in solving hydrological problems include: spectral ranges: 0.6, 0.3–0.9 μm, 10–12 μm; spatial resolution of 30–60 m, the image scale 1: 100000.1: 250000 [1]. For processing and analyzing data obtained from space images, MapInfo 12.0, MultiSpek W32, Ilwis 3.6 were used. Interpretation was performed by the visual-instrumental method using black-and-white, color spectrum-zone, color-synthesized materials of space information obtained from artificial earth satellites (Resource O1 No. 3, Cosmos, Landsat, Spot) in different seasons (spring-summer-autumn) from 1975 to 2018.

The methodological basis of this study is the ecotone concept according to which there are blocks (areas) experiencing impacts of the water body on the reservoir shore. According to [2], there are six main blocks: aqual – a water area with depths of more than 1.5-2.5 m (devoid of macrophytes); amphibian – a littoral area with periodic drying during the period of water drawdown of reservoirs, fluctuation – an annually flooded coastal area; dynamic – a non-annually flooded area; distant – a non-flooded area affected by shallow (up to 3-5 m) underlying groundwater, and marginal – an area affected by the water body transmitted through the microclimate of the previous blocks (transitional to zonal). Ecotone "water-land" systems of coastal reservoirs were studied by the method of topo-ecological profiling. Topo-ecological profiles were laid perpendicular to the water edge, from the reservoir into the coast to the zonal vegetation. Topo-ecological instrumental coast profiling involved location of test plots with a detailed study of the soil, vegetation, groundwater and elevations. All descriptions were accompanied by sampling for studying water salinity in water bodies and groundwater; morphological structures and salt composition; species composition and biological productivity of plant communities. Analysis of samples for mineralization and water composition of reservoirs was performed in the Kalmyk branch of the Russian Agricultural Academy n.a. A.N. Kostyakova in accordance with GOST 26449.1-85: the cation-anionic composition was determined by by the titrimetric method, dry residue – by the gravimetric method, pH – by the potentiometric method. Latin names of plant species were given according to S.K. Cherepanov [3]. The analysis of plant life forms is based on the approaches by I.G. Serebryakova [4]; ecological types of plants were identified by the method suggested by T.K. Goryshina [5]. When describing and classifying vegetation, the dominant method in the name of communities was used.

3. Research object

The research object is the Tsagan-Nur reservoir which was the largest in a chain of lakes located in the southern part of the Sarpinskaya depression created by the ancient riverbed of the Volga River in the Caspian lowland. The bed of the channel reservoir elongated from the north to the southeast; it has a length of 45 km and a width of 0.7 to 1.5 km. The average depth is 1.15 m. At the normal water level (NWL), the reservoir volume is 90.0 million m^3, the mirror area is 61.5 km^2. The main power of the reservoir is irregular water supply from the Volga river through the BP-1 channel. The main function of the reservoir is a drainage-waste water receiver whose waters were used for two administrative districts and commercial fishing. From 2008 to 2014, the maximum catch yield was recorded in 2012 (46.4 tons), and the minimum one – in 2013 (21.3 tons). According to V.G. Poznyak [6] and D.S. Petrushkieva [7], there were 34 fish species in the Tsagan-Nur reservoir. However, due to a decrease in the water surface area, the commercial catch is not carried out. In addition, since 1982, the Tsagan-Nur reservoir has been a core of the Sarpinsk regional wildlife sanctuary, a concentration site for many rare bird species (deer crane, bustard, little bastard, steppe eagle, white-tailed eagle, long-legged buzzard, etc.). The list of birds included 145 species [8].

4. Results and discussion

The study of the current ecological state of the Tsagan-Nur reservoir was carried out at three sites: in the upper part of the reservoir, in the central part and at the dam. According to the results of monitoring of remote sensing data, the maximum filling was observed in 1978 (55.4 km^2) [8]. Since 2012, the water surface area has been steadily decreasing. According to the data of space imagery in 2012, its area was

31.63 km^2, in 2015 – 20.2 km^2, and in August 2018 – only 6 km^2. A decrease in the water surface area of the reservoir is due to a decrease in the flow of drainage-waste water through the BP-1 channel caused by a reduced area of irrigated rice fields. Annual monitoring of water mineralization revealed a significant increase. The dam zone of the Tsagan-Nur reservoir is a salt reservoir, as the Volga river water flows through the BP-1 channel into the northern part of the reservoir; water reaching the southern dam part increases its mineralization. In this part of the reservoir, mineralization has significantly increased. In 2015, it reached maximum values (in April – 22.13 g/l, in September – 73.18 g/l), in spring of 2018, salinity was 70.24 g/l (Fig. 1) exceeding the average annual values (10–12 g/l) 7 times. In 2017, it was not possible to take samples in this part of the reservoir due to its transformation into a salt marsh. Since 2015, the reservoir has been divided into small two arms separated by islets; it continues to dry out.

In the central part of the reservoir and in the transient region, it was not possible to select surface waters due to strong desiccation.

In order to study the effect of the reservoir on adjacent (ecotone) territories, monitoring studies of the composition of the water-land ecotone system were carried out from 2008 to 2012. This article presents results of the 2017 autumn field research on the composition and structural-functional organization of the key section located on the left coast of the Tsagan-Nur reservoir at a distance of 1.5 km from the dam (47°22'37"N, 45°11'37"E). Four blocks of the ecotone system were studied and described: fluctuation, dynamic, distant, and marginal [9].

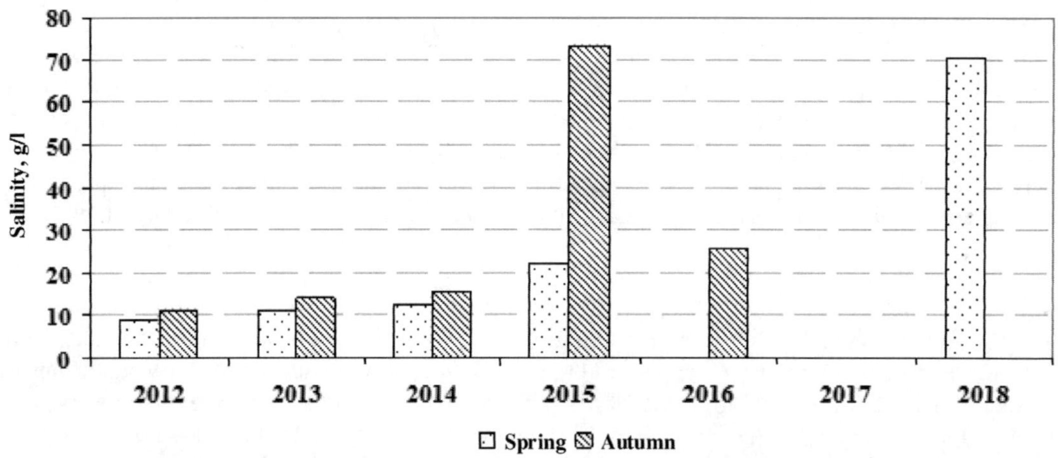

Figure 1. Changes in surface water salinity in the dam of the Tsagan-Nur reservoir.

The length of the fluctuation block was 250 m from the water's edge; the drying line was 110 m. The soil surface freed from water was gray, clayey, damp, takyr-like, devoid of vegetation. In the dehydration zone, the depth of groundwater (GW) was 0.7 m. Salinity of humic substances was 14.28 g/l, salinity was of sulphate-sodium-chloride type. Behind the strip of drying inland, the salt-bass community (*Bassia hirsuta* + *Salicornia perennans*) increased. The groundwater lay was at a depth of 1.3 m. The total projective coverage was 15 %. There are only two of the above listed species with a projective cover of more than 7 %. These are annual species that carry a high level of soil salinization. The air-dry weight of phytomass was 95 g/m^2.

The dynamic block, 150 m long, comprised two vegetation lanes. In the first lane, 65 m long, there were *Salicornia perennans* + *Bassia hirsuta* with a total projective cover of 95 %. The depth of groundwater was 1.6 m. HW mineralization was 16.02 g/l, the type of salinity was sodium sulphate-chloride. The community dominant was *Salicornia perennans* (PP 40 %), the subdominant was *Bassia hirsuta* (PP 35 %), other annual halophytes are *Suaeda salsa* and *Salsola soda*. The maximum height of the main mass of herbage is 40 cm. Biological productivity was 426 g/m^2. In the second lane of this unit, there was a petrosymonium-winch plant community (*Atriplex tatarica + Petrosimonia oppositifolia*) with a total projective cover of 65 %. Groundwater was deposited at a depth of 2.3 m. Salinity of humic substances was 12.36 g/l, the type of salinization was sulphate-sodium-chloride. The number of plant species was 13. The community dominant was *Atriplex tatarica*, the subdominant was *Petrosimonia oppositifolia* with an abundance sp_{1-2}. The remaining species are represented were less numerous: *Bassia hyssopifolia, Atriplex prostrata, Salsola soda, Suaeda salsa* and perennial herbs *Argusia sibirica, Vincetoxicum sibiricum. Artemisia santonica* grew from semi-shrubs; woody plant forms were represented by *Tamarix laxa* 20 cm in height. The air-dry weight of plant phytomass was 307 g/m^2.

The distant block began at a distance of 400 m from the water edge. In the first lane of vegetation, 80 m in length, the groundwater was located at a depth of more than 2.5 m. Petrosymonium-prosperous (*Suaeda salsa – Petrosimonia oppositifolia + P.brachiata*) communities grew here. The number of species was 11, the RPF of the community was 60 %. The dominant was *Suaeda salsa* whose abundance makes up half of the total projective cover of the grass stand (cop_1). Community subdominants were Petrosimonia species. *Salicornia perennans, Atriplex tatarica, Bassia hirsuta, Senecio vernalis, Polygonum aviculare, Suaeda acuminata* were observed. Perennial species were represented by Aeluropus littoralis, shrubs – by *Tamarix laxa* up to 40 cm in height. The biological productivity of the community was 92 g/m^2. In the second lane which was more than 50 m long, there were petrosymonium-santonin-full-tamarisk communities (*Tamarix laxa – Artemisia santonica + Petrosimonia brachiata*). Groundwater was deposited at a depth of more than 3 m. The number of species was 10, OPP was 60 %. The community was two-tier: the first tier was up to 3 m in height with a crowns density of 0.6. It was represented by *Tamarix laxa* shrubs. The second tier, up to 50 cm in height, was represented by *Petrosimonia brachiata* with an abundance of cop_{2-3} and Artemisia santonica with an abundance of sp_2. The abundance of other species is insignificant. Most of them are perennial herbaceous plants: *Puccinellia dolicholepis, Limonium caspium, Alhagi pseudalhagi,* and the long-rhizome xerophyte *Argusia sibirica.* Among the annual species, there were *Atriplex prostrata, Polygonum novoascanicum, Lepidium ruderale*. The air-dry weight of plant phytomass was 125 g/m^2.

The marginal block starts at 520 m from the water edge. Forbred-wormwood communities (*Artemisia lerchiana + A.santonica + A.pauciflora – Poaceta – Mixteherbosa*) grew here. The RPF of the community was 50 %, the number of plant species was 20. Soils and vegetation are of zonal nature: meadow soils are replaced by brown desert-steppe light loamy soils in combination with solonetz, there are more xerophilous species: *Arthmisia lerchiana, A. pauciflora, A.santonica, Potentilla bifurca grow from semi-shrubs, Alhagi pseudalhagi, Agropyron cristatatum, A.desertorum, Koeleria glauca, and APEMA, Apteratorum, aptortorum, A.desertorum, airoporhumi, and atiortorum.* Annual cereals were as follows: *Anisantha tectorum, Eremopyrum triticeum.* Herbs were represented by perennial species: *Achillea leptophylla, A.nobilis, Dianthus polymorphus, Kochia prostrata, Phlomis pungens*. The air-dry weight of plant phytomass was 58 g/m^2.

Floristic analysis of the ecotonic vegetation of the Tsagan-Nur reservoir showed that the species composition is represented by 135 species belonging to 26 families and 91 genera. The most numerous of them are Chenopodiaceae (31), Asteraceae (20), Poaceae (20) and Brassicaceae (13). Most of the plants are monocarpic herbs (54.3 %). Among annual monocarpic grasses, there are *Alyssum desertorum, Eremopyrum orientale, E. triticeum, Atriplex (Atriplex aucheri, A.micrantha, A.prostrata, etc.), Chenopodium album, Berteroa incana, Crepis tectorum, Lappula squarrosa, Lepidium perfoliatum.* Perennial herbs make up 33.3 % of the total number of plant species: *Achillea (A. leptophylla, A.nobilis), Bolboschoenus maritimus, Convolvulus arvensis, Dianthus polymorphus, Leymus ramosus, Carex stenophylla, Limonium (L. caspium, L. gmelinii , L.meyeri*), etc. There are 12 species of half-shrubs (8.9 %): *Artemisia austriaca, A.lerchiana, A.pauciflora, A.santonica, A.taurica, Camphorosma monspeliaca, Frankenia hirsuta, Halimione verrucifera, Kochia prostrata, Limonium suffruticosum, Halocnemum strobilaceum.* Trees are represented by shrubs; in this area, 2 species (1.5%) of Tamarix were observed – T.*laxa and T.ramosissima*. The vegetation of the ecotonic territory of the Tsagan-Nur reservoir is represented by 10 ecological types, of which euxerophytes predominate – 32.6

IOP Conf. Series: Materials Science and Engineering **663** (2019) 012039 doi:10.1088/1757-899X/663/1/012039

% (*Artemisia lerchiana, A. taurica, Atriplex aucheri*, etc.). Eumeophytes and transient xeromesophytes account for 19.3 % and 18.5 %, respectively. 11.1 % is accounted for by mesoxerophytes, 7.4 % – by egalophites, 5.2 % – by haloxerophytes. Halomesophytes make up 2.2 %, hygrophytes and halomesophytes account for 1.5 %, hydrophytes – 0.7 %.

5. Conclusion

Studies have revealed a significant increase in the degree of mineralization of surface waters of the Tsagan-Nur reservoir – up to 70.24 g/l which is seven times higher than average long-term data (10–12 g/l). Deterioration of water quality making water unsuitable for economic use was identified. Most of the reservoir (the central and transient zones) turned into a salt marsh bog. Salinization of groundwater in the conditions of the effusion type of the water regime increases the area of saline lands. Drying of reservoirs decreases biological diversity of flora and fauna, since in dry conditions, reservoirs are landscape reserves for conservation of natural biodiversity of vegetation, soil, animals and birds. It is necessary to eliminate reservoir regimes that contribute to the salinization of waters and adjacent territories, and replace them with more rational, non-adverse regimes.

References

[1] Garbuk S V and Gershenzon V E 1997 *Cosmic Systems for Remote Sensing of the Earth* (Moscow: A and B)

[2] Zaletaev V S 1997 The global network of aquatic-terrestrial ecotones, its functions in the biosphere and its role in global changes *Ecotones in the biosphere* (Moscow: RAAS) pp 77–90

[3] Cherepanov S K 1995 *Vascular plants of Russia and neighboring countries (within the former USSR)* (Leningrad: Science)

[4] Serebryakov I G 1962 *Ecological morphology of plants. Life forms of angiosperms and conifers* (Moscow: Higher School)

[5] Goryshina T K 1979 *Plant Ecology* (Moscow: High School)

[6] Poznyak V G 1987 Pisces *Animal World of Kalmykia* (Elista: Kalmyk book publishing house) 110 p

[7] Petrushkieva D S and Bugakov A A 2015 Ichthyofauna of inland commercial water bodies of the Republic of Kalmykia *Fish farming and fisheries* **11–12** 14–20

[8] Shapovalova I B 2017 The state of biodiversity of rare and endangered species of birds in inland waters of the steppe zone of southern Russia (Republic of Kalmykia) for 2008–2015 *Ecosystems: ecology and dynamics* **1(4)** 59–100

[9] Ulanova S S 2015 Ecological monitoring of the Caspian lowland reservoirs in the Republic of Kalmykia (on the example of the Tsagan-Nur reservoir) *Arid Ecosystems* **21.2(63)** 27–37

[10] Ulanov N E, Shaglinov P A and Lidzhieva N Ts 2017 Estimation of floristic and phytocenotic diversity of ecotones of Tsagan-Nur reservoir *In the World of Sci. Discoveries* **9(4–2)** pp 167–78

IOP Conf. Series: Materials Science and Engineering **663** (2019) 012040 doi:10.1088/1757-899X/663/1/012040

Possibility of using renewable energy sources in the Republic of Ingushetia

M I Malsagov, A R Elbazurov and G R Titova

National Research University Moscow Power Engineering Institute, 14, Krasnokazarmennaya ave., Moscow, 111250, Russia

E-mail: islamaga@mail.ru

Abstract. The article is devoted to an acute problem of the humanity - the scarce of energy sources, in particular in the Northern Caucasus. Therefore it is necessary to use renewable energy as a pure and undeveloped source. In this paper the authors explore the situation in Ingushetia. Nowadays the Republic has got a serious problem of electricity liabilities non-transparency of accounting system, quality and reliability of electricity supply. In the article it is possible to find general features of the region, the analysis of the current situation in the electric industry of the Republic, the power system characteristics, the electric industry development program and profitability of wind, water and geothermal energy analysis. The issue of switching to renewable energy is also considered.

1. Introduction

The civilization today can not be imagined without energy. Improving living conditions, growing life expectancy, needs satisfaction – all these require access to energy. Humanity is in constant search for new methods of obtaining energy, while increasing its consumption. Progress depends on industrial development, which depends on energy [1]. At the same time, we must not forget that energy is one of the serious factors that have an adverse impact on the environment.

The Republic of Ingushetia is an energy-deficient region [2]. The energy system of the republic is characterized by moral and physical deterioration of substations and power lines and, as a result, significant losses in the power grids with increase in electricity consumption by the population.

This growth is predicted in the future. All these factors affect the development of the Republic of Ingushetia and do not allow developing confidently in the industrial sector. One of the ways of energy development, as well as preserving the integrity of the natural heritage is using renewable energy sources.

2. General features of the region

The Republic of Ingushetia is located on the northern slopes of the foothills of the Greater Caucasus Mountain Range (in its central part) and on the small ridges adjacent to it – the Tersky, Sunzhensky and Skalist [3]. Length: from north to south – 144 km., from west to east – 72 km.

The climate of the Republic is continental, alpine, depending on the height above the sea level. The average temperature of January is –5 °C, July + 21 °C . The main river is Sunzha. There are also smaller rivers such as Assa, Fartanga, Armkhi, Guloykhi, Phatonka and Chemulga [4]. The total area of the river basin is 3073 km^2. There are significant reserves of water power; There is also a network of irrigation canals.

Ingushetia is an agrarian-industrial republic, 60% of the territory is occupied by agricultural land, half of which is arable. The industry of the Republic is underdeveloped.

3. Electric energy development program for the Republic of Ingushetia

In 2016 according to the Russian Government the decree subprogram under the title "Social-economic development of the Republic of Ingushetia for the period from 2017 to 2025" was issued. It is a part of state program "Nothern Caucasus Federal District development for the period until 2025". The main goal of the program is stabilization of social and economic situation in the Republic of Ingushetia providing long-term basis for economic growth and living standards improvement.

To achieve this goal it is necessary to solve the following problems: development of housing and public utilities as well as engineering infrastructure, overcoming technical underdevelopment, maintenance of industrial and agricultural development, development of electricity supply system.

Power self-sufficiency in particular the development of electricity and gas supply systems of the region is of key significance for sustainability of Ingush economy. The launch of important investment projects on creation of industrial and innovation clusters in the Republic make the solution of this problem especially acute. The priority direction is the development of electricity supply system which includes development of the integrated electric grid and generation in the Republic.

The development of generation is important because of the absence of generation facilities in the Republic. Construction of generation facilities will provide regional development according to the Republican focused program including the development of the advanced investment sites.

4. Solar energy development

Solar energy is ecologically pure as it has been coming to the Earth for billions of years and all the processes on the planet are accustomed to it. People must control and use it, preserving by that means the unique climate of the Earth.

The Republic of Ingushetia possesses huge resources of solar energy due to the large number of sunny hours and the level of solar radiation. As an example for our research we took a private house for a big family (more than 4 persons). At the first stage of designing of an autonomous solar power plant (ASPP) we make a list of all using equipment and calculate its energy input and voltage [5]. The recounted voltage of an electric user is also calculated so that to calculate the efficiency factor of an autonomous voltage inverter according to the formula (1):

$$P_{r,i} = \frac{P_{n,i}}{\eta} \tag{1}$$

where $P_{n,i}$ – nominal capacity of receiver i; $P_{r,i}$ – capacity of receiver recounted on the main distribution bus of ASPP; η – voltage inverter efficiency.

Then we evaluate how long the electric devises are used per day and calculate the daily demand on electric power making a load variance profile. Total capacities of loads in certain periods of time are calculated according to the formula (2):

$$P_{t,j} = \sum_{i=1}^{N} P_{r,i} \tag{2}$$

where N – number of receivers put into the network in j-m period of time

The output capacity of ASPP is determined as the maximum capacity of electric load for a summer daytime interval.

It is possible to get more accurate calculation of the demanded power level if we consider ASPPs of different capacities and state their generation level for each month, and the consumer will choose the capacity available to him on the basis of his requirements. To this purpose we have the data of METEONORM database about monthly insolation in the Republic present in Table 1.

Table 1. Data on insolation in the Republic from METEONORM database

Month	Jan	Feb	Mar	Apr	May	Jun	Jul	Aug	Sep	Oct	Nov	Dec
Insolation, KWh/m^2	39	60	95	127	160	162	170	146	112	75	41	30

After that we calculate the produced energy be means of solar modules, and write to the table 2.

$$W = kEP_w \quad (3)$$

where E – magnitude of insolation for the chosen month, P_w – capacity of the chosen module, k – ratio which equates 0.6 and 0.85 during summer and winter periods consequently. For the winter period the months with an average temperature below zero centigrade (November, December, January and February) were chosen.

Table 2. Energy generated by solar modules, kWh

	Month											
P_w, kW	Jan	Feb	Mar	Apr	May	Jun	Jul	Aug	Sep	Oct	Nov	Dec
5.5	180.9	278.2	312.5	417.5	528.3	534.3	560.0	480.5	368.0	247.8	190.7	139.3
6	197.4	303.5	340.9	455.4	576.4	582.8	610.9	524.2	401.4	270.4	208.1	152.0
6.5	213.8	328.7	369.3	493.4	624.4	631.4	661.8	567.8	434.9	292.9	225.4	164.6
7	230.3	354.0	397.7	531.3	672.4	680.0	712.7	611.5	468.3	315.4	242.8	177.3
7.5	246.7	379.3	426.2	569.3	720.5	728.6	763.7	655.2	501.8	338.0	260.1	190.0
8	263.2	404.6	454.6	607.2	768.5	777.1	814.6	698.9	535.2	360.5	277.4	202.6
8.5	279.6	429.9	483.0	645.2	816.5	825.7	865.5	742.6	568.7	383.0	294.8	215.3
9	296.1	455.2	511.4	683.1	864.5	874.3	916.4	786.2	602.1	405.5	312.1	228.0

Also the best possible banking angle of solar array was calculated for each month as shown in figure 1:

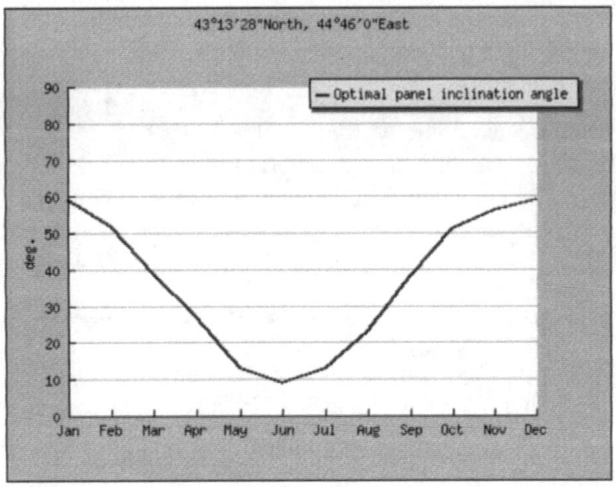

Figure 1. Optimum tilt angle of solar modules.

The best possible banking angle during the year was 33 degrees.

5. Water energy development

The own production of energy is one of the key tasks of the region. The Republic of Ingushetia possesses significant water power recourses which is about 1.5 bln kWh that is proportionate to the energy demands of the Republic and also wind and solar energy production capabilities. The main part of water power production capabilities of the rivers of Ingushetia (55 %) is located on the Assa River. According to the data of predesigning research of the previous years the construction of a hydroelectric power chain on the Assa River is highly recommended: Pervomayskaya WPP, Alkunskaya WPP, Targimskaya WPP, with total capacity of 40 MW each.

For this region one of the most efficient directions of developing non-traditional energy production is using the power of small water flows through micro and small WPP working within the range from 3 to 100 KW and to 5000 KW consequently. On the one hand it is explained by the significant capability of such flows and simplicity of their use, on the other hand the coming exhaustion of the big rivers' capability in this region.

6. Geothermal energy development

There is a good possibility of using geothermal energy in the Republic of Ingushetia as there are powerful sources of geothermal industrial waters with the temperature up to 160 above zero.

Though nowadays geothermal sources in the Republic mostly serve for thermal energy production, it possible to get electric energy from this source as well because the temperature of the geothermal waters is in the high-potential group [6].

Geothermal energy has a range of advantages: such as low cost which does not depend on climate and weather and has a high ecological compatibility. It is especially important because there are a lot of wildlife areas in Ingushetia and we can not undermine its natural identity.

7. Wind energy development

For the efficient utilization of wind energy we must have full information about the wind as a natural phenomenon and a source of energy [7]. Today we know a long-term average annual wind velocity in the region which serves as the initial feature of the wind intensity general level in the certain place. Based on it we can assess the possibility of using wind energy sources. The data on the long-term wind velocity are given in the Table 3.

Table 3. Data on the long-term wind velocity

Month	Jan	Feb	Mar	Apr	May	Jun	Jul	Aug	Sep	Oct	Nov	Dec	Year
Velocity, ms^{-1}	5.01	4.82	4.41	4.29	4.13	3.83	4.04	4.25	4.26	4.36	4.66	4.99	4.42

According to the evidence we come to the conclusion that the region possesses sufficient wind velocity to consider the point of a wind driven generator construction. It should be mentioned that we do not consider the construction of a wind driven generator higher than 10 m.

Considering a wind driven generator with the following characteristics in the table 4 as an example

The capacity of a wind driven generator is determined by the following expression:

$$N_{wg} = 0.125\rho\pi D_{vd}^2 \upsilon_{hv}(t)^3 \eta_w \eta_h \eta_g = 0.125 \cdot 1.226 \cdot 3.14 \cdot 4^2 \cdot 4.42^3 \cdot 0.45 \cdot 0.9 \cdot 0.95 = 255.8\ W \quad (4)$$

where ρ – atmospheric density; D_{vd} – vane diameter; υ_{hv} – air flow velocity at the height of the WDG tower at the moment of time t; η_w – windwheel efficiency factor; η_h – herringbone gear efficiency factor; η_g – generator efficiency factor.

Energy output by a WDG is calculated according to the formula:

$$E_o(T) = \sum_{j=1}^{N^r} \left[N_{wg} t(\Delta V_j^r) \right] T = 0.2558 \cdot 0.98 \cdot 8760 = 2196\ kWh \quad (5)$$

where N^r – number of wind velocity repeatability radiuses; T – monitoring period; $t(\Delta V^r_j)$ – wind velocity repeatability factor.

IOP Conf. Series: Materials Science and Engineering **663** (2019) 012040 doi:10.1088/1757-899X/663/1/012040

Table 4. Data on the long-term wind velocity

Capacity at 12 ms^{-1}	3500 W
Capacity at 9 ms^{-1}	2300 W
Capacity at 5 ms^{-1}	500 W
Generator starting torque	from 2.5 ms^{-1}
Range of generation altitude	3 to 25 ms^{-1}, more than 20 ms^{-1} defensive inhabitation is switched on
Number of vanes	3 items
Material of vanes	reinforced glass fiber with protective covering 3M, USA
Rotor diameter	4 m
Weight of the wind driven	Generator with vanes and tail unit: 130 kg
Vertical shaft	Bearing current collector
Working lifespan	Minimum 10 years

8. Results and Discussion

The research on this point showed that using renewable energy sources has a lot of advantages:

- Sustainable electricity supply which is absent in the most towns and villages of the Republic.
- Electric power of good quality (without voltage and frequency surges)
- Transparent accounting system of electric power
- Possibility of supplying power to the electric grid of the Republic.
- Releasing energy sources due to energetic self-sufficiency of consumers
- Decrease in equipment deterioration as a result of low-quality and unstable energy)
- Industrial production growth.

The only disadvantage is the necessity of huge investment at the first stage.

9. Conclusion

Taking into account the fact that Russia is making its first steps towards renewable energy development by legal acts as well as big power plants projects and the Republic of Ingushetia is also interested in this sphere , these energy sources must be developed that will improve living standards and the investment climate in the whole region guaranteeing economic growth.

Introduction of small generation would meet electric power demand in the Republic of Ingushetia and decrease the final cost of electric energy that is important for its development.

References

[1] Fortov V E and Popel O S 2012 *Renewable energy in the modern world* (Moscow: MPEI publishing house) p 9

[2] Popov E E 2017 *Scheme and development program of electric energy of the Republic of Ingushetia for 2018-2022* (Magas) p 7

[3] Dolgiyeva M B, Kartoev M M, Kodzoev N D and Matiev T H 2012 *History of Ingushetia* (Rostov on Don, Southern Publishing House) p 4

[4] *Social and economic development strategy of the Republic of Ingushetia for the period until 2030* 2015 (Magas, short version) p 10

[5] Okhotkin G P 2013 *The method of calculating power of solar power* (Moscow, Chuvash University Press) p 222

[6] Semenov V, Nikolsky A, Tomarov G and Shipkov A 2015 *Geothermal energy* (Moscow) p 131

[7] Burmistrov A A, Vissarionov V I and Deryugina G V 2009 *Methods for calculating renewable energy resources* (Moscow: MPEI publishing house) p 144

IOP Conf. Series: Materials Science and Engineering **663** (2019) 012041 doi:10.1088/1757-899X/663/1/012041

Environmental problems in the Kazakhstan coal industry and their solutions

T A Alimbaev[1], Zh S Mazhitova[2,4], B K Omarova[2] and Zh B Nurkina[3]

[1]Karaganda State University n.a. E.A. Buketov, 28. University city, Karaganda, 100028, Republic of Kazakhstan
[2]Astana Medical University, 49a. Mir st., Astana, 010000, Republic of Kazakhstan
[3]Sh. Ualikhanov Kokshetau State University, 76. Abay st., Kokshetau, 020000, Republic of Kazakhstan

E-mail: mazhitova_69@mail.ru

Abstract. This article discusses development of the coal industry in Kazakhstan after the 1990s and environmental problems caused by the coal industry. Along with industrial development, an increase in coal production, the level of environmental pollution is increasing. The authors say that transition to the market economy increased the economic potential of the republic, including coal companies, and caused an environmental crisis. The problem can be solved by maximizing the economically viable use of all types of waste and resources of the developed coal fields. These activities will solve geo-ecological problems, improve the environment through the greening of economic activities focusing on the needs of future generations. The methodological basis of the research is scientific objectivity which made it possible to analyze the subject taking into account economic changes in the industry.

1. Introduction
The coal industry in Kazakhstan and other republics of the former Soviet Union developed in hard conditions [1–4]. In 1991, the total coal production was 130.4 million tons, or 20.7 % of the total coal production in the territory of the former USSR. The industry had 26 mines, 7 deposits. Factories washed 27 % of all coal mined, including 98 % of coking coal. Coal was shipped to all republics of the former Union and Eastern European countries. The fundamental changes in the sociopolitical situation, acquisition of sovereignty, development of commodity-money relations, a decline in coal production required a search for new approaches in the coal industry. After the privatization of coal-mining enterprises, there has been a gradual recovery from the crisis. At present, the difficult period of coal production restructuring has been completed, positive results have been achieved, coal production volumes have increased.

2. Discussion
Transformations ensured the transition of the coal industry to market relations. Today, the coal industry produces 78 % of energy; there are opportunities to meet fuel needs of the household sector. There are 33 coal mining companies (5 foreign and 28 national ones). For ten years, they have invested more than 3 billion US dollars in the subsoil use of coal. In comparison with 1996, the volume of investment increased 10 times; in 2006, it amounted to 428.7 million dollars. Currently, coal production is 95–96

million tons per year. The expected volume of production will be 97 million tons. Taking into account the urgent need to ensure newly commissioned generating capacities with solid fuel, the Concept of development of the coal industry until 2020 has been developed. It involves an increase in coal production up to 145.6 million tons by 2020, or by 49.3 million tons, including coking coal – up to 24.3 million tons or by 11.4 million tons, thermal coal – up to 121.3 million tons or by 37.9 million tons, which meets the needs of domestic and foreign markets in coking and energy coals.

2.1 Problem statement

Coal-mining enterprises have a significant impact on the environment, as coal and surrounding rocks extracted and their processing products pollute it with dust, soot and gases. Methane and carbon dioxide create negative phenomena. Water from mines contains a large amount of salts and various solid impurities that pollute water intakes. Location of the rock in dumps cause land degradation. In order to reduce harmful effects on the environment, coal enterprises carry out very labor-intensive environmental protection works. The aging of mine equipment, hard mining and geological conditions due to the transition to deeper horizons decrease technical and economic performance and deteriorate the ecological situation caused by an increase in the anthropogenic impact on the environment. The coal industry is a basic component of the fuel and energy industry and the raw material base of metallurgy; it is one of the leading industries in the economy of Kazakhstan. As of 2016, Kazakhstan ranked eighth by coal reserves (25.6 billion tons, or 2.2 % of the world reserves, the BP Statistical Review of World Energy, June 2017) and tenth by the coal production volume (102.4 million tons, or 1.4 % of the world production volume).

In 2016, according to the British Petroleum company, Kazakhstan consumed 56.5 % of coal, 20.9 % of oil, 19.1 % of natural gas, and 3.3% of hydropower. The coal industry is one of the most important resource sectors of the economy of Kazakhstan. There are more than 300 coal deposits with geological reserves of 170.2 billion tons. More than 9/10 of all coal reserves are concentrated in the central and northern parts of the country. The largest basins are Ekibastuz (12.5 billion tons), Karaganda (9.3 billion tons) and Turgai (5.8 billion tons). 70% of coal reserves are mined in three giant fields (Bogatyr, Severny and Vostochny mines) in the Ekibastuz basin (Pavlodar region) and in four fields in Karaganda region (Borlinskoye, Shubarkol, Kushokinskoye and Saryadyr). The remaining coal reserves are mostly mined in the Karaganda basin (for the needs of local metallurgical enterprises). The coal industry is one of the largest sectors of the national economy which provides for the production of 74 % of electricity, full loading of coke-chemical production, satisfies needs of the population. Coal is widely used in heavy and mining industries, other industries related to mining. The share of metallurgy and other industries in the overall structure of coal consumption is comparable to the indicator characteristic of the household sector (approximately 20 % of the total consumption) [5].

2.2 Environmental problems of the coal industry in the Republic of Kazakhstan

The economic success of the industry is due to the commercial interest of coal companies which often violate environmental requirements. At present, among the real dangers, environmental issues are crucial due to technological impacts on the environment, including man-made transformations of the subsoil in the mining process. In the last quarter of the twentieth and the beginning of the twenty-first centuries, humanity faced serious environmental problems in mining operations. They are caused by man-made and natural processes. Large-scale and intensive activities lead to the accumulation of environmental problems, decrease economic capacities of the natural environment. At present, it is impossible to indicate a single source of energy whose use would not cause the degradation of the natural environment (NE).

In the republic, the main amount of harmful substances emitted into the atmosphere by extracting industries falls on the oil-extracting, coal and gas industries. Nearly 80 % of total emissions are sulfur dioxide, carbon oxides, nitrogen oxides and hydrocarbons. In addition, mining enterprises are the largest water consumers; they consume 30% of fresh water used in the country and up to 75 % of water used by industries. The coal industry has the largest discharge of polluted wastewater. Coal mining is accompanied by issuing and placing of 150 m^3 of host rocks and 100–120 m^3 of waste. With the extraction of one million tons of coal, up to 7.5 hectares of land in mines and up to 20 hectares in quarries are disturbed. When mining fields, the earth's surface subsides, the amount of subsidence varies from 12 to 20 m which causes losses of agricultural land. The development of coal deposits causes damage to all elements of the biosphere.

The solution of environmental problems, reduction of the load on the environment should be accompanied by creation of an underground system, a closed water system excluding the discharge of untreated water into external water bodies; systems for processing solid waste, reduction of alienation of land for their storage; creation of raw coal enrichment systems, dust suppression systems, a complex-mechanized technology for development of environmentally sensitive parts of the array, including near aquifers [6].

Analysis of environmental protection measures in individual mining regions shows that most technical solutions and measures are aimed at neutralizing and eliminating effects of production activities, rather than eliminating their sources. Reduction of the negative impact on the environment is carried out through partial restoration of the disturbed natural state which is achieved through land reclamation, clarification and purification of wastewater, dust and gas recovery by heat and power plants.

Environmental activities associated with achievement of established standards require significant costs. Water and air purification technologies, water treatment and gas trapping facilities are imperfect and ineffective [7]. Since there are no integrated treatment technologies, only a few dozen harmful substances are caught by treatment plants, and other are emitted into the air. These measures are ineffective, since they do not eliminate causes of negative phenomena which deteriorates the ecological situation in the coal mining regions. Without a significant amount of expenditures, it is impossible to eliminate harmful effects of coal mining.

The greening of a coal enterprise depends on technological processes of coal extraction, rational use and protection of the subsoil, complex use of waste to produce marketable products, elimination or reduction of environmental pollution, land violations, water circulation cycles and closed fuel and energy structures. Requirements are a system of restrictions (environmental and economic), technological directions and possible technical solutions for coal mining, rational use of natural resources and protection of the natural environment. During the extraction of minerals, a large number of mine, quarry and drainage waters polluting water bodies are released onto the surface. Phenols, aromatics and other pollutants in mine waters fall into the underground aquifer of groundwater and pollute it causing serious environmental and economic damage [8]. Currently, there is no universal technology for treating mine and other wastewater from soluble organic substances. Depending on the physical and chemical composition and technological properties of mine waters, various methods can be used to purify them.

2.3 Possible solutions to environmental problems in the coal industry of the republic

To solve environmental problems, drastic measures are required. It is necessary to develop low-waste production based on the integrated use of mineral and energy resources of mines and other enterprises of the coal industry. Extraction and enrichment of coal are integral parts of heat and electricity production. Therefore, a coal enterprise should be part of an energy enterprise. The final product is electrical and thermal energy. It is important to improve energy efficiency and environmental protection in the coal-energy enterprise in all the links of the technological chain: extraction and processing – energy production – energy consumption. Secondly, technological processes of the coal-energy enterprise will make it possible to use natural resources, create and apply waste-free and energy-saving technologies taking into account advantages of a through-production cycle. The organic combination "TPP – coal enterprise" will make it possible to reduce losses in the power supply network, utilize recovered energy of coal mining waste (methane, low temperature mine water, air flow, flue gases, recycled water, etc.) with a maximum efficiency.

In addition, it is necessary to use waste coal as a fuel. Since solid waste from coal combustion is much more in power plants than in coal-fired boiler houses, it is necessary to use it for laying the mined-out space in the mine which will make it possible to eliminate inevitable rehabilitation works. According to [9], technologies should be developed for processing and placing rocks in the mines without storage them on the surface. To reduce the volume of pumped and treated mine and quarry waters, appropriate technologies should be developed to reduce permeability of rocks, separation of the process and drainage water flows, prevention of surface waters from entering, dumping of highly mineralized waters in geological structures, etc. Implementation of these tasks can be achieved with through the integrated use of coal industry waste provided that waste should be considered as a product rather than a goal. Waste products should differ in technological stages of their formation; states of aggregation; valuable components; degrees of damage to the natural environment; costs [10].

Kazakhstan is developing a nationwide waste management program. The program is aimed at reducing waste, including coal products, processing accumulated industrial waste and land reclamation. Despite discussions about feasibility of valuation of natural resources, most countries recognize that natural resources are owned by the society, and the state is entitled to withdraw. The Statistics Division of the UN Secretariat summarized statistical data on natural resources in a number of countries and developed international recommendations of the environmental-economic accounting system (EEAS). The EEAS concept helps assess the relationship between the nature and the economy in the reproduction process taking into account the consumption of natural resources and the return of production wastes to the natural environment, preservation and improvement of the quality of the environment.

The study on water purification methods [11] shows that a sorption method can be applied using both natural and synthetic sorbents to purify natural and industrial wastewater from metals. This method can purify water from any contaminants. The process is easy to manage. At the same time, sorbents should be cheap. In combination with a sufficient depth of purification, they can be used to solve environmental protection issues.

Industrial activated carbons are very expensive, their number does not meet the demand. In the Republic of Kazakhstan, carbon sorbents are imported from the Russian Federation. The Chemical and Metallurgical Institute n.a. J. Abishev (Karaganda) developed a technology for producing carbon sorbents from low-sulfur, low-phosphorus Shubarkol stone long-flame coal characterized by low ash content (5–15 %) [12]. The sorbent fraction of 0-5 mm can be used instead of activated carbon in the process of wastewater treatment. Active carbon sorbents are produced from brown coal. Coal sorbents are worse than brown coal sorbents. The sorbent produced by the technology of the Chemical and Metallurgical Institute is unique, since it is not worse and cheaper than brown sorbents. The high content of volatile components in Shubarkol coal (about 45%) predetermines formation of a porous structure. The developed technology can be used to produce relatively cheap adsorbents.

3. Conclusion

The lack of a mandatory preliminary expert assessment of environmental consequences of the use of scientific and technological developments, effective economic levers that encourage economic managers to use natural resources rationally, departmental disunity in environmental protection, imperfect legal sanctions, localism and negligence of officials complicate the environmental situation Moreover, the lack of awareness of about the state of the environment, relevant knowledge and practical experience in this area, passive behavior of coal enterprises and citizens, the lack of ecological culture, belief in inexhaustibility of natural resources make the situation even worse. All these economic, legal, managerial and ideological factors are a base of the braking mechanism that impedes the optimization of environmental management. Unfortunately, a complex system of environmental and economic indicators has not yet been developed. Ways to optimize permissible limits of economic and environmental parameters taking into account mining effects have not been determined. Therefore, the most important problem for effective development of the coal industry is implementation of a program for comprehensive assessment of the environmental impact of coal enterprises. It should encourage them to develop an environmental management system.

IOP Conf. Series: Materials Science and Engineering **663** (2019) 012041 doi:10.1088/1757-899X/663/1/012041

References

[1] Kulikova M P and Balakina G F 2010 Ecological problems of using coal in the Republic of Tyva *Ecology and industry of Russia* **12** 37–9

[2] Schastlivtsev E L and Bragin V E 2007 *Geoecological problems in the coal-producing areas of Kuzbass Coal* **11(979)** 59–63

[3] Troshchenko V V 2009 Prospects and problems of the development of coal resources of Eastern Donbass *Geology, geography and global energy* **3(34)** 116–20

[4] Gubov A M, Permilov V A, Grigorieva I V, Zavozkin S Y and Sotnikov I Y High technology software web-tools to solve environmental problems of coal region *Sci. evolution* **2–1** 82–90

[5] Kim V 2017 State and Prospects of the Coal Industry of Kazakhstan *Mining J. of Kazakhstan* **9** 26–34

[6] *National Report on the State of the Environment and the Use of Natural Resources of the Republic of Kazakhstan for 2017* 2018 (Astana: Ministry of Energy of the Republic of Kazakhstan) 147 p

[7] Karenov R S 2016 Environmental problems of development of the coal industry in Kazakhstan and priority areas for their solution *Problems of law and economics* **2** 28–35

[8] Sokolov E M, Sheinkman L and Dergunov D 2013 Reduction of anthropogenic load of mine waters during the exploitation of coal deposits *Mining* **2(108)** 138–41

[9] Krasnoshtein A E and Zakirov D G 2009 Energy and environmental problems in the development of the coal industry and ways to solve them *Coal* **6** 69–73

[10] Klimov S L and Zakirov D G 2016 *Energy saving and problems of environmental safety in the coal industry of Russia* (Moscow: MGGU) 210 p

[11] Kim V, Siukhina V, Kuzgibekova X and Bogoyavlenskaya O 2014 Carbon sorbent from Shubarkol coal coal *Industry of Kazakhstan* **3(84)** 20–3

[12] Kim V A, Bogoyavlenskaya O A and Kudarinov S Kh 2007 A method for producing carbon sorbent from Shubarkol coal *Innovative patent no 20678* (Astana: KazGosINTI)

IOP Conf. Series: Materials Science and Engineering **663** (2019) 012042 doi:10.1088/1757-899X/663/1/012042

Use of the space segment of remote sensing for monitoring natural and anthropogenic emergencies in southern Russia

A V Mueva, V A Velegurin, V A Eviev, S A Sangadzhieva, Yu S Germasheva, K A Sarsen and D Isaev

[1] Kalmyk State University n.a. B.B. Gorodovikov, 11 Pushkin st., Elista, 358000, Russia

Email: mav_el@mail.ru

Abstract. The article is devoted to the problem of remote sensing of the Earth. The article analyzes prospects for using the space segment of remote sensing for monitoring emergency situations of natural and man-made character in southern Russia (using the example of the Republic of Kalmykia) to predict and control emergency situations and eliminate their consequences. The use of a space monitoring system for identifying landscape fires in arid territories was suggested.

1. Introduction

The spatial aspect of information about emergencies is crucial. It is always important to know the nature, size and location of threats which can affect people in the vicinity, industries and the environment.

2. The fight against emergencies has much in common with military actions

There is also an enemy of natural or man-made origin. There is a need to protect the population, plan operations and allocate human and technical resources. Emergency management has a shorter history than military affairs due to the fact that people did not recognize the importance of systematic work with emergencies. The scale of man-made disasters forced people to consider them as enemies, aggressive states in the neighborhood. Accordingly, the role of space remote sensing in dealing with emergency situations is growing [1, 2].

When controlling emergency situations, GIS created on the basis of space remote sensing play an important role. Emergency situations threaten people throughout the national territory. There are floods, mudflows and avalanches, hazardous production. In order to predict and control emergency situations and eliminate their consequences, it is necessary to know local features [3]. Emergency situations sometimes develop even more rapidly than military operations. Computer technologies are needed to work with spatial information.

The main task of the space monitoring system used for emergency situations is to fight against large-scale forest, steppe, peat and oil fires. The reason for its creation is increased degradation of the natural environment and intensification of natural processes. One of the reasons for the aggravation of this situation is climate warming [4] accompanied by a sharp temperature instability, an increase in the number and strength of hurricanes and other emergency situations caused by natural phenomena. Natural emergencies develop rapidly. Their scales grow nonlinearly in time.

Therefore, the effectiveness of fight against emergencies and their consequences depends on efficiency of rescue operations and compliance of operation plans with the actual state of the object or territory.

3. Problem statement

The start time of rescue operations depends on how quickly the emergency will be detected. In the vast and sparsely populated territories of Russia, operational detection can be ensured by means of aviation patrols or spatial monitoring.

The Air forest protection service has shown that monitoring of Kalmykia can be carried out from space. For example, the scale of fires and the extent of damage are not comforting. Regular observations of landscape fires are carried out only in zones of active protection of territories (two thirds of the total area) [5]. In the northern and eastern districts of Kalmykia (one third of arid territories), firefighting and fire accounting measures are not implemented. In the zone of active protection of arid territories, 500–1000 landscape fires are recorded annually on the area of 0.5–2.1 thousand hectares. This circumstance and large unprotected territories indicate a low level of fire protection of arid territories. Satellite imagery [6] allows for recording fires in protected and non-protected areas in operational and semi-operational modes. It is less expensive than aerial surveys. The space data on the state of arid territories are accurate. The method for detecting landscape fires is based on the use of space images and interpretation of images containing landscape fires in arid territories of Kalmykia. The geographical reference of the foci of combustion is calculated automatically from the orbital data of the corresponding Earth remote sensing satellites.

The results are displayed on the vector GIS map. If necessary, the vector map can be replaced by a raster topographic one [7].

4. Materials and methods

Studies have shown that the problem can be partially solved by using the available space information (CI) from the spacecraft NOAA (USA) [8, 9], as well as the resource satellite of the Resource Ministry of Natural Resources [10]. However, high efficiency of obtaining high-resolution data is impossible to achieve since the space monitoring system was initially focused on information from the Meteor and NOOA (USA) and the resource spacecraft "Resource" created for other purposes.

The All-Russian Research Institute for Civil Defense and Emergencies (VNII GOES) (Federal Center for Science and High Technologies) of the Emergencies Ministry of Russia has developed and improved the space monitoring system. However, they are dealing with the ground component of the system.

Methods have been developed for the thematic processing of space information to identify fire and flood conditions, air and water pollution, oil spills. Methods have been developed for rapid assessment of consequences of emergency situations for agriculture and forestry, damage caused by earthquakes. The method was adopted to the new spacecraft. However, to resolve contradictions between the temporal and spatial resolution of remote sensing data within the existing system is impossible.

Most of the work was performed by the laboratory which receives and processes space information (Elista). The laboratory can receive and process high-quality space images from all free access spacecrafts.

The most characteristic (priority) tasks of the system and requirements for monitoring information are summarized in Table 1.

5. Conclusion

The functions of the spacecraft for monitoring natural and man-made emergencies by means of low resolution make it possible to solve the following tasks aimed at reducing risks and consequences of emergency situations:

- monitoring of occurrence and development of forest, peat and steppe fires;
- monitoring of pollution of water areas and earth surfaces with oil products;
- monitoring of territories of canal and reservoir infiltration;
- monitoring of occurrence and development of flooding;

- monitoring of ecologically unfavorable territories that are on the verge of escalation into emergencies;
- monitoring of the onset of droughts and dynamics of their development;
- assessment of the scale of natural emergencies;
- monitoring of areas exposed to mudflow and landslide hazards.

Table 1. Information requirements for solving monitoring priorities

Priority problems, tasks	Information Requirements				
	Update frequency (t)	Update responsiveness (t)	Resolution (view / IR)		
			Spatial	Spectral	Radiometric
Identification of fires	24	0.25–2	10–20	5–40	0.1–0.5 0.1–0.2
destruction	24	0. 25–2	10	5–40	0.1–0.5
flooding	12	0. 25–1	30–100	5–40	0.1–0.5
Assessment of the degree of damage caused by accidents and disasters	3–6	1–3	1–2	5–40	0.1–0.5
Determination of the aircraft crash site	3–6	1–2	1–2	5–40	0.1–0.5
Determination of severity of accidents and the number of vehicles involved in the accident	6–12	1–3	10–20	2–6	0.1–0.5
Determination of the size of a pipeline accident zone	6–12	1–3	2–10	5–40	0.1–0.5
Pipeline Condition Monitoring	15	0.5–1	12	2–6	0.1–0.2

With regard to the identification of foci of landscape fires, the proposed method can be used in southern Russia. The results can be used to solve fire protection problems when firefighting equipment, etc. The technology for assessing landscape fire damage in arid territories of Kalmykia saved 3–5 % of the republic's budget.

It is necessary to expand the structure of the space monitoring system and include the atmospheric component as an additional element for remote sensing of the earth.

The Ministry of Emergency Situations uses systems created not on its order; when expanding the structure of the system, an order to create new media will be required. This requires investment.

At the present stage of development of SCM emergencies, its effectiveness was assessed only by the completeness and quality of the information received rather than by final results – prevented damage. The monitoring systems of other departments were assessed in accordance with the needs of these departments. To evaluate the options for the structure and composition of the aerospace monitoring system, it is necessary to create an apparatus for assessing their effectiveness. The creation of this device is a research issue.

References

[1] Kozoderov V V 2001 Information-dynamic aspects of the description of natural phenomena *Earth exploration from space* **5** 27–43

[2] Lavrova O Yu 2008 Modern problems of remote sensing of the earth from space *Earth and the universe* **5** 52–8

[3] Belyaev A I, Korovin G N and Lupyan E A 2005 Use of satellite data in the system of remote monitoring of forest fires MNR RF *Current problems of remote sensing of the Earth from space: Physical bases, methods and technologies for monitoring the environment, potentially dangerous objects and phenomena* (Moscow: GRANP polygraph) pp 20–9

[4] Karpunin Y N, Mitrofanov V D and Fedotov V P 2016 Modern approaches to the problem of improving the means of metrological support of promising equipment for remote sensing of

the earth *Astronautics and rocket production* **6** 94–101
[5] Petrov K M, Bananova V A, Lazareva V G and Unagayeva A S 2016 *Dynamics of the desertification processes of the northwestern Caspian region: physical-geographical and socio-economic aspects* (St. Petersburg: Himizdat)
[6] Akatkin Yu M, Efremov V Yu, Lupyan E A, Polishchuk A A, Proshin A A, Tolpin V A and Flitman E V 2008 Possibilities of remote work with the data of the remote monitoring system of agricultural lands of the Ministry of Agriculture of the Russian Federation *Current problems of remote sensing of the Earth from space: Physical bases, methods and technologies for monitoring the environment, potentially dangerous objects and phenomena* (Moscow: Polygraph Service LLC) pp 395–406
[7] Vedeshin L A 2009 On the use of space information in the interests of the socio-economic development of the regions of the Russian Federation *Exploration of the Earth from space* **5** 90–2
[8] Ershov D V, Korovin G N, Lupyan E A, Mazurov A A and Tashchilin S A 2004 Russian system of satellite monitoring of forest fires *Current problems of remote sensing of the Earth from space: Physical principles, methods and technologies for monitoring the environment, potentially dangerous objects and phenomena* (Moscow: Polygraph Service LLC) pp 47–57
[9] Ershov I I 2018 Software and hardware complex for remote calibration of electrical quantities measuring instruments *Automation and software engineering* **4** 29–33
[10] Manaev E F 2015 Remote sensing of the earth by a series of spacecraft Resurs-P *Perspectives of the project SCIENCE TIME* **4** 466–71

IOP Conf. Series: Materials Science and Engineering **663** (2019) 012043 doi:10.1088/1757-899X/663/1/012043

Sustainable use of natural resources in a hazardous landslide slopes of Cheboksary reservoir (river Volga)

I V Nikonorova, N F Petrov, A E Gumenyuk, V N Ilyin

Department of Physical Geography and Geomorphology, Chuvash State University, 15, Moscovsky ave., Cheboksary, 428015, Russia

E-mail: niko-inna@yandex.ru

Abstract. Often uncontrolled construction with insufficient consideration of the geotechnical and hydrogeological conditions of the territory, the increased man-made impact on the slope in the conditions of growing urbanization are leads to serious problems: landslide deformations and emergencies on the coast of the Cheboksary Reservoir, on the territory of the Cheboksary urban district and in the adjacent territories of Chuvash Republic. The increase in recreational load on coastal geomorphic systems also contributed to the activation of landslide, erosion, deflation and other dangerous relief-forming processes, and also affected the quality of water resources. The problem of assessing the stability of hazardous landslide slopes and slopes for civil, transport construction and recreational development of the coast of the Cheboksary reservoir is analyzed in this article. Minimizing the risk and damage from hazardous slope processes, including landslide, depends on the knowledge of the structure of the landslide geomorphic system and the slope array, harmonization of the interaction of the components of the designed object with the components of the geological environment (GE). On the basis of a professional analysis of engineering and geological factors, correct design models are proposed, with exact values of the slope stability coefficient (Cs) and rational design solutions leading to a reduction in natural and man-made risks.

1. Introduction

The problems of construction and territorial development of the coast of the Cheboksary reservoir are quite complex and require the solution of many issues on environmental management, on the safe functioning of such natural-technical systems.

Within the limits of Chuvashia on the Volga coast there are areas of two reservoirs of the Volga cascade: Cheboksary reservoir and Kuibyshevsky. The Cheboksary reservoir section is represented by both upstream and downstream. The lower backed bay is at the same time a zone of pinch-out of the backwater and the upper section of the Kuibyshevsky. The operation of the Cheboksary Reservoir began in 1981 at an intermediate mark of normal reservoir level NRL – 63 m (instead of 68 m). Since then, the banks are in the stage of adaptation to new conditions, the development of new equilibrium profiles, accompanied by the massive development of abrasion-landslide processes on the right bank and abrasion-scree on the left bank, and with a small share of abrasion-accumulative due to the relative youth of the reservoir. With the creation of the reservoir, its coastal areas began to be actively involved in the economic turnover and built up [1]. В качестве разделителя дробной и целой частей используется «точка» (в том числе на рисунках и в таблицах).

The high Volga right bank from Nizhny Novgorod to Volgograd has long been known as an area of intensive development of landslide processes, and is described in detail in numerous literature from the end of the 19th century on. (P.I. Krotov, A.P. Pavlov, I.S. Ragozin, G.S. Zolotarev, G.A. Golodkovskaya, A.P. Dedkov, etc.). Today, it is in this region with millionaire cities, with a high concentration of population and production potential of the country, the problem of the stability of the slopes is particularly acute due to fundamental changes in natural and man-made conditions, especially in connection with the creation and operation of a cascade of reservoirs. As a result, new known factors such as lithologic, stratigraphic, geomorphological, and erosional factors were added – increased bases of erosion, transformation of the hydrological and hydrogeological regimes of coastal areas, rising groundwater levels and changes in the weight of coastal massifs, including from building hazardous landslide slopes. The shortage of land resources and suitable areas, as well as architectural considerations, is forcing the builders of Cheboksary to actively engage in landslide slope surfaces with significant steepness, often affected by ravine erosion. There is an alignment, territory planning, backfilling of ancient and modern ravines and beams with technogenic soil (Fig. 1, 2).

Figure 1. The reconstructed Moscow Embankment (Volga River Embankment) in the city of Cheboksary

Figure 2. Embankment before reconstruction. One can see the cut-off tongue of the landslide-flow, which was put forward on the asphalt path

The rationale for the optimal economic development of the hazardous landslide slopes of the Cheboksary reservoir in order to provide the population of the Chuvash Republic with high-quality capital construction, social and industrial infrastructure will be devoted to this report. The obtained results are applicable for urban planning, recreational and transport development of the territory not only of the Chuvash Republic, but also of other territories under the conditions of intensive technogenic impact and influence of the functioning of reservoirs.

IOP Conf. Series: Materials Science and Engineering **663** (2019) 012043 doi:10.1088/1757-899X/663/1/012043

2. Materials and methods

The problem of assessing the stability of the slopes (Cs) remains in engineering geology (EG) and related sciences very relevant. This is evidenced by numerous publications and our own practice (since 1970) [2–14]. But the same information shows that against the background of the emergence of many certified design schemes and programs, the prevalence of the geotechnical approach in such surveys is preserved. This approach does not perceive the irreplaceable role of the geological component in solving the problem of determining the stability of landslide slopes. Unreasonably widespread method of circular-cylindrical sliding surfaces (CCSS), a complete replacement of domestic methods of calculation for foreign (Morgenstern-Price (1965–1971), Bishop (1955), Yanbu (1954), etc. [14]) without significantly improving the reliability of results, as well as reducing the demands on the quality of survey materials, the clogging of landslide studies by primitive algorithms, which underestimate the role of a specialist in the calculation process, have caused many emergencies. As a result, the problem of sustainability, instead of solving, is pushed into a dead end: most of the methods used for assessing Cs have retained the shortcomings of the geotechnical approach, on which K. Terzaghi, M.N. Goldstein, N.N. Maslov et al. paid attention [3, 5].

In addition to simplified design schemes that are far from the mechanisms of real landslides, the inability of many designers and even state expertise in the process of accepting objects can be attributed to the shortcomings of the geotechnical approach: a) to detect gross errors in exploration materials, especially geomorphological ones, which inevitably provoke erroneous design solutions and multimillion-dollar damages during the construction and operation of the object, b) their blind faith in the possibility of obtaining real design parameters of the shear strength of soils of the displacement zone by statistical processing of the results of laboratory tests according to the method of State Standard (GOST) 20522-96. The disadvantages include the drop in demand in the community of designers in qualified geological conclusions, and the growth in consumption of "machine" variants of geotechnical sections instead of sections built by an EG specialist. Below are considered the features of geotechnical and geological approaches to the problem of assessing the stability of slopes for construction purposes.

Much contribution to solving the problem of estimating Koo was made by M.N. Goldstein [3]: developing K. Terzaghi's thoughts on the relationship between soil mechanics and engineering geology, he was the first to point out the reason for such a state: "... it seemed to many that only purely mathematical difficulties restrain the progress of soil mechanics .... However, the practice soon showed all the groundlessness of such hopes, forcing us to pay special attention to the development of engineering geology (our emphasis)". The role of engineering geology is particularly increasing, according to M.N. Goldstein, in cases where "the complexity of the geological structure of the site and the heterogeneity of soil properties exclude the possibility of any kind of accurate calculation ... using methods of modern soil mechanics". In such cases, engineering geology should help "assess the degree of approximation of the performed calculations and indicate the order of possible deviations of the results of calculations from reality ...". From the foregoing it is clear that M. Goldstein among specialists in landslides, including geologists, came closest to understanding the source of unreliable Cs. But another conclusion follows from his position – the time of hopes for solving stability problems only on the basis of a mechanical-mathematical approach has obviously passed. Geotechnics is likely to have exhausted its capabilities in this area, and further progress should be expected from the geological component of the Cs problem. But the trouble is that this truth is far from being realized by many geologists and prospectors.

With existing regulatory documents, the results of the thoughtless execution of the designer's will (how many wells and where to drill, how deep, how many monoliths to take, what tests to perform) are quite satisfactory for the examination, without bearing any serious responsibility for the quality of the computational models and calculations. Unfortunately, mainly geotechnics, not sufficiently sophisticated in geology and exogenous geodynamics are make models and calculations, As a result, the problem Cs remains, as it were, in a state of suspense between geotechnics and geology. There are sufficient up-to-date data (materials from surveys of design institutes, laboratory studies of soils at Moscow State University, etc.), indicating that errors in Cs estimates are made due to the inexperience

IOP Conf. Series: Materials Science and Engineering **663** (2019) 012043 doi:10.1088/1757-899X/663/1/012043

of specialists. The amateurishness of surveyors and designers has now become the scourge of modern landslide, the main cause of the most expensive emergencies.

The choice of a rational method of calculation for mass use in design and survey organizations we have decided in favor of the method of "leaning slope" as the most correct, simple, convenient and quite reliable for determining Cs slopes and Pl (landslide pressure) for designing anti-landslide measures (ALM). The method is considered in detail by I.O. Tikhvinsky (1981). As part of our algorithm, it allows to get Cs and Pl on the borders of all structural elements and blocks of landslide geomorphic systems, fairly reliable data on the role of various factors in the slope stability, at different positions of the surface water level (SWL), under loadings and unloading compartments taking into account the soil resistance to compression in front of the frontal edge of the flat displacement unit (ПС) of the entire landslide system. n relation to the design formula of a leaning slope, we have chosen an algorithm and a table of source data that requires constant work of the prospector himself at all stages of the sustainability assessment procedure – from drawing up a structural calculation model using his own data from a field landslide mapping and filling the table of source data to communicating with a computer during the calculation, consciously changing the strength characteristics for obtaining the peak and residual parameters of soils of the displacement zone. By this, our algorithm is significantly different from most algorithms that operate according to the principle of the so-called "black box" almost without the participation of a surveyor. Geotechnics are usually engaged in calculations, and as a result, it often turns out that the calculation model is poorly connected with the geological structure and the structure of the forecast or real landslide.

It is important to note that the quality of the assessment of the Cs slope is determined not so much by the number of applied methods, but by the quality of the computational model and the accuracy of the shear strength parameters of soils of the displacement zone, discussed below.

3. Results and discussion

Drawing up correct design models, i.e. such models, which reflect the structure of real landslides, their mechanisms, are compiled according to certain principles. The main condition for the compilation of such models is the availability of own field detailed mapping materials of the landslide area with the definition of the genetic type of slopes, their condition and boundaries, types of landslides on the mechanism and structure in accordance with modern concepts of the structure and component-element composition of landslides. Below in Figure 3 shows the structural model of a multi-block landslide, with the main components of landslide systems that are part of the correct design models.

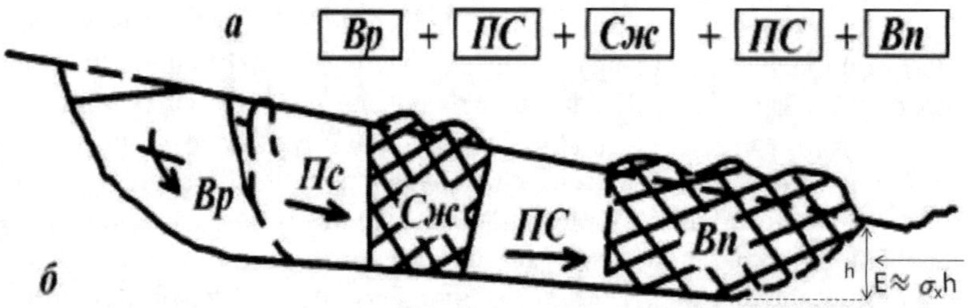

Figure 3. A single-level slip landslide, consisting of 5 blocks: a – functional kinematic scheme; b – structural-functional graphic model. Вр – head rotation unit, ПС – flat displacement block, Сж – internal plastic compression block, Вп – lingual tongue block, E – compression resistance at the front edge of the tongue

The correctness of the computational model is a direct consequence of the competence of the researcher of slope processes and phenomena, including landslides, in matters of their genetic diversity, their structure and composition, their mechanisms of nucleation and separation from the array, the characteristics of their movement and stop, their natural classification.

The choice of design parameters of the averaged shear strength of soils of the displacement zone according to the results of reverse calculations is carried out differentially depending on the genetic type of the slope, the absence or presence of landslides of various types within it, their sizes, age and preservation, on survey tasks. Success is determined by the degree of geometric conformity of the calculated model to the original: the former for old landslides, the modern one for current landslides, and the future for forecast ones. The need for such indicators of strength arises after a direct calculation of the stability according to laboratory tests of soils and obtaining a result that diverges from the optimum (for example, Cs> 1.0 for a landslide slope, or Cs <1.0 for a clearly stable slope). To design a ALM on a landslide slope, it is necessary to have indicators of both peak strength, at which the landslide body is separated from the massif, and residual strength, at which the landslide body movement stops. In multilevel landslides each tier can have its own strength parameters. Reverse calculations should be performed on the same models on which the software is designed. Peak parameters are obtained on retrospective models, reflecting the pre-creep condition of the slope, and residual ones – on models that depict the post-creep state of the slope, taking into account its modern profile and changes in its denudation processes after the landslide body disappears. These secondary changes can be quite significant, especially on the banks of reservoirs, rivers, lakes and seas.

Other recommendations in the problem of assessing sustainability will also be relevant. The landslide capture depth of an array is determined by drilling and sounding data, and additionally by geometric constructions of the kinematics of the system blocks (especially rotation blocks (Bp) and reset (Cб). The latter method is especially effective for slip landslides with head blocks Bp, which are less than 25–30 m. According to our data [15], blocks in a tier play different functional roles, and the magnitudes of the shear and holding forces of the tiers depend on the type of their active head units – rotation (Bp), plane displacement (ПС) or reset (Cб). In fig. 3 it can be seen that between the Bp and ПС blocks, in the process of movement, compressive stresses arise at the bottom of the displacement surface, and tensile stresses occur in the upper part. As a result, when plastic soils between them can be formed shaft extrusion due to squeezing plastic clay from the bottom up to the surface, and for solid, semi-solid soils – crack fracture and graben-like structures by lowering the wedges of the soil separated from the edges of the crack. In the frontal part of a landslide, the block Вп is often formed not by pushing or pushing the ground to the surface, as is usually shown in the design diagrams, but by compressing the soils of the frontal part of the tier according to the "barrel" principle (remember the coopers who made wooden barrels) and abrupt progressive movement of similar compression zones in the direction of displacement of the system by progressive involvement in the movement of new portions of the slope. The shear forces of the head block are maximum for block Bp, minimum for ПС. In almost all cases, when the mechanism of separation of a landslide from the massif, especially at its beginning, is almost close to the mechanism of slip landslides, it is advisable to determine Cs using the methods of limiting equilibrium, developed, like the method of "leaning slope", in relation to landslides with a "fixed" sliding surface.

If it is necessary to assess the stability of non-slide slopes (talus, erosion, scree, etc.), with a natural Cs more than 1.0 (1.1–1.15), such predicted landslides are preferable by the mechanism, dimensions and shapes of the displacement surface, which are set landslides in the area surveyed in similar landscape conditions. The mechanism of forecast landslides is established according to survey data. The mechanism of forecast landslides is established according to survey data. In this case, the question of the possibility of the formation of particular landslides due to technogenic changes in the hydrogeological and soil conditions of the massif is being resolved. Technological factors of sustainability are also considered – the limits of possible undercuts and powders in the vertical planning of the construction site, the stability of the projected vertical benches and slopes, recommendations for improving their sustainability.

4. Conclusion

Minimizing the risk from hazardous processes depends on the accuracy of forecasting the conditions for reliable operation of the natural-man-made system. To obtain the optimal result, it is enough to have one method for solving the problem on the correct model with the calculated characteristics of the soils of the displacement zone, refined by inverse calculations.

There are numerous new data on the landslide phenomena of recent years, which have arisen as influenced by the regime of the Cheboksary hydroelectric station, and the urban economy. They were studied for the purpose of building up landslide slopes. In their study, typifications and classifications, their own methodological developments were applied, starting with clarifying the concepts of "landslide", "landslide process" and "landslide mechanism" and ending with the principles of systematics and nomenclature of simple and complex landslide systems, their typification and classification. We recommend to increase the stability of the slopes of the Cheboksary reservoir with a number of ALM: 1) vertical leveling of the slope (cutting off the head active blocks of landslides, which will lead to unloading of the head parts of the landslide tiers), 2) regulation of runoff, 3) adjusting the frontal part of the landslide systems with counter banks.

Among a number of principles of applied landslide, developed in the course of the research, we substantiate the principles of structural mapping of landslide systems and their elements, recognition of landslide taxons according to their diagnostic features in nature, making kinematic computational models for designing rational composition of anti-landslide activities. The classification of landslide systems should be based on: signs of separation of the landslide body from the massif; signs of movement of the landslide body; signs of the structure of the landslide body. They express the essence of landslide bodies, the type of displacement mechanism, without the determination of which sustainable environmental management in the hazardous landslide slopes of the Cheboksary reservoir is impossible.

References

[1] Gumenyuk A and Nikonorova I 2018 Landscape study of Cheboksary and Kuybyshev reservoirs coasts for recreational using *Managing risks to coastal regions and communities in a changing world* (Academus Publishing) pp 1–1. DOI: 0.31519/ conferencearticle_5b1b93ab223f57.36658580 Retrieved from: https://academuspub.com/ en/nauka/conference_article/1982/view (Date of access 24.08.2018).

[2] Sharpe C F S 1938 *Landslides and related phenomena: a study of mass-movements of soil and rock* (New York: Columbia University Press)

[3] Terzaghi K 1950 Mechanism of landslides. In Paige S *Application of Geology to Engineering Practice* (New York: Geological Society of America) pp 83–123

[4] Ketchum B H 1972 The water's edge: critical problems of the coastal zone In *Coastal Zone Workshop* 22 May – 3 June (Woods Hole, Massachusetts) (Cambridge: MIT Press)

[5] *Landslides. Analysis and Control* 1978, ed. Schuster R, Krizek R (Washington D.C.: National Academy of Sciences, Springer-Verlag) Translation from English 1981, ed. Zolotarev G S (Moscow: Peace) pp 32–85

[6] Cruden D M 1991 A simple definition of a landslide *Bulletin of the Int. Assoc. of Engineering Geology* **43(1)** 27–9

[7] Cruden D M and Varnes D J 1996 Landslide types and processes. In Landslides: Investigation and mitigation *Serial: Transportation Research Board Special Report* **247** 36–75

[8] *Hazardous exogenous processes* 1999 ed. Osipov V I (Moscow: GEOS) 290 p

[9] Kay R and Alder J 1999 *Coastal Planning and Management* (London: E & FN Spon)

[10] Highland L and Bobrowsky P T 2008 *The Landslide Handbook – a guide to understanding landslides* (USGS Circular) 1325

[11] Van Den Eeckhaut M, Poesen J, Govers G, Verstraeten G and Demoulin A 2007 Characteristics of the size distribution of recent and historical landslides in a populated hilly region *Earth and Planetary Science Letters* **256(3)** 588–603 Retrieved from: https://doi.org/10.1016/j.epsl.2007.01.040

[12] Harp E L, Keefer D K, Sato H P and Yagi H 2011 Landslide inventories: theessential part of seismic landslide hazard analyses *Engineering Geology* **122(1–2)** 9–21 Retrieved from: https://doi.org/10.1016/j.enggeo.2010.06.013

IOP Conf. Series: Materials Science and Engineering **663** (2019) 012043 doi:10.1088/1757-899X/663/1/012043

[13] Jackson L E Jr, Bobrowsky P T and Bichler A 2012 *Landslide. Identification, Maps and Mapping* – Canadian Technical Guidelines and Best Practices related to Landslides: a national initiative for loss reduction *Geological Survey of Canada* 7059
[14] Porter M and Morgenstern N 2013 *Landslide Risk Evaluation* – Canadian Technical Guidelines and Best Practices related to Landslides: a national initiative for loss reduction *Geological Survey of Canada* 7312 21 p
[15] Petrov N F, Nikonorova I V and Nikitina O V 2017 *Structural landslide study: aspects of classification* (Cheboksary: Chuvash State University) 218 p

IOP Conf. Series: Materials Science and Engineering **663** (2019) 012044 doi:10.1088/1757-899X/663/1/012044

External condition of transport clusters as a criterion of environmentally oriented management of sustainable development

T A Emelianova[1], D V Novikov[1,2], A V Novikov[1], N G Konokotin[1]

[1] State University of Land Use Planning, 15 Kazakov Sreet, Moscow, 105064, Russia
[2] prof.novikov@hotmail.com

Abstract. The article presents the most important purpose of the formation of transport clusters in Russian economy. The authors touch upon the issue of the increase in the market level of logistics infrastructure and provision of land management of high quality using environmentally oriented sustainable development strategy. The analysis of strategic development prospects by industry sector of key shippers is presented.

1. Introduction

One of the most important purposes of the creation of transport clusters in Russian economy is to increase the market level of logistics infrastructure and ensure land management of high quality using environmentally oriented sustainable development strategy. At the same time, it is necessary to understand that cluster members commit to implement basic agreements and legal documents, the totality of which forms the institutional environment of any cluster.

All the market principles, the development of which is presented by internal agreements, are applied to the economic activities of participants of transport cluster. According to the authors, it is necessary to use the principles and rules of WTO interaction in an adapted form in order to create an institutional environment for transport cluster that supports the principles of environmentally oriented sustainable development management.

Taking into account the configuration for the requirements of the activities of cluster members, these principles may have the following form:

- The provision of the most favored nation treatment for other cluster members;
- The provision of conditions for the attraction and use of products from the external environment;
- Market methods of counterparty selection
- The refusal to use toxic technologies in economic activity;
- The resolution of disputes through negotiations involving domestic arbitrators.

2. Materials and methods

The inclusion of an economic entity into a transport cluster implies compliance with the indicated principles and mutual concessions when in the course of joint business operations by cluster members. As a rule, for all participants, the inclusion into a cluster expands the possibilities of integration into a single logistics system, opens up new opportunities for interaction with other business entities, facilitates

IOP Conf. Series: Materials Science and Engineering **663** (2019) 012044 doi:10.1088/1757-899X/663/1/012044

the access of manufactured products to the market, and increases the flow of direct investment in sustainable development programs.
However, despite the fact that the transport cluster has the ability to use its own guidelines and principles of sustainable development, all cluster members, including central companies, must develop their own competitive advantages. Environmentally-oriented sustainable development management assumes that all cluster members must take into account the territorial peculiarities of doing business:

1. Significant land area and a large number of constituent entities of the Federation (86 entities) compared to other countries - the USA (50 states), Mexico (31), Brazil (21), Venezuela (20), Germany (16 states), Austria (6 states, 2 territories), India (25 states, 6 union territories).
2. The increasing differentiation of the development of the territories of the Russian Federation: the maximum gap in the level of economic development is 8 times, in the cost of living - 5 times. There is also a gap in the distribution of human and natural resources, in the level of urbanization of territories, the intensity of land use and the density of logistics infrastructure.
3. The irregular distribution of freight and passenger traffic flows, manifested in a higher concentration of traffic in large cities (Moscow, St. Petersburg) and regions with a pronounced raw material orientation.

Studying the intensity of passenger and freight flow in every region (Figure 1, Table 1), it is possible to conclude that the highest rates belong to regions that have objective conditions for using transport infrastructure. In particular, the largest transit center for passengers in Russia is Moscow, which has always been a key passenger hub for all types of transportation.

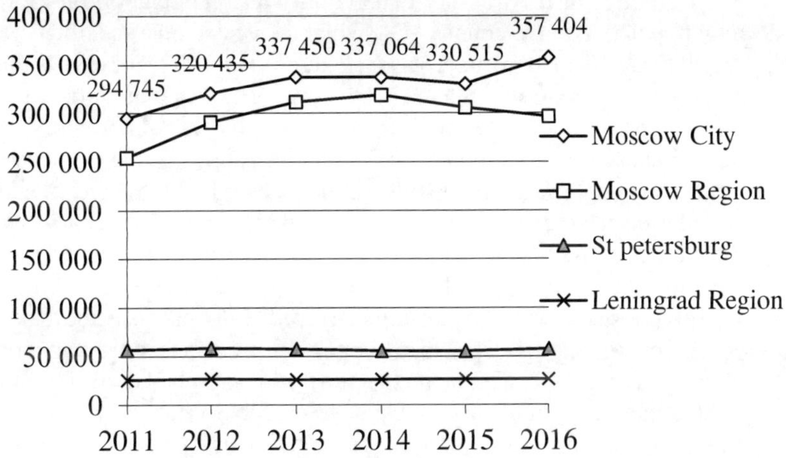

Figure 1 – Passengers sent, thous. peop.

At the same time, it is advisable to pay special attention during the formation of priority directions for the sustainable development of transport cluster to those territories for which the logistics infrastructure as the basis of transport cluster is the main activity. Such territories are in a high-risk zone, since they do not have product differentiation, which entails increased vulnerability to a specific type of activity, which forms the economic, social and environmental parameters of the current state of the territory. It is advisable to pay special attention to issues of sustainability of development for the territories of transport clusters stationing that satisfy the following conditions:

- More than 25% of economically active population work in business entities that are participants of transport cluster
- The economic structure of the territorial product is created by more than 50% of central companies of transport cluster.

Table 1 – Shipped goods, mln. t

	2011	2012	2013	2014	2015	2016
Kemerovo region	222.4	226.6	233.6	242.6	242.7	249.7
Chelyabinsk region	62.0	76.0	73.1	74.4	66.9	66.2
Sverdlovsk Region	62.1	65.0	61.4	61.0	57.7	59.1
Krasnoyarsk Territory	52.7	53.2	47.7	45.2	49.3	51.3
Belgorod Region	56.8	57.1	54.2	54.4	53.5	51.0
Irkutsk Region	59.3	59.3	57.6	53.6	50.6	49.1
Perm Territory	40.1	38.9	39.0	42.2	41.2	41.9
Orenburg Region	34.8	37.7	35.2	36.6	35.4	35.0
Krasnodar Territory	40.0	39.0	35.1	31.3	31.7	31.9
Republic of Karelia	23.7	26.7	26.9	28.3	27.3	29.0
Murmansk Region	28.2	28.3	27.5	26.3	26.3	28.7
Leningrad Region	36.9	37.7	35.5	31.5	27.8	28.5
Republic of Bashkortostan	29.4	30.5	30.1	31.5	28.4	28.4
Lipetsk Region	22.1	26.1	25.9	25.3	25.9	24.9
Samara Region	23.2	23.0	23.7	23.9	23.6	20.9
Khabarovsk Territory	20.7	22.3	21.3	21.3	21.6	20.4

According to the data from Russian Statistics Agency, it is possible to conclude that as of the end of 2016, there were 335 territories in the Russian Federation with a population of 15.9 million people (about 11% of the total population of the country) [1], which depended on a specific type of activity and needed to develop a sustainable development program. Therefore, if transport cluster operates in such a territory, these environmental circumstances must be taken into account during the creation of a sustainable development program.

The nature of the problems and the extent of the difficulties caused by the influence of the external environment on the functioning of transport clusters can be assessed after the global economic crisis of 2008-2009, which undermined the economic stability of all national economies due to strict specialization and difficulties in risk management. This event necessitated the intervention of authorities and the provision of various forms of support to all categories of stakeholders.

The main instruments of state support were direct subsidies to improve the balance of regional budgets, the development of logistics infrastructure, support for producers, professional retraining of the population in order to reduce social tension in the territorial labor markets. In this situation, the environmentally oriented management of sustainable development of transport cluster can be provided with additional resources, the use of which requires the corresponding economic, social and environmental potentials.

It is necessary to note that, for business entities, the impact of external crises is usually short-term, and the consequences of inclusion into a cluster are characterized by long-term sustainable impact on the internal environment, which significantly affects not only the nature of production relations, but also the requirements for technologies, management system and performance criteria. Therefore, the use of only economic indicators during the development of a program of environmentally oriented sustainable development management seems to be inadequate.

In addition, the analysis of the external environment for the formation of environmentally oriented management of sustainable development of transport cluster should include the analysis of strategic development prospects in the context of the industry of key shippers.

Accordingly, within the framework of the issue of the state of the external environment for the implementation of environmentally oriented management of sustainable development of transport cluster, it is advisable to take into account the willingness of cluster participants to the occurrence of risk situations, the transformation of the product structure of economy, technological innovations that change the way business is conducted, and legal requirements that establish boundaries and rules of land use.

In this context, experts suggest dividing economic entities according to their degree of readiness for adaptation in accordance with environmental conditions into three groups:

1. The most prepared entities are competitive, characterized by high growth rates of labor productivity and a significant share of exports in product portfolio.
2. Mid-prepared entities are characterized by stable growth rates of labor productivity, a constant share of exports in product portfolio, and are competitive in national and international markets.
3. Least trained entities have rather low labor productivity, the decrease in the share of exports and the increase in the share of imports in product portfolio, with low competitiveness» [2].

Most potential and existing participants of transport cluster should make quite definite efforts to increase the activity of innovation and strengthen their role in market competition.

3. Results

The analysis of the external environment during the choice of a priority direction for the sustainable development of transport cluster involves a whole range of macroeconomic and industry risk factors, including those produced by various global events: sanctions, international agreements and the emergence of new competitors. Accordingly, not only the success of the business operations of specific participants, but transport cluster as a whole depends on the relevance of the chosen priority area of sustainable development to the opened prospects.

4. Results

In case when an economic entity becomes a member of transport cluster, it receives additional opportunities to increase its competitiveness. In addition, the use of a cluster approach to increase competitiveness in modern conditions is possible in several ways:

- organizational;
- marketing;
- ecological;
- economic;
- financial.

For example, during the use of organizational option, a business entity creates competitive advantages due to its position in the technological chain of a cluster, realizing the technological potential and the intellectual potential of employees. The marketing option is the development of consumer potential of the market, as well as the reduction of the elasticity of supply and demand. Ecological option suggests that due to the participation in the cluster, land use intensity is reduced, and technologies are used in activities to reduce the negative impact on the environment.

The economic option is reasoned by the availability of access to partners, that is, it allows reducing the transaction costs of the economic activity of an economic entity. The financial version of the approach assumes that the strategic efforts of a business entity are focused on the quality of a product, and other cost areas play a less important role due to work in the cluster.

In addition it is necessary to note that it is the appearance of a synergistic effect as a result of the interaction of economic entities that distinguishes a cluster from scientific and production associations existing in controlled economy, when the role and contribution of each participant are taken into account, but the combined effect is not always considered as a necessary result of the interaction.

It is the cluster that creates the conditions under which the demand for sustainable development and business activity determines the feasibility of transforming ways of long-term coordination of the interests of government with the interests of business representatives when not the all types of economic activity and the spheres of national economy are highly competitive.

References

[1] Russian State Statistical Service 2018 *Socio-economic situation in Russia* (Moscow: Goskomstat) URL: http://www.gks.ru/free_doc/doc_2017/social/osn-12-2017.pdf.

IOP Conf. Series: Materials Science and Engineering **663** (2019) 012044 doi:10.1088/1757-899X/663/1/012044

[2] Tomsk City Administration 2018 *Study of the impact of the inclusion of Russia into the World Trade Organization (WTO) on the socio-economic situation of Tomsk region and certain types of economic activity* Analytical Report URL: http://old.tomsk.gov.ru/export/sites/ru.gov.tomsk/ru/economy_finances

Increasing the efficiency of using renewable energy sources for energy supply of autonomous highland objects

Yu S Petrov, M K Khadikov, I Yu Zorina and A K Muzayev

North Caucasian Institute of Mining and Metallurgy (State Technological University), NCIMM (STU), 44, Nikolaev st., Vladikavkaz, 362021, Russia

E-mail: alexander.muzaev@mail.ru

Abstract. Ways of increasing the efficiency of using renewable energy sources in the autonomous power supply of highland facilities are considered. Descriptions of autonomous multifunctional energy complexes that simultaneously use converters of various types: sunlight, wind and water flows into electrical energy, patented in NCIMM (STU) are given. The prospects of using renewable energy sources for autonomous power supply of highland facilities are substantiated.

1. Introduction

Renewable energy sources (RES) are currently finding wider and more diverse applications [1–3].

One of the most popular and effective applications of RES is autonomous energy supply of low-power facilities and private residential households. This is particularly true for remote highland settlements that are not provided with centralized power transmission lines.

2. Articulation of issue

This energy drawback of highland settlements is fully compensated by the presence (and, therefore, the ability of using) of powerful and almost constantly acting natural energy flows: air (wind), water (mountain rivers) and sunlight. The latter play a special role in providing individual objects with energy. Suffice it to say that in some countries the so-called "solar houses" are already in operation. These are houses in which energy for domestic needs is fully generated from sources of solar panels.

Figure 1 shows a diagram of the conversion and use of energy in autonomous generating complexes operating on RES. The first operation is to capture a part of the total energy flow for its further transformation.

A significant difficulty in the use of RES is the dispersion of natural energy flows (for example, the small specific energy of solar radiation, i.e., energy per unit area of the irradiated surface).

To increase the power generated by the primary converters to the required consumed value, various methods and devices are used. For example, to increase the total power of solar panels, the total irradiated surface area is increased (for which roofs are often used, for example). The concentration of solar flows occurs using special devices – reflectors, directing the sun's rays on the working surface of solar panels.

Both methods can be used to power private houses, for instance, mountain settlements.

In the conditions of mountainous territories, mountain rivers are an effective source of energy raw materials.

Possibilities of using RES for energy supply of remote highland settlements.

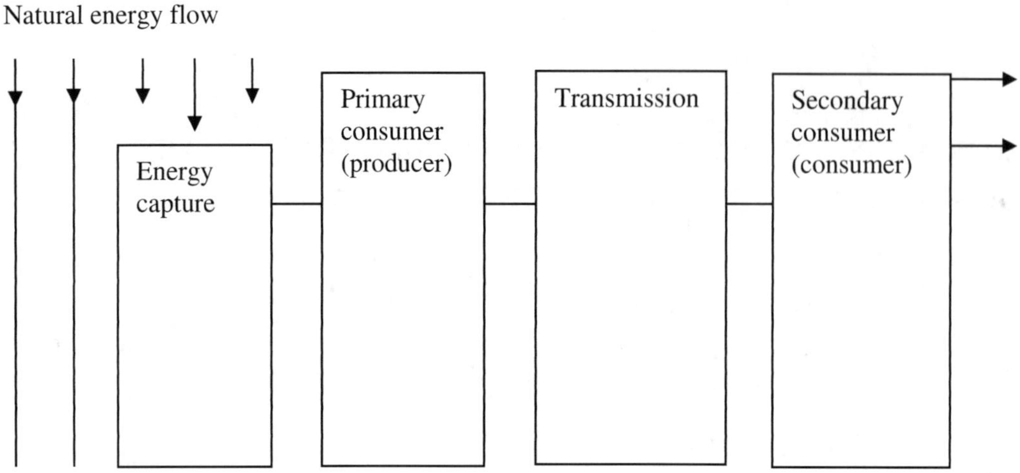

Figure 1. The diagram of the conversion and use of energy in generating complexes

The part of the natural energy flow received by the primary consumer (producer) is even used to convert its energy into energy convenient for further use (electric, thermal). Energy is transferred to the consumer and as a result of its use (second converter), the energy state of the consumer changes, as well as energy dissipates into the surrounding production.

Micro Hydro Power Stations can be installed directly on mountain rivers (for instance, cantilevered HPS) or on branches taken away from them (hose HPS).

If the use of solar generators related to a daily change in solar radiation, then the use of hydro generators in the mountains related to a seasonal change in the water level in rivers, which should be taken into account when designing and operating micro HPS.

Wind is constantly blowing in the mountain gorges, which provides conditions for the development of wind energy, despite certain instability of the wind flow.

To make the picture of the mountain energy flow complete, it is important to note the possibility of using in some cases geothermal energy sources.

3. Materials and methods

Natural energy flows have different energy potentials, different affordability, and various environmental characteristics. For example, for North Ossetia-Alania, the total (estimated) hydropower potential is 22723.4 million kWh per year; solar radiation and wind energy potential are 16,000 and 1.666, respectively.

The output characteristics of real generating plants operating on RES are, to one degree or another, random. This largely characterizes the autonomous energy systems, designed, for instance, to work in high altitude conditions.

Changes in wind speed, solar radiation intensity, volume and pressure of water in rivers in mountainous conditions occur quite often and do not have a strict regularity. As a result of this, mathematical models describing the processes in RES plants should be statistically distributed and be analyzed by appropriate methods.

In the most general case, the energy state of a system using renewable energy sources can be expressed by the energy balance equation

$$\sum S_i = \sum S_{помр} + S_{нк} \quad (1)$$

where $\sum S_i$ is the total power of the transducer generating component of the overall system; $\sum S_{npmp}$ is the total power of the consumed system; $S_{нк}$ is power transmitted to the storage device.

With excess generated power, part of it is transferred to the storage device and it works in consumer mode ("+" sign). With a lack of generated power (compared with that required for the normal operation of consumers), it comes from storage devices to the consumption system, and the power has a "–" sign. Formally, the summand "–" can be transferred to the left side of the equation with the "+" sign, which corresponds to the operation of the storage device in generator mode, i.e. in power transfer mode to the consumer.

Equation (1) is written for mean values over a certain time. From a mathematical point of view, expressions written for instantaneous power values will be more correct

$$\int_0^t \sum^S p_i(t)dt = \int_0^t \sum^S p_n(t)dt + \int_0^t {}^S p_{НК}(t)dt \tag{2}$$

Equation (2) is statistically distributed, since all terms under integral sign are random functions.

To calculate autonomous multifunctional power plants, it is necessary to work and analyze the equation (2). As you know, there are two main ways to work statistically distributed equations: the transition to the mathematical expectations of the members of the equation, and the Monte Carlo method. For practical purposes, we can use equation (1) for average power values, given that the generated S_i and consumed S_n power are explanatory variables, and the accumulated power $S_{нк}$ is a dependent variable, determined by the values of generated and consumed power.

4. Results and discussion

Two types of devices are being developed at NCIMM (STU): devices for converting one type of natural energy flows into a convenient one for further use [4–6], as well as multifunction devices that simultaneously convert the energy of various types of natural energy flows [7–9].

The wind wheel proposed by Professor Yu.S. Petrov [4], allows rotation of the panels in one direction regardless of the direction of the wind, which is essential for areas with changing parameters of the wind flow vector. The machine is designed for small capacities and application for power supply of autonomous low-energy facilities, for instance, in mountain conditions.

Solar battery proposed by M.K. Khadikov [5], differs in small dimensions and high efficiency in the use of sunlight in a limited installation volume. It is supposed to be used in the system of autonomous energy supply, for instance, objects of highland settlements.

Professor I.D. Alborov and PhD student I.Yu. Zorina proposed an original design of the solar collector, which allows increasing the efficiency of thermal converters of this type by about 2 times. The machine has a wide range of performance and can be operated in conditions of autonomous power supply of objects of various capacities. The most promising use of the machine is to power individual high altitude facilities.

A series of inventions relates to machines of a hybrid type, using the conversion of radiant and wind energy in power supply at the same time. The simultaneous conversion of natural energy flows of various types (in this case, wind and sunlight) in the power supply has a number of positive qualities. Firstly, the machine capacity is increased; secondly, the reliability of power supply is increased (at least one generating stream is gradually eliminated); thirdly, the area occupied by the hybrid installation is reduced in comparison with the total area that would occupy two sections separately; In addition, the environmental load on the environment is reduced (compared to using each converter separately).

Currently, NCIMM (STU) is developing samples of autonomous multifunctional energy complexes – (AMEC), with a view to their subsequent use in mountain conditions.

One of the existing factors hindering the use of AMEC is the need for an energy storage device of sufficient capacity. The role of the accumulating mechanism is to absorb (store) the excess of generated

IOP Conf. Series: Materials Science and Engineering **663** (2019) 012045 doi:10.1088/1757-899X/663/1/012045

energy and compensate (if necessary) for its lack in the consumption system. The problem of creating an energy storage device that fully meets modern requirements has not yet been resolved.

5. Conclusion

In NCIMM (STU), work is underway, for instance, to create an integrated heat-electric storage device, which, according to its characteristics (calculations and experiments show), can quite satisfy the AMEC requirements of average power.

The use of AMEC is quite promising, especially in mountain conditions. The use of the energy of natural energy flows circulating in the surrounding space, in particular, on the basis of AMEC, will not only simplify the energy supply to high mountain regions, but also significantly ensure the environmental sustainability of the mountain region.

References

[1] *Alternative energy sources and energy conservation* 2014 (Moscow: Science and Technology) 320 p

[2] Elistratov V V *The use of renewable energy: a training manual* 2008 (St. Peterburg: Publishing house of the Polytechnic University) 224 p

[3] Vasiliev Yu S, Elistratov V V et al 1995 *Renewable energy sources and hydraulic accumulation: a training manual* (St. Petersburg: Publishing House Polytechnic University)

[4] Petrov Yu S 1998 Patent no 2105190, published on 02/20/98 Bull. no 5

[5] Khadikov M K 2017 Patent no 172079, published on 06/28/2017, Bull. no 19

[6] Alborov I D and Zorina I Yu 2017 Patent no 170208, published 04/18/2017, Bull. no 11

[7] Petrov Yu S, Sakhansky Yu V, Zorina I Yu and Ilikoev G V 2014 Patent no 148781, published 20/12/2014, Bull. no 35

[8] Petrov Yu S, Sakhansky Yu V, Zorina I Yu and Ilikoev G V 2016 Patent no 158761, published 01/20/2016, Bull. no 2

[9] Petrov Yu S, Khadikov M K, Sokolov A A and Muzaev A K 2019 Patent no 188444, published 04/12/2019, Bull. no 11

IOP Conf. Series: Materials Science and Engineering **663** (2019) 012046 doi:10.1088/1757-899X/663/1/012046

Spectral approach to the estimation of the time series of the power change in the substation metering units and its application in the control modes of substation electricity metering systems

V N Pichugin[1], A A Soldatov[2] and A M Pinaev[1]

[1]Chuvash state university is named after Ilya Nikolayevich Ulyanov, 15, Moskovsky ave., Cheboksary, 428015, Russia
[2]PAO "MRSK of Volga"-"Chuvashenergo», 4/4, I. Yakovlev ave., Cheboksary, 428015, Russia

E-mail: vladimir_iii@mail.ru

Abstract. The wide possibilities of modern interval metering devices installed at substations of distribution grid companies are not fully used today. In the process of measuring the electrical parameters of the network meters, there may be abnormal modes of electricity metering substations, resulting from intentional or negligent actions of staff, failure of existing equipment or parts thereof. The considered circumstances lead to large economic losses of electric power enterprises, expressed in the underestimation of electricity. One of the effective ways to solve this problem is the use of intelligent techniques to identify incorrectly operating metering stations of the substation network, as well as the development of an optimal model of an automated multifunctional information and measuring system for industrial use and software implementation. The purpose of the study – the development of methods of control and diagnosis of normal operation of the substation electricity metering system, determined by compliance with the reliability of electricity metering substation information and measurement systems. On the basis of the offered method the algorithm of control and diagnostics of modes of operation of the information-measuring equipment of system of the account of the electric power of substations is offered. The software for a two-level multifunctional measurement system combining the functionality of an automated electricity metering system, a system for monitoring and diagnosing the modes of operation of the information and measuring complex of substations, and a system for monitoring the quality of electricity has been developed. The results of the research were used in the development of recommendations for the construction of automated information and measurement systems based on existing power metering systems of substations for distribution grid companies. The developed technique and the corresponding algorithm of control and diagnostics of abnormal operating modes of information-measuring complex of substations allow to quickly determine the technical condition of the metering system of substations, to plan corrective measures to eliminate violations.

1. Introduction

The balance method, reviewed Zhelezko Y.S. [1], in control of maintenance of normal modes of work of substation systems of the account of the electric power is used by power engineers quite often. At the same time, at the moment there is no clear idea of the possibility of its use in automated control systems of the above modes, which is confirmed by the lack of control modules and diagnostics of information-measuring systems of electricity metering as part of automated control systems of power facilities, reviewed Kochneva E.S. [2]. For automatic measurement of the unbalance power in the substation and report on the work of substation metering system, reviewed Pazderin A.V. [3], requires that the reference values of the quantity under consideration imbalance of power was obtained at a single time and date that can be implemented with function "latch" the instantaneous States of the energy consumption that characterize the current volt-ampere the status of metering of electricity, reviewed Mironovskiy L.A. [4].

2. Materials and methods

The statistical processing of experimental data of full-scale field tests and mathematical description of the random law of change in the magnitude of the substation unbalance of instantaneous power, as well as its computer simulation.

3. Development of methods to identify faulty information measuring complex by calculating the power in the substation metering stations

Simultaneous calculation of power on all metering devices of the substation and calculation of unbalance of instantaneous power is made by the formula:

$$HБ_W = \frac{W_{vv} - W_{ot}}{W_{vv}} \times 100\% \tag{1}$$

where $HБ_W$ – the relative value of the actual unbalance of power at the substation, W_{vv} – the sum of the capacities in the metering stations of the substation inputs and in the communication lines of substations, when electricity is supplied through them, W_{ot} – the sum of the capacities in the metering stations of the outgoing connections, in practice, usually does not exceed ±2,2 % of the substation.

In fact, the calculation of the value 1 in time resembles a non-harmonic periodic oscillatory process, reviewed Soldatov A.A. [5], which can be represented as a result of the addition of a certain number of harmonic oscillations of changes in the power differences recorded by the information-measuring units of electricity metering of the input and output connections of the substation.

For a three-phase network, the sum of the products of the values of phase currents, voltages and the cosine of the angle between them, taking into account the transformation coefficients, determines the amount of electric power. A change in any of the parameters under consideration leads to a change in the value of the total power fixed by the electricity metering device. At the same time, a sharp change in the state of the value 1 should be understood as an abnormal measurement or a miss, and the repetition of an abnormal character on several consecutive measurements indicates a deliberate interference in the work of electricity metering devices or a technical malfunction in which the values of the current, voltage or cosine of the angle are underestimated, due to which the substation accounting of electricity becomes unreliable.

It is required to describe the random process of changing the value of the substation unbalance of instantaneous power for the possible determination of the boundaries of the confidence interval of the measured unbalance of instantaneous power, within which the normal modes of operation of substation electricity metering systems are supported.

To carry out the appropriate control, it is required to have software and hardware computing facilities reviewed Pichugin V. N. [6], capable of implementing instantaneous power data acquisition from all substation information and measuring power metering units, with further processing of the received information.

For the period from Dec 2018. to February 2019. at different times of day produced n=2202 measurements substation imbalance of power. At daily intervals t_i=24, the average values of the unbalance of the power y_i for the entire period under consideration were determined by the values given in table 1.

Table 1. Average power unbalance y_i

τ_i	The number of the measurements for the time interval	Average $\overline{y_i}$
1	94.00	-0.18
2	82.00	-0.19
3	81.00	-0.21
4	70.00	-0.19
5	72.00	-0.19
6	84.00	-0.085
7	113.00	-0.057
8	106.00	-0.058
9	95.00	-0.13
10	105.00	-0.028
11	100.00	-0.077
12	112.00	-0.067
13	104.00	-0.097
14	96.00	-0.097
15	95.00	-0.068
16	76.00	-0.067
17	66.00	-0.073
18	71.00	-0.020
19	62.00	-0.077
20	62.00	-0.12
21	102.00	-0.13
22	146.00	-0.14
23	108.00	-0.15
24	100.00	-0.20
Subtotal	**2202.00**	**-0.11**

When calculating the average values for each hour interval specified in the table, the power unbalance data were used, which were characterized by small average deviations from each other. This means that there is no shift that distinguishes one day of measurement from the other, and there are only cyclic fluctuations associated with the magnitude of the unbalance of power. In turn, the value of the unbalance of power depends on the error of information-measuring metering units, which in turn depends on the current load of the current transformers of a separate information-measuring metering unit. The average value imbalance for all time intervals τ_i performed to reduce the level of the random component $\varepsilon(\tau)$ time series. Consider the value $\overline{y_i}$ as a time sample of a series, investigating the structure of the hourly cyclic oscillations of the daily change in the unbalance of power at the substation. Make a matrix of values $\left\{\tau_i, \overline{y_i}\right\}$, i=1,2,…,24 and build a graph of the values of the time series $\overline{y_i}$. By figure 1 you can see the broken oscillatory character of the value change $\overline{y_i}$, in this connection, to determine the structure of this time series, you can use the methods of harmonic analysis.

Find the coefficients of the Fourier series by formulas (2)–(4):

$$a_0^* = \frac{1}{n} \cdot \sum_{i=1}^{n} y_i \tag{2}$$

$$a_k^* = \frac{2}{n} \cdot \sum_{i=1}^{n} y_i \cos\left(\frac{2\pi k}{T}\tau_i\right) \tag{3}$$

$$b_k^* = \frac{2}{n} \cdot \sum_{i=1}^{n} y_i \sin\left(\frac{2\pi k}{T}\tau_i\right) \tag{4}$$

IOP Conf. Series: Materials Science and Engineering **663** (2019) 012046 doi:10.1088/1757-899X/663/1/012046

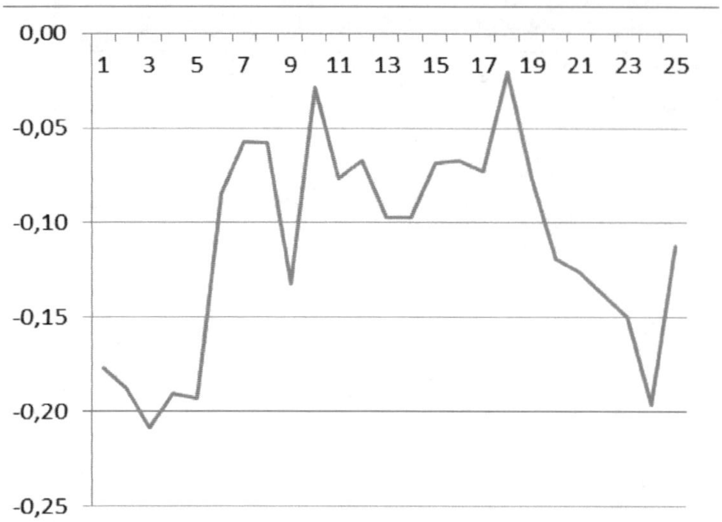

Figure 1. Figure values $\overline{y_i}$

For T = 24, Δτ = 1, n = 24 it is advisable to calculate a0, a1*, a2*, b1*, b2*, an/2*=a12*, bn/2*= b12*. The symbol "*" means that the coefficients are calculated by discrete values of the time series.

Value

$$S_k = a_k^2 + b_k^2 \tag{5}$$

characterizes the energy contribution of the k-th harmonic to the function y(τ). We determine the spectral composition of the function y(τ) as a function of the value Sk of the harmonic number k, for which we consider table 2.

Table 2. Spectral composition of the function y(τ).

k	1	2	3	4	5	6	7	8	9	10	11	12
a_k	-0,056	-0,021	0,012	0,01	-0,011	-0,012	0,0027	-0,00091	-0,0025	-0,0066	-0,0097	0,015
b_k	-0,026	-0,023	0,0039	0,0033	0,0057	-0,014	0,0061	0,0088	-0,0099	0,0054	-0,0079	0,0000011
S_k	0,0038	0,00099	0,00015	0,00019	0,00015	0,00032	0,000044	0,000079	0,00011	0,000072	0,00016	0,00023

Since, a0=-0.11, we get S0=0.013.

By conducting further studies for individual substations, it is possible to obtain the values of the functions y(τ). Having established the maximum possible threshold of change y(τ), it is possible to offer as a criterion of observance of reliability of the substation account of the electric power a condition of not exceeding of the considered size.

4. Conclusion

In the course of research, an appropriate algorithm for simulation of changes in the operating modes of the substation electricity metering system based on the experimental statistical base was developed. On the basis of provisions and conclusions the software of the automated multipurpose information and measuring system is developed and realized. The results of the study can be used as a basis for the development of tools for detecting the unreliability of electricity metering on intersystem power lines. Based on the recommendations developed in this paper, methods can be developed to improve the efficiency of measurement and control systems of electricity metering.

IOP Conf. Series: Materials Science and Engineering **663** (2019) 012046 doi:10.1088/1757-899X/663/1/012046

5. Acknowledgments
The authors Express their gratitude for the comprehensive support and advice of colleagues of PJSC "IDGC of Volga"-"Chuvashenergo".

References

[1] Zhelezko Y S 2003 Nedouchet elektroenergii, dopustimyye nebalansy i ikh otrazheniye v normativakh poter' *Elektricheskiye stantsii* **11** 18–22

[2] Kochneva E S 2014 Vyyavleniye nedostovernykh izmereniy elektricheskoy energii s pomoshchyu aposteriornogo analiza *Vestnik Samarskogo gosudarstvennogo tekhnicheskogo universiteta Ser. Tekhnicheskiye nauki* **2(42)** 32–40

[3] Bartolomey P I and Pazderin A V 2004 Nablyudayemost raspredeleniya potokov elektricheskoy energii v setyakh *Izvestiya vuzov. Problemy energetiki* **9–10** 24–33

[4] Mironovskiy L A and Slayev V A 2010 *Algorithms for evaluating the result of three measurements* (St. Petersburg: Professional Publ.) 192 p

[5] Soldatov A A and Evdokimov Y K 2017 Postroyeniye mnogofunktsionalnoy avtomatizirovannoy sistemy i algoritmov kontrolya i diagnostiki rezhimov raboty sistem ucheta elektroenergii elektrosetevykh podstantsiy *Pribory i sistemy. Upravleniye. Kontrol. Diagnostika* **3** 1–10

[6] Pichugin V N, Fedorov R V, Nemkova M P and Veryaskina M A 2015 Automation of the software package' Relay protection and automation service' useful in energy industries. *Ecology, Environment and Conservation* **21** (Suppl.) AS1–AS5

IOP Conf. Series: Materials Science and Engineering **663** (2019) 012047 doi:10.1088/1757-899X/663/1/012047

Application of man-induced raw material resources as a way to solve ecological problems

D K-S Bataev[1], V Kh KHadisov[1,2], A Z Abukhanov[2] and M Sh Salamanova[1,2]

[1]Kh. Ibragimov Complex Institute of the Russian Academy of Sciences, 21a, Staropromyslovskoe shosse, Grozny, 366002, Russia
[2]State petroleum technological university in Grozny named after M. D. Millionshikov, 30, Kadyrov street, Grozny, 364021, Russia

E-mail: madina_salamanova@mail.ru

Abstract. The article deals with the issue of man-made resources utilization in the form of ceramic crushed bricks and production fault with fine and big fillers for small objects production from the ceramic-concrete. This wasteless technology of wall materials production allows saving the environment. There is a peculiarity of preparing forming ceramic-concrete compositions as well as in structures of these compositions. There are formulas of light-weight concretes based on ceramic brick filler and parameters necessary for quality and durable small-piece materials. Such characteristics as density, thermal conductivity, cold resisting property and compression strength have been studied as exactly these characteristics are under special attention. Developed compositions of light ceramic-concretes had rather low thermal conductivity within 0.49–0.98 Wt/(m·°C), ceramic-concrete type by cold resisting property was within F50 to F100 limits, density changed from 1300-1700 kg/m^3, compression strength was 41 MPa. Using crushed bricks in light concretes allows reducing the cost by 30–35.

1. Introduction

Nowadays there are a lot of ecological problems in the world, so issues of rational use and effective conservation of natural resources are rather acute for the high standards of living in every country. It is necessary to note that resource saving of the natural potential means introduction of wasteless and low waste technologies, based on the secondary use of raw material resources, that allows supporting environmental safety [1, 2, 5, 8].

A construction industry has an important role in resource- and power-saving, as it efficiently uses man-made (secondary) raw material which does not require spending on search and pit development, processing and transportation from other regions. These cheap man-made resources are buildings' and structures' demolition wastes, production faults as crushed bricks, in order to receive secondary filler big and fine. The issue arises what is the amount of these waste, as statistics [3, 4, 6, 9, 10] shows, annually our country cumulates more than 22 mln tons of the man-produced construction wastes, 60 % of this resource is ceramic crushed bricks and concrete (fig. 1).

2. Problem statement

The article gives results of studies on production of light ceramic-concretes with the use of secondary filler in the form of ceramic crushed brick (CCB) and production fault bricks (PFB). Ceramic concrete is used for wall materials production, so it must have low density and thermal conductivity, cold resisting property and compression strength to 30 MPa. In the compositions under study special attention is given to porous filler, as ceramics nature, grain-size composition and properties influence strength parameters of the concrete brick.

Figure 1. Emergency removal of brick buildings

The composition of the light concrete was done by the special methodology, as ceramic filler demands a lot of water and influences structure and strength formation of the ceramic-concrete. In the composition under study there are hard, liquid and gaseous stages, and mixing water is redistributed among them in a specific way and changes technological and rheological characteristics of a formation concrete mix. While preparing forming mixture, added mixing water is instantly absorbed by the ceramic brick filler, fine and big, that promotes formation of the primary capillar-porous structured joint. But in some time, absorbed water in its pores and micro flows gradually release water into the hardening, getting strength concrete system, as a result rather hard structure of the concrete brick is forming [7, 11, 14, 18].

3. Materials and methods

The forming concrete mixture had the components ratio given on figure 2a, the fine filler from ceramic crushed brick had Abram's fineness module 2.5 and fractional composition shown in figure 2b. In some compositions natural silica sand of Chervlensk origin was used. It refers to the class of fine ones with Abram's fineness modulus 1.8.

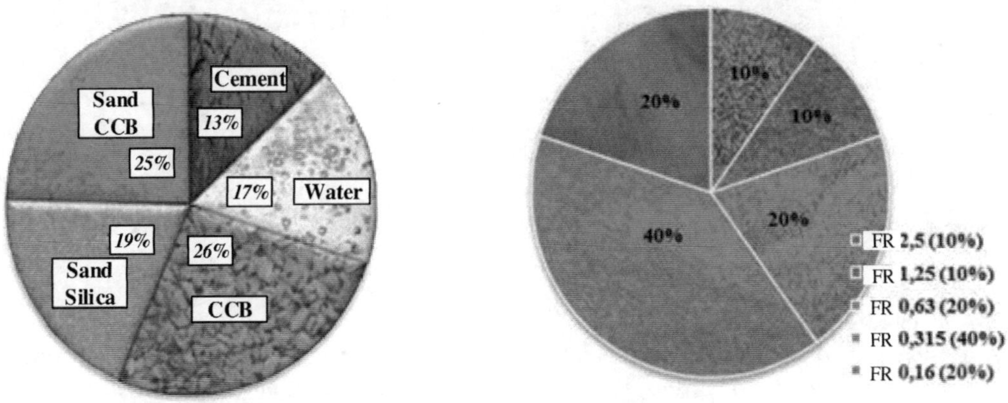

Figure 2. The ratio between concrete mixture components (a) and a fracture composition of the fine filler from crushed brick (b)

Superplasticizing addmixtures GP-9 with the dosage 1.5 % from the binding component and C-3 in the amount 2 % from the binding were used while studying the influence of plasticizers on the technological parameters of the ceramic-concrete mixture. Portlandcement without additives CEM I 42.5 H from SUE "Chechencement" was used as a main binding component. Its characteristics correspond to GOST 31108-2003.

Formulas and properties of samples from light ceramic-concrete are in table 1 and figure 2.

Table 1. Formulas and properties of light ceramic-concretes

№	Materials expenditure, kg for 1 m³						W/C	concrete density, kg/m³	$R^{28}_{compression}$, MPa,	Thermal conductivity, Вт/(м·°С)	Cold resisters property
	C	CCB	S	W	C-3	GP 9					
Common light ceramic concrete											
1	240	820	540	290	-	-	1.10	1674	11.4	0.48	50
2	330	745	520	299	-	-	0.93	1703	20,9	0.52	50
3	315	790	560	230	6.2	-	0.75	1735	24,7	0.53	50
4	315	800	530	238	-	4.7	0.78	1744	24,5	0.53	50
5	405	755	500	230	8.0	-	0.59	1757	35,9	0.52	75
6	405	745	115+ +405*	225	8.0	-	0.57	1835	41.1	0.97	100
Coarse pored light ceramic concrete											
7	255	975	-	282	-	-	1.1	1312	7.3	0.35	25
8	410	915	-	285	-	-	0.80	1397	17.3	0.38	25
9	405	952	-	266	8.0	-	0.66	1415	18.9	0.36	35

Note: composition 6 is a mixture of two kinds of sands (115 kg is CCB sand, and 400 kg is dense silica sand of Chervlensk origin).

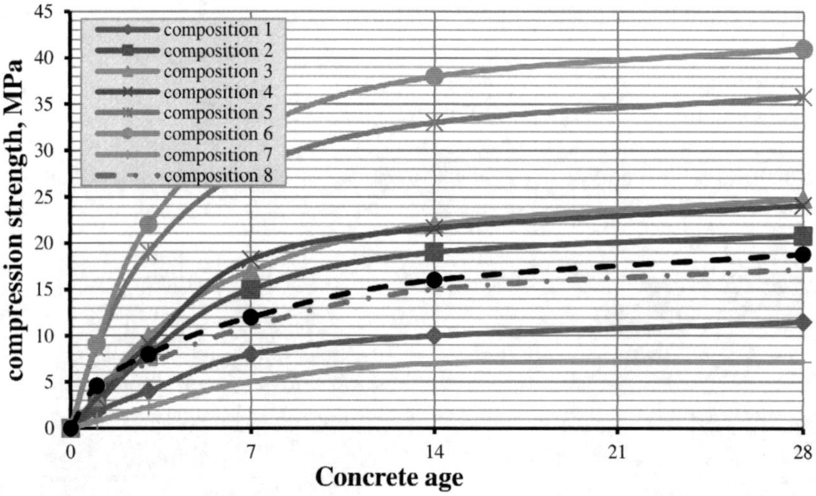

Figure 3. Strength development of ceramic concrete samples.

4. Results discussion

By the results we can notice that there is dependence between the density of the light ceramic concrete and the density of the cement rock itself. It is necessary to take into account the amount of the mortar stage is only 40–44 % from the volume of the whole composite [12, 13, 16–19]. Cement component is present in greater amount in the space between pores, cavities, caves whose sizes are in millimeters. We can see that in composition 7 minimal parameter is in thermal conductivity, and the most expensive component – cement – is minimal, the weight decreases significantly, thus, coarse-pored ceramic concrete can be used for non-constructional walling products.

In the coarse pored concretes there is a deficit of the binding component, the structure is more porous, cement covers the filler surface in a thin layer, and inter-grain space is almost not filled. As the result there are latent closed cavities and pores, which reduce thermal conductivity.

The study of the thermal conductivity of the light concretes with the use of ceramic brick filler was conducted with specially prepared samples 250x250x30 mm in size (figure 3) with the apparatus for determining thermal conductivity ITP-MG4.

In the formulas of light ceramic concretes with the use of fine ceramic artificial sand the thermal conductivity parameters change within 0.48–0.97 Wt/(m·°C), and if compared with the common heavy concrete, whose thermal conductivity is 1.5–1.87 Wt/(m·°C), it proves the efficiency of the suggested developments. Sample test for cold-resisting property, showed that the studied compositions are characterized by the type of cold resisting property from F50 to F100.

It is necessary to note that the best cold resisting property parameter is in type 6 – F100 type, there was natural silica sand in this composition, and may be it improved the cold resisting property. It is important to note that the presence of dense silica sand (400 kg) had a negative influence on thermal conductivity and density of the sand.

Better values in thermal conductivity are in coarse pore light ceramic concretes within 0.35–0.38 Wt/(m·°C), the density lowers till 1310–1410 kg/m^3, but a big amount of open pores and cavities has a negative influence on the cold resistant property.

Figure 4. Macrostructure of the light concrete sample based on ceramic brick filler 250x250x30 mm in size

In the compositions of the light concrete, the attention is paid to secondary ceramic brick filler, that is why adhesion of the mortar phase with the porous filler is a determining factor. It is especially important because having high values of strength characteristics in the contact zone, there will be a better, unloaded condition of the matrix binding system, and it will function as a whole, increasing the strength of the composite. To receive higher adhesion strength concretes of big fillers from hard and strong rocks are used. The combination of these fillers with the concrete matrix will mutually compensate depleted strength characteristics of every stage – mortar phase or filler, which will provide significant strengthening of the composite, due to unloading of the matrix system and high filler characteristics. In this case the concrete will have a "reinforcing effect" of the composite with the filler from the inside.

Studying the destruction mechanism of the samples of the projected light concrete (fig. 4) we can note that in the compression test, the cube destructed along the filler, and not in the zone of contact of the cement rock with the ceramic filler. It evidences that the adhesion strength of the filler with the cement rock and the mortar part are significantly higher, than the strength of the filler from the crushed brick itself.

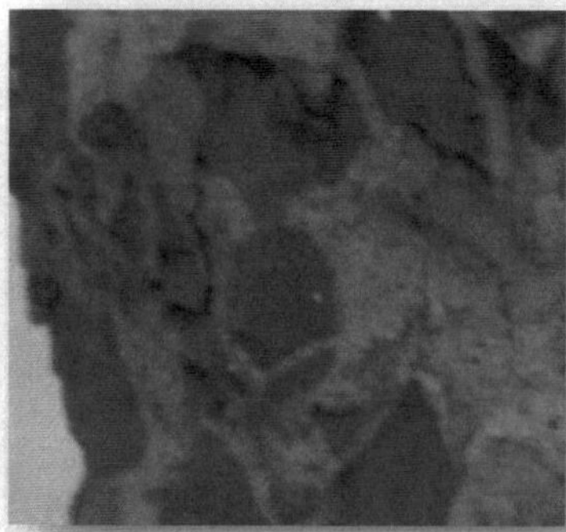

a) b)

Figure 5. Ceramic concrete sample after the test for the compression strength. a) the "clip" phenomenon; b) the contact zone of filler junction with the mortar part

Energy dispersive microanalysis of the studied light ceramic concrete sample (composition 5), with an electronic microscope Quanta 3D 200 i with the integrated system of microanalysis Genesis Apex 2 EDS from EDAX showed the following microphotos at scaling by 1000, 2500, 5000 times and the chemical composition given in table 2. The results prove dense structure, the contact zone is nearly invisible (fig. 5). In the zone of contact there are evenly distributed all structural oxides.

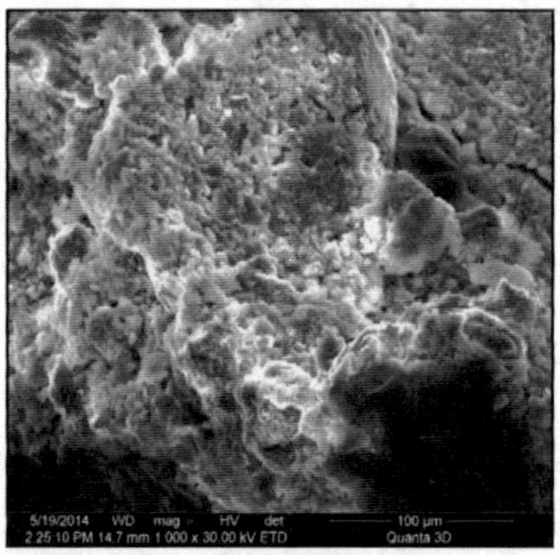

a)

b)

c)

Figure 6. Microphotos of the contact zone of the CCB filler with the cement rock: a) scaling in 1000 times; b) scaling in 2500 times; в) scaling in 5000 times

It has been determined [1,3,7,8,19] that the parameter of the adhesion strength in the contact zone "cement rock – filler" and composite strength in general are greatly influenced by the nature and properties of the filler. The following factors influence the concrete strength:

1. Form and surface roughness of the filler grain, its purity;
2. Chemical and mineralogical filler composition;
3. Strength parameters of the cement rock;
4. Chemical and mineral additives which improve adhesion strength;
5. Microdefects in the structure at the contact zone;
6. Water content of the concrete to the testing moment.

Table 2. Oxide composition in the contact zone of the mortar binding part and the filler from CCB

Oxides	Values at every part		
	1	2	3
Na_2O	2.55	0.72	0.62
MgO	2.44	0.87	1.48
Al_2O_3	8.34	11.33	8.85
SiO_2	27.33	27.09	28.18
SO_3	4.06	2.77	5.25
K_2O	0.57	0.57	1.29
CaO	51.68	51.85	44.57
TiO_2	0.25	0.26	0.41
Fe_2O_3	2.79	4.54	9.35

5. Conclusion

The following conclusions can be drawn:

- The properties of crushed bricks after building demolishing in the form of CCB have been studied;
- Man-made secondary fillers for the formulas of light and coarse pored ceramic concretes have been suggested;
- The formulas of light ceramic concretes with the use of CCB as fine and big fillers have been developed;
- The processes of structure formation and strength development as well as the properties of light ceramic concretes have been suggested.

It has been determined that using secondary man-made raw material is a topical issue for our region and for the whole world, as the ecological issue and ways of resource and power saving are acute nowadays.

References

[1] Murtazaev S, Salamanova M 2015 High-Strength Concretes Based on Multicomponent Binding and Fractionated Aggregates from Waste from Processing of Subsurface Rocks *Sustainable development of mountain regions* **1(23)** 23–8

[2] Hillemeier B, Buchenau G, Herr R, Huttl R, Klubendorf St and Schubert K 2006 *Spezialbetone, Betonkalender* **1** 534–549 (Ernst & Sohn)

[3] Murtazaev S-A YU, Salamanova M SH, Bisultanov R G and Murtazaeva T S-A 2016 High-Quality Modified Concretes with the Use of a Binder on the Basis of a Reaction-Active Mineral Component *Construction materials* **8** 74–80

[4] Zhang S, Lu D and Xu Z 2015 Effect of dolomite powders on the hydration and strength properties of cement mortars *Proc. XIV Int. Cong. on the Chemistry of cement* (Beijing, China) 320 p

[5] Nocun-Wczelik W, Szybilski M and Zugaj E 2015 Hydration of Portland cement with Dolomite *Proc. XIV Int. Cong. on the Chemistry of cement* (Beijing, China) 320 p

[6] Tironi A, Scian A N and Irassar E F 2015 Hydration of ternary cements elaborated with limestone filler and calcined kaolinitic clay *Proc. XIV Int. Cong. on the Chemistry of cement* (Beijing, China) 320 p

[7] Lukutcova N, Kulesh I, Macaenko A, Postnikova O, Nikolaenko A and Tuzhikova M 2014

Environmental safety improvement for decorative fine concrete using man-made glauconite sands *Construction and reconstruction* **1(51)** 79–84

[8] Udodov S A, Chernykh V F and Chernii D V 2008 Application of porous filler in finishing compositions for foamed concrete *Dry construction mixtures* **3** 70

[9] Nesvetaev G, Koryanova Y and Zhilnikova T 2018 On effect of superplasticizers and mineral additives on shrinkage of hardened cement paste and concrete *MATEC Web of Conf. 27 Ser. "27th R-S-P Seminar, Theoretical Foundation of Civil Engineering (27RSP), TFoCE 2018"* p 04018

[10] Stelmakh S A, Nazhuev M P, Shcherban E M, Yanovskaya A V and Cherpakov A V 2018 Selection of the composition for centrifuged concrete, types of centrifuges and compaction modes of concrete mixtures *Physics and Mechanics of New Materials and Their Applications (PHENMA 2018)* ed. Kim Yun-Hae, Parinov I A and Chang S-H P

[11] Soldatov A A, Sariev I V. Zharov M A and Abduraimova M A 2016 Construction materials on the bases of the liquid glass. In the collective works: Acute issues of construction, transport, machine engineering and technosphere safety *Proc. of IV annual scientific-practical conf.* (North-Caucasus federal university named after N I Stoyanov) pp 192–5

[12] Jeknavorian A, Roberts L, Jardine L et al. 1997 Condensed Polyacrilic Acid-Aminated Polyether Polymers as Superplasticizers for concrete *Proc. Fifth CANMET//ACI Int. Conf.* (Rome, Italy) pp 173–4

[13] Ohta A, Sugiyama T and Tanaka Y 1997 Fluidizing Mechanism and application of Polycarboxylate-Based Superplasticizers *Proc. Fifth CANMET//ACI Int. Conf.* (Rome, Italy) pp 173–19

[14] Dosho Y 2007 Development of a Sustainable Concrete Waste Recycling System "Application of Recycled Aggregate Concrete Produced by Aggregate Replacing Method" *J. of Advanced Concrete Technology. Japan Concrete Institute* **5(1)** 27–42

[15] Yanagibashi K, Yonezawa T, Iwashimizu T, Tsuji D, Arakawa K and Yamada M 2004 A new recycling process for coarse aggregate to be used concrete structure *Environment-Conscious Materials and Systems for Sustainable Development. Proc. of RILEM International Symposium* (Tokyo) pp 137–43

[16] Lesovik V S, Alfimova N I and Trunov P V 2014 Reduction of energy consumption in manufacturing the fine ground cement *Res. J. of Applied Sci.* **9(11)** 745–8

[17] Lesovik V S, Zagorodnyuk L K, Mestnikov A E, Kudinova A I and Sumskoi D A 2015 Designing of mortar compositions on the basis of dry mixes *Int. J. of Applied Engineer. Res.* **10(5)** 12383–90

[18] Bataev D K-S, Uzaeva A A, Uzaeva S A, Uzaev M A and Tuzurkaeva T A 2018 The Study of Shrinking Deformations of Repair Compositions on Barkhan Sands *Int. Symp. on Engineer. and Earth Sci. (ISEES 2018) Advances in Engineering Research* vol 177 pp 254–7

[19] Khadisov V Kh and Saidumov M S 2013 Light concretes with the use of ceramic crushed brick and faulty production *Problems of innovative biosphere friendly social-economical development in construction, housing and utility and road complexes: Materials of the 3-d Int. science-practical Conf.* (9–10 April 2013, Bryansk) in 2 volumes, vol 1, ed. Board: A V Alekseitsev, N P Lukutsova, V S Yanchenko and M A Senushenkov (Bryansk, Bryansk State Engineering Technological Academy) pp 189–94

IOP Conf. Series: Materials Science and Engineering **663** (2019) 012048 doi:10.1088/1757-899X/663/1/012048

Wastes recycling in the construction

S-A Yu Murtazaev[1,2], Z Kh Ismailova[2] , M Sh Salamanova[1,2] and S Kh Ibragimov[2]

[1]Kh. Ibragimov Complex Institute of the Russian Academy of Sciences, 21a, Staropromyslovskoe shosse, Grozny, 366002, Russia
[2]State petroleum technological university in Grozny named after M. D. Millionshikov, 30, Kadyrov street, Grozny, 364021, Russia

E-mail: madina_salamanova@mail.ru

Abstract. For many years fuel and energy and ore-dressing industries have been the sources of industrial waste which, if treated correctly, can be reused as secondary raw material resources accumulating earlier wasted investments and power costs. The article introduces possible ways of recycling both in our country and abroad. Basing on this, ways of reusing man-produced wastes in Grozny as bottom ash waste and chippings from rock crushing into spalls have been described. Study results showed that secondary man-produced raw material can compete with and substitute non-renewable natural resources as their chemical composition and quality parameters prove their unique nature, also it is possible to take into account solution of ecological problems. Developed formulas of multi-component bindings based on finely dispersed fly-ash and concretes with the use of enriched artificial aggregates promoted recycling of secondary raw materials and improving ecological situation in the region.

1. Introduction

Our country has a rich natural resource potential, but despite the variety and abundance of resources the issue of their rational use and effective preservation is acute. To support environmental safety and balance in nature it is necessary to reuse natural raw materials [1, 2].

Statistics showed that to receive various industrial goods about 1/3 of extracted raw material resources is used, and 2/3 are coproducts and waste [3], which results in ecological disbalance, and pollution of valuable lands.

That is why, waste reuse as a full value raw material base allows solving many economic tasks, such as saving natural resources, preventing water, soil, air pollution, increasing productivity and inventing new composite materials [3, 4, 9, 10].

European countries have high results in it. For example, the Netherlands have 100 % of material recycling. Besides, that country has a sustained ecological tendency of improving construction industry based on introduction of continuous non-waste industry with recycling man-made raw material (fig. 1) [6, 7, 15–18].

2. Problem statement

Nowadays, there is a lot of fundamental research [3, 8, 4, 11], allowing utilization of slurry, cinder and ashy waste, wastes of ore-dressing and processing enterprises, refinery waste and so on for the production of nearly all construction materials (fig. 2).

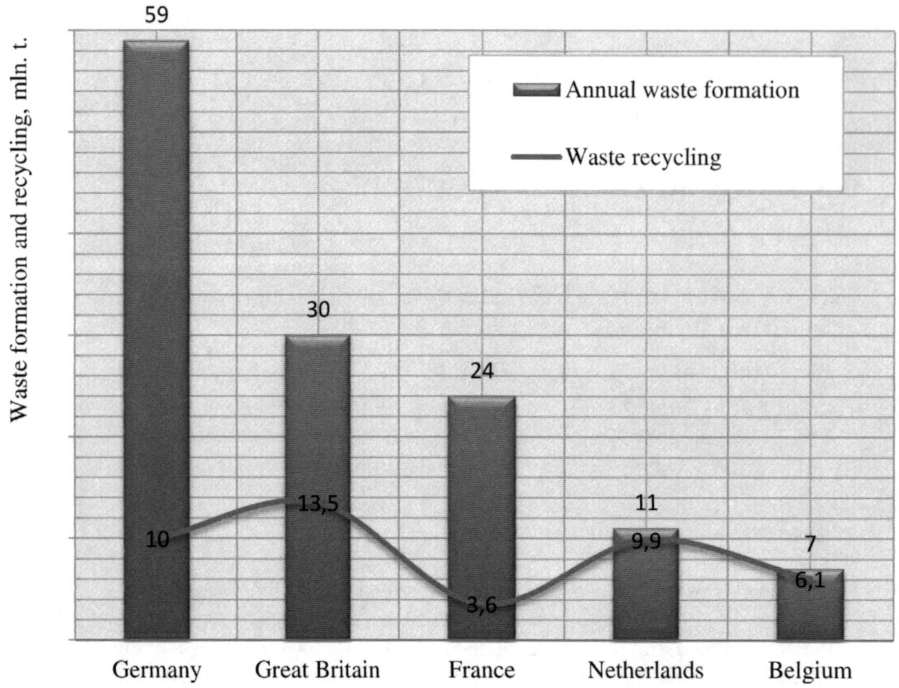

Figure 1. Formation and recycling of secondary resources in European countries [2]

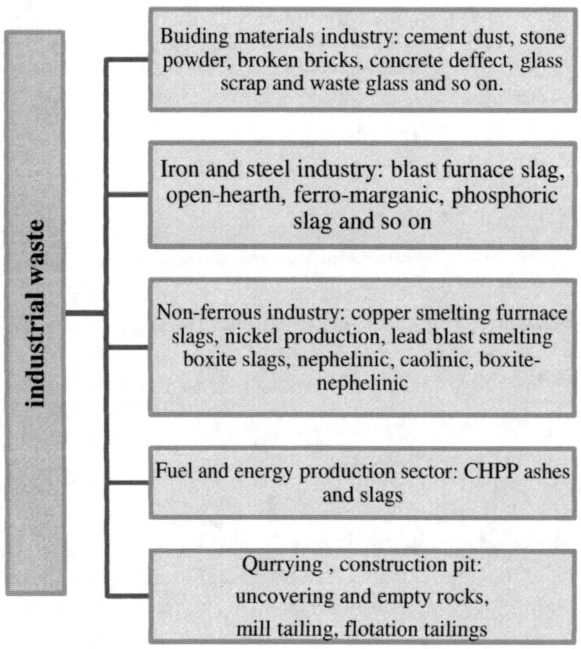

Figure 2. Industrial wastes used in construction material production

Basic list of construction materials received from industrial wastes is as follows: Portland cement clinker, Portland cement with mineral admixtures, blast furnace cement, slag alkali cementing materials, mixed cement free bindings, aggregates for concrete, slag cotton; slag-glass ceramic, autoclave hardening bindings; nepheline cement; materials for soil stabilization, castable, heat insulating materials and so on.

Thus, industrial waste management in combination with local raw material resources is perspective and profitable in construction material production, in particular, in developing multi component bindings and concrete on their basis.

In the Chechen Republic nowadays there are large accumulations of man-made wastes from fuel and energy industry in the form of bottom ash waste which take huge areas of hundreds hectares of valuable land plots and pollute the environment of the region (fig. 3). At the regional plants of nonmetallic construction materials up to 350 thousand m^3 of waste accumulate as stone dust, stone crumbs, and stone milling chippings. Huge areas of valuable lands are used to store the waste, though they could be used for agriculture, but to transport and store the waste significant means from the production work are used [4, 5, 9].

a) b)

Figure 3. Industrial waste: a) bottom ash waste; b) crushed stone chippings

3. Materials and methods

The work presents results of long term study of bottom ash waste properties and rock crashing waste for further use in developing high quality multi component bindings and concretes on their basis. In tests Portland cement SUE "Chechencement" CEM I 42.5 H, was used as a main binding component which has constant chemical and mineralogical compositions. Its basic characteristics are in table 1, chemical analysis in % by mass: SiO_2 = 17.45; Al_2O_3 = 3.88; Fe_2O_3 = 3.72; MgO = 1.12; CaO = 71.56; SO_3 = 0.76; TiO_2 = 0.33; K_2O = 1.07; Na_2O = 0.11.

Table 1. Basic cement properties

Manufacturing plant and kind	Specific area m^2/kg	Nytroxilene, %	Density kg/m^3	Time of set, hour – min		Supporting pressure, MPa, 28 days and nights	
				beginning	end	pressure	bending
"Chechncement" CEM I 42.5 H	330	25	3100	2-15	3-40	52.6	6.2

For production of multi component binding bottom ash waste was used which meets the requirements of AUSS 25818-91 “Fly ash of heat power plants for concretes. Technical requirements”. The main components of bottom ash waste are ashy microspheres which have such properties as: low density, adequate strength, ideal spherical form, high adhesion with polymer materials, high water hermiticity, low heat loss [6, 9–12].

Studying ashy microspheres from TPP in Grozny with a scanning electronic microscope (fig. 4) showed that they are finely-dispersed particles of grey color, spherical and shiny smooth surface. Also, some unevenness of different structure and size has been determined, covered porous shells in some microspheres have been determined.

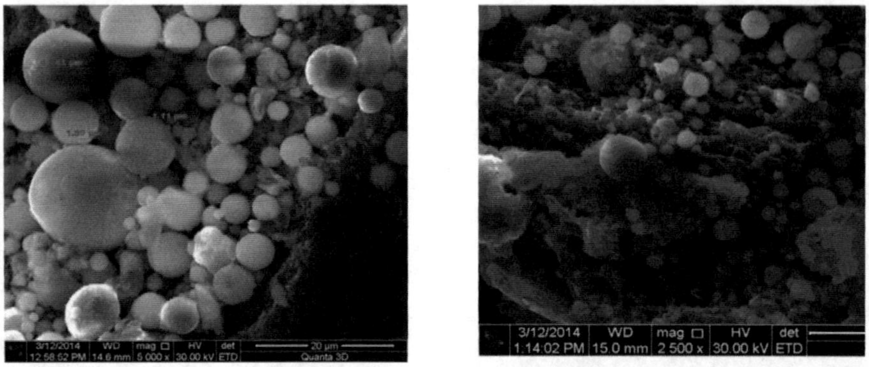

Figure 4. Microphoto of ashy microspheres of bottom ash waste from TPP in Grozny

Chemical composition of ashy microspheres in % by mass: MgO = 1.49; Al_2O_3 = 23.89; SiO_2 = 62.88; K_2O = 0.48; CaO = 1.7; Fe_2O_3 = 7.95; TiO_2 = 0.11; SO_3 = 0.06; ignition loss= 0.9.

Further, formulas of multi component bindings (table. 2) have been developed which were produced by combined 30 minutes grinding of Portland cement, bottom ash waste and dry super plasticizing admixture Polyplast SP-1. Specific area dependence of the received binding filled with bottom ash waste on the length of grinding and strength development of the new cements is given in figures 5, 6.

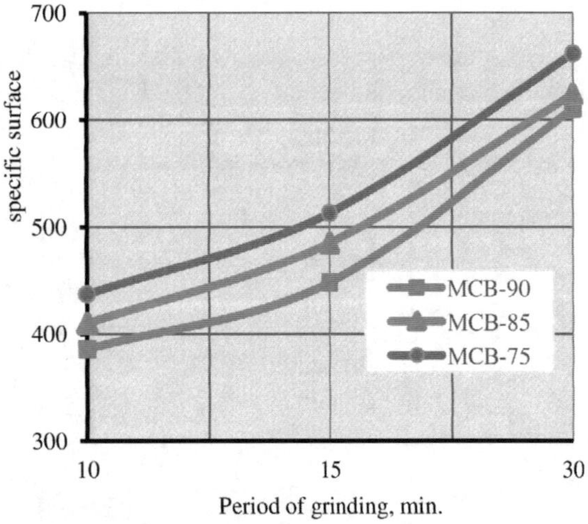

Figure 5. Dependence of the surface area of multicomponent bindings on the length of grinding MCB

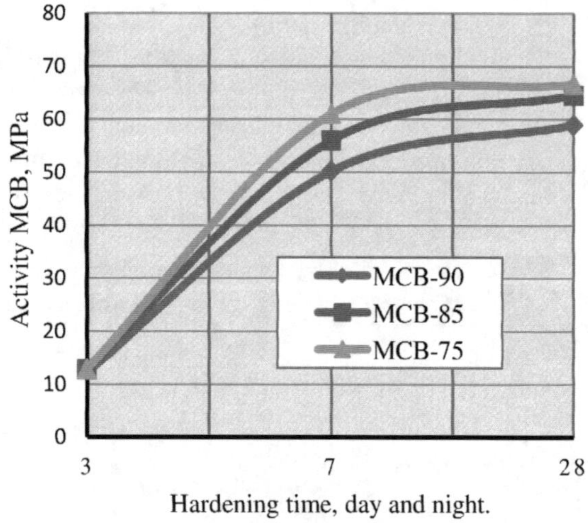

Figure 6. Strength development of multicomponent bindings

Table 2. Formulas of multicomponent bindings

№ composition	Activation method	Symbolic designation	Binding components content, %		
			PC	Ash microspheres	CR-1
2	Combined vibroactivation during 30 minutes	MCB-90	89	10	1.0
3		MCB-85	84	15	1.0
4		MCB-75	74	25	1.0

4. Results discussion

The results showed that the surface area of the bindings under study increases smoothly with longer period of grinding, but it is necessary to take into account that MCB-75 has the highest granulation factor, possibly it is caused by the character of ash microspheres which are softer than those of Portland clinker [1, 3]. Activity of multicomponent bindings is one of the most important parameters for receiving hard and durable composites.

Compositions of bindings MCB-75 and MCB-85 demonstrated good results which prove that ash microspheres combined with a plasticizing agent promoted full hydration of clinker minerals, moreover, active components of the bottom ash are additional source of the additional hydrated calcium silicate [1, 3].

There is no medium and big natural sand in the republic and application of fine sand results in excess expenditure of cementing components and increase of water demand for the concrete mixture. That is why, to enrich them crush wastes of gravel-sand rocks from Argunsk occurrence, such as chippings were used [4, 5, 9]. Visual survey of the chipping grains showed that these products have sharply angular form and contoured surface of the particles, and grains mostly have fractions of 2.5–1.25 mm and particles less 0.16 mm, that provides better binding in the contact zone of artificial sands with the cementing component and improves the concrete strength. As for the cost, the filler from by-passed stone is much cheaper (up to 6–10 times) than natural sands and their application decreases the prime cost of 1 m^3 concrete by 10 % [1,6].

Taking that into account, to receive better filler we combined fine sand of Chervlenskii occurrence $M_{chipings}$ = 1.7 and waste chippings of the gravel crushing into spall from Argunsk open pit M_{chips} = 3.8 at the ration 1:1 (it was determined experimentally).

To receive the high quality concrete the cubiform crushed stone from the granitic-diabasic rock from Alagirsk occurrence with fractions mixture 5–10 mm and 10–20 mm in the ration 70:30, spalls strength M1200 have been used. Superplasticizing admix Glenium, whose action is caused by the combination of electrostatic and spatial effect, achieved by lateral water repelling molecule chains of polycarboxylate ester was used for mixture plasticization and improving concrete characteristics [4, 18].

Table 3 shows experimental compositions and basic characteristics of high quality concrete with the suggested multicomponent binding MCB-75, fortified finely grained filler based on by-passed stone chippings and the superplasticizing admix Glenium.

Table 3. Compositions and characteristics of high-strength concrete

no.	Expenditure, kg per 1m³				Chipp/ Cement	additive expenditure, %	water absorption, %	concrete density, kg/m³	Concrete compression strength at days, MPa	
	MCB-75	FFF	S	W					7	28
1	480	870	880	168	0.35	1.3	3.7	2320	33.1	42.4
2	480	870	840	168	0.35	1.3	3.5	2350	49.6	65.2
3	460	840	870	161	0.35	1.6	2.8	2390	46.1	63.9
4	500	860	870	175	0.35	1.4	2.1	2380	50.9	66.7
5	520	840	900	182	0.35	1.8	2	2400	53.2	71.9

Note: FFF is fractional fine filler: mixture of fine sand from Chervlensk occurrence M_{chips} = 1.7 and crush waste chippings from Argunsk open pit $M_{chippings}$ = 3.8 in the ratio 1:1; S – spalls; W – water

Basing on the results of the experiment it can be noted that complex use of the components: ashy microspheres, finely fractionated filler with sharply angular form and contour surface of chipping screening grains, providing better binding with the cement rock and superplasticizing admix Glenium promotes structure and strength formation of high-quality concretes.

5. Conclusion

Thus, basing on the use of multicomponent bindings from ash microspheres and fractioned fillers made from rock processing wastes formulas of highly strong concretes with better physico-mechanical properties have been suggested. The results prove practicability of industrial waste utilization, which helps to free valuable lands used for industrial waste storage.

References

[1] Murtazaev S-A U and Salamanova M SH 2015 High strength concrete with graded aggregate from rock processing waste *Sustainable development of mountain territories* **1(23)** 23–8

[2] Hillemeier B, Buchenau G, Herr R, Huttl R, Klubendorf St and Schubert K 2006 *Spezialbetone, Betonkalender* **1** 534–49 (Ernst & Sohn)

[3] Murtazaev S-A U, Salamanova M SH, Bisultanov R G and Murtazaeva T S-A 2016 High-Quality Modified Concretes with the Use of a Binder on the Basis of a Reaction-Active Mineral Component *Construction mater.* **8** 74–80

[4] Zhang S, Lu D and Xu Z 2015 Effect of dolomite powders on the hydration and strength properties of cement mortars *Proc. XIV Int. Cong. on the Chemistry of cement* (Beijing, China) 320 p

[5] Nocun-Wczelik W, Szybilski M and Zugaj E 2015 Hydration of Portland cement with Dolomite *Proc. XIV Int. Cong. on the Chemistry of cement* (Beijing, China) 320 p

[6] Tironi A, Scian A N and Irassar E F 2015 Hydration of ternary cements elaborated with limestone filler and calcined kaolinitic clay *Proc. XIV Int. Cong. on the Chemistry of cement* (Beijing, China) 320 p

[7] Lukutcova N, Postnikova O, Nikolaenko A, Macaenko A and Tuzhikova M 2014 The improved environmental safety of decorative and fine-grained concrete on the basis of the use of technogenic glauconitic sands *Building and reconstruction* **1(51)** 79–84

[8] Udodov S, Chernykh V F and Chernyi D V 2008 Application of porous filler in finishing compounds for foamed concrete *Dry construction mixtures* **3** 70

[9] Nesvetaev G, Koryanova Y and Zhilnikova T 2018 On effect of superplasticizers and mineral additives on shrinkage of hardened cement paste and concrete *MATEC Web of Conf.* vol 27 "27th R-S-P Seminar, Theoretical Foundation of Civil Engineering (27RSP), TFoCE 2018" p 04018

[10] Stelmakh S A, Nazhuev M P, Shcherban E M, Yanovskaya A V and Cherpakov A V 2018 Selection of the composition for centrifuged concrete, types of centrifuges and compaction modes of concrete mixtures *Physics and Mechanics of New Materials and Their Applications (PHENMA 2018)* Abstracts & Schedule, ed by Yun-Hae Kim, I A Parinov and S-H Chang 337

[11] Soldatov A A, Sariev I V, ZHarov M A and Abduraimova M A 2016 Construction materials based on liquid glass *Topical issues of construction, transport, machine-building and technosphere safety Mat. of IV Annual sci.-pract. conf. in North-Caucasus federal university*, ed in chief N I Stoyanov pp 192–5

[12] Jeknavorian A, Roberts L, Jardine L et al 1997 Condensed Polyacrilic Acid-Aminated Polyether Polymers as Superplasticizers for concrete *Proc. Fifth CANMET//ACI Int. Conf.* (Rome, Italy) pp 173–4

[13] Ohta A, Sugiyama T and Tanaka Y 1997 Fluidizing Mechanism and application of Polycarboxylate-Based Superplasticizers *Proc. Fifth CANMET//ACI Int. Conf.* (Rome, Italy) pp 173–19

[14] Dosho Y 2007 Development of a Sustainable Concrete Waste Recycling System "Application of Recycled Aggregate Concrete Produced by Aggregate Replacing Method" *J. of Advanced*

IOP Conf. Series: Materials Science and Engineering **663** (2019) 012048 doi:10.1088/1757-899X/663/1/012048

Concrete Technology. Japan Concrete Institute **5(1)** 27–42
[15] Yanagibashi K, Yonezawa T, Iwashimizu T, Tsuji D, Arakawa K and Yamada M 2004 A new recycling process for coarse aggregate to be used concrete structure *Environment-Conscious Materials and Systems for Sustainable Development. Proc. of RILEM Int. Symp.* (Tokyo) pp 137–43
[16] Lesovik V S, Alfimova N I and Trunov P V 2014 Reduction of energy consumption in manufacturing the fine ground cement *Res. J. of Appl. Sci.* **9(11)** 745–8
[17] Lesovik V S, Zagorodnyuk L K, Mestnikov A E, Kudinova A I and Sumskoi D A 2015 Designing of mortar compositions on the basis of dry mixes *Int. J. of Appl. Engineer. Res.* 10(5) 12383–90
[18] Bataev D K-S, Uzaeva A A, Uzaeva S A, Uzaev M A and Tuzurkaeva T A 2018 The Study of Shrinking Deformations of Repair Compositions on Barkhan Sands *Int. Symp. on Engineering and Earth Sci. (ISEES 2018) Advances in Engineer. Res.* vol 177 pp 254–7

IOP Conf. Series: Materials Science and Engineering **663** (2019) 012049 doi:10.1088/1757-899X/663/1/012049

Vegetal fuel as environmentally safe alternative energy source for Diesel engines

A K Apazhev, Y A Shekikhachev, V I Batyrov, R A Balkarov, Kh B Kardanov, Kh L Gubzhokov and A L Bolotokov

Mechanization and enterprises power supply department, Kabardino-Balkarian State Agrarian University named after V. M. Kokov, 1c, Lenin Aveю, Nalchik, 360030, Russia

E-mail: shek-fmep@mail.ru

Abstract. Oil crisis and sharp increase of oil and petroleum market prices, resource limitation fastened the search for alternative energy sources. Among alternative energy sources renewable and especially received from vegetable biomass (fuel alcohol, vegetable oils) are of the greatest interest. With the account of this the article gives the results of studying the peculiarities of the bio-Diesel operational cycle and possibilities to adapt existing Diesels to biofuel. It has been determined that substitution of the Diesel fuel (DF) with rape-seed oil (RSO) and biofuel increases maximal and intermediate injection pressure, earlier start and greater injection duration, loss in frequency and oscillatory amplitude in the pressure pipeline after completing the injection. When RSO is used there appear increased inside lacquer film on the blowgun; which results in the needle valve immobility down to hanging up; needle valve immobility leads to the time lag of inlet and lengthening of the injection, increase in maximal injection pressure. There is quality loss of spraying the fuel, increase in non-uniformity and medium volume diameter of fuel drops, as well as penetration of the fuel jet into the air medium. Higher viscosity of the biofuel increases the injection rate due to leakage decrease in plunger pairs of diesel fuel injection pumps (DFIP).

1. Introduction

Ethylic alcohol (EA), vegetable oils made from crops, and methyl alcohol (MA) both from wood and its wastes are the best in economic terms [1, 2]. Fuel alcohol and vegetable oils can be used to substitute partially or fully traditional motor fuels (petroleum and Diesel). Biofuel from VO and alcohols are of especial interest and can fully substitute petroleum product fuels. It is also important that biofuel allow saving oil fuels and improve ecological qualities of engines.

Energy value of vegetable oils per kilogram is 37…40 mJ, while Diesel fuels typically have 42…46 mJ. Despite insignificant difference in energy value, vegetable oil density (0.91…0.94 kg/l) is higher than that of Diesel fuel (0.82…0.86 kg/l).

Vegetable oils have greater kinematic viscosity (7.5…10 times), that is why they dissipate worse. Cetane number of the vegetable oils on average is 16% lower than of the Diesel fuel (DF).

Basing on the combination of the physico-chemical and economical parameters RSO can be recommended as a basis for biofuel.

There are three trends of studying biofuel production from RSO and its application:

- possibility to use pure RSO or in combination with DF(RSO+DF);

IOP Conf. Series: Materials Science and Engineering **663** (2019) 012049 doi:10.1088/1757-899X/663/1/012049

- RSO biofuel with EA or MA additives (RSO+EA, RSO+MA);
- biofuel production by processing RSO into complex ethyl- and methylesters (REE, RME).

Nowadays most developed countries conduct researches on RSO biofuel production. France has produced ecologically pure fuel based on RSO which is named "Diestro" [1, 2]. "Diestro" was used in tractors and cars. The results showed up to 50 % decrease in sooth emission.

The economic alliance "EurobioDiesel" (France, Germany, and Italy) is developing a new kind of fuels from RSO and glycyl alcohol monoethers. Glycyl alcohol monoethers are received from sunflower and soy oil.

"John Deer" (the USA) tested biofuel Diesel made from oil, butanol and ethanol. The biofuel included oil 53.3 %, ethanol 13.3 %, butanol 33.4 %, tricthylanine 3.3 % and lanolin acid. It has been determined that operating on the biofuel Diesel capacity remained the same despite decreased by 19 % combustion heat in comparison with DF [3].

"Deitz" company Diesel (Germany) F3L413W had good results. It operated on the mixture of 40 % DF, 40 % RSO, 19 % water and 1 % of emulsifier agent. According to the expertise to use this composition on a common scale it is necessary to solve some problems: sooth prevention on the piston head, a head and a valve, an injection nozzle tip; better pulverization of a more viscous fuel [4].

Streyr WD209 Diesel has been tested in Australia which was a combination of rape seed oil with Diesel fuel in different proportions.

The issue of producing vegetal biofuel is also studied in our country [5]. The studies have showed that when Diesel D-240 operates on biofuel (75% RSO+25% DF) there is no decrease in its capacity and fuel efficiency; there is a decrease in the concentration of harmful substances with exhaust fumes: carbonic oxide by 55 %, hydrocarbons by 2…5 times. The results of the study prove applicability of RSO as a component of the composite fuel from the point of view of the economy and ecology.

2. Materials and methods

The physical parameters of DF, biofuel composites based on RSO, DF and alcohols were determined with common laboratory devices and facilities [6-9]. Density was measured with a densitometer graduated in 0.001 g/sm^3, kinematic viscosity with a capillary glass fluiditymeter, surface tension with a Rehbinder gauge, flash temperature with a TTH-1, sulfur content with a lamp method. Experimental tests were done on the motorless and motored testing facilities with fuel injection equipment assemblages of a Diesel engine D-240 and SMD-62. Hydraulic characteristics of the high pressure fuel system elements (HPFS) were determined on the testing facilities KI-22201 with the add-on device KI-15713.

HPFS section adjustment was at fuel temperature in the pump head 40°C with pressure 0.1 MPa. Then a geometrically effective plunger stroke and geometric commencement of fuel delivery with spillage under pressure 2.4 MPa were determined. Geometrically effective plunger stroke was determined with an indicator graduated in 0.01 mm, spillage pressure was controlled with a manometer graduated in 0.2 MPa. The tests were conducted on the testing facilities "Hartridg-1100" with a fuel pump (FP) UFP-5. The facility allows stepless changing of the drive shaft rotation frequency from 0 to 4200 min^{-1}, controlling pressure in the FP pump head, supporting test temperature specifications with the automatic temperature controller. After adjusting DFIP section a cyclic inlet and an onset angle of the fuel injection were determined with the stroboscoping light of the testing facility.

3. Results discussion

Physico-chemical properties of the fuel in the engine greatly determine its power, economy and ecological parameters. A final decision on biofuel usage as an alternative to the traditional DF must be done on the results of motor tests on the test facilities and operational tests.

To choose the most perspective and cheap energy material as a base for biofuel we compared the most important characteristics of different vegetable oils and commercial DF (Table 1).

Table 1. Physico-chemical properties of the vegetable oils and Diesel fuel

Parameters	Oil			Diesel fuel
	rape	soya	cotton	
Fuel lower heating value, kJ/kg	37300	36400	35400	42500
Fuel density, kg/m^3	920	920	920	840
Viscosity at 20^{0}C, mm^2/s	75	57	72	6
Cetane rating	32-36	36-39	35-38	45
Wijs	97	130	105	6
Acidity, mg Potassium hydroxide/100 ml	0.04	0.70	0.09	5.0
Surface tension, dine/cm	35.8	35.8	35.5	26…30
Temperature, ^{0}C:				
blushing	-9	-4	-1	≤0
hardening	-18	-9	-3	≤-7
colliquation		-15	+(5…11)	≤0
filtrability	+15	+4	+14	≤0

The motorless and motor tests of the biofuel composites test samples based on RSO, DF and alcohols (ethyl – EA, methyl – MA) (tables 2, 3), peculiarities of the biodiesel operational cycle have been studied as well as possibilities to adapt existing Diesel engines to biofuel.

Table 2. Elementary composition of basic and composite fuels

Fuel	Elementary fuel composition			
	C	H	O	S
Diesel fuel (DF)	0.870	0.126	0.004	0.002
Rape seed oil (RSO)	0.780	0.100	0.120	0.000
Ethanol (EA)	0.522	0.130	0.370	0.000
Methanol (MA)	0.375	0.125	0.500	0.000
75% RSO+25% DF	0.802	0.106	0.091	0.000
50% RSO+50% DF	0.825	0.113	0.062	0.000
90% RSO+10% EA	0.754	0.103	0.145	0.000
70% RSO+30% EA	0.700	0.110	0.194	0.000
90% DF+10% MA	0.820	0.126	0.054	0.0018
RSOE	0.770	0.110	0.110	0.006

Table 3. Basic and composite fuel characteristics

Fuel	Fuel lower heating value, kJ/kg	Density, kg/m^3	Cetane rating	Fuel viscosity at 20 ^{0}C, mm^2/s	Weight coefficient, kg/kg
Diesel fuel (DF)	42500	840	45	6	14.35
Rape seed oil (RSO)	37300	920	36	75	12.00
Ethanol (EA)	26500	790	9	2	8.55
Methanol (MA)	19665	795	5	0.55	6.52
75% RSO+25% DF	38600	900	38	38.2	12.56
50% RSO+50% DF	39900	880	40	39.8	13.20
90% RSO+10% EA	36356	902	33	39.5	11.65
70% RSO+30% EA	34383	890	27	27.2	10.97
90% DF+10% MA	40216	835	41	5.4	13.56
RSOE	37000	884	52	7.6	12.70

Table 4. Changes in engine power depending on the combined fuel composition

Fuel	Calorific efficiency, mJ/kg	Engine capacity, kW	PC, %
Diesel fuel (100%)	42.8	19.9	32.0
DF+RSO (75%+25%)	41.3	19.4	32.4
DF+RSO (50%+50%)	39.9	19.2	31.7
DF+RSO (25%+75%)	38.5	18.9	32.4

It has been determined that the decrease of Diesel capacity with vegetable oil in comparison with DF was below 10% (table 4) [10].

Experimental tests of Diesel D-240 and SMD-62 operating with RSO and composition of RSO with DF (75% + 25% DF) showed that Diesel D-240, with the combustion chambers similar to the chamber "CCRDI" adapts better to RSO and biofuel in comparison with the Diesel engine SMD-62, with the direct injection.

Operating on biofuel compositions Diesel keeps its operational qualities. Diesel capacity and efficiency decrease proportionally to power capacity and cyclic fuel inlet.

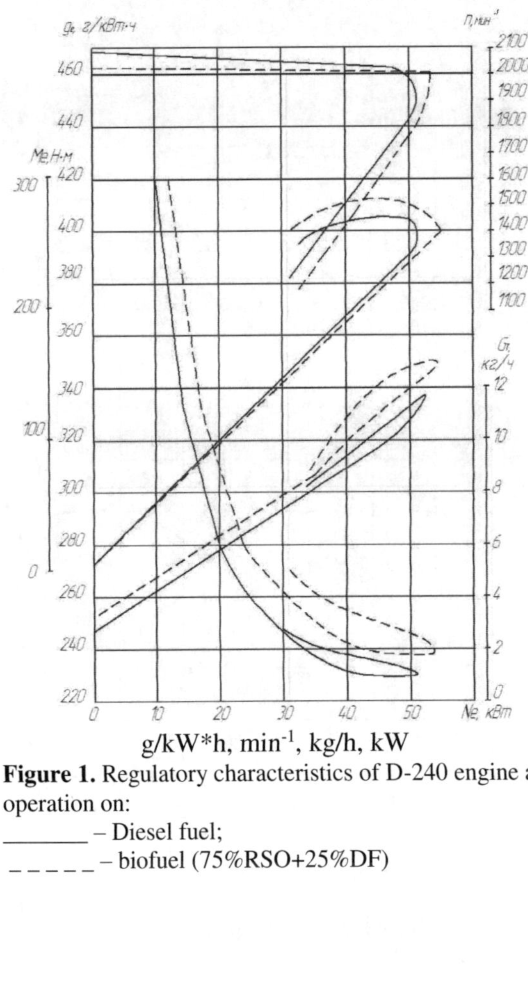

g/kW*h, min^{-1}, kg/h, kW

Figure 1. Regulatory characteristics of D-240 engine at operation on:
_______ – Diesel fuel;
_ _ _ _ _ – biofuel (75%RSO+25%DF)

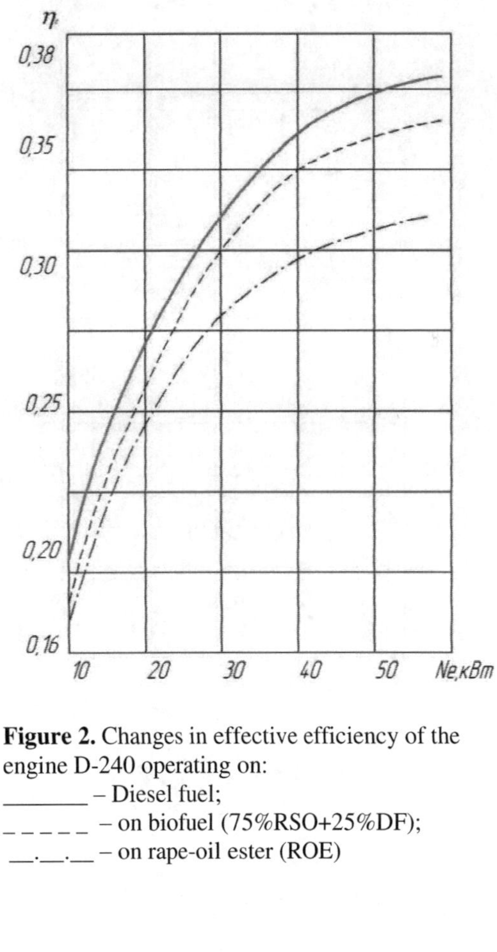

Figure 2. Changes in effective efficiency of the engine D-240 operating on:
_______ – Diesel fuel;
_ _ _ _ _ – on biofuel (75%RSO+25%DF);
__.__.__ – on rape-oil ester (ROE)

Shown in figure 1 regulatory Diesel characteristics D-240 operating on DF and biofuel composite (75%RSO+25%DF) prove that at constant regulations of fuel injection equipment running torque reserve coefficient when operating on biofuel is higher (11.5 %), than with DF (9.88 %). Diesel D-240 peak capacity when operating on biofuel composite is 2 % lower (56.8 kW) comparing to DF (58.0 kW). Regulatory irregularity degree when operating on biofuel (3.87 %) and DF (2.79 %) differ insignificantly.

Usage of RSO as fuel, or biofuel based on RSO combined with DF or alcohols introduces significant changes into the burning process. Biofuel have less Cetane rating that is less autoignition. That is why the ignition lag increases. As studies show [10], despite the ignition lag increase, Diesel operating on biofuel, have lower pressure rise rates and less intensity of heat generation.

Sluggish combustion at the first (kinetic stage) complicates fuel burning in the main (diffusive stage) and increases its duration. It increases the total period of burning and worsens Diesel indicated efficiency.

Quality analysis of the biofuel burning process is proved by the data of Diesel D-240 tests on the test operating on different fuels (figure 2).

At augmented intermediate rating of Diesel D-240 the peak capacity when using biofuel is higher by 7.69 % at $n = 0,85n_{max}$ (figure 3) and by 4.34% at $n = 0,7n_{max}$ (figure 4). The lowest fuel consumption rate when applying biofuel is higher by 3% at $n = 0,85n_{max}$ and by 7.93% at $n = 0,75n_{max}$.

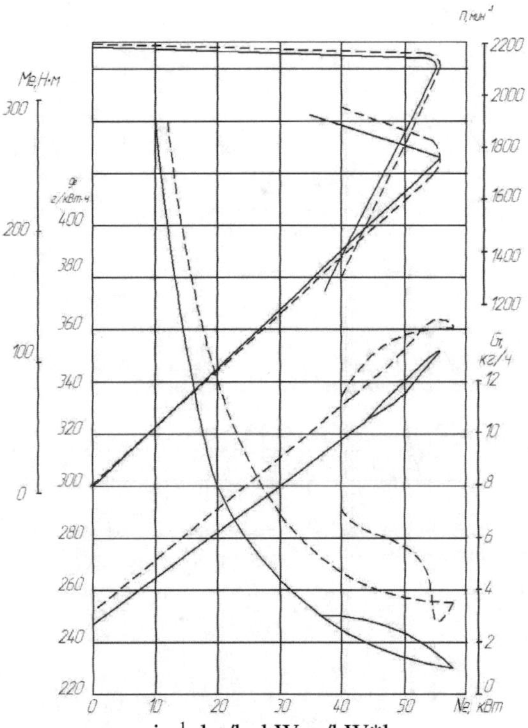

min^{-1}, kg/h, kW, g/kW*h

Figure 3. Partial regulatory engine parameters D-240 operating on:
_______ – Diesel fuel;
_ _ _ _ _ – biofuel (75%RSO+25%DF)

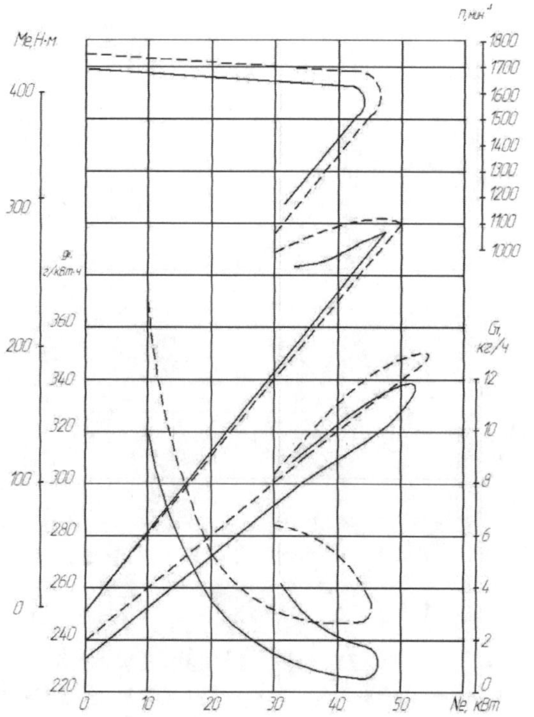

Figure 4. Partial regulatory engine parameters D-240 operating on:
_______ – Diesel fuel;
_ _ _ _ _ – biofuel (75%RSO+25% DF)

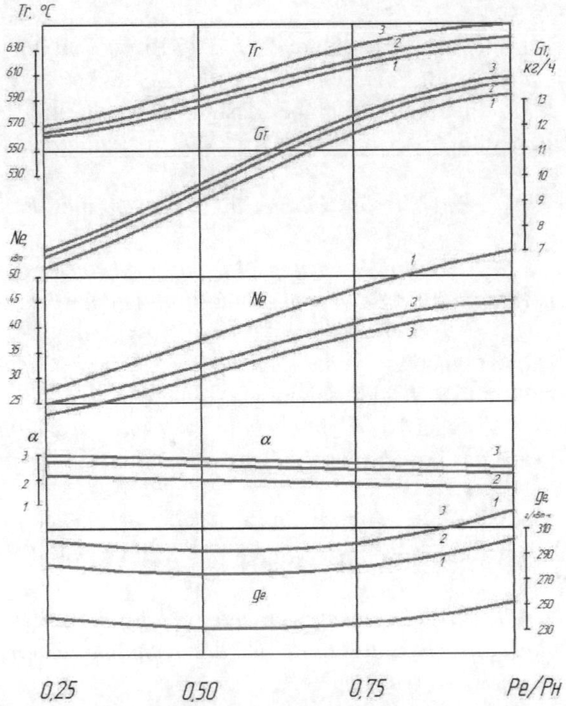

Figure 5. Load characteristics of Diesel D-240 operating on different fuels (2200 min^{-1}):
1 – Diesel fuel (DF);
2 – biofuel (90% RSO+10%EA);
3 – biofuel (70%RSO+30%EA)

Application of biofuel blends based on RSO and ethylic alcohol (EA) keeps Diesel functional properties at some worsening of power and economic parameters which is proportionally lower to the calorific value of the applied fuels (figure 5). At the same time Diesel toxic parameters significantly improve: sooth content decreases by 50 %, nitric oxide by 12 %.

4. Conclusion

The following conclusions can be made:

1. Fossil fuel shortage can be solved by biomass fuel. Rape seed oil is recommended as a base for biofuel.
2. The comparative analysis of physico-chemical and cost parameters of different biofuel blends allows recommending the blend RSO with ethylic alcohol as the most perspective biofuel. Additives can be recommended up to 30 % of the volume. Further increase of the alcohol additive decreases nitric oxide effect in the Diesel exhaust fumes and increases combustion period.
3. Operating on the biofuel Diesel keeps its functional properties. Diesel capability and efficient performance here scale down depending on energy intensity decrease and cyclical fuel inlet. To rehabilitate power and economic parameters it is necessary to increase cyclical feed and lead angle of the fuel injection.
4. EA additive into RSO allowed significant increase (more than 2 times) and RSO viscosity decrease. But, biofuel blends are steel highly viscous and overcome DF viscosity by 5…7 times, that needs additional heating of the biofuel or external heating of the injector.
5. Toxicity components content (N_{ox} and soot) with biofuel decreases significantly in comparison with their content in Diesel EF operating on DF (soot content decreases by 50 %, nitric oxide by 12 %).

References

[1] Koichev V S, Gritsay D I, Kobozev A K and Batyrov V I 2016 Perspective biofuel blends in Diesel engines *Scientific idea* **5** 191–6

[2] Koichev V S, Kobozev A K, Shvetsov I I and Gazizov I I 2018 *Alternative fuels for modern tractor engines Currentissues of the science-technical progress in AIC* (Stavropol': Stavropol'

State Agrarian University) pp 341–7

[3] Koichev V S, Kobozev A K, Shvetsov I I, Gritsai D I and Gerasimov E V 2017 Biofuel mixtures: perspective motor fuel *Res. J. of Pharmaceutical, Biological and Chemical Sci.* **8(5)** 642–6

[4] Koichev V and Mosikyan K 2016 Influence of combustion chamber design for power and fuel efficiency of gasoline engines running on natural GAS *Bull. of National Agrarian university in Armenia* **3** 44–6

[5] Koichev V S, Gazizov I I and Parashenko T A 2013 *Using biofuel in Diesel engines Farm Machinery operator* **7** 8–9

[6] Kobozev A K, Shvetsov I I and Koichev V S 2016 *Nozzle testing and adjusting techniques Currentissues of engineering education* (Stavropol': Stavropol' State agrarian university) pp 36–43

[7] Kobozev A K, Shvetsov I I and Koichev V S 2016 *Outlet valve checking procedure Current issues of engineering education* (Stavropol': Stavropol' State agrarian university) pp 43–6

[8] Kobozev A K, Shvetsov I I and Koichev V S 2016 *Plunger pair checking procedure Current issues of engineering education* (Stavropol': Stavropol' State agrarian university) pp 47–51

[9] Kobozev A K, Shvetsov I I and Koichev V S 2016 *Checking methods of Diesel fuel injection pump installation Current issues of engineering education* (Stavropol': Stavropol' State agrarian university) pp 51–6

[10] Batyrov V I, Koichev V S and Bolotokov A L 2016 *Diesel power and ecology parameters, operating on mixture and composite biofuel Current issues of the science-technical progress in AIC* (Stavropol': Stavropol' State Agrarian University) pp 260–7

IOP Conf. Series: Materials Science and Engineering **663** (2019) 012050 doi:10.1088/1757-899X/663/1/012050

Principles of cadastral valuation and usage of water reservoirs in the Republic of Kalmykia

E A Staselko, O V Erdniev, T A Balinova, U S Germasheva, S A Snagadjieva and A V Baryshev

Kalmyk State University named after B.B. Gorodnikov, Suseev street, 4, d. Elista, 358000, Russia

E-mail: staselkoelena@mail.ru

Abstract. The article deals with the principles of the cadastral valuation in the area of water reservoir construction site. Rules of using water resources, technical maintenance and water reservoir improvement are given. Special attention is paid to the land-use planning in the area of the water reservoir construction. Basing on the results of the pre-design and land-use planning, a number of land plots and active areas of water utilization scheme and participants who receive profit from the water reservoir have been determined. These data can be used to determine the cadastral value and the land tax.

1. Introduction

During national cadastral valuation estimated area zoning is fulfilled. An estimative zone is a part of lands similar for the intended purpose, type of the pragmatic use, and close in cadastral value land plots. Depending on the territorial value of the estimated zones their boundaries coincide with the land plots boundaries with the account of the existing development and land utilization, linear type facilities, as well as with the boundaries of cadastral districts or blocks. A map (scheme) of estimative zones is done on the results of the evaluation and a cadastral value of the area unit is established within the zones boundaries.

2. Problem

There are following water reservoirs in the Republic of Kalmykia: Arshan-Zelmen, Gorodovikovskoe, Laganskoe, Nugra, Proletarskoe, Ulan-Erginskoe, and CHograiskoe. The total area of the flood land in the Republic of Kalmykia is 1231980 ha. There are 37 water inlets on the area of Ketchenerovskogo, Maloderbetovskogo, Oktiabrskogo, Sarpinskogo, TSelinnogo, IUstinskogo, Iki-Burulskogo, IAshkulskogo, and Priiutnenskogo regions. 8 water reservoirs were built in 2011, and Elistinskoe is one of them. Its water will be used for irrigating agricultural lands, and there will be a tourist-recreation site on its banks. The water reservoir is 14 km away from Elista, in the Gashun-Sala Clough. Its area is about 600 ha, its useful storage is 35 mln m^3 of water. The characteristic levels and storages of the water reservoir are in figure 1 [1].

These project is important for the steep region as Kalmykia has little water sources especially in the central part of the republic.

Water storage usage must be done according to the rules of water resources utilization and rules of engineering instruction and water reservoir improvement by the Decree of the Government of the Russian Federation on 22 April 2009 no. 349 [2–4].

IOP Conf. Series: Materials Science and Engineering **663** (2019) 012050 doi:10.1088/1757-899X/663/1/012050

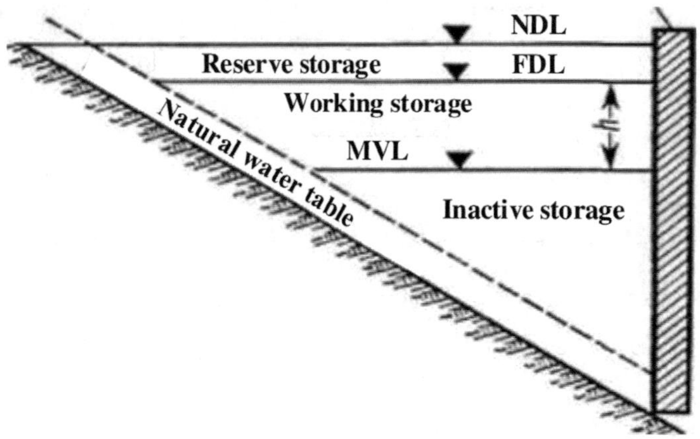

Figure 1. Characteristic levels and reservoir capacities.
NDL is the highest banked-up water level, which the dyke can perennialize at normal operation of all facilities;
FDL is the highest banked-up water level which can be held for a short time during flood routing for the safety of the facilities;
MSL is the minimal permissible level in a normal operating environment

Water will be taken from the CHernozemelskii irrigation channel by the underground method. A pump house is built for this. It will feed water through the water-pipe. Besides, the facility includes the cutting-off dam and dike 19 m high (fig. 2, 3).

(a) (b)

Figure 2. Early works (a) – site preparation for the construction of Elistinskii water reservoir; (b) – bottom solidifying works at Elistinskii water reservoir

(a) (b)

Figure 3. Elements of the water supply system to the water reservoir: (a) – Elistinskii water reservoir dam; (b) – Elistinskii water reservoir pipe

3. Materials and methods

While constructing water reservoirs vast lands are flooded that is why they are reserved for state needs. Lands can be reserved for state needs not more than for 7 years [5, 6], and the state cadastral valuation of lands must be done once in 5 years [7].

Land use planning must be done on the water reservoir construction site before it has been built [8, 9]. Succession of works at local land management for water reservoir construction is given in works by S.N. Volkov [10]:

- The location of all lands affected by the water reservoir and waterworks facilities is determined, their boundaries with households/farms and flood zones, underflood, shallow waters, construction sites etc. Distribution of the water reservoir itself which takes the main part of the granted lands is determined within the flood boundaries (on the flood level), which are on the project plans in scales 1:25 000 or 1:10 000, land plots are marked which are taken by the water reservoir and hydroelectric generating station. Possible rincons are determined as well as their economic use.
- The areas are calculated (total and agricultural) and explications are made to every landed property (land-use) after determining the boundaries of the granted lands (flooding level). Project explications are done by the land categories for every region, subject of Federation and the water reservoir itself.
- Plot on the boundaries of the flooded lands on the project plans which are not excluded from the landed property (land-use). New construction is prohibited there; the character of the agricultural land-use is changed as well as their productivity. The area of the flooded lands is calculated in the project explication. The area of every land-plot (land-use) is calculated and the project explications of the rest of lands are done.
- The location of the populated areas of industrial centers, bridges, roads, communication lines and other objects which cannot be located at the former area due to flooding, underflooding, bank erosion, landslides, worse malarias conditions, institutionally-managemental reasons is impossible or loses their commercial value. It is decided whether it is necessary to protect some objects. Shallow waters, their boundaries and areas are determined. The issue of their economic use or engineering protection is solved according to the construction rules 104.13330.2016 [11].
- The economic loss for the land owners, land users and land lords is determined. When the economic loss is determined due to early termination of their obligations to the third parties, work costs for land rehabilitation, lost profits are calculated. For land lords – the rent payment for rent lease for the period after withholding the land plot, the right cost for the rent lease of the land plot at the market.
- Proposals are made for the removal and use of the fertile soil layer on flooded areas for the operational project.
- The proposals are made for major conditions of the land grant. They are: reconstruction of the affected transport links, measures to protect natural resources and the environment, cultural and historical objects protection, determining the terms of water reservoir resources use and so on.
- The proposals are made to reorganize existing land-property (land-use), manufactures, resettlement, and their costs. Here flooding and underflooding of agricultural lands and settlements is taken into account, as well as disjointedness of the land properties and land-uses.

4. Results discussion

The data received during the pre-design and land use planning allow determining the number and used areas of the water utilization system parts, who receive the profit from the water reservoir and make their evaluation. Area data can be used to calculate the cadastral value and the land tax.

The flooded area is mostly covered with agricultural lands, that is why the specific indicator of the cadastral value, used for calculations on this kind of water utilization is equal to the average value of the specific indicators of the agricultural land value for administrative areas where the waterworks facilities are placed. The indicators for every region are given in table 1.

Table 1. Specific indicators of the cadastre values of the agriculture lands in the flooded municipal regions

name of the region	Specific indicator of the land cadastre values, rub/ha
Ketchenerovskii	669.60
Maloderbetovskii	1630.63
Okt'abr'skii	427.27
Sarpinskii	3681.47
Tselinnyi	4091.44
Ustinskii	421.73
Iki-Burulskii	2195.75
Yashkulskii	640.60
The average specific indicator of the cadastre land value, rub./ha	1719.81

State cadastre evaluation of lands is done to determine the cadastre value of the land plots for different purposes. Cadastre value is the market cost of the property determined during the state cadastre evaluation the method of mass evaluation, or the market cost, determined for every object individually when it is impossible to determine the market cost by mass methods [12]. When determining cadastre value the objects of evaluation are property objects enlisted in the State Immovable Property Cadastre on the date of cadastre evaluation.

Land evaluation is done with the account of land, city-planning, forest, water and other cadastres [13]. It results are registered in the register or classified code of data on the registered immovable property (land plots, buildings, constructions, unfinished objects and buildings), as well as data on the State Boundary of the Russian Federation, and boundaries between the subjects of the Russian Federation, municipal boundaries, boundaries of population areas, target areas and use-restricted zones.

5. Conclusion

It has been determined that during utilization the Elistinskii water reservoir will further turn into the complex of hydro technical facilities for systematic contentment of the requirements of all parts of the water resources utilization system. It is conditioned by the deficit of water resources in the Republic of Kalmykia.

Land resources and water objects need systematic and regular quality and quantity classification. Here special attention should be paid to the changes in the ameliorative state of lands, the groundwater depth level, as well as indicating of changes in the object cost.

References

[1] Tesalovsky A A 2014 *Methods of cadastral evaluation of land reserved for the construction of reservoirs for complex purposes: dissertation for the Candidate of technical Sciences degree* (St. Petersburg) 187 p

[2] *Water code of the Russian Federation* Federal law of 03.06.2006 no 74-FZ

[3] *On approval of Regulations on development, coordination and approval of rules for the use of reservoirs, including model rules for the use of reservoirs* Resolution Of the government of ecology of the Russian Federation on 22.04.2009 no 349

[4] *On the approval of Guidelines for the development of rules for the use of reservoirs* Order of the Ministry of natural resources and ecology of the Russian Federation on 26.01.2011 no 17

[5] *Land code of the Russian Federation* Federal law on 25.10.2001 no 136-FZ

[6] *Tax code of the Russian Federation* Federal law on 31.07.1998 no 146-FZ
[7] *On state cadastral evaluation of land* Federal law dated 03.07.2016 no 237-FZ
[8] *On state registration of real estate* Federal law dated 13.07.2015 no 218-FZ
[9] *On land management* Federal law on 18.06.2001 no 78-FZ
[10] Volkov S N 2001 *Land Management* (Moscow: Kolos)
[11] *Engineering protection of the territory from flooding and flooding* Order of the Ministry of construction and housing and communal services of the Russian Federation on 16.12.2016 no 964/PR On approval of SP 104.13330.2016
[12] Polikarpov A M, Mirzoeva A Je and Ovchinnikova A S 2015 Problems of the cadastral system in the Russian Federation Education, economy, society **3–4(49–50)** 125–9
[13] Bykowa E and Sishchuk J 2015 The state real estate cadaster in Russia – problems and perspectives ZFV – Zeitschrift fur Geodasie, Geoinformation und Landmanagement **140(1)** 22–6

Assessment of snow cover pollution of the mining area

Ya T Suyundukov[1,2] , I N Semenova[1,2], R F Khasanova[1,2], Yu S Rafikova[1], M B Suyundukova[1,2], G R Ilbulova[2] and E E Papyan[2]

[1]Institute of Strategic Studies of the Republic of Bashkortostan, 1, Kutuzov St., Sibay, 453837, Russia, Republic of Bashkortostan
[2]Bashkir State University, 21, Belov St., Sibay, 453837, Russia, Republic of Bashkortostan

E-mail: rezeda78@email.ru

Abstract. The purpose of the work is to assess snow cover pollution of Sibay city with sulfates and heavy metals. The paper presents snow cover study conducted in January-February 2019 in accordance with the requirements of generally accepted regulatory and procedural documents [11–12]. The site near the pond in the northern part of the city was as a conditional background. The paper describes the analysis of samples for the content of heavy metals carried out by atomic absorption spectrophotometry, the sulfate concentration measured by the photometric method. The study used pH-meter to determine the pH. To calculate the total pollution indicator the study used obtained results. Snow sulfate pollution was at the admissible level. The highest content of sulfates was in the area of industrial enterprises (the excess of background indicators was in the range of 5.1–11.1 times). The study revealed polyelement pollution of snow with such metals as (Fe, Mn, Cu, Zn) in the zone of quarry emissions. Snow cover next to the pit dumps has a moderately dangerous pollution category. Copper and zinc make the main contribution to pollution. Thus, the activity of mining enterprises contributes to the accumulation of elevated concentrations of metals in the atmospheric surface layers, which negatively affects the ecological and hygienic situation in the region.

1. Introduction

Numerous ethnic groups have long inhabited Zauralskaya zone of the Republic of Bashkortostan, located in the southern Urals. For a long time, the main branch of the economy was pasture cattle breeding. At the beginning of the last century, people discovered various minerals, mainly non-ferrous metal ores, in the depths of this region, after which intensive development of their deposits began. Since then, the living conditions of the population have changed dramatically and they have adversely affected environment. The development of mining production reflected not only in the structure of the regional economy, in the ecological condition of the environment, but also in the medical and demographic situation: many diseases of people also depend on ecological situation [1–4].

The city of Sibay with a population of more than 60 thousand is the largest settlement of the Trans-Ural zone of the Republic of Bashkortostan, which in essence is the cultural, educational and industrial center of the region. The Sibay branch of Uchalinsky mining and refining plant (UMRP) (previously called BMSK), founded in 1948, and is engaged in mining and enrichment of polymetallic ores. The main objects of the mining and processing plant, which are sources of pollution of the environment, are the quarry, the processing plant and the tailings dam. In addition, a large thermal power station has been operating in the city for more than 10 years. The long-term exploitation of mining and other industrial enterprises has led to the formation of technogenic provinces in the region, which, overlapping with the natural geochemical background, further aggravated the ecological situation. One of the environmental problems of the city is contamination with various pollutants, first, heavy metals (HM). According to Roshydromet, presented in the State report "On the state and environmental protection of Russian Federation in 2017", Sibay city was included in the list of Russian Federation with a moderately dangerous category of soil pollution by heavy metals for 2007-2017 [5]. The report noted that the priority heavy metals are copper, cadmium, zinc and plumb. The total indicator of soil pollution in one-kilometer zone belongs to the dangerous category. Long-term studies of a number of authors have confirmed this fact [6–9].

The tense of environmental situation of the Trans-Ural zone of the Republic of Bashkortostan affected the health of the population. An increased level was in its territory compared to the average republican indicators of circulatory system diseases, respiration and digestion, skin and nervous system in adults and diseases of the respiratory organs, digestion, endocrine system and metabolism, nervous system and sensory organs in children [10].

Of particular relevance to Sibay city is the study of environmental risks acquired in the last months of 2018 due to emissions of toxic substances from the Sibay open-cast mine into the atmosphere of the city. This event received a vivid media coverage and caused serious public unrest. At this time, citizens of Sibay complained of headache, sore throat, cough, and tears. There was an excess of the admissible concentration limit (ACL) of sulfur dioxide in the air. For the rest of the other studied parameters there were no exceeding. This fact aims to conduct studies of the state of the atmospheric air using non-standard methods, which, along with generally accepted indicators, would perhaps more fully allow for assessing the environmental risks to the population from mining facilities.

There are various methods for assessing the state of atmospheric air, but one of the most accessible is monitoring of snow cover. This method is a relatively cheap and informative indicator of pollution by emissions from industrial enterprises and vehicles in winter period. Snow has a high sorption capacity and is a carrier of not only wet, but also dry precipitation, and therefore gives an objective assessment of all atmospheric pollution during the winter period. Due to the natural processes of concentration of pollutants in the snow, the content of pollutants in it is one of the most significant criteria for assessing the ecological state of the atmosphere [10].

The purpose of this study was to assess the ecological state of snow cover in Sibay based on the accumulation of heavy metals and sulfates.

2. Materials and methods

The paper presents snow cover study pollution conducted in January-February 2019 in accordance with the requirements of generally accepted regulatory and procedural documents [11-12]. Researchers have taken snow samples manually using a plastic cylinder with a diameter of 10 cm and a height of 20 cm in plastic bags. Sampling was out with visually clean and flat areas of snow cover; each plot area was of 10 sq.m. At each site by the envelope method, researchers have taken 5 samples, which then have mixed into 1 combined sample. The study comprised a total number of 11 composite samples. The site next to pond in the northern part of the city was as a conditional background. Then researchers delivered snow samples to the laboratory and laid out of the packages in plastic containers (pots) for natural thawing. Analysis of melt water samples for HM content was carried out by atomic absorption spectrophotometry (spectrophotometer Contr AA, Analytik Jena).By the photometric method (KFK-2 photometer) researchers measured the sulfate concentration and used pH-meter "HI 2211 HANNA" to determine the pH.

Polyelement pollution causes big problems when rationing the degree of pollution of the environment. At present, it is customary to use a formal calculation of total pollution factors (Zc). Sanitary and hygienic standards do not only approve the calculation of total pollution factors , but also conventionally classify into classes of danger : $Zc= \sum K_{ki} - (n-1)$, where $K_{ki}= C_{idet} / C_{iback}$; C_{idet} –

IOP Conf. Series: Materials Science and Engineering **663** (2019) 012051 doi:10.1088/1757-899X/663/1/012051

determined content of the i-th toxicant in the soil ; C_{iback} – the value of the ground content of the i-th toxicant ; n – total number of toxicants used in the calculation [13].

Critical values that characterize the total pollution of Zc by danger are as follows: when Zc< 16 pollution is admissible; when 16 < Zc< 32 – moderately dangerous; when 32 < Zc< 128 – high dangerous; when Zc> 128 – extremely dangerous.

3. Results and discussion

3.1. Sulfates cover

First, the researchers studied aerotechnogenic substances containing sulfur in snow samples. As part of sulfur dioxide, which adversely affects the health of people and animals, sulfur is a priority pollutant of the atmosphere. However, sulfurous anhydride loses this status when adsorbed by soil and oxidized to sulfate. Sulfates can migrate in the form of soluble salts and be temporarily deposited by a soil-absorbing complex as part of mineral salts, or absorbed by soil biota. The proportions between sulfate pools depend on the soil properties, all sulfates can participate in the prolonged fixation of sulfur in the soil organic matter. The contribution of sulfate to the change in pH of the soil solution depends on the buffer of the soil capacity [14].

Sulfur compounds, like all pollutants, are removed from the atmosphere in two ways: dry deposition and wet – with precipitation. Mainly sulfates are to be out with wet removal. The amount of sulfates in sediments, as a rule, correlates well with their content in the air [14].

Despite the high content of sulfur dioxide in the atmospheric air of Sibay city, pollution with snow sulfates was admissible. The highest content of sulfates was in the area of industrial enterprises – the Bashkir mine control department (BSMCD) (ground exceedance by 11.1 times), the Drilling reagent plant (8.1 times) and quarry dumps (5.1 times).

Table 1. Assessment of acidity and sulfate content in snow samples

Sampling location	Hydrogen indicator, pH	Sulfateion, mg/dm^3	Concentration rations of sulfateions relative to the ground	Pollution category
Ground (next to the pond, district Fisherman's House)	6,5	5,35	1,0	admissible
Pervomayskaya st.,41 / 1	6,8	6,58	1,2	admissible
Quarry dumps	7,6	27,16	5,1	admissible
Small town Zoloto	6,4	13,7	2,6	admissible
District Fisherman's house (residential buildings)	7,1	8,64	1,6	admissible
Mayakovsky park	8,0	11,93	2,2	admissible
Next to Drilling reagent plant	8,0	43,21	8,1	admissible
District Vostochnuy, Kusimov st.	8,1	13,17	2,5	admissible
Next to Bashkir mine control department (BSMCD)	8,2	59,26	11,1	admissible
Nurimanov st.	8,6	17,28	3,2	admissible
Small town Gornuy	8,7	20,16	3,8	admissible

3.2 Assessment of snow cover pollution with heavy metals

As a result of the investigations performed, the researchers have revealed polyelement pollution of snow in the zone of the quarry emissions with metals (Fe, Mn, Cu, Zn).

As a result of the study, the researchers found that snow cover in the small towns Zoloto and Gornuy had admissible pollution category. Nurimanov street, next to quarry dumps, had moderately dangerous category of pollution. Copper and zinc made the main contribution to pollution.

Table 2. The content of HM in snow samples (concentration ratios with respect to the ground)

Sampling location	Cu	Zn	Fe	Mn	Zc	Pollution category
Village Turkmenovo(ground)	1,0	1,0	1,0	1,0	1,0	admissible
Small town Zoloto	3,7	3,9	1,4	1,3	7,3	admissible
Nurimanov st. (next to the quarry dumps)	10,9	9,3	2,9	2,3	22,3	moderately dangerous
Small town Gornuy	3,4	1,3	0,8	2,1	4,5	admissible

4. Conclusion
Thus, the activity of mining enterprises contributes to the accumulation in the surface layers of the atmosphere of elevated concentrations of heavy metals that adversely affects the ecological and hygienic situation in the region.

5. Acknowledgments
The publication was prepared in the framework of a research project №18-413-020004 p_a supported by the Russian Foundation for Basic Research (RFBR) and the Government of the Republic of Bashkortostan.

6. Conflict of interest
The authors of the article confirm the absence of a conflict of interest.

References
[1] Novikov S M and Ivanenko A V 2009 Assessment of damage to the health of the population of Moscow from the effects of suspended substances in ambient air *Hygiene and sanitation* **6** 41–3
[2] Chudinova O.N. 2008 *The Impact of technogenic air pollution on the health of the people of Transbaikalia (on the example of the city of Ulan-Ude)* Abstract of diss. ... cand. biol. sciences (Ulan-Ude)
[3] Health aspects of air pollution – answers to follow-up questions from CAFÉ 2004 *Report on a WHO working group meeting* 15–16 January 2004 (Bonn, Germany)
[4] Dorn O Yu 2007 *Biochemical and immunological factors of the pathogenesis of gastropathy in patients with occupational chronic dusty pulmonary pathology* Abstract of diss. ... cand. med. sciences (Novosibirsk)
[5] *On the state of natural resources and environmental protection of the Republic of Bashkortostan in 2016: State report in 2017* (Ufa: Ministry of Nature Management and Ecology of the Republic of Bashkortostan)
[6] Opekunov A Yu, Opekunova M G, Somov V V, Mitrofanova E S and Kukushkin S Yu 2018 Effect of the development of the Sibay deposit (Southern Urals) on the transformation of the flow of metals in subordinate landscapes *Bulletin of Moscow University, Series, Geography* **1**, 14–24
[7] Opekunova M G, Somov V V and Papyan E E 2017 Soil pollution in the area of influence of the mining enterprises of the Bashkir Trans-Urals *Soil Sci.* **6** 744–58
[8] Semenova I N, Rafikova Yu S, Suyundukov Ya T and Biktimerova G Ya 2016 Regional Peculiarities of Microelement Accumulation in Objects in the Transural Region of the Republic of Bashkortostan *Springer Int. Publish.* (Switzerland)
[9] Nikonov V N, Shamsutdinova L V and Zvereva T I 2015 Natural and anthropogenic geochemical anomalies of heavy metals in the Bashkir Trans-Urals as an aspect of the system analysis of geopathogenic zones *Ural Ecological Bull.* **2** 3–7
[10] Semenova I N, Rafikova Yu S and Drovosekova I V 2013 *Population of the city of Sibay in*

conditions of technogenic pollution: demography, state of health, features of microelement status (Ufa: Gil, Bashkir encyclopedia)
[11] Vasilenko V N, Nazarov I M and Friedman Sh D 1985 *Monitoring of snow pollution* (Leningrad: Hydrometeoizdat)
[12] GOST 17.1.5.05-85 *Nature Conservation. Hydrosphere. General requirements for sampling of surface and seawaters, ice and precipitation* 1985 (Moscow: Standards Publishing House)
[13] Gl 5174-90 *Guidelines for assessing the degree of air pollution of populated areas with metals on their content in snow cover and soil* 1990 (Moscow: Standards Press)
[14] Saet Yu E, Revich B A and Yanin E P 1990 *Environmental Geochemistry* (Moscow: Nedra)
[15] Maslova I Ya 2008 Impact of sulfur-containing aerotechnogenic substances on some agrochemically significant processes and soil properties *Agrochemistry* **6** 80–94

IOP Conf. Series: Materials Science and Engineering **663** (2019) 012052 doi:10.1088/1757-899X/663/1/012052

Primary succession of plants of technogenic dumps of the Kabardino-Balkarian Republic

A Ya Tamakhina, A A Ahkubekova, A A Gadieva and R A Tiev

Kabardino-Balkarian State Agrarian University named after V M Kokov, 1v, Lenina ave., Nalchik, 360030, Russia

E-mail: aida17032007@yandex.ru

Abstract. The article presents data on the regularities of the flow of primary succession on the terraces of the tailing dump of the Tyrnyauz tungsten-molybdenum combined plant. In the accumulative ecotope of old-age terraces of the lower level of the dump, grass-wormwood and herb-wormwood communities dominate with a projective cover of up to 60 %. In the transeluvial ecotope of middle-aged terraces, there are ruderal-natural grass-forb grass communities with a projective cover of 37 %. In the eluvial ecotope of young terraces of the upper level, there are monodominant explent communities with a projective cover of 13 %. Old age terraces have the greatest similarity with the zonal flora and have characterics of relatively high species richness and saturation. With an increase of succession length, there are changes in the dominant life forms (from terophytes to hemicryptophytes), a decrease in the proportion of species with wide ranges (pluriregional, holarctic, and palaearctic), an increase in the number of stenobionts and hemistenobionts. A number of species growing in extreme conditions of ecotopes of middle and upper terraces are characterized by the absence of deviations from ecological regimes and indifference to elevated concentrations of Mo and W. High toxicity of the substrate constrains the speed of transition from initial to subsequent stages of primary succession. To restore vegetation, it is necessary to carry out both biological and technical remediation.

1. Introduction

An important factor determining the direction of succession and the duration of its individual stages is the initial state of the substrate and the ecotope as a whole. Significant heterogeneity of the surface of technogenic landscapes causes a high contrast of the modes of insolation, water and mineral nutrition, in which regeneration biogeocenoses are forced to develop [1].

Chemical pollution to a greater extent than the type of community and the group of soils affects the species saturation and the projective cover of the species. [2]. The features of the succession process on toxic substrates are the high duration of the pioneer stage of succession, the insignificant coverage of vegetation for a long time, the poverty of the floristic composition, the small number of dominant species, the simplified spatial structure, the predominance of species-assectators [3–5]. Ecologically plastic ruderal species with wide ranges and mechanisms of protection against high concentrations of toxic elements represent plant communities of pioneer stage [6–14].

On the territory of the Kabardino-Balkarian Republic (KBR), there is JSC "Tyrnyauz tungsten-molybdenum combined plant" (TTMCP), which has referred to the tails of the concentrating plant and has accumulated over 80 million m^3 of tailings of the tungsten, molybdenum and copper, bismuth concentrates with high toxicity and high solubility in groundwater. The average content of main metals in the surface layer of industrial waste is (g/t): W – 420, Mo – 145, Zn – 290, Cu – 69, Sr – 315, Rb – 118 [15].

Due to the relatively poor study of the characteristics of the flow of primary succession on a toxic substrate, the purpose of the study was to monitor the composition of the plant community in dumps from non-ferrous metal enrichment of № 1 TTMCP tailings.

2. Materials and methods

In the course of monitoring the vegetation cover of the tailing dumps (2016–2018), a mountain steppe at a distance of 3–4 km up the valley of the r. Baksan (ground) and ecotopes at different levels of the terraces of the western slope of the tailing dumps have become the object of the study. The climatic conditions of the tailing areas are close to arid ones with an average annual precipitation of about 350 mm. The climate adjacent to the tailing areas is temperate with an average annual precipitation of 600 mm. The soils are dark chestnut, the reaction of the upper horizons (0–40 cm) is neutral and slightly alkaline (pH = 6.9–7.5), the total nitrogen content is 0.26 %, and water-soluble salts are 0.05–0.06 %.

The alluvial landscape of early-age upper terraces (Tu) has high density of the soil, a strongly alkaline reaction of the substrate (pH = 9.3–9.5). The transit transeluvial landscape of middle-aged terraces (Tm) is distinguished by a more loose and moist soil covered with a continuous thin layer of young soil up to 2 cm thick with an alkaline substrate reaction (pH = 8.3–8.7); the accumulative landscape of the lower old-age terraces (Tl) – wet soil with a primary soil layer up to 5 cm and a less alkaline reaction of the substrate (pH = 7.9–8.2).

In the course of the monitoring, there was a geobotanical survey of the terraces of the tailing and the ground site, and botanical descriptions were drawn up with an indication of the values of the projective cover of the species during the maximum development of grass stand (July-August). In total, the reseachers have compiled 40 descriptions of the pioneer group flora and 8 descriptions of the background landscape . The size of accounting sites was 5 x 10 m2. The similarity of the plant communities of the terraces and the background area was estimated by the Jacquard coefficient (Kj), the degree of uniform distribution of the abundance - by the Pielou index (E), the species richness - by the Shannon – Weaver index (H). For the analysis of species diversity, curves of "dominance – diversity" were used.

Ecological assessments of species were determined by amplitude and point ecological scales. According to the values of climatic (It clim.) and soil (It soil) of tolerance indices, the researches divided herbaceous plant species into stenobiont (SB), hemistenobiont (HSB), mesobiont (MB), hemieubribous (HBB) and eurybiotic (EB)[16]. For the ecological-geographical analysis of vegetation, reports on the flora of the KBR were used [17]. Life forms of plants were determined by Raunkier's classification. The content of Mo and W in the soil and plant samples was determined by the method of mass spectrometry with inductively coupled plasma. The degree of resistance of individual plant species to heavy metals was assessed by the coefficient of biological absorption (CBA) and the translocation coefficient (TC).

3. Results and discussion

In the communities of old-age terraces there are grass-paniculate-wormwood and herb-paniculate-wormwood communities dominated by *Artemisia scoparia* Waldst., herb-wormwood dominated by *A. austriaca* Jacq. and *A. marschalliana* Spreng. Due to the high floristic saturation, the total projective cover of herbaceous plants reaches 60 %. On the middle-aged terraces, a ruderal-natural grass and forb community there is a projective cover of 37 %. Among the cereals are most abundant *Festuca valesiaca* Gaudin, *Elytrigia repens* (L.) Nevski, *Melica transsilvanica* Schur, *Setaria viridis* (L.) P. Beauv., *Achnatherum caragana* (Trin.) Nevski. Herbs represented with xerophilic and ruderal species (*Verbascum thapsus* L., *Cichorium intybus* L., *Achillea millefolium* L., *Pyrethrum partenifolium* Willd., *Scabiosa bipinnata* K. Koch, *Tussilago farfara* L. and others). 13 % of monodominant communities were found on young terraces with herbaceous plants *Tussilago farfara* L. and *Cirsium vulgare* (Savi) Ten. Bushes of sea buckthorn (*Hippophae rhamnoides L.*), planted for the purpose of primary reclamation, are in a depressed state and dry massively.

The floristic composition of the studied ecotopes includes 71 species of vascular plants. The plant community of old-age terraces is characterized by the most equal distribution of abundance by taxa, relatively high species richness and richness compared to middle-aged and young terraces. The total projective cover of Tl is 1.3–2.8 times higher than that of Tm and Tu (table 1).

Table 1. Characteristics of plant cover of phytocenoses at different stages of succession

Terraces	Species richness, pcs.	Species saturation, pcs. / m^2	The total projective cover , %	E	H
Tu	19	6.5±1.8	39.0	0.52	2.23
Tm	53	18.6±2.4	84.1	0.65	3.49
Tl	62	24.1±3.2	108.8	0.76	4.50

The shape of the "dominance – diversity" curves indicates an increase in biodiversity and an increase in the dominance of species in the successive age range due to a decrease in the stress load of ecotops. The low S-shaped curve of the "dominance – diversity" curve of Tl is close to Preston's log-normal distribution (most species of this ecotope have an average abundance; there are few dominant and isolated species). The rank distribution of species of ecotope Tu is approaching the geometric distribution of Motomura, which is a characteristic of poor low-species communities in the early stages of succession. The rank distribution of the ecotope types Tm is characterized by a low compliance of the geometric model, approaching a lognormal one.

Chemical contamination of the substrate leads to a decrease in the coefficients of similarity between communities and an increase in taxonomic diversity by reducing the species richness of the types and families and increasing the proportion of monotypic taxa [2]. Old age terraces of TTMCP have the greatest similarity in floristic composition with middle-aged terraces ($K_{jl-t} = 0.53$) and ground ecotope ($K_{jl-g} = 0.29$). The smallest similarities with the background ecotope have early-age terraces ($K_{ju-g} = 0.13$). Compared with the monitoring data of 2003 ($K_{jl-t} = 0.45$, $K_{jl-g} = 0.35$, $K_{ju-g} = 0.08$), an increase in the similarity of the floristic composition of old and middle-aged terraces, as well as early-age terraces and background, was found. Reducing the similarity of old-age terraces and background ecotope is associated with the suppression of the life potential of zonal species in conditions of high toxicity of the substrate and, as a result, weak germination of seeds and / or loss of the ability to form them. A number of authors have noted an increase in the frequency of mutations and anaphase anomalies, a change in the mitotic index of the root meristem cells and a delay in cell division at the stages of mitosis under the influence of tungstates, molybdenum and other heavy metals [18, 19].

A characteristic feature of the primary succession on a technogenic substrate is the regular changes in the dominant life forms – from terophytes to hemicryptophytes [20]. The initial stages of vegetation restoration on the terraces of TTMCP are mainly formed by hemicryptophytes and terophytes. With an increase in the duration of succession, the proportion of terophytes decreases by 23.17 %, and that of chamephytes – by 2.5 %, hemicryptophytes and phanerophytes – increases by 14.3 and 6.45 %, respectively (table 2).

Table 2. Characteristics of plant cover of phytocenoses at different stages of succession

Life form	35–50 years (Tl)		25–34 years (Tm)		15–24 years (Tu)	
	pcs.	%	pcs.	%	pcs.	%
Phanerophytes	4	6.45	2	3.77	0	0.00
Chamephytes	5	8.06	4	7.55	2	10.53
Geophites	3	4.84	2	3.77	0	0.00
Hemicryptophytes	35	56.45	29	54.72	8	42.10
Terophytes	15	24.20	16	30.19	9	47.37
Species richness	62	100.00	53	100.00	19	100.00

The flora of the ground area and the pioneer communities of the terraces is represented by 17 geo-elements. Extensive ranges (Pluriregional, Holarctic and Palaearctic) have more than a third of species (34.95 %). The most widespread in the vegetation cover are the Palearctic, Caucasian and South Siberian geo-elements (26.52; 21.70 and 14.47 %, respectively). The participation of species with a wide ecological amplitude (46.67 and 42.31 %, respectively) increases in the plant communities of Tu and Tm compared with more favorable conditions of Tl (35.49 %). In the pioneer communities of Tm and Tu, the depletion of the geographic spectrum of species was observed, represented by 13 and 8 geo-elements, respectively.

In the succession age series, mesobiont and hemieubribo species are most prevalent. The proportion of species with a low potential ecological valency (SB, HSB) during succession increases from 9.5 to 32.2 %, while species with a wide range of adaptation to adverse edaphic conditions (HBB, EB) decreases from 40 to 31 %. The specific weight of mesobionts with intermediate ecological valence is highest in the early stages of succession (50.6 %), in the later stages its decrease to 35 % was noted.

The proportion of species with deviations from environmental comfort factors increases in the age succession series. The greatest deviation from comfortable conditions at all levels of the terraces is marked by edaphic factors (table 3).

Table 3. The proportion of species with deviations of environmental factors from comfort, %

Environmental factors	Tl	Tm	Tu
Thermal mode	14.5	9.4	-
Continentality	12.9	5.7	-
Soil moisture	12.9	13.2	10.5
Soil acidity	41.9	30.2	15.8
Soil nitrogen availability	45.2	35.8	15.8
Soil salt regime	45.2	41.5	42.1

Species with no deviations from ecological regimes (*Artemisia scoparia* Waldst., *Cichorium intybus* L., *Taraxacum officinale* F.H. Wigg., *I. germanica* L., *Phleum phleoides* (L.) H. Karst., *Plantago major* L., *P. media* L., *Ambrosia artemisiifolia* L., *Melilotus officinalis* (L.) Pall., *Scabiosa bipinnata* K. Koch, *Ephedra procera* Fisch. et Mey., *Echium vulgare* L.), noted in the plant communities of middle-aged and early-age terraces, show no visible signs inhibition of ultrahigh Mo and W concentrations in the substrate.

In the tailing substrate, the average content of Mo and W was 112 and 230 μg/g, respectively. The translocation coefficients of the species studied with respect to Mo and W (CT>1) indicate the triggering of the protective function of the roots, which accumulate most of the elements entering them. The ability of plants to the biological absorption of Mo (CBA-1) indicates the possibility of their use in phytoremediation of soils contaminated with molybdenum. Biological absorption of W does not depend on the gross content of the element in the substrate (Table 4). High concentrations of tungsten in the roots indicate that the element is easily accessible to plants in alkaline environmental conditions, favorable for the transition of tungsten into mobile and easily accessible forms of migration for plants

[21].

In general, in the succession age series of technogenic ecotopes with a toxic substrate, a regular change of the abiotic S-model of plant community organization (patients of extreme conditions) at the initial stage of succession (young terraces), R-model (segetal and ruderal groups) on middle-aged terraces and transition is observed to CRS models on lower terraces [22]. However, the speed of this process is strongly inhibited.

According to the results of monitoring of vegetation cover, it was established that the dam overgrowth proceeds according to the type of primary succession. In accordance with the polyvariant model of primary succession [4], the pioneer stage (first stage) still continues at the upper level of terraces (Tu), which is characterized by the settlement of pioneer plants (one- and two-year grasses with short root system, one- and two-year grasses with perennial long root system), low species diversity and general projective cover. On the terraces of the middle level (Tm), the second stage of pioneer succession continues (transitional "cereal" stage). In the transit transeluvial ecotope of the middle part of the slopes, grass-forb and forb-grass communities are noted, they contain long-grass grasses, high-grain cereals; species diversity and total projective cover increase. In the transeluvial-accumulative ecotope of the lower old-age terraces, a third (transitional "shrub" stage) is observed, which is characterized by the gradual formation of natural phytocenoses and shrub layer, which contributes to the accumulation of leaf litter and the formation of litter. The rate of transition from initial to subsequent stages of primary succession is constrained mainly by the high toxicity of the substrate.

Table 4. Content in aboveground (a) and underground (u) plant phytomass (C, μg / g dry matter), biological absorption (CBA) and translocation (CT) of Mo и W

Indicators	*Artemisia scoparia*	*Phleum phleoides*	*Melilotus officinalis*	*Taraxacum officinale*	*Inula germanica*	*Echium vulgare*
Ca (Mo)	32.12±0.12	37.94±4.01	74.08±7.83	43.66±4.62	49.11±3.08	53.23±4.75
Cu (Mo)	63.46±0.30	84.72±4.73	118.47±10.25	94.46±6.44	132.56±11.47	87.54±10.61
CT (Mo)	1.96	2.23	1.60	2.16	4.55	2.23
CBA (Mo)	1.18	1.11	1.76	1.35	11.73	12.38
Ca (W)	0.34±0.11	0.46±0.05	0.51±0.07	0.29±0.05	0.43±0.08	0.38±0.09
Cu(W)	0.59±0.08	0.63±0.09	0.94±0.10	0.67±0.05	0.84±0.30	0.96±0.85
CT (W)	1.73	1.37	1.24	2.31	1.95	2.53
CBA (W)	0.008	0.007	0.006	0.005	0.011	0.010

According to the results of monitoring the vegetation cover of the TTMCP tailing dump, it was established that the dam overgrowth proceeds according to the type of primary succession, aimed at the gradual replacement of ruderal plants with species of natural flora and increase in species diversity. With an increase in the succession, changes in the dominant life forms, a decrease in the proportion of species with wide ranges, an increase in the number of hemievribionts and mesobionts were noted. Species with no deviations from ecological regimes growing in ecotopes of young and middle-aged terraces are characterized by indifference to chemical pollution of the habitat and are promising for phytoremediation of soils contaminated with molybdenum. Due to the extremely low self-healing potential of vegetation on the toxic substrate of the TTMCP tailing dump, it is necessary to carry out both biological and technical reclamation for the restoration of vegetation cover.

References

[1] Goleusov P V 2012 Features of the reproduction of the resource characteristics of herbaceous phytocenoses in anthropogenically disturbed ecosystems of the forest-steppe zone *Sci. statements of Belgorod State University Ser. Natural Sci.* **3(122)18** 124–30

[2] Zhuykova T V, Meling E V, Kaygorodova S Yu, Bezel V S and Gordeeva V A 2015 Features of soils and grassy plant communities in the conditions of technogenesis in the Middle Urals *Ecology* **3** 163–72

[3] Gusev A P 2015 Assessment of the potential of self-healing of vegetation in man-made landscapes

IOP Conf. Series: Materials Science and Engineering **663** (2019) 012052 doi:10.1088/1757-899X/663/1/012052

(by the example of the south-east of Belarus) *Rus. J. of Applied Ecology* **3** 8–12

[4] Sumina O I 2014 Primary successions in quarries as a full-scale model for studying the processes of formation of terrestrial ecosystems *Theoretical and applied ecology* **1** 40–4

[5] Reid N B and Naeth M A 2005 Establishment of a vegetation cover on tundra kimberlite mine tailings: 1. A Greenhouse Study *Restoration Ecology* **13(4)** 594–601

[6] Korznikov K A 2016 Vascular plants of spontaneously overgrown areas of the alluvial gold deposit development (Langeri River, Central Sakhalin) *Tomsk State University Bull. Biology* **4(36)** 62–78

[7] Zalikhanov M Ch, Kolomyts E G, Sharaya L S, Tsepkova N L and Surova N A 2010 *Alpine ecology in models* (Moscow: Science) 487 p

[8] Tamakhina A Y, Dzakhmisheva I Sh, Beslaneev E V and Gadieva A. A. 2018 Bioaccumulation of Heavy Metals by Medicinal Plants of the Inula Genus *J. of Pharmaceutical Sci. and Res.* **10(5)** 1263–6

[9] Gumaelius L, Lahner B, Salt D and Banks J A 2004 Arsenic hyperaccumulation in gametophytes of P. vittata: a new model system for analysis of arsenic hyperaccumulation *Plant Physiol.* **136** 3198–208

[10] Meagher R B 2000 Phytoremediation of toxic elemental and organic pollutants *Current Opinion in Plant Biology* **3(2)** 153–62

[11] Mendez M O, Maier R M 2008 Phytostabilization of Mine Tailings in Arid and Semiarid Environments – An Emerging Remediation Technology *Environ Health Perspect* **116(3)** 278–83

[12] Baker A J M and Brooks R R 1989 Terrestrial higher plants which hyperaccumulate metallic elements – A review of their distribution, ecology and phytochemistry *Biorecovery* **1(2)** 81–126

[13] Kochian L V, Pence N S, Letham D L D, Pineros M A, Magalhaes J V, Hoekenga O A and Garvin D F 2002 Mechanisms of metal resistance in plants: aluminum and heavy metals *Plant and Soil* **247** 109–19

[14] Assunção A G L, Schat H and Aarts M 2003 Thlaspi caerulescens, an attractive model to study heavy metal hyperaccumulation in plants *New Phytologist* **159(2)** 351–60

[15] Gurbanov A G, Bogatikov O A, Vinokurov S F, Karamurzov B S, Gazeev V M, Leksin A B, Shevchenko A V, Dolov S M and Dudarov Z I 2015 Geochemical assessment of the environmental situation in the region of the Tyrnyauz tungsten-molybdenum combined plant (Kabardino Republic of Balkaria, Northern Caucasus): sources of environmental pollution, impact on the state of the territory and rehabilitation measures *Reports by the Academy of Sci.* **464(3)** 328–33

[16] Zhukova L A, Dorogova Yu A and Turmukhametova N V 2010 *Ecological scales and methods for analyzing the ecological diversity of plants* (Yoshkar-Ola: Mari State University Publishing House) 368 p

[17] Shkhagapsoev S X 2015 *Vegetation cover of Kabardino-Balkaria* (Nalchik: "Tetragraf" LLC) 352 p

[18] Reutova N V, Dreeva F R, Reutova T B and Shevchenko A A 2015 Study of the genotoxic effect of tailings of the mining and processing plant, heavy metals *News of the Kabardino-Balkarian Sci. Center of the Russian Academy of Sci.* **1(63)** 112–6

[19] Gazizova N I, Petrova N V and Karimova F G 2013 Effect of tungstate on the growth of pea roots and protein phosphorylation on tyrosine *Plant physiology* **60(6)** 819–27

[20] Shpilevskaya N S. 2017 Primary succession on technogenic sandy substrate (using the sand mining quarry as an example) *Bull. of Polesye State University Ser. of Natural Sci.* **1** 26–30

[21] Yurgenson G A, Solodukhina M A and Smirnova O K 2009 To the problem of biological absorption of tungsten, molybdenum and arsenic by plants on the example of wormwood Gmelina *Bull. of the Transbaikalian Center of the Rus. Academy of Natural Sci.* **1** 16–21

[22] Mirkin B M and Naumova L G 2015 The concept of the plant community: history and current state *J. of General Biol.* **76(1)** 63–76

IOP Conf. Series: Materials Science and Engineering **663** (2019) 012053 doi:10.1088/1757-899X/663/1/012053

Ecological and economic aspects of negative impact of transport in large cities

T M Teziev[1], O G Beriev[2], E K Samrailova[3], B S Cidaev and S Ch Savkhalova[1]

[1]North Ossetian Republican Organization of All-Russian Trade Union of Education, 2, Dimitrova str., Vladikavkaz, 362040, Russia
[2]Geophysical Institute of Vladikavkaz Scientific Center of the Russia Academy of Sciences, 93a, Markova str., Vladikavkaz, 362002, Russia
[3]Russian University of Transport (MIIT), 9, Obrazcova str., Moscow, 127994, Russia
[4]North Caucasian Mining and Metallurgical Institute (State Technological University), 44, Nikolaeva str., Vladikavkaz, 362021, Russia

E-mail: tteziev@yandex.ru

Abstract. This paper discusses a variety of methodological approaches to economic and environmental assessment of the negative impact of transport in urban areas. The role of the ecological crisis, which is growing in large cities, is noted. There is a trend of a sharp increase in the number of vehicles in the cities. The need for a comprehensive solution of economic and environmental problems based on the systematic collection, analysis and synthesis of environmental information has been formulated. The foundation for economic and environmental assessment of the negative impact of transport in urban areas is the use of a simulation model based on software package AnyLogic PLE. It is noted that the solution of environmental problems of urban transport is a set of program measures of different nature with the aim to minimize the load on the biosphere. The proposed economic mechanisms and criteria were used in economic and mathematical models, which determined the state of urban ecological systems, and made it possible to make necessary assessment of their sustainability.

1. Introduction

Environmental protection is one of the most urgent problems of our time. Scientific and technical progress, which contributes to new technological systems, ultimately leads to the fact that the harm caused by man to nature, acquires dangerous dimensions that threaten the very biological basis of the existence of our society. Globalization processes of the modern world in all aspects are already in the middle of the last century brought to the fore the problem of environmental safety. Moreover, this problem was often vivid at the regional level in a highly urgent context with the effects of cultural, economic, and military opposition to communities.

The social need for qualitative ecological knowledge, practical application of the results of the environmental studies, was formed in the context of the global environmental crisis caused by anthropogenic factors.

The deterioration of the ecological situation undermines social needs and contradicts the interests of the population, causing social and economic tensions at the regional and national levels. Without managing social and ecological processes and lack of tasks to reduce the environmental capacity of the

IOP Conf. Series: Materials Science and Engineering **663** (2019) 012053 doi:10.1088/1757-899X/663/1/012053

economy, our society can make the environment unsuitable for human existence, rather than healing the environment. It can bring destructive social processes that can interrupt the onward development of civilization.

For many decades, the economic development of Russia to a certain extent was carried out at the expense of the environment and human health. The environmental crisis, which manifests itself in varying degrees, arose as a result of the neglect of environmental safety of society and coincided with an economic slowdown in development [1].

2. Main sources of pollution

Air emissions from various polluting sources are a powerful factor of the negative impact on humans and the surrounding environment.

Contemporary environmental problems are caused by a lag of economic thoughts, neither the classics nor the theoreticians of the XX century. The expenses of society for the restoration of natural environment are not taken into account in the schemes of public reproduction. The attempts are being made to develop fundamentally new concepts of development.

The modern economic world is built on theoretical schemes of the classics of bourgeois and communist political economy as amended by scientists of the XX century. This type of development is called the *technogenic type of economic development*. Its characteristic features are the absence of social reproduction of the economic mechanism of reproduction for consumed natural resources and changeable conditions; the use of artificially created means of production without regard to environmental restrictions; the rapid and depleting use of non-renewable natural resources; overexploitation of renewable resources at a rate exceeding their recovery.

Technogenic type of production is characterized as nature-intensive or nature-depleting.

There are many models of a man-made type of development. There are two generalized models:

1. Frontal economy;
2. The concept of environmental protection.

Frontal economy is based on the principle of unlimited natural resources. With the growth of productive forces, population size, deep destabilization of the environment has become apparent; the cost (of resources) of life has sharply increased. Now every 24 hours 35 thousand people die from hunger, 20 thousand of them are children under 5 years old.

The concept of environmental protection. Some developed countries (100 countries) began to create the bodies for nature protection from the negative impact of production. However, the production formula remained the same, focused on expanding reproduction, and the most important the economic component. The environmental component is considered as a factor to reduce economic efficiency of social production.

Negative consequences of the technogenic type of economy are land desertification (aridization), deforestation; lack of natural raw materials; greenhouse effect; depletion of the ozone layer; acid rains; shortage of fresh water, ocean pollution, extinction of many species of flora and fauna. If people do not solve these problems, it will lead to the extinction of man as a species [2].

The main polluters are emissions from industrial enterprises and vehicles. Atmospheric pollution occurs mostly in cities than in the countryside. This is due to the fact that there are much fewer industrial enterprises and vehicles in rural areas.

The main sources of pollutants that enter the air are man-made, namely industry and motor vehicles. Emissions of pollutants into the atmosphere are mainly volatile gaseous chemicals. As the results of environmental studies, air emissions and atmosphere pollution are one of the most powerful and permanent factors of negative impact on humans and the surrounding environment.

In large cities, industry is represented by such fields as non-ferrous metallurgy, mechanical engineering and metalworking, electric power industry, fuel industry, timber and woodworking industries, and the production of building materials. A large and significant contribution to air pollution is also made by vehicles, the number of which is constantly increasing mainly due to large cities.

IOP Conf. Series: Materials Science and Engineering **663** (2019) 012053 doi:10.1088/1757-899X/663/1/012053

As a result of industrial activitities, the atmosphere is polluted by carbon monoxide, lead, sulfur dioxide, nitrogen oxides, hydrocarbons, including benzo (a) pyrene, photooxidants formed under conditions of intense solar radiation.

At the present stage of overcoming the economic crisis and improving the quality of people's life, the state of the environment and environmental benefits in the system of social reproduction should be more accurately and specifically evaluated. The preservation and reproduction of environmental goods are perceived as the most important part of the process of economic development, ensuring the growth of well-being, improving the quality of work and life of people to the level of developed countries; maintaining the reproduction of healthy generations; recovery of mining sites and recycling of waste for their recycling. With modern development of urbanized areas, technogenic components (especially motor vehicles) have a very negative impact on the environment and human health [3].

The concentration of a significant part of the population in urban areas has led to the need for the earliest possible solution to the problem of protecting people's health and preventing the occurrence of environmental disasters. Now it is becoming more and more obvious the interrelation of production, natural and social processes in society. There is a merging of objects of human economic activity, its habitat and social and economic factors into the complex interconnected systems, developing according to peculiar, not sufficiently studied laws. The health of the population is 25...50 % dependent on lifestyle, 20...40 % on the environment, 15...20 % on genetic factors, and only 10 % on the activities of health services. If we adhere to the position that a disease is a disturbance in a person's normal life due to the conditions of its existence in the environment, then it is an environmentally caused phenomenon and, therefore, can be used as an indicator of environmental quality, and various aspects of it. A large group of diseases is directly related to the natural environment. These are pathological conditions caused by local geochemical features, climatic factors, regional biocenological characteristics, variability of pathogens of infectious diseases in a specific host (for example, influenza virus), etc. [2]

Another large group of pathological processes is associated with *production activities*, with a limited zone of influence and exposure to a small contingent of people directly working in adverse conditions (for example, it has been established that significant concentrations of dioxins in humans have a suppressive effect on endocrine, hematopoietic and immune system functions). The third group occurs under the influence of *social causes* and *intrapopulation interaction* (mainly, it is a psychogenic pathology: cardiovascular, endocrine diseases, mental disorders). This pathogenic factor may be ethnic (for example, dietary patterns — the distribution of raw food supply of fish and meat) and genotypic (genetically determined differences in oxidation of alcohol) characteristics of populations.

Finally, an extensive group of diseases, as well as an increase in overall morbidity, often without sufficient factual justification, are associated with *intense pollution* of water and air (up to 60 % of the Russian population lives under conditions of a constant excess of MAC in air) and soil as a result of human production. The water treatment system is not effective to prevent waterborne diseases (in particular, caused by poliomyelitis viruses, hepatitis, etc.). Each disease is characterized by its own characteristics (spatial distribution, prevalence in the population, spectrum of nosological forms), but all of them are causally related to environmental factors. The primary task in ensuring the health of the population and increasing the duration of active life of people is a deep study of the correlation links between medical-biological and anthropogenic factors.

A powerful complex of harmful and highly toxic environmental factors that have arisen because of the scientific and technological revolution influence on a modern person. At present, 4 million toxic substances are registered in the external environment and their number increases by 6 thousand annually. Over the last decades alone, more than 1 million tons of nickel, about 1 million 1 cobalt, more than 600 thousand tons of zinc were emitted into the atmosphere, 1.5 million tons of arsenic and the same amount of silicon. A particularly large amount of harmful substances enters the environment in industrialized countries, in large cities. The main anthropogenic sources of the increase in carbon dioxide in the atmosphere are energy and transport, running on organic fuel.

The situation in the modern world is that not a single nation, not a single country can single-handedly emerge from an impending ecological catastrophe. It is not even visible to them, except in its local manifestations.

Industrial development has created not only a special world of technology, but also the "urban world". If in 1800 only 5 % of the population lived in the cities (50 million people), then by 2020 the number of citizens should have increased by more than 10 times and reach 53 % (3.4 billion people). Although cities with dynamically developing large-scale industrial and energy-intensive production as a whole occupy only 0.5 % of the Earth, they have a destructive effect on nature. Explosive urbanization (after 1950 about 2.3 billion new citizens on the planet) enabled not only to concentrate population and technology in cities, but radically transform human activities, change the quality indicators of the whole way of life, people's lifestyle [2].

In recent years, the population has had the opportunity to purchase personal vehicles under affordable credit conditions or to rent it by leasing. Moreover, there are many private business carriers.

A significant growth of vehicles has led to a decrease in the average speed on the carriageway, traffic jams, traffic accidents, increased release of toxic substances into the atmosphere, increased noise and vibration from passing vehicles, increased electromagnetic radiation from radio electronic road systems, and pollution of the urban area and the adjacent aquatic environment with petroleum products, as well as problems with parking for motor vehicles, etc. It is obvious that the growth of motor transport to the city highways created acute economic and environmental problems that require effective comprehensive scientific and practical solutions [3].

There is no mechanism to control emissions of motor vehicles, which would be based on economic instruments of environmental incentives. Such a toolkit could be based on an economic and environmental assessment of the negative impact of transport in urban areas [1].

The relevance of such tools is difficult to overestimate due to the need to implement a systematic and integrated approach to planning and implementation of environmental protection measures, taking into account all the major environmental, economic and social factors in traffic organization in different time and seasonal intervals. Administrations of large cities are taking measures to prevent harmful emissions into the environment; however, their number does not decrease.

Therefore, a complex solution of economic and environmental problems is vital based on systematic collection, analysis and summarization of environmental information. It is necessary to investigate economic aspects that clearly define measures to protect urban areas from dangerous concentrations of harmful substances by transport, optimizing the choice of solution options based on criteria of environmental and economic efficiency, formation of programs and mechanisms that implement the chosen solution. Separate implementation of individual measures of environmental improvement in the processes of functioning of urban transport is contrary to the current environmental situation, determined by the extreme scale of the impact of vehicles on the urban biosphere [3].

Since traffic flows in urban traffic create environmental pollution, it is possible to determine measure effectiveness to protect this environment at a streaming level [7]. Optimization will affect both individual elements of the urban transport system and the measures themselves that limit environmental damage from the scale of the system being optimized.

Within the framework of economic and environmental safety of motor transportation system, it is proposed to model the levels of environmental pollution by motor transport using the theory of traffic flow [5]. The main characteristics of the traffic flow, which can significantly affect the efficiency of vehicles in urbanized areas, include the speed and intensity of movement of vehicles [8].

Therefore, the forecasts on speeds and intensities of urban traffic are very relevant in the process of developing measures to improve the performance of vehicles, improve the road network and the environment in cities.

Fuel consumption by motor transport can be directly used as a criterion for the quality of urban traffic, due to the fact that this indicator is associated with efficiency, speed indicators and indirect release of toxic components along with exhaust gases.

Fuel costs can be translated into value terms, which is a function of changes in the dynamic state of traffic and the level of costs to improve urban traffic [5–6].

Protection against noise and vibrations from urban traffic flows is provided by the construction of anti-noise screens along urban highways, tunnels, bypass racks, equipment, sound-absorbing elements of town houses, etc. When exceeding the permissible noise level by motor transport, health indicators of the population deteriorate and labor productivity drops.

3. Simulation model

The basis for economic and environmental assessment of the negative impact of transport in urban areas is a simulation model based on AnyLogic PLE software package, which will display the movement of vehicles in various traffic conditions [4].

The initial data for simulation:

- a map of urban area with streets defined for the carriageway (Table 1);
- traffic flows with speeds from minimum to maximum and traffic intensity from minimum to maximum;
- dependence of the emissions of harmful substances (carbon monoxide, sulfur oxide, lead oxide, mercuric oxide, etc.) per unit of time on the speed and intensity of movement of vehicles;
- values of maximum permissible norms for emissions of harmful substances per unit of time;
- intensity of natural recovery in certain areas of the environment from emissions of harmful substances that exceed the maximum permissible norms.

Figure 1. Traffic flows in urban areas modeled in AnyLogic PLE [4]

As a result of modeling, all possible situations are assessed with the dynamics of exceeding the maximum permissible norms of emissions of harmful substances into the environment of urbanized territories and the dynamics of subsequent natural restoration of the environment from pollution.

Further, it is necessary to determine the implementation cost for various measures of economic, legal, administrative, social and other nature, compensating for the damage from the maximum permissible pollution by harmful substances or preventing such pollution due to restrictions on the use of motor vehicles in urban areas moving on fuel from oil products [1].

An example of the negative impact of transport on the environment and human health can be the fact that, on average, with mileage of 15 thousand km per year, each car burns 2 tons of fuel and about 26–30 tons of air, including 4.5 tons of oxygen, which is 50 times more people need.

At the same time, a car emits into the atmosphere (kg/year): carbon monoxide – 700, nitrogen dioxide – 40, unburned hydrocarbons – 230 and solids – 2–5. In addition, many lead compounds are emitted due to the use of mostly leaded gasoline. The studies have shown that in houses located near the main road (up to 10 m), residents suffer from cancer 3-4 times more often than in houses remote from the road at a distance of 50 m [7].

The total implementation cost of various measures to compensate for environmental damage can be divided into road users in the form of environmental municipal tax on polluting vehicles.

To reduce the level of air anthropogenic pollution of the urban basin and its negative effects, current environmental protection methods should be improved to implement new organizational and economic policies to protect the urban biosphere and stimulate economic and social development.

Qualitative and quantitative changes in the road transport system of cities will occur when using the economic mechanism to manage this system to reduce the cost to eliminate pollution effects.

The solution for the environmental problem of urban transport could be a set of program measures to minimize the burden on the biosphere. An important part is a program for the effective organization of urban traffic, progressive methods of automatic regulation at intersections to reduce the number and duration of stopping vehicles, eliminating traffic jams and jams, and increasing the average speed of traffic, also requires the transfer of pedestrian crossings or above the roads [3, 5, 6].

With centralized traffic control by means of traffic lights and sign regulation, it is possible to achieve the increase in the average speed of traffic [6]. At the same time, in the future, there is a gradual transition of urban transport to non-petroleum-based fuels: natural gas, electricity, which will eliminate the problem urgency of environmental pollution.

Perspective automation of urban traffic should include full control of each car from the initial to the final point of its movement based on automatic (unmanned) control algorithms, including Internet application for traffic, the formation of an optimal route coordinated with electronic routes of all road users according to the minimum route indicators and limited travel time. Thus, the planned intensity of traffic will be distributed along urban highways, taking into account their throughput capabilities and the required average traffic speed.

The economic component for the regulation of environmental transport safety during its evolutionary development should have a system of economic incentives for environmental activities; the improved system of carriage charges for pollution; prices for goods and services in cities, formed on the basis of environmental requirements; stimulation of processes of centralized traffic control with the growth of environmental vehicles [1].

Each new change in the ecological status of the urban transport system must be tested on a simulation model to achieve the required conditions of economy and environmental safety for road traffic [6].

The proposed environmental and economic approaches and models, by the example of reducing automobile noise, are aimed at optimizing environmental protection measures of the urban biosphere to create acoustic conditions acceptable to the population.

The developed economic mechanisms and criteria were used in economic and mathematical models, which determined the current state of urban ecological systems after which the necessary assessment of their sustainability was made.

The constructed economic and mathematical model to assess the economic efficiency of the developed environmental measures was based on the economic indicators of absolute and relative efficiency, taking into account a discount factor, which determined forecast characteristics for a set of measures to protect the urban air environment from harmful emissions.

In the process of model application of known statistical methods to predict the state of the urban air environment, the interrelations of the main economic factors that caused changes in the city environmental and economic indicators were identified, evaluated and investigated. A set of environmental protection measures for environmental rehabilitation of the city's motor transportation system has been developed with minimal additional costs for this system functioning [9].

4. Conclusions

The main emission sources in the atmosphere of large cities are industrial enterprises and road transport. Emissions from stationary sources into the atmosphere of megacities are growing over the last period.

The proposed economic and mathematical model can predict pollution indicators of the urban air basin. Based on the correlation analysis, this model is capable of generating a forecast of the state of the urban air environment on the results of the obtained forecasts in the city's environmental activities.

It is necessary to combine economic, organizational, legal and educational measures, enhance the role of local authorities and administration in environmental protection, public environmental organizations, and personal initiative of citizens based on general public awareness of the environment, planned design and construction of large economic objects, education to respect nature and improve the ecological culture of the population.

References

[1] Zolotovskaya N K 2000 *Economic regulation of the level of negative impact of transport in urban areas* PhD thesis ... in economy sci. (Rostov on Don) 148 p

[2] Samrailova E K 2006 *Social and environmental problems of inter-regional conflicts* Monograph (Moscow: Publishing house "Paleotype") pp 7–29

[3] Bityukova V R 2002 Principles and methods of integrated assessment of the ecological state of the urban environment *Problems of urbanization at the turn of the century* ed A G Makhrova (Smolensk: Oikumena) pp 189–98

[4] *AnyLogic PLE* Retrieved from: https://www.anylogic.ru/blog/besplatnaya-versiya-anylogic-dlya-obrazovatelnykh-tseley/ (date of access 09.11.2018)

[5] Gasnikov A V, Klenov S L, Nurminsky E A, Kholodov Ya A and Shamray N B 2010 *Introduction to mathematical modeling of traffic flows* ed by A V Gasnikova (Moscow: Publishing House of MIPT) 363 p

[6] Markuts V M 2008 *Traffic flows of highways and city streets. Part 1 Intensity and safety of car traffic, traffic intersection capacity. Practical application* (Tyumen) 108 p

[7] Beriev O G and Teziev T M Dynamics of emissions in the atmosphere of RNO-Alania for the period 2006–2016 Bulletin MANEB **23(4)** 11–9

[8] Zhivoglyadov V G 2005 Theory of traffic and pedestrian traffic *Bull. of universities. Nor.-Caus. Region* 1082 (Rostov on Don: Publishing house)

[9] Galachieva S V, Sokolov A A, Sokolova O A and Makhosheva S A 2018 Assessment system for sustainable development of regional economic complexes of mountain territories *Sustainable development of mountain territories* **3(37)** 329–36

IOP Conf. Series: Materials Science and Engineering **663** (2019) 012054 doi:10.1088/1757-899X/663/1/012054

Avalanche and mudflow situation in the Republic North Ossetia-Alania

O G Beriev[1], T M Teziev[2], E K Samrailova[3], B S Cidaev[4] and S Ch Savkhalova[2]

[1]Geophysical Institute of Vladikavkaz Scientific Center of the Russia Academy of Sciences, 93a, Markova str., Vladikavkaz, 362002, Russia
[2]North Ossetian Republican Organization of All-Russian Trade Union of Education, 2, Dimitrova str., Vladikavkaz, 362040, Russia
[3]Russian University of Transport (MIIT), 9, Obrazcova str., Moscow, 127994, Russia
[4]North Caucasian Mining and Metallurgical Institute (State Technological University), 44, Nikolaeva str., Vladikavkaz, 362021, Russia

E-mail tteziev@yandex.ru

Abstract. Snow avalanches and mudflows cause great damage to federal and domestic roads, economic facilities and, most importantly, they are dangerous to people's lives within the mountainous territory of the Republic of North Ossetia-Alania. Avalanches represent the greatest danger on the northern section of the Transcaucasian highway from the village of Buron to the Northern portal of the tunnel through the main Caucasus ridge, which is 28.3 km long. The number of avalanches is directly dependent on the amount of snow that fell during the winter. During the season of 2011–2012 due to the large amount of snow that fell, there were 371 avalanches (including 270 forcible avalanches and 101 avalanches by self-rising). During the anomalously low snow season of 2013–2014, there were only 13 avalanches, which came off. Therefore, that is the lowest index since the avalanche period of 1986–1987. During the study period, the road was closed from 1 to 17 times per season, which also depends on the number of avalanches. Mudslides pose a great danger to the population of the republic, roads and objects of the national economy. There are mudslides after heavy precipitation on the territory of the republic.

1. Introduction

Snow avalanches and mudflows are one of the hazards in the mountainous areas and pose a greater threat to the lives of people and economic objects.

In the Republic of North Ossetia-Alania, avalanches on the northern section of the Transcaucasian highway from the village of Buron to the Northern portal of the tunnel through the main Caucasus ridge, which is 28.3 km long, represent the greatest danger.

To study avalanche and mudflow situations on the territory of North Ossetia-Alania, the materials of the annual reports of the Ministry of Emergency Situations of the Republic from the 2009–2010 season to the 2016–2017 season [1–8] were used. The data on avalanche situations are presented in Table 1.

The number of avalanches is directly dependent on the amount of snow that fell during winter. During the season of 2011–2012 due to the large amount of snow that fell, there were 371 avalanches (including 270 forcible avalanches and 101 avalanches by self-rising). During the anomalously low snow season of 2013–2014, there were only 13 avalanches, which came off. It is the lowest index since the avalanche period of 1986–1987. During the study period, the road was closed from 1 to 17 times per season, which also depends on the number of avalanches.

2. Formation of avalanches

Foehn winds have a significant effect on avalanche formation. The distribution and intensity of foehn winds in Greater Caucasus varies widely. The number of days with foehn winds is observed at settlement Nizhniy Zaramag – 17, Kobi – 15, and in the high-mountain settlement Kazbegi – 58. Thus, the number of days with foehn winds in the mountains increases dramatically when the height increase.

The melting of snow at foehn winds and avalanches are not so much due to the increase in temperature as the duration of foehn winds. The duration of foehn winds in the study area varies from several hours to 10–14 days. Foehn winds particularly frequent here in late February - early March.

The main factor in the accumulation of large amounts of snow on the leeward slopes in this area is the prevailing southwestern wind.

Table 1. Avalanche situation in Republic of North Ossetia-Alania from 2009–2010 to 2016–2017

Years	Avalanches, which came down			Maximum snow height (cm)	When road was closed during the season (days)
	total	self-rising	forcibly		
2009-2010	240	190	50	102	17
2010-2011	234	189	45	108	6
2011-2012	371	270	101	107	10
2012-2013	216	129	87	100	9
2013-2014	13	-	13	6	1
2014-2015	169	114	55	143	3
2015-2016	200	103	97	106	11
2016-2017	127	103	24	91	3

During the winter, there were 45 days with southwestern wind, which was blowing at the speed of 5–12 m/c at the altitude above 2500 m above the sea level. Wind gusts, reaching up to 20 m/c, are usually in December, January and February. Below the altitude 2000 m above the sea level, wind speed and its direction vary as well as the terrain peculiarities. Vegetation has a great influence on wind transfer of snow, for example, dense forest in Tsei gorge contributes to the accumulation of snow masses, while rare shrubs in the upper reaches of other gorges cannot significantly affect the nature of the wind transfer of snow.

The period of formation of sustainable snow cover as well as changes in its structure is greatly influenced by air temperature [9].

The threat of snow avalanches also exists on another important highway - the Georgian Military Road near the Cross Pass (Upper Lars checkpoint). In accordance with Russian and Georgian agreements, when there is a threat of an avalanche, the traffic through the pass is limited or terminated completely. In this case, there are up to several hundred heavy vehicles that are harmful to the environment due to long-running engines on the territory of the republic

Avalanches bring not only material damage to highways, but also they take people's lives. In 2008, a snow avalanche caused the death of four people in the area of Zaramagskaya hydroelectric power station.

The task to ensure avalanche safety in the northern part of TransKAM from the village of Buron to the Northern portal of the tunnel through the main Caucasian ridge is assigned to the Roki avalanche unit (RVPLO) under the North Ossetian militarized unit of the FASM of Roshydromet. Its base and location are in the village of Nar and on the North portal. There is also a weather station on the North portal that transmits weather reports and forecast avalanche dangers. A special road service, equipped with special road equipment, carries out snow clearing after avalanches.

Mudflows pose a great danger to the population of the republic, roads and objects of the national economy. As a rule, there are mudflows after heavy precipitation on the territory of the republic

North Ossetia-Alania is one of the most mudflow areas of Russia. Mudflows happen annually due to the contrasted relief, and significant precipitation [10]. In the mountainous part of the republic, there are 145 mudflow basins. The total area affected by mudflows is approximately 2165 square kilometers, which is 25 % of the total area of the republic. The average area of the mudflow basin is 29.4 square kilometers. The total length of the mudflow channels is approximately 792.7 kilometers; the average value is 11.4 kilometers. The largest area affected by mudflows in North Ossetia is in the valley of the Ardon river (937 square kilometers) [11].

Mudflows on the territory of North Ossetia can have not only a natural character, but also the so-called "hand-operated", for example, in case of emergency discharge of water from the reservoir of the Zaramagskaya hydroelectric power station [12]. Zaramagskaya hydroelectric power station itself also represents a great danger due to its location in a seismic zone [13]. During the study period, the following mudslides were noted.

In March 2010, Dzaurikau-Tskhinval gas pipeline, which supplies South Ossetia with Russian gas, was damaged by mudflows that came down on TransKAM. A powerful snow mud flow of more than 1000 cubic meters came down at 3:00 a.m. in the area of Nizhny Zaramag and covered the territory of Zaramag gas measuring station (GIS). As a result, the structures of the metering unit were damaged and two of the three sections of parallel technological branching of the GIS with a diameter of 159 mm each appeared under the obstruction. The destruction and ignition of the main gas pipeline with a diameter of 420 mm did not occur. There was no interruption in the supply of natural gas.

At 11:40 pm on July 16, 2011, 1 km below the Victoria Hotel, Tseyiskoe Gorge, Alagirsky District, North Ossetia-Alania, 2 mudflows descended with the roadway overlapping; there were no victims and injuries. Energy and water supply was not disrupted.

In 2012, the following mudflows were observed on the territory of the republic:

- the right side of the river Mamisondon, a nameless water stream 1 km above the river Khalatsa caused deformation of the technological road through the gas pipeline L=20 m, V – 5000 m^3;
- the river Kuatedon - road blocking on the Mamison L=20 m, V-3000 m^3;
- on June 18, the river Kasaikom, 71 km of TransKAM – road blocking L=150 m, V–2000 m^3;
- on August 23, the river Kasaikom, PK 52 on TransKAM – road blocking L=250 m, V – 1500 m^3; the river Andorrag, PK 438 on TransKAM – temporary overlap of the road section L=50 m, V – 1000 m^3;
- on July 6, the river Andorrag (the left tributary of the Ardon river in Mizur) - TransKAM overlap L=50-70 m, V – 1000 m^3;
- on August 20, Kolota water stream (the right tributary of the Fiagdon river) - Harisjin-Hilak road blocking L=30 m V–10 000 m^3; the river Lamardon (the left tributary of the Gizeldon river) – the bridge deformation at Lamardon settlement, V-5000 m^3; the river Fardon (the right tributary of the river Genaldon) – a mudflow descended into Saniban lake with no damage, V-10000 m^3; highway Gizel – Koban, PK 110,5 (the area of a new bridge) – deformation of a shore protection dam L=40 m, V – 1200 m^3; water stream Dzhanukadzhin PK 65,5 along the Georgian Military Road, V – 2000 m^3.

In 2013, the following mudflows were observed on the territory of the republic:

- on July 4, water stream Dallagkom (the left tributary of the river Zakka), V – 1000 m^3; on July 5, Kozidon river basin, starboard, 2 beams 2,5 km from the river mouth, V – 1000 m^3;
- on July 5, the river Kozidon, starboard, 2 beams 3,1-3,3 km from the river mouth, V – 2000 m^3; on July 8, the valley Kaikomkondon, starboard ~ 3,5 km from the river mouth, V – 1500 m^3;
- on July 5, the river Zemegondon, the left side ~ 2950 m from the river mouth , V – 2000 m^3; on July 1-2, the river Kasaikom, TransKAM (71 km), V – 5000 m^3 with overlap and deformation of the road, which length is 200 m;

- on July 19, the river Kasaikom, TransKAM (71 km), V – 1000 m^3 with overlap and deformation of the road, which length is 200 m;
- on July 21, Wills beam, the right side of the river Skazdon valley, V– 9000 m^3;
- on July 21, nameless beam to the north of Wills beam, the right side of the river Skazdon valley, V – 1000 m^3;
- on August 26, the river Tsatadon (headwaters) V – 9000 m^3; on September 24, water stream North Kasaik (the right source of the river), V – 1500 m^3.

In 2014, the following mudflows were examined on the territory of the republic North Ossetia - Alania:

- on June 6, Darg-Shuardon (the right tributary of the river Dzamarashdon, 3 km from the river mouth), V – 1500 m^3,
- on June 23, beams on the right side of the river Mamisondon (gas pipeline Dzuarikau - Tskhinval, 81,4 – 81,6 km), V - 200 m^3, 500 m^3, and 1000 m^3, the threat of deformation of the pipeline, the overlap of the technological road, which length is 50 m;
- on June 25, Bezimyannaya beam (Dzuarikau - Tskhinval gas pipeline, 83 km), V-3000 m^3, the overlap of the technological road, which length is 50 m;
- on June 27, the river Labagom (Dzuarikau - Tskhinval gas pipeline, 62 km), the overlap of the technological road, which length is 20 m in spring time;
- on September 2, beam on the left side of the river Genaldon, which is 8 km above the river mouth of the Kanidon, V – 2000 m^3 in spring and summer time; on September 2, beam on the left side of the river Genaldon 8,7 km higher of the river mouth Kanidon, V – 2500 m^3;
- on September 23, the river Vilsadon, 1,5 km from the river mouth, V – 2000 m^3 in summer time; on September 23, the left tributary of the river Vilsadon, 1,9 km from the river mouth, V – 1500 m^3;
- on September 25, beam on the right side of the river B. Dur-Dur, 0,4 km higher from the river mouth M. Dur-Dur, V – 2000 m^3, damage of forest land;
- on October 11, water stream Kolota (the right tributary of the Fiagdon river), 40 m of the road to Hilak mineral springs were destroyed;
- on October 11, the river Lartsikom (the right tributary of the Midagrabin river),V – 10000 m^3, the road to Midagrabinsk waterfalls, which is 100 meters long, was destroyed.

On April 15, 2015, as a result of a mudflow of 1500 m^3 on the river Bezimyannay (the right tributary of the river Mamison), 1 km above the river Khalatsa, there was the overlap of a 50 m long road; traffic on TransKAM was stopped for 20 hours; on June 14, the result of a mudflow of 3000 m^3 - the left side of the river Zakka, beam between the brooks Toa and Uallagk, there was the overlap of technological serpentine roads about 150 m long; on June 20, as a result of a mudflow of 1500 m^3 the southern outskirts of the settlement Mizur, 56 km of Dzuarikau-Tskhinval gas pipeline, the technological road 300 m long was deformed; on August 6, as a result of a mudflow from 4000 to 5000 m^3 on the river Kasaik, the right tributary of the river Ardon, there was an overlap of TransKAM with a length of 100 m, the traffic was stopped for 16 hours;

On August 6, as a result of a mudflow of 5000 m^3 on the river Kolota (the right tributary of the Bugultadon river, 2.5 km from the river mouth), there was an overlap, which is 15 m long.

On September 26, the river Bartui (the left tributary of the Karaug river) there was a mudflow of more than 10000 m^3; on September 30, as a result of a mudflow on the Skazdon river, which was 50000 m^3 the re was a deformation of a ski slope of 500 m, the water intake was destroyed, the right flank was 30 m long; on November 6, as a result of a mudflow at the settlement Badashta (the right tributary of the Ardon river, the south of the settlement Nuzal), TransKAM was blocked, the traffic was stopped for a day, the gas pipeline was deformed.

On June 15, 2016 at 7 p.m., after intense precipitation during two days, on the river Kasikak (the right tributary of the Ardon river, 71st km of TransKAM) there was a low-density mudflow (V - up to 2000 m^3) with predominance of water component. Pipe water passage was immediately clogged with debris, after which the stream broke out onto the road and rushed down right along the roadway, and deformed the asphalt and filled it with stone material for more than 200 m. Water gradually flowed down to the lower slope of the road, where there was a site for storing technogenic soil, on which large gullets were formed with the demolition of loose detrital material in the river Ardon.

In 2017, mudflows in the republic were represented by the following events:

- on July 5, the river Kasaikom (the right tributary of the river Ardon, 71 km of TransKAM), a mudflow (V-3000 m^3) blocked TransKAM L=100 m c with erosion man-made dump;
- on July 10, water stream Wills (the right tributary of the river Skazdon), there was a mudflow, V - 8000 m^3. As a result, the ski trail was damaged and blocked by detritus L=100 m. On the same day, there was a mudflow on the river Orakhkom (the left tribute of the river Tseidon), V- 5000 m^3, which destroyed hiking trail to the glacier L=300 m;
- on August 5, starboard of the river Urukh opposite the settlement Kussu, a mudflow, V - more than 10000 m^3 blocked the river bed Urukh. There was forest destruction in the flood plain, S=1 hectare;
- on October 20, Kolot beam (the right tribute of the river Fiagdon along 12 km), a mudflow, V - 5000 m^3, destroyed the highway Fiagdon – Khilak L= 40 m.

To provide anti-avalanche and mudflow hazards, it is necessary to use modern methods for detection of avalanche-prone and mudflow foci, including drones and other equipment. In places of constant avalanches and mudflows, we should build anti-avalanche galleries and tunnels on the roads.

3. Conclusions

Avalanches and mudflows in the mountainous territory of the Republic of North Ossetia-Alania is a threat to the lives of people, federal and inter-republican roads and economic facilities.

To prevent damage and ensure the safety of people from snow avalanches and mudflows, it is necessary to introduce modern methods of monitoring dangerous natural processes, as well as build avalanche galleries and tunnels on dangerous sections of roads.

4. References

[1] *Materials for the annual state report "On the state of protection of the population and territories of the Republic of North Ossetia-Alania from natural and man-made emergencies in 2010"* 2010 (Vladikavkaz) 196 p

[2] *Materials for the annual state report "On the state of protection of the population and territories of the Republic of North Ossetia-Alania from natural and man-made emergencies in 2011"* 2011 (Vladikavkaz) 194 p

[3] *Materials for the annual state report "On the state of protection of the population and territories of the Republic of North Ossetia-Alania from natural and man-made emergencies in 2012"* 2012 (Vladikavkaz) 197 p

[4] *Materials for the annual state report "On the state of protection of the population and territories of the Republic of North Ossetia-Alania from natural and man-made emergencies in 2013"* 2013 (Vladikavkaz) 218 p

[5] *Materials for the annual state report "On the state of protection of the population and territories of the Republic of North Ossetia-Alania from natural and man-made emergencies in 2014"* 2014 (Vladikavkaz) 197 p

[6] *Materials for the annual state report "On the state of protection of the population and territories of the Republic of North Ossetia-Alania from natural and man-made emergencies in 2015"* 2015 (Vladikavkaz) 248 p

[7] *Materials for the annual state report "On the state of protection of the population and territories of the Republic of North Ossetia-Alania from natural and man-made emergencies in 2016"* 2016 (Vladikavkaz) 244 p

[8] *Materials for the annual state report "On the state of protection of the population and territories of the Republic of North Ossetia-Alania from natural and man-made emergencies in 2017"* 2017 (Vladikavkaz) 245 p

[9] Zalikhanov M Ch 1974 *Snow avalanches and prospects of development of the mountains of North*

Ossetia (Ordzhonikidze: Publishing house "IR") 148 p

[10] Kondratyeva N I 2017 *Changes in mudflow activity from West to East on the sulfuric slope of the greater Caucasus. Geological and geophysical studies of the deep structure of the Caucasus: Geology and Geophysics of the Caucasus: modern challenges and research methods* (Vladikavkaz: VSC RAS, GIF) pp 594–602

[11] Kondratieff N In, Adzhiev A H, Bekkiev M Y, Gadoeva (Georgieva) M M, Perov V F, Seinova I B and Gucunaew L V 2015 *Inventory of mudflow hazard in the South of European part of Russia* (Moscow; Nalchik: Feoriya; OOO Printing house) pp 55–6

[12] Zaporozhchenko E V, Kamenev N S, Nikulin A S et al 2012 Baddon Basin (Republic of North Ossetia-Alania): mudflow processes natural and man-made *Proc. of the Int. seminar "Dangerous natural processes in the mountains: the lessons of the Karmadon disaster"* (on the 20th of September 2002-the anniversary of the disaster on the glacier Kolka) (Vladikavkaz) pp 24–6

[13] Hazaeva F M 2014 The impact of the cascade of "Zaramag" on the mountain landscapes of the RIS-Alania *Sustainable development of mountain areas* **4(22)** 70–4

Impact of the expropriation of the territories of natural reserves for mining purposes on protected populations

M D Vasilyev, NV Vasilyeva, OI Matveeva and Yu I Trofimtsev

[1]Institute of Mathematics and Informatics, M.K. Ammosov North-Eastern Federal University, 58, Belinsky str, Yakutsk, 677013 , Russia

E-mail: trofimtsev@mail.ru

Abstract. The paper discusses the impact of allocation of resource reserves for the development of mineral deposits on protected populations. The main method of assessing this effect is mathematical modeling. The results of the study were obtained using a model based on two parabolic partial differential equations (diffusion equations). The equations describe the dynamics of a population in a bilocal habitat, when the boundary of the protected area divides the area into two parts–protected and unprotected. At the same time, the species are free to pass from one part of the habitat to another. The model also takes into account the diffusion movement of species across the habitat and cross-diffusion, which means both the competition of species for available resources and their possible coexistence. Numerical calculations demonstrate that with a proper assessment of the impact on the natural potential of the reserve, it is possible to maintain stable densities of protected populations in both parts of the habitat, but at a slightly lower level. The case of high man-made loads on protected populations leading to their degeneration is also considered. The work shows that the presence of a sufficient food resource has a beneficial effect on the density of the protected and unprotected parts of the population.

1. Introduction

Diamond mining is one of the main industries of the Republic of Sakha (Yakutia). After the accident at the Mir mine, which will begin to be restored only in 2024, production fell by 10–11 percent. Therefore, placer deposits are becoming increasingly attractive for industrial development. A significant increase in their production is associated with the activities of LLC Almazy Anabara. Among placer deposits, the alluvial type is most common, developed in small and medium rivers. The length of the placers varies from kilometers to several tens of kilometers, while width varies from several tens to hundreds of meters [1].

The territory of Yakutia is more than three million square kilometers, of which 28.5 percent belong to protected areas. This is also due to the low population density, about 0.3 people per square kilometer. In Yakutia, a harmonious system of specially protected natural areas (PAs) has been developed, including natural parks, sanctuaries, wildlife refuges, resource reserves, protected landscapes and natural monuments, as well as green areas of residential areas [2]. A fundamentally new form of protected areas has been introduced: a resource reserve containing elements of the fourth and seventh categories of protected areas of the International Union for Conservation of Nature. The introduction of reserves is dictated by the extensive network of protected areas, which affects the way of life of indigenous peoples. Resource reserves include zones of absolute peace, sacred sites, licensed resource extraction areas,

IOP Conf. Series: Materials Science and Engineering **663** (2019) 012055 doi:10.1088/1757-899X/663/1/012055

seasonal natural resource extraction areas, traditional nature management areas, protected landscapes and nature monuments.

2. Problem statement

Placer diamond mining in Yakutia is moving farther north to the Arctic Circle expropriating the territories of resource reserves. An example of such a progression is the Verkhne-Munskoye diamond deposit, which is located in the west of Yakutia, 170 km from the city of Udachny in the territory of the Muna resource reserve. At present, four kimberlite pipes have been explored: Zapolyarnaya, Deimos, Novinka, and Komsomolskaya Magnitnaya. Today it is the largest investment project of the ALROSA company. The Verkhne-Munskoye field will bring about 1.8 million carats of diamonds per year, and its reserves are sufficient to continue production for more than 20 years until 2042. The official discovery of the Verkhne-Munskoye diamond deposit was held on October 31, 2018. The reduction of the territory of the resource reserve affects the protected animal populations, which include moose, wild deer, sable and Siberian weasel. The aim of the work is to assess the impact of mining on the density of protected populations using mathematical modeling of their dynamics.

3. Methods for study

We consider the dynamics of the population in a bilocal area, which is specific in the presence of protected and unprotected parts and a generalized resource conducive to the existence and development of the population. The study uses a mathematical model, which is a system of parabolic partial differential equations and describes arbitrary (diffusion) distribution of the population in a two-dimensional area [3]. The model that takes into account migration flows caused by the uneven distribution of parts of the population in the range is as follows:

$$\begin{cases} \dfrac{\partial u}{\partial t} = \nabla(\varepsilon_1(u,v)\nabla u - p_1 u\nabla v - q_1 u\nabla r) + f_1(u,v), \\ \dfrac{\partial v}{\partial t} = \nabla(\varepsilon_2(u,v)\nabla v - p_2 v\nabla u - q_2 v\nabla r) + f_2(u,v). \end{cases}$$

Here $u(x,y,t)$ and $v(x,y,t)$ are population density, divided into two parts–unguarded and protected–respectively. Migration flows are described by diffusion coefficients $_1(u,v)$, $_2(u,v)$ and directional migration parameters p_1, p_2. The resource function is set as $r(x,y)$. The natural increase in the density of parts of a population is determined by functions f_1 and f_2:

$$f_1(u,v) = m_1u + d_1(v \quad u) \quad c_1u^2 \quad b_1,$$

$$f_2(u,v) = m_2v + d_2(u \quad v) \quad c_2v^2.$$

In functions m_1 and m_2 are the coefficients of population growth outside the protected area and inside it; c_1 and c_2 are the coefficients of competition within the population in and outside the protected area; d_1 and d_2 are the coefficients of exchange of individuals between the protected area and the rest of the habitat. The term b_1 is interpreted as a quantity that describes external impact on the unprotected part of the population caused by mining operations.

The diffusion coefficients $_1(u,v)$ and $_2(u,v)$ are positive values depending on the density of populations:

$$_i(u,v) = \ _{i1} + \ _{i2}uv, \ i = 1,2.$$

Here $_{11}$, $_{12}$, $_{21}$, $_{22}$ are taken as second-order matrices with non-negative elements; the case of linear diffusion is obtained at zero $_{12}$, $_{22}$. The introduction of non-linearity into migration processes is explained by the existing competition between species of different parts of the population for resources, which increases the dispersion of species across the habitat. It is also associated with the Allee effect which states that the accumulation of species increases the competition between them for resources and living space, but increases survivability of the group as a whole. This implies that for the development of the population, the limiting factors are both "overpopulation" and "under-population" [4, 5].

The treatise considers habitat $\Omega = [0, l_1]\ [0, l_2]$, on the boundaries of which conditions were set that suggest that species of the population do not leave the habitat, and that their numbers on the borders are zero. The alienation of the protected area was described by changing parameter l_2. The system of equations was solved numerically [6–9].

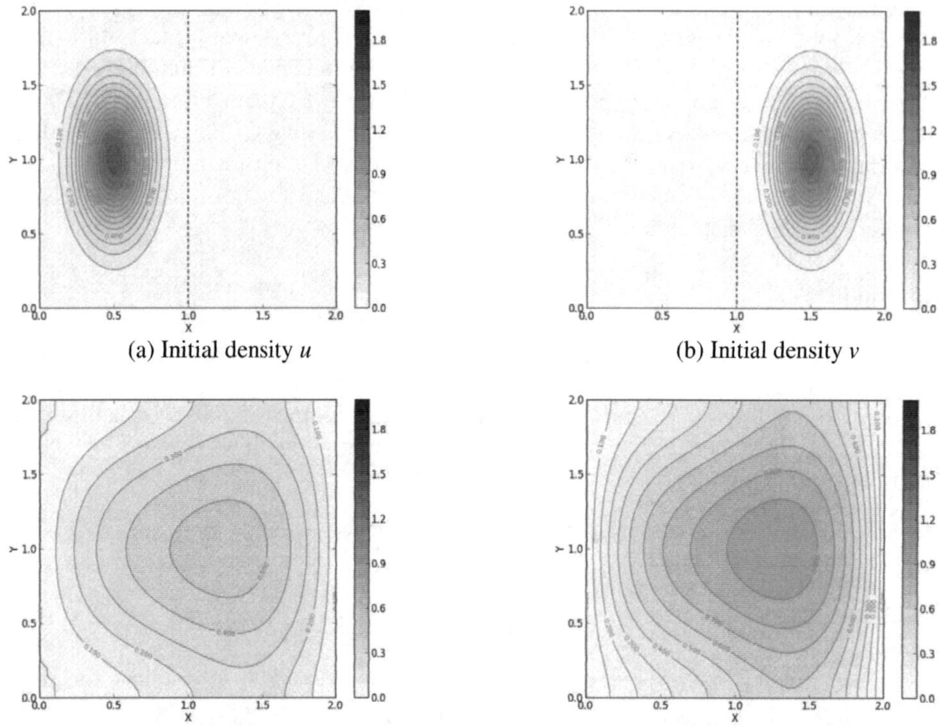

(a) Initial density u (b) Initial density v

(c) Limiting density u (d) Limiting density v

Figure 1. Case of the same areas of the protected and unprotected parts of the habitat

4. Discussion

The calculations showed that a decrease in the area of the protected area causes a much smaller effect than expected. Moreover, if a technogenic impact is small, they stabilize at a certain, but lower initial level. This can be explained by the presence of the protected part of the reserve. Fig. 1 shows the initial and limiting values of the densities of both parts of the population with the same areas of the protected and unguarded parts of the habitat. The limiting values of densities are 0.5 for unprotected and 0.9 for protected parts. When the protected area decreases, the limiting values stabilize at a lower level—0.4 and 0.8, respectively (Fig. 2). A further decrease in the area of the protected part practically does not change the limiting densities of both parts of the population (Fig. 3).

In our model, there is a fold-type bifurcation [10], which leads to the degeneration of the entire population, which is shown in Fig. 4. It occurs at high man-made loads, i.e. with a large value of b in the model. At moment of time t = 3, with equal areas of the habitat parts, non-zero densities of the protected and unprotected parts of the population preserve. If the area of the protected part of the habitat is reduced, then at the same time point, the unprotected part of the population no longer exists (Fig. 5). Let the main food resource of individuals is concentrated on the unguarded part of the habitat (Fig. 6). With a large anthropogenic impact and equal areas of the habitat parts, this leads to a small but non-zero density of the unprotected part of the population (Fig. 7). The expropriation of a part of the reserve for mining results in a very small but non-zero density of the unprotected part of the population (Fig. 8).

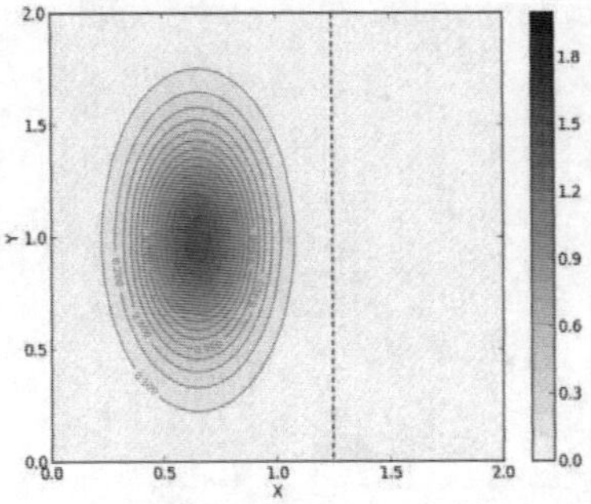

(a) Initial density u

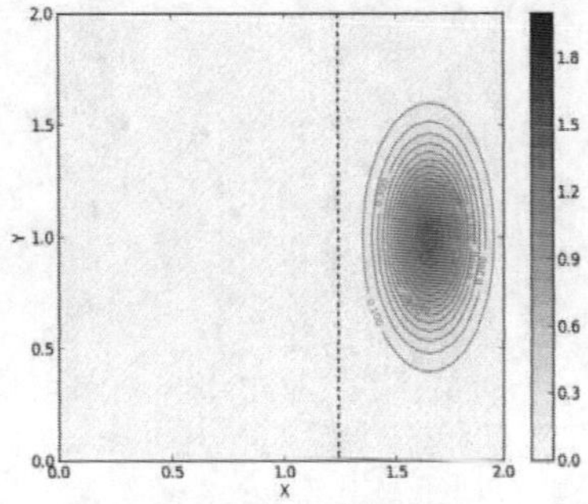

(b) Initial density v

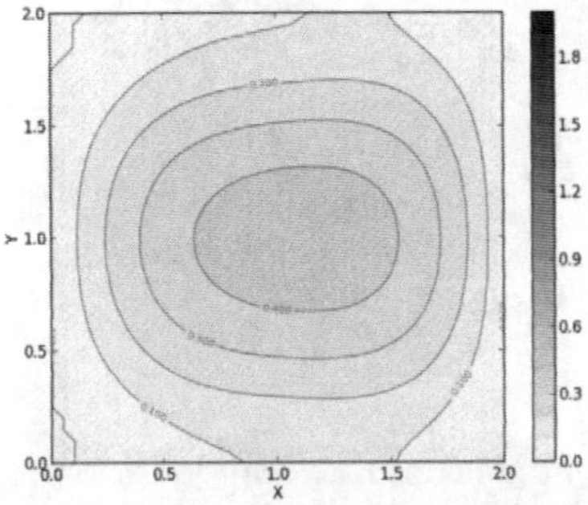

(c) Limiting density u

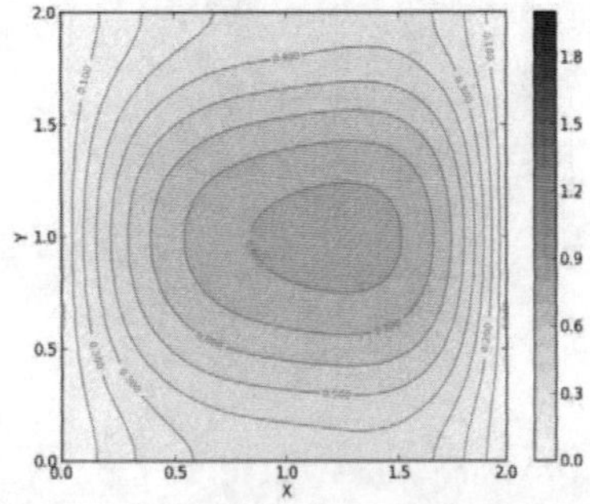

(d) Limiting density v

Figure 2. Case of the ratio of the areas of the protected and unguarded parts of the habitat of 5:3

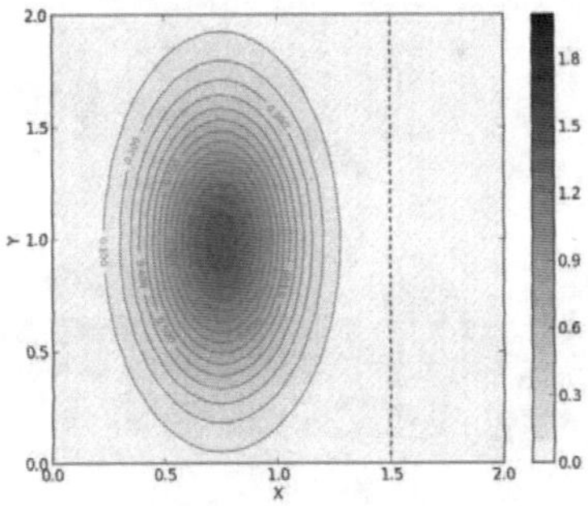

(a) Initial density u

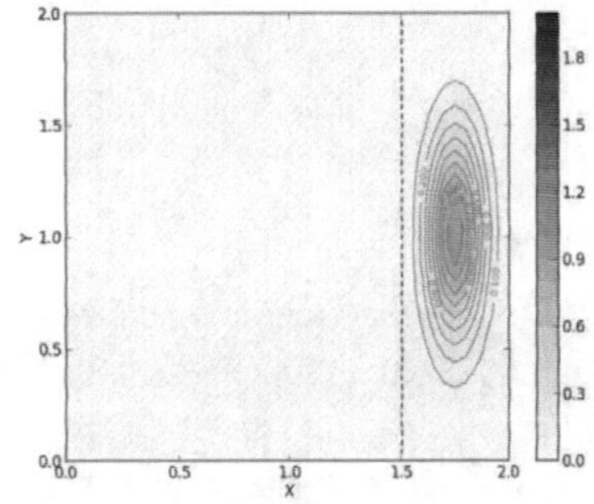

(b) Initial density v

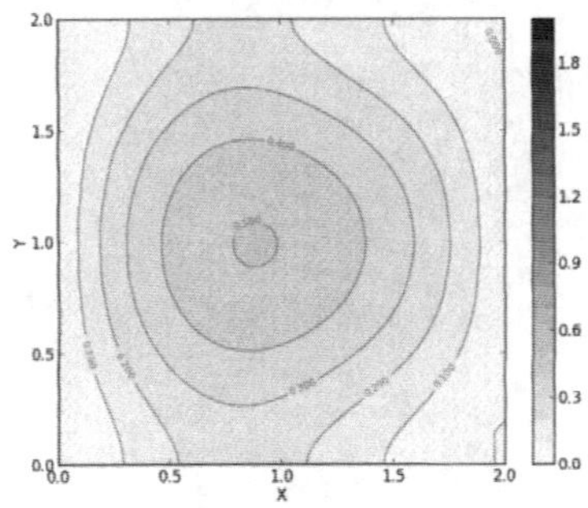

(c) Limiting density u

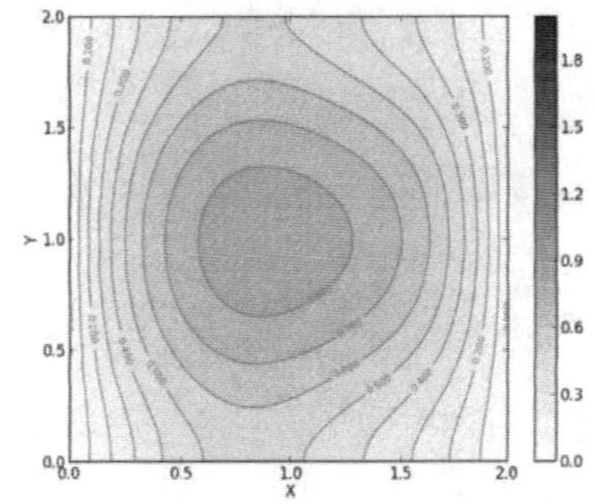

(d) Limiting density v

Figure 3. Case of the ratio of the areas of the protected and unguarded parts of the habitat of 3:1

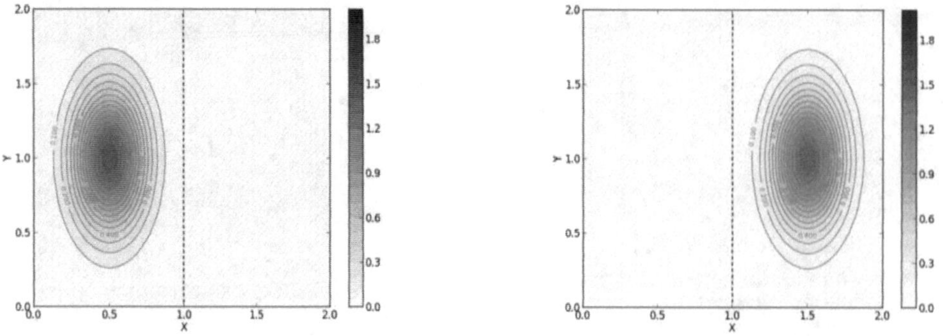

(a) Initial density u (b) Initial density v

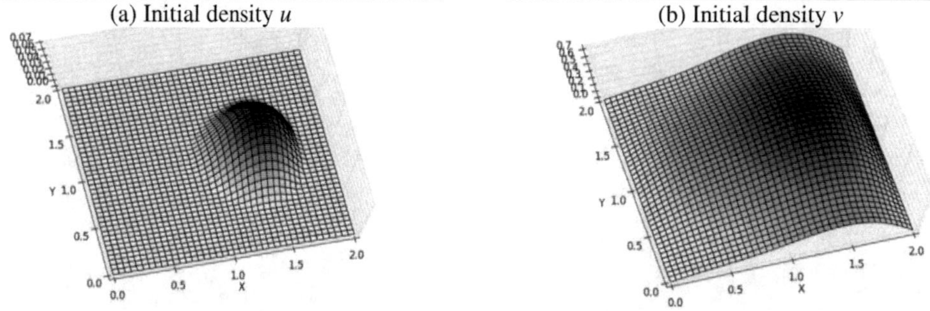

(c) Intermediate density u, $t=3$ (d) Intermediate density v, $t=3$

Figure 4. Case of the same areas of the protected and unguarded parts of the area in the presence of bifurcation in the system

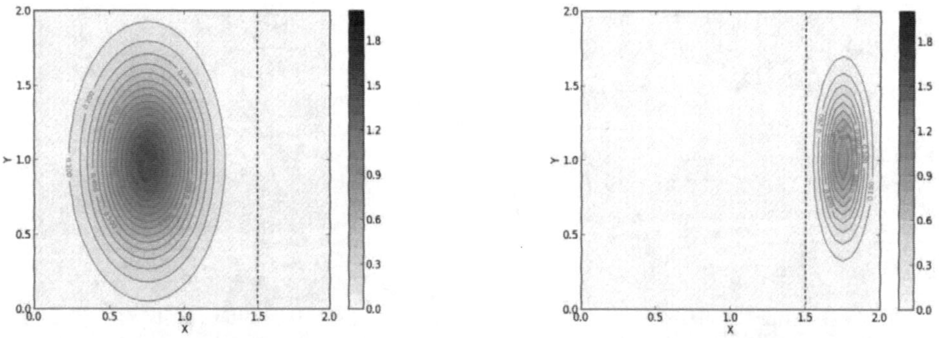

(a) Initial density u (b) Initial density v

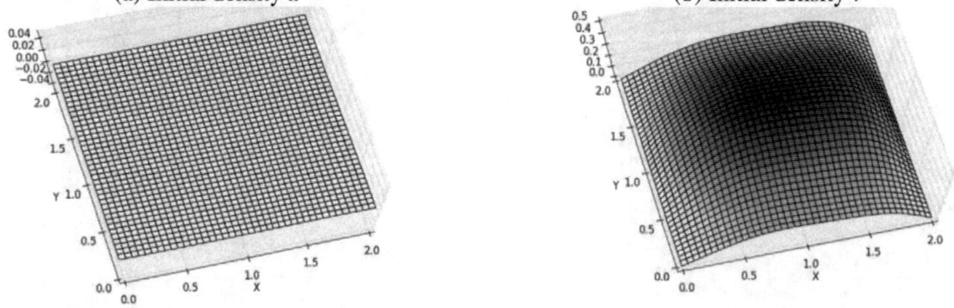

(c) Intermediate density u, $t=3$ (d) Intermediate density v, $t=3$

Figure 5. Case of the ratio of the areas of protected and unprotected parts of the habitat of 1:3 in the presence of bifurcation in the system

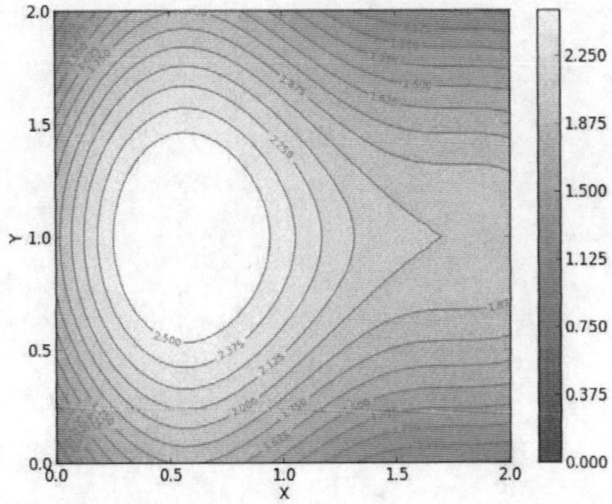

Figure 6. Setting generalized resource, the most favorable zone is in the unguarded part of the habitat

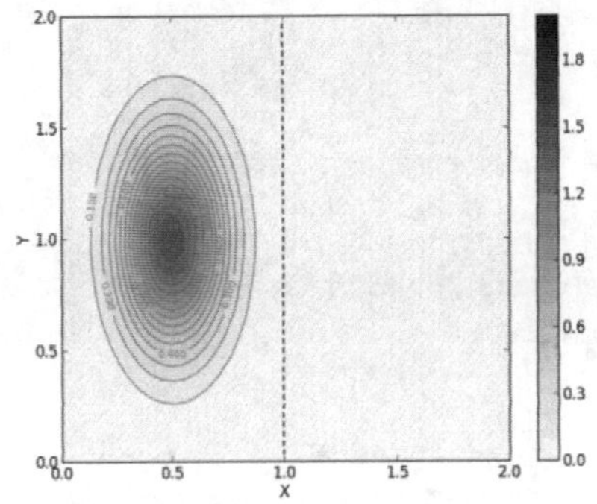

(a) Initial density u

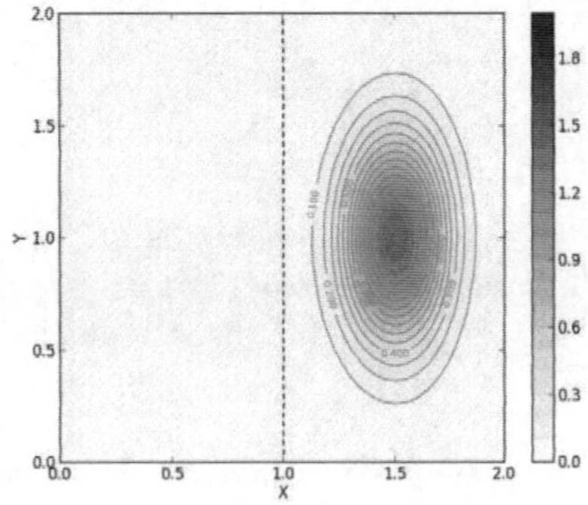

(b) Initial density v

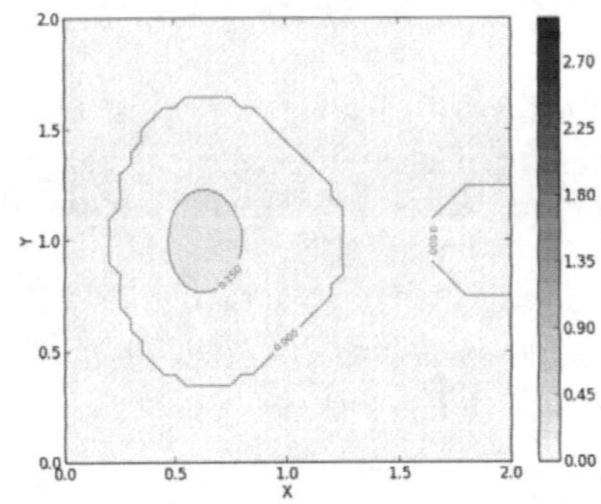

(c) Intermediate density u, $t=10$

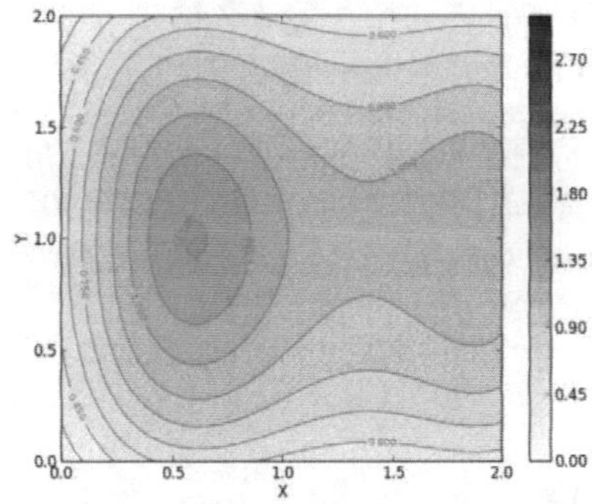

(d) Intermediate density v, $t=10$

Figure 7. Case of the same areas of protected and unprotected parts of the habitat with the introduction of a generalized resource

5. Conclusions

When solving problems of the economy, it is necessary to take into account the industrial impact on the environment and to solve environmental problems simultaneously with the economic ones. The transfer of a part of the lands of a natural reserve for the extraction of natural resources does not lead to a large negative impact on the protected populations only with a correct assessment of the natural potential.

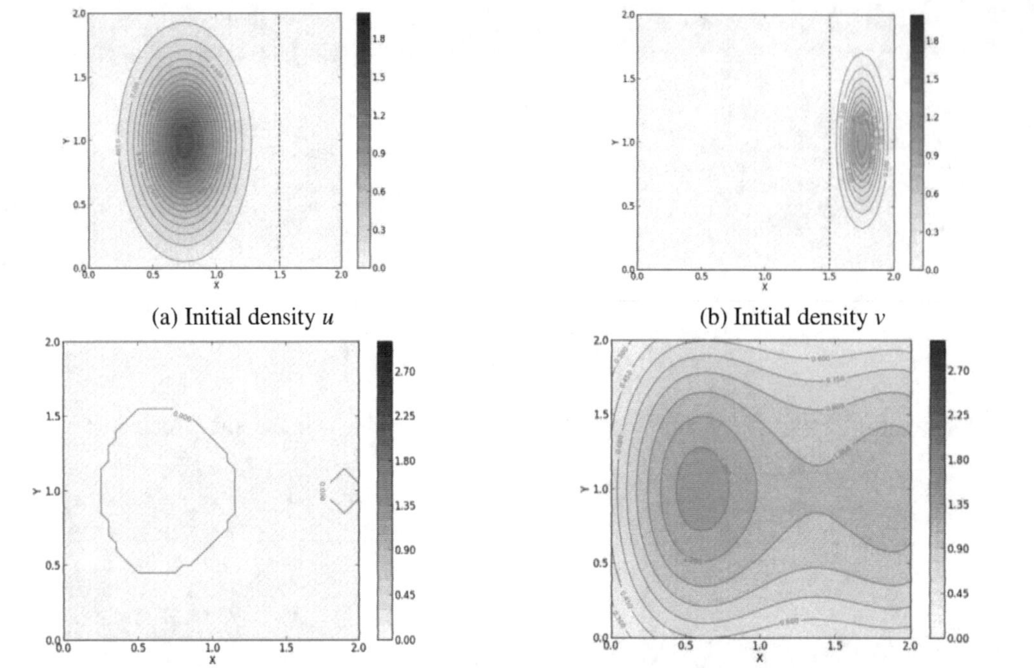

(a) Initial density *u* (b) Initial density *v*

(c) Intermediate density *u, t=10* (d) Intermediate density *v, t=10*

Figure 8. Case of the ratio of the areas of protected and unprotected parts of the habitat of 1: 3 with the introduction of a generalized resource

References

[1] Argunov K P 2005 *Diamonds of Yakutia* (Novosibirsk: Nauka)

[2] Fridovsky V Yu and Prokhorov VA 2008 *Security of the Republic of Sakha (Yakutia)* (Novosibirsk: Nauka)

[3] Knabner P and Angermann L 2003 *Numerical methods for elliptic and parabolic partial differential equations* (New York: Springer-Verlag)

[4] Tulumello E, Lombardo M C and Sammartino M 2014 Cross-diffusion driven instability in a predator-prey system with cross-diffusion *Acta Appl Math* **132** pp 621–33

[5] Dubey B, Das B and Hassain J 2002 A predator-prey interaction model with self and cross-diffusion *Ecol. Model* **141** 67–76

[6] Vasilyev M D and Trofimtsev Yu I A diffusion model of protected population on bilocal habitat with generalized resource *Proc. of the 8th Int. Conf. on Mathematical Modeling* AIP Conf. Proc. 1907.030014-12 Retrieved from: https://doi.org/10.1063/1.5012636

[7] Vabishchevich P N 2013 *Additive operator-difference schemes (splitting schemes)* (Moscow: Krasand)

[8] Vabishchevich P N 2010 *Numerical methods: Computational practice* (Moscow: Librcom book house)

[9] Samarsky A A and Vabishchevich P N 2009 *Numerical methods for solving convection-diffusion problems* (Moscow: URSS)

[10] Vasiliev M D 2017 Numerical study of a two-dimensional model of a protected population in the bilocal area *VEFU Bull.* **1(57)** 44–61

IOP Conf. Series: Materials Science and Engineering **663** (2019) 012056 doi:10.1088/1757-899X/663/1/012056

Heat pumps as alternative energy sources: characteristics, state and prospects for use

T F Tsgoev, R A Tebloev, O B Dzhatiev and L A Zagalova

North-Caucasian Mining and Metallurgical Institute, 44, Nikolaeva str., Vladikavkaz, 362021, Russia

E-mail: tsgoevt@inbox.ru

Abstract: Currently, heating and hot water supply of urban facilities is provided, as a rule, by centralized heating systems. The source of thermal energy in such systems are city combined heat and power stations (CHPs) – which combine generation of electricity and heat – or district boilers. The benefits of centralized heat supply are widely recognized. However, their use also has drawbacks and limitations. The construction of extended heating mains to remote objects, as well as to objects in areas with low building density, is associated with significant capital investments and large heat losses along the mains. Their total operating expenses are significant. Serious problems also arise in the reconstruction of existing facilities and the construction of new ones in populated urban areas with dense buildings. The current tariffs for thermal energy, combined with the cost of connecting to the urban heating networks, make it increasingly necessary to think about alternative methods of heating. In addition, a large amount of fuel is burned at CHPs and boiler houses, which adversely affects the environment. A modern rural house is mainly heated from autonomous boilers, private heat generators, or stove heating. Solid fuel is traditionally used as energy in the stoves. The stove usually has a firebox (where fuel is burned), channels through which flue gases are transported, and a chimney that removes flue gases to the atmosphere.

1. Introduction

At present, Russia – as well as the whole world – is facing two interconnected problems: saving fuel and energy resources, and reducing environmental pollution. In the conditions of depletion of organic fuel reserves and a sharp increase in the cost of developing new deposits, it is becoming increasingly irrational to burn coal, gas and oil products in millions of low-power boiler houses and individual furnace units, causing a large amount of harmful emissions into the atmosphere and significant environmental degradation situation in cities and the world [1]. When stocks of traditional energy sources, such as oil, gas and coal, inexorably decrease, their cost is high enough, and their use leads to the formation of the greenhouse effect on the planet, an increasing number of countries in their energy policy, turn their eyes to alternative energy sources.

2. Problem statement

To solve the problem of limited fossil fuels, researchers around the world are working to create and put into operation alternative sources of energy.

Alternative energy sources are environmentally friendly renewable resources which conversion yields electrical and thermal energy used for human needs. Such sources primarily include the energy of wind and sun, the waters of rivers and seas, geothermal energy, as well as biofuels, derived from the biological mass of animal and vegetable origin.

In addition to traditional types of alternative energy, Recycle (a project on an environmentally friendly lifestyle) chose the most interesting and environmentally friendly energy sources of the future [2]. Among them:

1) Energy from the turnstiles. Several research centers in the world had the idea to use the flow of people through the turnstiles as an innovative energy generator. The Japanese company East Japan Railway Company decided to equip each turnstile at the railway stations with generators. The installation operates at a train station in Shibuya district of Tokyo: piezoelectric elements are embedded in the floor under the turnstiles, which generate electricity from pressure and vibration, which they receive when people step on them. Similar is being introduced in China and Holland.

2) Energy from road humps. The concept of power generation with the help of the so-called road humps began to be implemented first in Great Britain, then in Bahrain. The invention is a road ramp generator consisting of two metal plates, slightly rising above the road. Under the plates, an electric generator is laid that generates current whenever a car passes the ramp.

3) Energy from a soccer ball. The principle of the accumulation of energy by the Soccket ball is fairly simple: the kinetic energy generated from hitting the ball is transferred to a tiny pendulum-like mechanism that drives the generator. The generator produces electricity that accumulates in the battery. The stored energy can be used to power any small electrical appliance, for example, a desk lamp with an LED.

4) Energy of volcanoes. One of the main developments in the development of volcanic energy belongs to American researchers from the pioneer companies of AltaRock Energy and Davenport Newberry Holdings. The test subject was a sleeping volcano in Oregon. Salt water is pumped deep into rocks the temperature of which, due to the decay of the radioactive elements present in the crust of the planet and the hottest mantle of the Earth, is very high. When heated, the water is converted into steam, which is supplied to a turbine that generates electricity.

5) Energy from human heat. The principle of thermoelectric generators operating on temperature difference has been known for a long time. But only a few years ago, technology began to allow the heat of the human body to be used as an energy source. A group of researchers from the Korea Advanced Institute of Science and Technology (KAIST) has developed a generator built into a flexible glass plate. Such a gadget will allow fitness bracelets to be recharged from the heat of the human hand, for example, during jogging when the body is very hot and contrasts with the temperature of the environment.

6) The energy of walking on paving slabs. Each of the busiest streets in the cities undergoes up to 50,000 steps a day. The idea of using pedestrian flow for the useful transformation of steps into energy was implemented in a product developed by Laurence Kemball-Cook, director of a British company Pavegen Systems Ltd. The engineer created paving slabs that generate electricity from the kinetic energy of walking pedestrians.

7) Energy from a bicycle. To recharge a player, phone or tablet, it is not necessary to have an outlet nearby. Sometimes it is enough just to work the pedals. For example, Cycle Atom, an American company, released a device that allows charging an external battery while cycling and subsequently recharging mobile devices.

8) Energy from wastewater. One of the pioneers of the idea was Professor Bruce Logan of the University of Pennsylvania. The general concept is very difficult for a non-specialist to understand and is built on two pillars: the use of bacterial fuel cells and the installation of so-called reverse electrodialysis. Bacteria oxidize organic matter in wastewater and produce electrons in this process, creating an electrical current.

However, at the present stage of development of science and production forces we consider the main type of alternative energy is to use low-grade heat of the earth's natural resources (air, water and soil) to generate thermal energy using heat pumps (TN).

3. The principle of operation and types of heat pumps

The topic of using heat pumps is becoming increasingly important not only because of the increase in the cost of energy, but also as climate conditions change. Today, this method of heating, as well as cooling buildings in the summer, is even considered as one of the methods to combat global climate change on our planet [3].

Heat pumps are energy efficient structures that use renewable energy of the earth's natural resources – heat from the soil, water, and air (Fig. 1).

By regulating pressure by the expansion valve, a such coolant flow into the evaporator is set which provides a definite calculated boiling temperature; on boiling, the coolant takes the heat provided by the collector from environment. Gas formed from the coolant is pumped into the compressor where it is compressed and, being heated, ejected into the condenser The condenser is the heat-rejecting element of a heat pump. Heat is rejected into water in the heating circuit with gas cooling down and turning back into liquid. Coolant is decompressed in the expansion valve and returns into the evaporator. The working cycle begins all over.

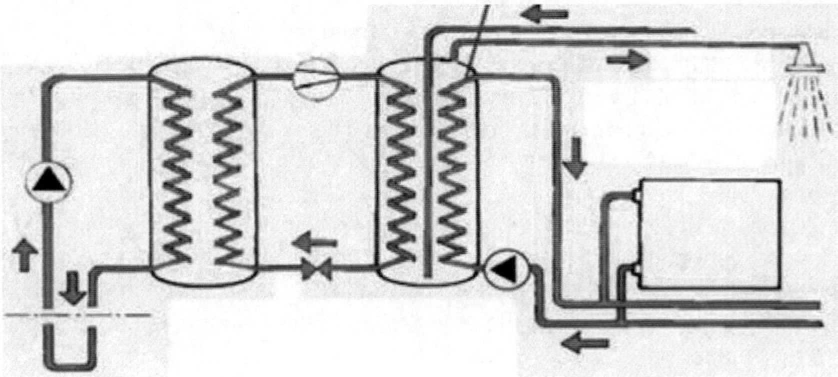

Figure 1. Heat pump operation principle

The device consists of:

- external contour with a heat carrier;
- internal circuit with a coolant;
- evaporator;
- compressor;
- capacitor.

The working principle is as follows. Non-freezing liquid moves through the external contour of the heating system and absorbs the heat of the environment. In the pump, this liquid gives off about 5 degrees to the refrigerant and continues to circulate. The refrigerant boils (at minus 10 °C) and turns into gas; the compressor compresses the gas, which increases the temperature. Once in the heat exchanger, this gas gives off heat to the internal heating circuit, cools itself, turning back into liquid and returning to the evaporator.

Like a refrigeration unit, a heat pump consumes a certain amount of energy to realize the thermodynamic cycle (compressor's drive).

Depending on the type of a heat pump, the method of their installation differs. For example, for an air-water pump, drilling of wells and earthworks will not be required.

The system also uses freon. The external contour of the heat pump can absorb energy from different environments: earth, water, air. Labor costs for its creation depend on the type of pump and its configuration. It is most difficult to arrange a ground-water pump in which the outer contour is horizontally located in the ground, since this requires large-scale earthworks. If there is a reservoir near the house, it makes sense to make a water-water heat pump. In this case, the outer contour is simply lowered into the pond.

There are different classifications of heat pumps.

IOP Conf. Series: Materials Science and Engineering **663** (2019) 012056 doi:10.1088/1757-899X/663/1/012056

1. According to the operational functions of heat pumps, we conditionally distinguish four main categories:

- Heat pumps used only for heating, i.e. ensuring a comfortable temperature in the room, as well as heating sanitary water.
- Heat pumps used for both heating and air conditioning throughout the year. The most common are reversible air-air machines. Heat pumps use hydraulic circuits to distribute heat and cold, and they can provide both operating modes at the same time.
- Integrated systems based on heat pumps that provide heating of premises, cooling, hot sanitary water and, sometimes, disposal of exhaust air. At the same time, the water can be heated either by superheat removal supplied from the gas compressor, or by a combination of superheat removal and using the regenerated heat of the condenser. Using only the selection of heat overheating is advisable when only space heating is required.
- Heat pumps designed exclusively for heating sanitary water. Most often, they use ambient air as a heat source, but they can equally use exhaust air.

Heat pumps vary depending on the source of heat extraction:

- geothermal: the unit uses heat from the soil, underground or surface groundwater. Information the costs of this unit and installation works of geothermal heat pumps for heating is fairly large; however, the solution is advantageous in terms of operating costs. This is due to the fact that the temperature of the earth at a depth is constant even in the winter period and is quite high, which means that the electricity costs will be lower;
- water: a device with the collector arranged in a reservoir below the depth of frost as rings or a zig-zag. The lowest-cost option. The drawback are the requirements for the volume of water and depth for a particular region;
- air heat pump for heating uses air as the heat source. Air heat pumps are not effective everywhere, as they depend on wind energy. They are mounted in a windy spot;
- using secondary heat: for them, the sources of heat are sewage and other water resources.

4. Characteristics of heat pump operation

A heat pump is characterized by the following parameters: *thermal power* Q_{CON} (thermal energy released during condensation of the vapor of the working substance in the condenser per unit time); *cooling capacity* Q_{EVAP} (thermal energy transferred to the working substance when it evaporates in the evaporator per unit time); *power* P_{COM} spent on pumping the working substance, and the *coefficient of energy transformation* μ_{HPU}.

The *energy efficiency* of a heat pump unit (HPU) is estimated by the transformation coefficient, which is the ratio between the received thermal power in the condenser per unit time and the power consumption P_{COM} by the compressor's drive:

$$\mu_{HPU} = \frac{Q_{CON}}{P_{COM}} \tag{1}$$

The transformation coefficient is greater than one, and it is the higher, the smaller the amount of additional energy expended in the compressor. Usually it equals 2...4, i.e. per 1 kWh of energy expended on the drive, 2...4 kWh can be obtained by using a low-grade energy source.

The higher the μ_{HPU}, the more efficient the heat pump is. The transformation coefficient depends on:

- temperature difference between a low-grade heat source (LPHS) and a high-potential heat source (HPHS); the higher it is, the lower the coefficient of performance (COP),
- thermodynamic properties of the refrigerant,
- features of the thermodynamic cycle,
- technical perfection of the heat pump design.

Depending on these factors, the COP ranges from 2.5 to 5.

IOP Conf. Series: Materials Science and Engineering **663** (2019) 012056 doi:10.1088/1757-899X/663/1/012056

The efficiency of heat pumps is usually characterized by the value of the dimensionless energy transformation ratio (Ktr), determined for an ideal Carnot cycle by the following equation:

$$K_{tr} = \frac{T_{rej}}{T_{rej} - T_{src}} = \frac{T_{cond}}{T_{cond} - T_{evap}} \quad (2)$$

where T_{rej} is the temperature potential of the heat rejected to a heating or heat supply system, K; T_{src} is the temperature potential of a heat source, K; T_{cond} the condensation temperature of the working fluid; T_{evap} the evaporation temperature of the working fluid.

The transformation coefficient of a heat pump, or heat-pump heat supply system K_{tr} is the ratio between the useful heat rejected into the consumer's heating system and the energy spent on the system operation. It is equal to the amount of useful heat obtained at the temperatures T_{rej} and T_{src}, per unit of energy spent on the drive of the heat pump or heat supply system.

5. State of heat pump application abroad

The production of heat pumps in each country is primarily focused on meeting the needs of its domestic market. Air-air reversible heat pump installations for heating and summer air conditioning are the most common in the USA, Japan and some other countries, while in Europe water-water and water-air pumps prevail. In Sweden and other Scandinavian countries, the availability of cheap electricity and the widespread use of centralized heating systems have led to the development of large heat pump installations. In the Netherlands, Denmark and other countries of this region, the most accessible type of fuel is gas, and therefore, gas-driven and absorber-driven heat pumps are rapidly developing [4].

Millions of heat pump installations are being exploited in the United States and more than half of them in the housing and utilities sector. The most common are reversible air-air heat pump installations with electric drive for year-round air conditioning in rooms. More than 50 companies produce heat installations, 30% of newly built houses, such as cottages, are equipped with heat pump installations.

The systems of heat supply of residential and public buildings from heat sources of low and high levels such as soil-water are rapidly developing. Highly efficient technologies and technical means of selection of the heat of the soil have been developed. There is an effective system of fines (for CO2 emissions from fuel combustion) and incentives for using low-grade heat sources for heat supply.

In Sweden, since the beginning of the 80s, the development of heat pump installations has been very intensive. This country is characterized by the use of large installations with a thermal power of more than 30 MW. The source of low-grade heat is mainly purified wastewater, sea water and wastewater from industrial enterprises. Among these heat pump installations, the largest are located in the cities of Malmö (40 MW), Uppsala (39 MW) and Örebro (42 MW).

The most powerful (320 MW) Stockholm installation using the water of the Baltic Sea as low-grade heat source. This unit, located on barges attached to the shore, cools the sea water in winter from 4 to 2 °C. The cost of heat from this installation is 20% lower than the cost of heat from boilers. The amount of heat generated by heat pump installations in Sweden is already about 50%.

In Germany, there are hundreds of thousands of heat pump installations that are used in water- and air-based heating and air conditioning systems. Electric heat pumps are predominant. In addition, they use hundreds of high-power units driven by diesel and gas engines. Sources of heat are air (external and exhaust), soil, water, etc. [4].

In Switzerland, the first heat pump installations were built back in the 30s. Nowadays, tens of thousands of heat pump installations are in operation, mostly of a small thermal capacity. Large plants have been built for operation in systems of centralized heat supply. The largest one is in Lausanne with a thermal capacity of 7 MW with an electric drive [5].

6. State of heat pump application in Russia.

The experience in the implementation of heat pumps in our country is still modest, but there are conditions for their wider application. Nevertheless, in Russia the process of applying heat units has already begun. Of course, the total capacity of the installed equipment is immeasurably small compared to the leading countries, but many public buildings in Perm, Kaliningrad, Tuapse, Samara, Penza, Moscow and Leningrad regions are already heated by this energy-saving technology.

In Russia, the development and implementation of heat pumps is carried out by the Siberian Branch of the Russian Academy of Sciences (Institute of Thermal Physics), Irkutskenergo OJSC. Heat pumps are manufactured in Novosibirsk, Moscow, and Nizhny Novgorod [5].

But, unfortunately, in most regions of the country, the works in this direction have practically stalled. For instance, in the territory of North Ossetia-Alania, no thermal stations operate. Considering the importance of using alternative types of fuel by the example of heat pumps, the staff and students of the North-Caucasian Mining and Metallurgical Institute have initiated a project on application of alternative types of energy using a soil-water heat pump for single-storey buildings in rural areas of the country [5].

The project proposed the use of a geothermal heat pump. The parameters of the primary circuit of the heat pump unit were calculated, as well as the total length of the collector pipes, the total area of the site under it and the calculations of technical and economic indicators. The results showed that the annual volume of generated heat energy – as a result of the implementation of proposed idea – was 669.6 Gcal/year with a capital investment in the construction of the installation of 1,792.81 thousand rubles. Prevented environmental and economic damage in this case amounts to 1,171.67 thousand rubles, and the profit from the sale of products is 867.27 thousand rubles. As a result, the payback period for these calculations is only 2.1 years.

7. Conclusions

The accumulated long-term experience of practical operation of heat pump heating systems abroad and in Russia, feasibility studies and design studies of their implementation in real small and large construction sites located both in dense urban areas and in rural areas, indicate wide possibilities of effective application of heat pump systems and ensure noticeable economic, energy saving and environmental effects due to their utilization. Additional potential for improving the efficiency of heat pump use also lies in the possibility of their implementation not only for heating and hot water supply, but also for air conditioning, including monitoring and managing indoor air humidity and in a number of technological processes. This is done using reversible heat pumps that allow changing the direction of the heat flow.

References

[1] Heat pumps. Use in residential buildings for heating, hot water supply, air conditioning and ventilation 2011 (Tver) Retrieved from: https://refdb.ru/look/1051428-p3.html

[2] Ten alternative energy sources that you knew nothing about Retrieved from: https://recyclemag.ru/article/10-neobychnyh-alternativnyh-istochnikov-energii

[3] Tsgoev T F and Tebloev R A 2019 State and prospects for the use of heat pumps in the regions of Russia on the example of North Ossetia-Alania Int. United Academy of Sci. Collection of sci. papers based on *Proc. of the XII International Sci. and Pract. Conf.* (10 December 2018) Part I pp 79–84

[4] Khakimullin B R and Bagautdinov I Z 2006 Foreign experience in fuel pump operation *Int. Sci. J. Innovation Sci.* **4**

[5] *European heat pump market* Retrieved from: https://www.abok.ru/for_spec/articles.php?nid=6506

[6] Semenov B A and Solovyov V A *Problems and features of the use of ground heat pumps for autonomous heating* Retrieved from: https://cyberleninka.ru/article/n/problemy-i-osobennosti-ispolzovaniya-gruntovyh-teplovyh-nasosov-dlya-avtonomnogo-teplosnabzheniya-obektov-v-tsentralnyh-regionah

[7] Kalnin I M, Lazarev L Ya and Savitsky A I *Energy-saving, environmentally friendly technologies for heat supply of industrial and residential premises* Retrieved from: http://ekip-tnu.ru/doc1.shtml

IOP Conf. Series: Materials Science and Engineering **663** (2019) 012056 doi:10.1088/1757-899X/663/1/012056

[8] Yeskendirov A N *Heat pumps as an effective way to save energy* Retrieved from: http://kazatu.kz/assets/i/science/sf14_energo_121.pdf
[9] *Guidelines for the use of heat pumps with the use of secondary energy and non-traditional renewable energy sources* Approved and put into effect by the Decree of the Moscow City Architecture Committee 31 January 2001 no 8
[10] Tsgoev T F and Gutonova Z H 2013 Heat pump systems as alternative ways of green heating *Proc. of sci.-techn. (sci.-pract.) conf. NTK-2013* (Vladikavkaz: North Caucasus Institute of Mining and Metallurgy (State Technological University)) pp 16–21

IOP Conf. Series: Materials Science and Engineering **663** (2019) 012057 doi:10.1088/1757-899X/663/1/012057

The problem of miscarriage in a region with a high content of heavy metals in the soil

T I Tsidaeva[1], A A Cheldieva[1], N D Kastueva[1], F M Gagloeva[2], K G Tomaeva[1], N G Salbieva[1], M A Kirgueva[1] and E G Plieva[1]

[1]North-Ossetian Medical Academy of the Ministry of Health of Russian Federation, 40, Pushkinskaya str., Vladikavkaz, 362025, Russia
[2]Republican Clinical Hospital of the Ministry of Health of North-Ossetian Republic–Alania, 39, Barbashov str., Vladikavkaz, 362002, Russia

E-mail: t.tsidaeva@rambler.ru

Abstract. Reducing the incidence of miscarriage in order to improve the demographic situation is currently one of the most pressing problems. The article presents the results of a comparative statistical analysis of reporting forms of federal statistical observation No. 13 “Information about pregnancy with an abortive outcome”, No. 32 “Information about medical care for pregnant women, parturient women and puerperas” with an insert for it “Information about the regionalization of obstetric and perinatal care in maternity hospitals (wards) and perinatal centers ", data from the State Report "On the state of sanitary and anti-epidemic well-being of the population in the Republic of North Ossetia-Alania in 2018" and statistical yearbooks of the Republic of North Ossetia-Alania. It was revealed that in the Republic of North Ossetia-Alania from 1991 to 2018, due to the introduction of modern methods of diagnosis and treatment of women suffering from miscarriage, as part of the state's policy in the field of women's and children's health, a significant decrease in infant and perinatal mortality occurred at practically the same frequency of a miscarriage. The latter was the basis for the assumption about the negative impact on the reproductive function of women of a high content of heavy metals in the soil, in particular cadmium and lead. It is concluded that it is necessary to conduct active monitoring of the content of heavy metals in the serum of women of reproductive age.

1. Introduction

For almost 30 years (from 1991 to 2018), thanks to the implementation of state policy measures in the field of maternal and child health aimed at improving the quality and accessibility of medical care for pregnant women and newborns in the Republic of North Ossetia-Alania, significant progress has been made in reducing infant mortality [1, 2] (Figure 1).

The infant mortality rate, which was 16.8 per 1000 live births in 1991, dropped to 4.6 by 2018, i.e., almost 3.7 times, which corresponds to the all-Russian trend [3].

Perinatal mortality rate, significantly increased in 2012–2013 in connection with the transition to new criteria for birth registration, began to decline from 2014, and in 2018 it was 6.40 versus 12.09 in 2013.

It should be noted that a greater contribution was made to the reduction of perinatal mortality by one of the components of this indicator – early neonatal mortality (6.91 ‰ in 2012 and 2.3 ‰ in 2018). The stillbirth rate over the same period changed slightly (4.88 ‰ in 2012 and 4.20 ‰ in 2018).

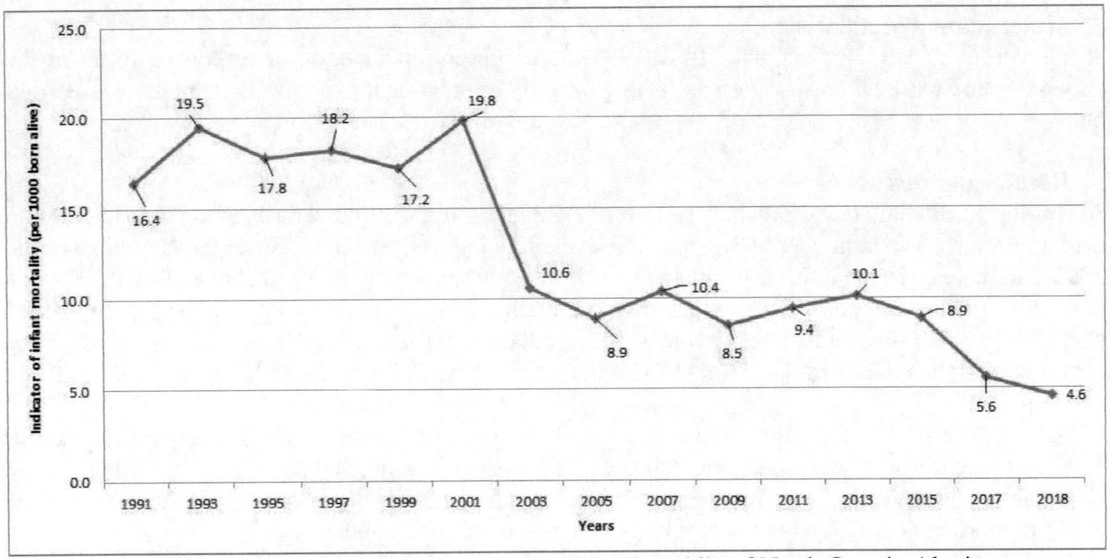

Figure 1. Dynamics of infant mortality in the Republic of North Ossetia-Alania

At the same time, throughout all the analyzed years with a relatively stable mortality rate of the entire population, the birth rate has fluctuations [4]. This indicator was decreasing from 1991 until 2001; from 2002 to 2014, there was a steady increase in the birth rate, which again fell into decline in 2015 (Figure 2).

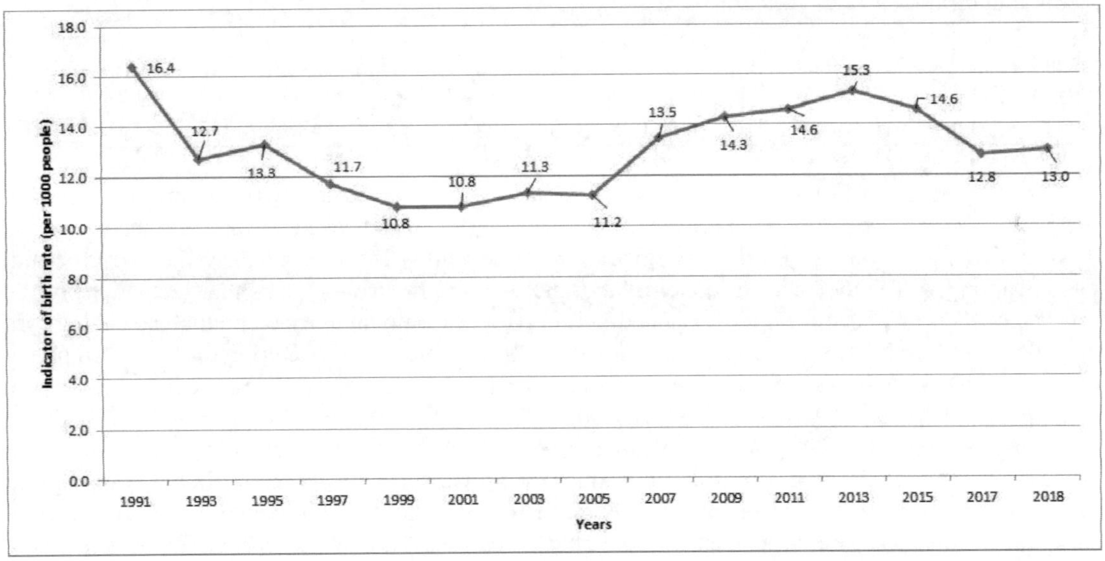

Figure 2. Dynamics of birth rate in the Republic of North Ossetia-Alania

During the specified period of time, in RNO-Alania the number of induced abortions has sharply decreased and continues to decrease annually. Reducing the number of abortions per 1000 women of childbearing age, however, does not lead to an increase in the birth rate.

In order to find reserves for improving the demographic situation, indicators characterizing reproductive losses were analyzed.

2. Materials and Methods

A retrospective analysis of the data, from 1991 to the present, included in the reporting forms of the Federal State Statistical Observation and recognized by international expertise as reliable, was carried out.

3. Results and Discussion

According to official statistics, since 1991 in the Republic of North Ossetia-Alania there has been a distinct decrease in the number of pregnancies (excluding induced abortion). Since the beginning of the 2000s, an increase in the number of pregnancies has begun (excluding induced abortion). By 2009, the number of pregnancies increased compared with 2002 by 41.3 %. In 2011–2013 the largest number of pregnancies (excluding induced abortion) was recorded with a maximum of 11,242. Since 2014, the birth rate has started to decline again, the number of all pregnancies by 2018 has decreased by 23 %, reaching 9,109.

At the same time, the rate of miscarriage per all pregnancies (11.1 % in 1991) reached 14.5 % in 1998 and practically did not change over the last 10 years (11.1 % in 2009 and 11.2 % in 2018).

Table 1 provides data on the frequency of preterm birth in the republic over the past 10 years, as well as the proportion of children born with a body weight from 500 to 2500 g.

Table 1. Dynamics of the number of children born weighing 500–999 g, 1000–1499 g and 1500–2499 g

	2009	2010	2011	2012	2013	2014	2015	2016	2017	2018
Born alive prematurely, including birth weight of	489	533	572	605	770	634	584	536	447	454
500-999 g	42	32	27	26	25	43	35	40	25	20
1000-1499 g	39	51	64	54	194	69	78	59	52	44
1500-2499 g	408	450	481	525	481	522	471	437	370	390

The data presented in the table demonstrate an increase in the frequency of preterm birth and an increase in the proportion of premature newborns since 2009, which peaked during the transition of the Russian Federation to new criteria for registration of live births recommended by the World Health Organization [5]. The dynamics of preterm labor presented in the table repeats the fluctuations of the total number of pregnancies in the indicated period. However, the number of premature babies born alive decreased from 2013 to 2018 by 41%, and the birth rate only by 23%. During the described period of time, clinical protocols, recommendations, and the procedure for providing medical care in the obstetrics and gynecology profile were introduced (with the exception of assisted reproductive technologies) approved by order of the Ministry of Health of the Russian Federation as of November 12, 2012 No. 572n "On approval of the procedure for providing medical care profile "Obstetrics and Gynecology" (except for the use of assisted reproductive technologies)" [6], which certainly improved the quality of medical care for women with possible miscarriage. The lack of correlation between these phenomena in the implementation of all standards of medical care was the basis for searching for an explanation of this fact and suggested that this imbalance may be due to the influence of an unfavorable environment.

The body of pregnant women is known to be most sensitive to the effects of various anthropogenic substances that pollute the environment, which is associated with increased gas exchange, a decrease in the rate of metabolic processes and degraded ability to neutralize toxic substances. The harmful effects of heavy metals that accumulate in the pregnant woman's body also affect the embryo/fetus. The degree of damage to the ovum directly depends on the duration of the half-life of metal pollutants, their ability to penetrate the uteroplacental barrier and the ability to exert an embryo and fetotoxic effect.

According to the Center for Hygiene and Epidemiology in the Republic of North Ossetia-Alania [7], annually soil samples reveal an elevated content of heavy metals, in particular, cadmium, lead, etc.

Table 2. Characteristics of the soil condition (according to 2018)

Chemical Name	Indicator (gross content)	
	Limit [mg/kg]	Actual content [mg/kg]
Cadmium	2.0	6.1
Lead	130.0	320.1

As follows from Table 2, the gross content of toxic substances recognized by scientists from different countries – that have an adverse effect on the reproductive system, such as cadmium and lead, is almost three times higher than the limits.

Cadmium is a highly toxic element, the main source of which in the soil is industrial and waste water. This metal is characterized by a long biological half-life. Prolonged intake of even low doses of cadmium can be accompanied by dysfunction of various organs and systems, including reproductive, manifested in infertility, miscarriages, complications of pregnancy and childbirth, congenital malformations of the fetus [8].

High lead content in the soil is observed near the metallurgical industry, roads, in wastewater. With prolonged exposure to lead on human body suffers a decrease in resistance to infectious agents, toxic effects on the brain, kidneys, and blood [9, 10]. Manifestations of negative effects on the reproductive system are infertility, miscarriages, complicated pregnancy and childbirth, fetal malformations [11, 12].

4. Conclusions

Thus, on the basis of a retrospective analysis of statistical data characterizing the most important medico-demographic indicators in the Republic of North Ossetia-Alania in 1991-2018, it was found that with an increase in the quality of medical care for pregnant women, there was no decrease in the incidence of miscarriage.

Considering that the health of citizens is determined by many factors and, first of all by environmental pollution by anthropogenic chemicals [13]—which accounts for at least a third of the diseases in the population [14]—we found it possible to associate the lack of positive changes in the rate of miscarriage in the soil of heavy metals.

In the framework of the implementation of the Parma commitments, WHO identified experts as cadmium, arsenic, mercury, lead, etc., as priority reproductive toxicants [8].

According to foreign scientists, increasing the concentrations of cadmium and lead in the serum contributes to the habitual loss of pregnancy [15].

This preliminary study confirmed the relevance of its further development with the active monitoring of concentrations of cadmium and lead in the serum of pregnant women for the development of effective measures to prevent miscarriage.

References

[1] Tsidaeva T I 2002 Maternal and Child Health Programs, a reserve for reducing perinatal mortality *Proc. IV Ros. Forum "Mother and Child"* (Moscow: MIK) 136 p

[2] Tsidaeva T I 2004 *Miscarriage in the Republic of North Ossetia-Alania: medical and social aspects, organization of medical care, and prevention* Extended abstract of dissert of Doctor of Sciences (Medicine) (Moscow: Sci. Center of Obstetrics, Gynecology and Perinatology)

[3] Baibarina E N, Shuvalova M P, Sorokina Z H, Lenyushkina A A and Tsymlyakova L M 2011 Transition of the Russian Federation to international criteria for birth registration of children: view of the health care organizer *Obstetrics and gynecology* **6** 4–8

[4] *Statistical Yearbook of the Republic of North Ossetia-Alania (Statistical compendium of the Federal State Statistics Service)*

[5] Malyshkina A I, Pesikin O H and Kuligina M V 2014 Organization of perinatal care for very early

IOP Conf. Series: Materials Science and Engineering **663** (2019) 012057 doi:10.1088/1757-899X/663/1/012057

preterm labor in the Central Federal District of the Russian Federation *Obstetrics and Gynecology* **6** 74–8

[6] *Order of the Ministry of Health of the Russian Federation* as of 12 November 2012 no 572n "On approval of the procedure for providing medical care in the profile" obstetrics and gynecology (except for the use of assisted ATSDR/EPA reproductive technologies)

[7] *State report "On the state of sanitary and anti-epidemic well-being of the population in the Republic of North Ossetia-Alania in 2018"*

[8] WHO. European Regional Office *Indicators based on biomonitoring exposure to chemical pollutants. Meeting report* (Catania, Italy) 19–20 April 2012 p 43 Retrieved from: http//www.euro.who.int__data/assets/pdf_file/0014/171221/e96640r.pdf.

[9] *ATSDR/EPA Priority List for 1999: Top 20 Hazardous Substances* Agency for Toxic Substances and Disease Registry, U.S. Department of Health and Human Services. http:/www.atsdr.cdc.gov/99list.html; & U.S. CDC, National Center for Environmental Health, National Report on Human Exposure to Environmental Chemicals 2001 Retrieved from: http:/www.cdc.gov/nceh/dls/report/Highlights.htm

[10] *CHE Toxicant and Disease Database. Collaborative on health and the environment* Retrieved from:http://database.healthandenvironment.org accessed 16 October 2007

[11] Couloures K and Vasan K 2011 Prenatal lead poisoning due to maternal exposure results in developmental delay *Pediatr. Int.* **53(2)** 242–4

[12] Kireeva Yu V and Shubina O S 2008 Lead content in the mother-placenta-fetus system *Successes of modern natural sci.* **6** 193

[13] Olden K and White S L 2005 Health-related disparities: influence of environmental factors. *Med. Clin. North Am.* **89(4)** 721–38

[14] *Fourth National Report on Human Exposure to Environmental Chemicals* 2009 Department of Health and Human Services 519 p Retrieved from: http//www.cdc.gov/exposurereport

[15] Ajayi O O, Charles-Davies M A and Arinola O G 2012 Progesterone, selected heavy metals and micronutrients in pregnant Nigerian women with a history of recurrent spontaneous abortion *Afr. Health Sci.* **12(2)** 153–9

IOP Conf. Series: Materials Science and Engineering **663** (2019) 012058 doi:10.1088/1757-899X/663/1/012058

Natural risks in the Kabardino-Balkar Republic`s landscape zones

A B Uzdenova [1,3], F E Kanametova[2], O O Dakhova[2], N V Tatarenko[2], R G Miskarova[1], A S Otarova[1] end Z Zh Gergokova[1]

[1] Nature Research High Mountain Geophysical Institute, 2, Lenin str., Nalchik, 360030, Russia

[2]Kabardino-Balkarian State University named after H. M. Berbekov, 173, Chernyshevsky str., Nalchik, 360004, Russia

Abstract. One of the most serious obstacles of a natural character for the development of recreation and tourism are natural hazards and processes. The high-mountainous part of the republic is especially subject to such processes, where most of the recreational objects are concentrated. Of the entire spectrum of hazardous natural processes, mudflows, avalanches, landslides, karst, subsidence, slides, debris, flood and flooding are the most widespread in the republic. In the mountainous part because of strongly dissected relief with steep slopes and considerable depth of incision of narrow river valleys, avalanches, mudflows, landslides and debris, slides, glacial activities are most active. The flat part and the foothills have their own set of natural hazards, among which floods, flooding, subsidence, soil erosion, and droughts prevail. These areas are often exposed to hailstorms. The entire territory of the Kabardino-Balkar Republic is an area of seismic activity. Spontaneous natural processes and phenomena have the most significant impact on the live of the population of the republic and the functioning of recreational facilities located in the zone of their occurrence. Each demonstration of risks in recreational areas leads to undesirable consequences, from dissatisfaction with rest to death of people and wealth.

1. Introduction

Natural risk is the danger perceived by a subject (person) of a natural process or its anthropogenic counterpart in the implementation of any activity [1].

Hazardous natural processes of both natural and man-made are spontaneous events that unexpectedly demonstrate themselves as powerful destructive forces, usually not controlled by human. All these processes are very different in terms of formation factors, features of demonstrations, duration of flow, which can be from several minutes (avalanches) to several hours (mudflows), days (landslides), months and years (channel and soil erosion). Arising independently, many of these processes and phenomena can interact with each other, enhancing the common effect [2].

The diversity of the natural conditions of the republic determines the level of natural hazards for its individual parts. In winter, avalanche activity is active in mountainous areas, mudflow, landslide, landslide and scree processes, and karst are active in summer time. Here also floods are usually formed, reaching then to the plats.

IOP Conf. Series: Materials Science and Engineering **663** (2019) 012058 doi:10.1088/1757-899X/663/1/012058

2. Statement of the problem

Destructive spontaneous processes are a serious problem for the recreational development of the territory, the solution of which is connected with the development of rational methods and measures of protection, and also the identification of safe areas for perspective development of building and laying of tourist routes [3].

3. Materials and methods

On the basis of cartographic data [4] for all landscape-climatic zones of the Kabardino-Balkar Republic [4], the area of exposure of the territory to various types of natural hazards for farming and living conditions of the population was calculated.

The research was conducted on the map Buslik A.V. from the Atlas of the Kabardino-Balkar Republic. The author divided the territory of the Kabardino-Balkar Republic according to the level of risk of natural hazards into 6 zones:

- high – avalanches, mudflows;
- medium – soil erosion, mudflow, hail, floods;
- low-flood;
- a combination of high and medium levels – mudflows, soil erosion, floods, hail, earthquakes;
- combination of medium and low levels – soil erosion, landslides;
- no risk.

We have carried out a generalization of the map, by combining the zones combining the level of risk with the zone with the average level of risk, since these zones contained the same types of danger (Fig. 1).

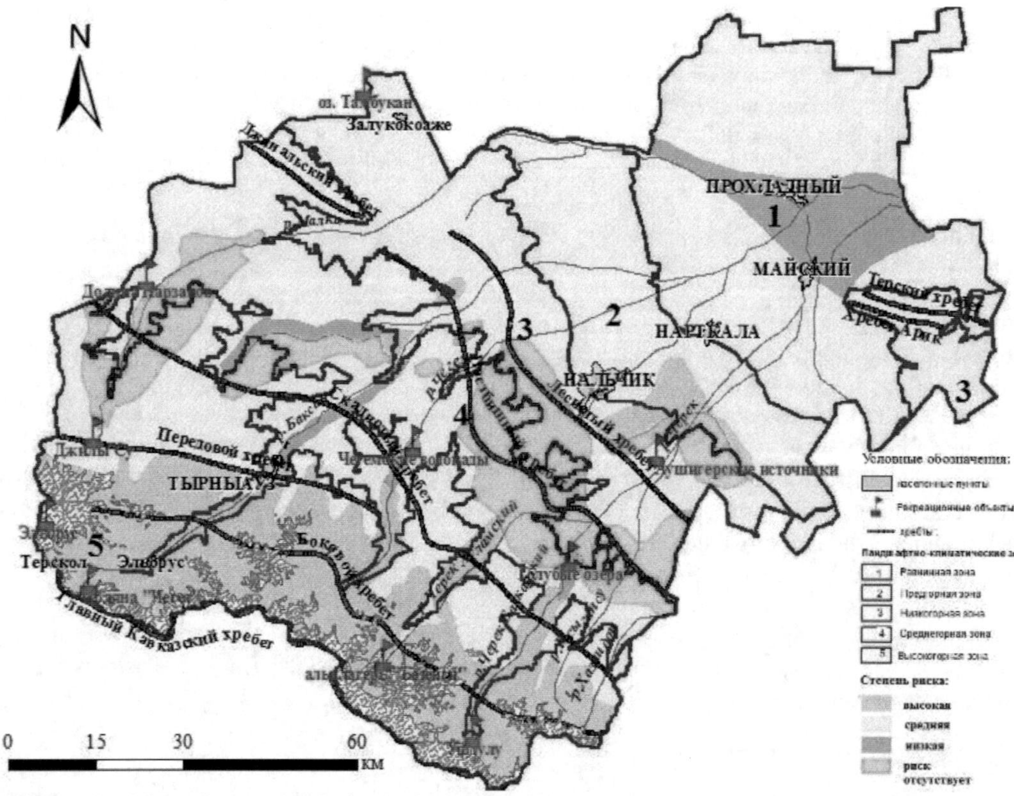

Figure 1 . Map of natural risks in the landscape-climatic zones of the Kabardino-Balkar Republic

We have calculated the area of various degrees of risk for each landscape-climatic zone with the help of GIS technology (Table1).

Table 1. The areas of various degrees of risk for landscape-climatic zones

Landscape-climatic zones	Risk level			
	High	Medium	Low	No risk
Flat	-	83	17	-
Piedmont	6	92	-	2
Low-mountain	3	69	1	27
Mid-mountain	8	67	5	20
High-mountain	56	35	-	10

To identify the most favorable landscape-climatic zones according to the level of risk, we used the formula (1)

$$\mathrm{CP} = \sum_{i=0}^{3} \boldsymbol{i} \cdot \boldsymbol{S_i} \qquad (1)$$

i *– level of risk*
S_i *- the fraction of the area with risk level i in the zone under consideration.*

The sum of points for all categories of risk that threaten the area, characterizes the average level of risk for the territory at all. For the classification of a territory by this indicator, each type of risk was assigned points:

Score	***Risk level***
• 0	No risk
• 1	Low
• 2	Medium
• 3	High

4. Discussion of the results

The research obtained the following data:

• Landscape-climatic zones	Score
• Flat	1.8
• Piedmont	2.0
• Low-mountain	1.5
• Mid-mountain	1.63
• High-mountain	2.38

An analysis of natural risks in the Kabardino-Balkar Republic showed that the entire territory of the Republic is a subject to various dangerous natural processes.

According to the results of the research, the territory was ranked in descending order of the natural hazards point risk.

The highest risk is the high-altitude zone.

Such dangerous natural processes as mudflows and avalanches that take the most disastrous forms are prevalent in the region. Destructive spontaneous processes in the Elbrus region are a serious problem for the recreational development of the territory, the solution of which is connected with the development of rational methods and measures of protection, and also with the identification of safe areas for perspective development of building and laying of tourist routes. In the recreational areas exposed to hazardous natural processes (in the area of Chegetskaya glade, Glade Azau, etc.), a large number of hotel complexes and other recreational facilities are built and are being built, and often with no permits for it [3].

Many of the existing routes and areas and newly created mountain-tourist complexes in the Elbrus region are under threat of dangerous natural processes, and some have already experienced the effects of these phenomena. So, the dangerous snow avalanche processes repeatedly destroyed the recreational complexes on the territory of Elbrus region. All recreational facilities built on Chegetskaya glade for the last 10–15 years are in the zone of avalanche danger [3].

In the valley of the river Adyrsu mudflows destroyed Climbing camp "Stal" and "Molnia". (1940), "Dzhailyk" (1983). In the valley of the river Adylsu by powerful blows of breakthrough mudflows in 1958 and 1959. mountaineering camps «Elbrus» and «Shhelda» were exposed. In 1967 mudflow along the river Kyrtyk was destroyed and damaged many houses in the village of Verkhnij Baksan. Several times (in 1966, 1967, 1983) under the influence of destructive mud-stone flows village Gubasantisu, many buildings and structures of the Baksan Neutrino Observatory, a highway leading to the Elbrus region was damaged. Less powerful demonstrations of mudflow activity occurred in many areas and caused damage of the main road connecting the lowland regions of the republic with the Elbrus region [7].

On 1.09.2017, a breakthrough of the Bashkar glacier lake occurred in the headwaters of the Adyl Su gorge. The breakthrough was accompanied by serious damage to the Prokhladny-Baksan-Elbrus highway. Alpine camp Adylsu was damaged and the Green Hotel was washed away next to the Bashkar glacier.

Of the 43 mudflow pools river Chegem, three of them - characterized by the volume of removal of more than 500 thousand m3 [6]. Two of them are located in the upper reaches on the left side of the valley, and one - from the starboard, it carries a threat for Bulungu. Mudflows of the Bulungu-Su river, which in the estuary part flows through the outskirts of the village of Bulungu, are especially dangerous. Mudflows here are threatened by the Basil, highway and agricultural land. Less powerful debris flows are less than 500,000 m3, but more than 100,000 m3, it means that the mudflow danger is strong, it threaten the villages of Verkhniy Chegem, Eltyuby, agricultural land, hydraulic structures, and roads.

Piedmont and flat zones are subject of medium risk. The main natural hazards are floods, hail, waterlogging, soil erosion, karst, earthquakes. About 760 km2 of the territory of the Kabardino-Balkar Republic are subject to flooding, where the cities of Prokhladny, Maisky, Terek are located, agricultural lands and objects of the agro-industrial complex are concentrated, and a significant part of the population lives in the republic [3].

Due to the peculiarities of the relief and climatic conditions almost annually in the warm period, the territory of the Kabardino-Balkar Republic is exposed to thunderstorm processes. On average in the republic for a year about 70 days are noted with heavy rains and thunderstorms. These phenomena are most active from the third decade of April to the end of September.

Flooding and the flood caused by them are a great danger. One of the most devastating floods was the June 2002. On the territory of the Kabardino-Balkar Republic, were destroyed 552 and damaged 3276 houses, 80 bridges, 340 km of roads were damaged, gas supply was disrupted in 12 settlements, 5 mudflow trays, 75 km of shore protection structures, 156 km of water lines, 66 water intakes, agricultural crops were destroyed on a large area. The hydrographic network of flood impact is represented in the republic by the Terek River and its tributaries of various orders. Almost the entire territory of the river Terek, that is, about 83 % of the Kabardino-Balkar Republic area. The natural flooding zone with a total area of 760 km2 is confined to the valleys of the Malka, Baksan, Chegem, Nalchik, Cherek and Terek rivers and is characterized by a high groundwater level with a low aeration zone, represented by loam and loamy sand.

In the areas of these natural risks there are sanatoriums, outflows of mineral waters, where a large number of recreants gather.

Low-risk and low-mountain areas are subject to low risk. There is a large area of zones where there is no risk (48 %). The risk is practically absent in the recreational areas of the city of Nalchik, in suburban areas. Also safe are the areas of Aushigersky hot springs and Blue Lakes and the Narzanov Valley in the west of the republic.

The main hazards are soil erosion, hail and floods. Mudflows are widespread in the west of the middle mountain zone. Thus, the mudflow along the Birdalysu river with a maximum recorded volume of more than 500,000 m^3, which was destroyed many times from 1909 to 2006. popular resort in the area of mineral springs Gili-su [7].

Table 2. Natural hazards affecting recreational facilities in different years

Date and time of Emergency Situations	Recreational facility	Brief Description of Emergency Situations
1940	Climbing camp "Stal" and "Molnia".	In the valley of the Adyrsu river, the "Stal" and "Molnia" climbing camps were destroyed by mudslides.
1958-1959	Climbing camps "Elbrus" and "Shhelda".	In the valley of Adylsu, climbing camps "Elbrus" and "Shhelda" were subjected to powerful blows from breakthrough mudflows.
9.01.1987	Glade Cheget	The avalanche blocked the entire Cheget glade, reached the Cheget hotel complex, destroyed the economic buildings, squeezed out the glass in the restaurant building, and the snow wave effect was detected on the upper floors of the hotel
12.01.1987	Upper river Baksan (Elbrus region)	A massive avalanche occurred as a result of prolonged intense snowfall. There were snow blockages on the roads, numerous cliffs of communication lines and power lines. Part of the rest in tourist hotels was evacuated
18.02.1998	Mountain Cheget	Snow avalanche. Damaged ski trail.
14.02.2000	Elbrus region	With the ascent of the mountain peak in the upper river. Shkhelda in a snow-ice avalanche group of 7 climbers hit. They all died.
17.04.2000	Tourist base Chegem	A snow avalanche between the village of Bulungu and the tourist base "Chegem." Road blocked
19.06.2000	Bezengi gorge	Rockfall. Cherek district, Bezengi gorge. The group of climbers from Samara has suffered
10.02.2004	Mount Cheget	As a result of an avalanche from the slope of Mount Cheget, 7 snowboarders died.
03.03.2005	Elbrus region , Terskol	Several avalanches came off, as a result of which the café on the Glade Azau was partially destroyed
11.08.2006	Mountain Elbrus. Gil-Su	There was a breakthrough of the adjacent lake Birdjalıran, as a result of which a water-stone mudflow was formed. In the area of the resort Dzhily-Su 50 thousand m3 were deposited. In the resort area in the floodplain of the river pools with mineral water, the foundations of the new bathroom pavilions, a road bridge and a section of the road were destroyed. There were no human victims.
11.02.2007	Adyr-Su gorge	As a result, avalanches killed 3 people
26.01.2009	Chegem waterfalls	The collapse of the rock. A dammed lake was formed that flooded the area.
1.09.2017	Elbrus region	Breakthrough Lake Bashkar. The breakthrough was accompanied by serious damage to the Prokhladny-Baksan-Elbrus highway. Alpine camp Adylsu was damaged and the

IOP Conf. Series: Materials Science and Engineering **663** (2019) 012058 doi:10.1088/1757-899X/663/1/012058

		Green Hotel was washed away at the foot of the Bashkar glacier.
24.03.2018	Glade Cheget	An avalanche came down to the parking lot.

On July 21, 2011, an intensive downpour (54 mm – sediment layer) caused mudflows and a rock block collapse, which led to rupture in several places of the gas pipeline, debris on the road in the section with Nizhny Chegem and v. Khushto-Syrt length of about 3 km and damage of private buildings on the site "Small Waterfalls". There were no casualties [8].

Landslides cause significant direct damage in their area of action, destroying roads, houses and structures, damaging pipelines, telephone and electrical networks, causing agricultural land [9]. The most dangerous are the landslide processes that are developing on the territory of cities. Relocation of significant rock masses as a result of landslides can lead to disastrous consequences and become a natural disaster [10,11]. A total of 140 landslides (landslide sites) have been registered in the Kabardino-Balkar Republic. Most of them are located in the pools of the Chegem, Baksan, Cherek rivers. Three landslides marked in the pool Malka.

Based on the literature data, a table 2 of natural hazards has been created that affected recreational facilities in different years.

The obtained data served as the basis for mapping the distribution of tourist and recreational areas of the Kabardino-Balkar Republic according to the level of danger of natural processes (Fig. 2).

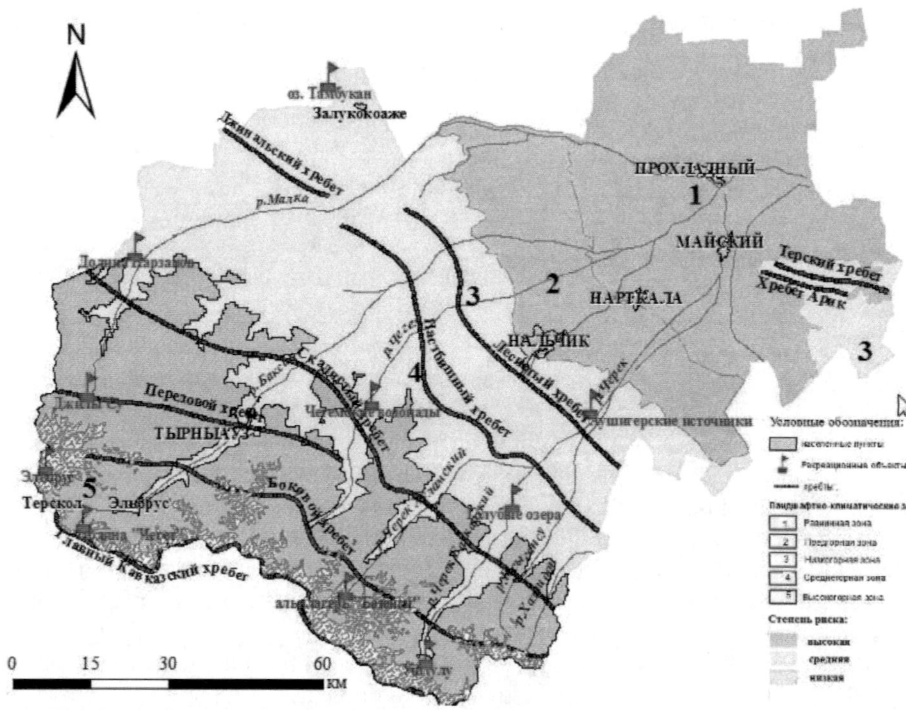

Figure 2. The basis for mapping the distribution of tourist and recreational areas of the Kabardino-Balkar Republic according to the level of danger of natural processes

5. Conclusion

The events of recent years show that the modern economic development and use of recreational potentially natural hazardous territories in the republic are not focused on the necessary observance of elementary rules and regulations ensuring the protection of people, objects, and territories. In the development of recreational areas, elementary precautions are often not observed - hazardous and potentially dangerous areas are developed or even settled. Most emergencies on the territory of the

IOP Conf. Series: Materials Science and Engineering **663** (2019) 012058 doi:10.1088/1757-899X/663/1/012058

Elbrus region, caused by the ddemonstration of dangerous natural processes and phenomena, occur in areas of active recreational activities and population living.

References

[1] Kuzmin B S 2007 Risk and danger in natural and man-made disasters *IX All-Russian Conf. with the participation of foreign scientists "Modern methods of mathematical modeling of natural and anthropogenic catastrophes"* (Barnaul) pp 43–7

[2] Aleksandrova A Yu 2008 *Evaluation methods* (Moscow: Knorus) 214 p

[3] Razumov V V, Shagin S I and Kolychev A G 2012 Natural hazards in the recreational zones of the Elbrus region *Georisk* **1** 28–38

[4] Buraev R A 1997 *Atlas of the Kabardino-Balkarian Republic* (Moscow)

[5] Uzdenova A B and Kanametova F E 2018 Recreational landscape assessment of the Kabardino-Balkar Republic *Trends in the development of sci. and ed.* **45** 123–9

[6] *Cadastre of avalanche-mudflow danger of the North Caucasus* 2001, ed. by Academician RAS M Ch Zalikhanov (St. Petersburg: Gidrometeoizdat) 112 p

[7] *Cadastre of mudflow danger in the south of the European part of Russia* 2015 (Nalchik: Printing House) 148 p

[8] Uzdenova A B and Kondratieva N V 2016 Estimation of the level of danger of natural processes in tourist and recreational areas of the Kabardino-Balkar Republic *ITportal* **1(9)** Retrieved from: http://itportal.ru/science/economy/otsenka-stepeni-opasnosti-prirodnykh/

[9] Crozier M J 1999 Slope instability: landslides. In: Alexander D, Fairbridge R W (eds.) *Encyclopaedia of Environmental Science* (Dordrecht) pp 561–2

[10] De Marchi B, Scolobic A, Delli Zotti G and Del Zotto M 2007 *Risk Construction and Social Vulnerability in an Italian Alpine Region* Task 11: Risk Perception, Community Behaviour and Social Resilience, Case Study (Italy)

[11] Staffler H, Pollinger R, Zischg A and Mani P 2008 Spatial variability and potential impacts of climate change on flood and debris flow hazard zone mapping and implications for risk management *Natural Hazards and Earth System Sci.* **8** 539–58

IOP Conf. Series: Materials Science and Engineering **663** (2019) 012059 doi:10.1088/1757-899X/663/1/012059

Adapted methodology for comprehensive contamination assessment of bottom sediments on the content of various forms of heavy metals

V S Valiev, D V Ivanov, I I Ziganshin, A B Aleksandrova, R R Khasanov, D E Shamaev and V V Malanin

Research Institute for Problems of Ecology and Mineral Wealth Use of Tatarstan Academy of Sciences, AS RT, 28, Daurskaya str., Kazan, 420087, Russia

E-mail: podrost@mail.ru, adabl@mail.ru

Abstract. As a result of long-term studies of bottom sediments of lakes, rivers, ponds and reservoirs of the Republic of Tatarstan, considerable factual material has been accumulated, which enabled the translation of quantitative estimates into qualitative characteristics. This necessitated a comprehensive assessment and generalization of the exceedance of the thresholds set for all metals, similarly to the specific combinatorial water pollution index (SCWPI). At the same time, the authors emphasize the features of such a comprehensive assessment of bottom sediments, and also suggest methods and algorithms for solving emerging problems. We use the method of frequency characteristics of data series of metal content (Cd, Pb, Co, Cu, Ni, Zn, Cr, Mn, Fe) in bottom sediments, score the relevant characteristics, generalize the point estimates and bring them to the appropriate classes of purity. As a result, a method for the integrated assessment of contamination of bottom sediments was developed and tested, which solves the problem of the integral assessment of the quality of bottom sediments in terms of a complex of pollutants using the example of heavy metals. The method is based on the approach of the SCWPI. Therefore, its results are comparable to combinatorial assessments of water quality, which will allow deriving a uniform assessment of the water body as a whole.

1. Introduction

Management of environmental management and environmental quality involves accounting and assessing the indicators of the state of all phase components in this environment: the atmosphere, hydrosphere, lithosphere and biota. The most important link accumulating and redistributing the flows of matter and energy is the hydrosphere formed by a complex system of water bodies [1, 2].

Bottom sediments formed due to sedimentation of inorganic and organic material suspended in water play a significant role in shaping the chemical composition of surface water bodies. Therefore, assessing the degree of contamination of water bodies involves not only determining the quality of their aquatic environment, but also a generalized characteristic of bottom sediments [2, 3].

The peculiarity of bottom sediments, as an indicator of the state of a water body, is that they are the last link in the input of substances into water bodies and watercourses, thereby integrating the geochemical features of the catchment areas, man-made discharges and water bodies [4].

IOP Conf. Series: Materials Science and Engineering **663** (2019) 012059 doi:10.1088/1757-899X/663/1/012059

2. Problem statement

In Russia, the control of the composition of bottom sediments of water systems is carried out within the framework of the Unified State System for Monitoring the Environment. Despite the fact that control of the composition of bottom sediments is provided, it can hardly be effective in the absence of clear criteria for the quality of bottom sediments, which combine the whole range of pollutants into a single indicator. The problem of qualitative assessment of bottom sediments is aggravated by the diversity of their composition, which, nevertheless, can be reduced to two main factors: the particle size distribution and the content of organic matter [5–8]. Therefore, a comprehensive integrated assessment of contamination of bottom sediments is currently relevant for a complex of pollutants, in particular, heavy metals.

3. Materials and methods

The work uses the data of long-term studies of bottom sediments of lakes, rivers, ponds and reservoirs of the Republic of Tatarstan [9, 10]. To work out the established assessment methodology, the content of metals (Cd, Pb, Co, Cu, Ni, Zn, Cr, Mn, Fe) was considered as pollutants in bottom sediments, taking into account the granulometric composition of bottom sediments and the content of organic matter in them. The analysis of the long-term data series allowed us singling out the statistical background, the actual threshold of the normal regional content of various forms (mobile and gross) of the nine scrutinized metals in bottom sediments [11].

We have generalized the exceedances of the thresholds by all metals, similarly to the specific combinatorial index of water pollution (SCWPI).

The system of formalized indicators in our proposed method uses the frequency characteristics of data series about the content of different metals in bottom sediments, scoring system for these characteristics, compiled scores and reduced figures to the appropriate purity classes, similarly to the methodological guidelines RD 52.24.643-2002 developed by the Hydrochemical Institute of the Federal Service of Russia for Hydrometeorology and Environmental Monitoring (Roshydromet).

The contamination level of bottom sediments is estimated by a relative characteristic calculated on the basis of the actual concentrations of the aggregate pollutants and the corresponding background values (standards). The estimate is based on the aggregation of information about the detection frequencies of concentrations exceeding the background values and the multiplicities of this excess. The combination of the level of contamination of bottom sediments with various metals and the frequency of detection of these cases makes it possible to obtain a complex characteristic summarizing a number of indicators in the form of a particular purity class.

The additivity of the action of various metals with their simultaneous presence is realized by summing individual indicators that evaluate the contribution of each of them separately.

In addition to the system of formalization of the estimated parameters of the SCWPI, for the compatibility of both assessments, a similar final gradation of purity classes was used, implemented in the form of five basic and three complementary gradations.

At the same time, instead of linear interpolation for scoring, we used nonlinear one (polynomial and logarithmic), as it more truly encompasses the breakdown of ranges. For example, Fig. 1 shows the dependence of S_α scores (in the terminology of [12]) on the exceedance occurrence frequency of. At the same time, we used a 5-point scale of ranks similar to SCWPI: <1, 1-10, 10-30, 30-50,> 50, with a ranking from 0 to 4.

Another important difference from the classical method of calculation is the difference in the scale evaluating the multiplicity of exceedances (Appendix G [12]). The scale of scores that take into account the multiplicity of exceedances in relation to bottom sediments, in our opinion, is too stretched. At the same time, the authors of the technique of the SCWPI suggest researchers, in some cases, to set the gradations themselves, depending on the nature and toxicity classes of the substances under consideration. Explicitly, such an amendment to the RD [12] is present only for oxygen, and it is emphasized that the proposed scale is calculated for substances of the 2-class of hazardousness.

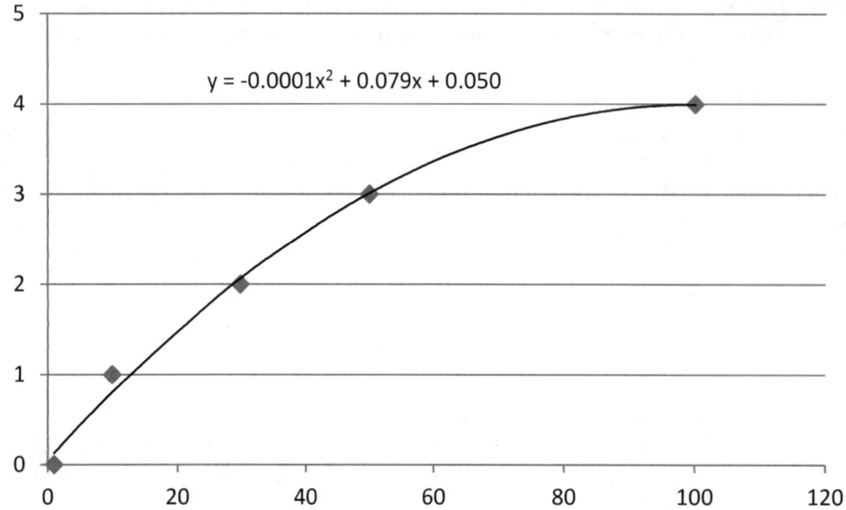

Figure 1. Polynomial dependence of the scoring S_α on the exceedance frequency

Due to the fact that we assess the quality of bottom sediments relative to background concentrations of pollutants, it was decided to estimate the scale (similar to Appendix G [12]), taking into account not only the hazard classes of various metals, but also the actually observed ranges of their concentrations in bottom sediments. As a result of analyzing the frequency characteristics of variational series of our own observations, the boundaries of the scale were the averaged frequency characteristics of the variational series of metal contents in bottom sediments, reduced to their median. For example, the rounded values of the ratios of the maximum values of the 95% confidence interval observed over all the variation ranges of concentrations to their medians form the limiting (most extreme) range of the scale, starting from 20.

4. Discussion

As a result of the study, the following base scale was determined, which is adjusted by a coefficient that takes into account the hazard class of a particular metal being evaluated:

- <1 (0 points)
- 1-3 (1 point)
- 3-9 (2 points)
- 9-20 (3 points)
- > 20 (4 points)

The formed scale is also logarithmic, which is clearly shown in Fig. 2.

Hazard class adjusts the scale as follows:

Scale rank multiplicity = Rank multiplicity × (0.5 × Hazard class).

That is, the current gradation of the exceedance multiplicity scale is calculated for hazard class 2. For the first class, the ranges of its gradations will be, respectively, 2 times (× 0.5) lower, and for grade 3 1.5 times higher.

Among the metals we considered, the first class (highly hazardous) substances include Cd, Pb, and Zn; the second class (moderately hazardous) includes Co, Ni, Cu, Cr; the third class (little dangerous) makes up Mn. When calculating the gradations of the scale of Fe exceedances, we also refer to the third class, although this metal is not considered dangerous.

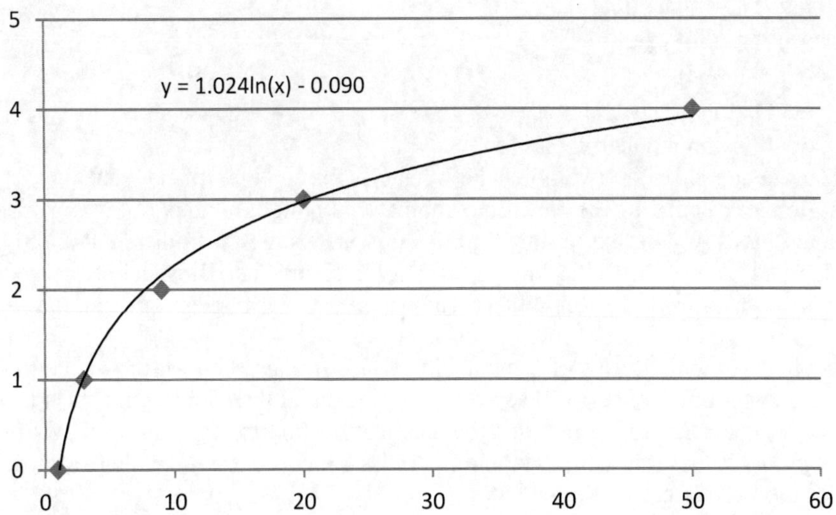

Figure 2. Logarithmic dependence of the scoring S_α on the exceedance frequency

Since most of the heavy metals in water bodies are bound with organic or inorganic ligands, the bottom sediments of water bodies can theoretically adsorb both metal ions and their complex compounds with dissolved organic and inorganic substances. A special role in the binding of metal ions into a durable chelate complex is played by humic substances, widely represented in pelitic fractions (particles less than 0.01 mm). Thus, active sorption by bottom sediments of metals depends on the share of pelitic fractions in them. Respectively, the content in bottom sediments of the fraction of pelitic fractions above 30 % forms their special capacity, theoretically capable of accumulating metal concentration 2–3 times higher than bottom sediments with a low content this fraction.

At the same time, when summarizing estimates considering a series of observations as a whole, the total proportion of samples with a high content of fine particles is important rather than the absolute content of pelitic fractions in a particular sample. For such an assessment, we have introduced another factor correcting the background indicator itself against which frequencies and multiplicities are calculated. The correction is a linear relationship: $K_{pelit} = 0.48 \times D + 0.52$; where D is the proportion of observations with a particle content of <0.01 mm more than 30 % expressed as fraction of 1. The dependence is established as a result of analyzing the frequency characteristics of concentration series of various metals in bottom sediments with different particle size distribution. In addition, the calculated background values also rely on an indicator such as a pelitic fraction content of more than 30 %, suggesting higher background concentrations for such bottom sediments.

Accounting for critical indicators of pollution, in relation to bottom sediments, is also somewhat modified and does not come from $S_\alpha \geq 9$, accounting for 56.3 % of the maximum possible value (9 out of 16), but from $S_\alpha \geq 12$ (75 %), which we think is more justified if we consider that we are talking about critical pollution.

Since the value of the specific combinatorial index is the weighted average of all quotients of S_α, it is convenient to present this indicator in a normalized (reduced to fractions 1) form, simply dividing it by 16 (the maximum possible value of S_α). Thus, the analyst more clearly determines the proportion of current pollution relative to the maximum possible value.

The final grades of grade points, adapted to the assessment of sediments, are as follows:

With the values of the specific combinatorial index (not normalized, that is, not reduced to 16):

- <1 (Clean bottom sediments, class 1 cleanliness)
- 1-3 (Slightly polluted, class 2)
- 3-5 (Moderately Polluted, class 3a)
- 5-6 (Heavily polluted, class 3b)

- 6-8 (Dirty, class 4a)
- 8-9 (Dirty, class 4b)
- 9-10 (Very dirty, class 4c)
- 10-11 (Very dirty, class 4d)
- > 11 (Extremely Dirty, class 5).

In this gradation, there are two points between classes, and one subclass each.

The system of calculating the specific combinatorial pollution index of bottom sediments (SCWPI-BS), adapted by us, was tested on three groups of water bodies with different anthropogenic load: 1) Forest lakes and lakes of the Raifsky area of the Volga-Kama State Biosphere Reserve; 2) Lakes in rural areas of various regions of the Republic of Tatarstan; 3) Water bodies of Kazan located directly within the city.

The content of mobile forms of 9 metals in surface (up to 20 cm) layers of bottom sediments was assessed, the upper limits of regional standards were taken as threshold levels [11] (Table 1). The upper limits were accepted due to the fact that the classical method for calculating SCWPI does not operate with background values, but with maximum permissible concentrations, the excess of which uniquely determines the toxic effect.

Table 1. Summary table of concentrations of gross and movable forms of metals recommended for use in calculating SCWPI-BS

Metal	**Rivers and water reservoirs**	**Lakes**
	Gross forms	
Cd	0.92	0.62
Pb	18.8	25.2
Co	15.5	13.8
Cu	31.1	37.4
Ni	63.9	45.7
Zn	74.9	103.2
Cr	44.7	39.8
Mn	925.0	773.5
Fe	26227.0	26746.7
	Movable forms	
Cd	0.42	0.38
Pb	3.70	7.34
Co	0.68	0.81
Cu	1.45	3.27
Ni	2.14	3.31
Zn	7.49	19.4
Cr	1.22	1.04
Mn	318.3	350.6
Fe	351.5	252.5

The calculation of SCWPI-BS for three groups of reservoirs showed the adequacy of the results obtained to expert assessments carried out by experts. The results of the calculation are presented in Table 2.

Recall that the indicated indices S are normalized and reduced to fractions 1, therefore, when compared with ULAIR, they can be multiplied by 16. Kaverage is calculated similarly to the SCWPI method.

Table 2. The result of SCWPI-BS calculation for three groups of reservoirs with different anthropogenic load

Group	$S_{combinatorial}$	$S_{specific}$ (SCWPI-BS)	$K_{average}$	Purity class
1	0.55	0.06	2.47	Clean 1
2	1.84	0.20	11.3	Moderately polluted 3a
3	4.15	0.46	21.0	Dirty 4b

5. Conclusions

Thus, the accepted and widely used methodology for assessing combinatorial water pollution can be adapted for assessing the quality of bottom sediments.

The system of formalized indicators of the adapted method uses the same frequency characteristics of the data series: scoring system for these characteristics, compilation of the scores and bringing them to the appropriate purity classes as the original guidelines of RD 52.24.643-2002. However, the heterogeneity of the composition of bottom sediments should be taken into account, as well as the distribution of metals in different layers, their hazard class, and the calculations should be adjusted using appropriate coefficients.

References

[1] Ke X, Gui S, Huang H, Zhang H, Wang C and Guo W 2017 Ecological risk assessment and source identification for heavy metals in surface sediment from the Liaohe River protected area, China *Chemosphere* **175** 473–81

[2] Li R, Chai M and Qiu GY 2016 Distribution, fraction, and ecological assessment of heavymetals in sediment-plant system in mangrove forest, South China Sea *PLos One* **11** e0147308

[3] Lili L et al 2019 Assessment of the Geographical DetectorMethod for investigating heavy metal source apportionment in an urban watershed of Eastern *China Sci. of the Total Environment* **653** 714–72

[4] Dauvalter V and Kashulin N 2018 Assesment of the ecological state in the arctic freshwater system based on concentrations of heavy metals in the bottom sediments *Geochemistry international* **56** 842–56

[5] Tomilina I I and Komov V T 2002 Bottom sediments as an object of toxicological studies (review) *Biology of inland waters* **2** 20–6

[6] MacDonald D D, Ingersoll C G and Berger T A 2000 Development and Evaluation of Consensus-Based Quality Guidelines for Freshwater Ecosystem *Archives of Environmental Contamination and Toxicology* **39** 20

[7] Royer G B and Burgos R W 2002 The roles of natural organic matter in chemical and microbial reduction of ferric ion *Sci. of the Total Environment* **307** 167–78

[8] Wasserman J, Oliveira F and Bidarra M 2003 Cu and Fe associated with humic acids in sediments of a tropical coastal lagoon *Organic Geochemistry* **28** 813–22

[9] Valiev V S, Ivanov D V, Ziganshin I I, Shamaev D E, Malanin V V and Marasov A A 2016 Metals form distribution in lakes sediments of natural and urban areas of the republic of Tatarstan *Proc. of the Karelian Res. Centre of the Russian Academy of Sci.* vol 9 pp 57–67

[10] Ivanov D V, Khasanov R R, Malanin V V and Zakonnov V V 2015 The spatial division of metals in sediments of the Kuibyshev reservoir *Water: chemistry and ecology* **82** 81–6

[11] *Development of regional standards for the background content of pollutants in bottom sediments of water bodies of the Republic of Tatarstan* Report on the State Contract no 18ME-33C as of 2 November 2018 (Kazan: Institute of Environmental Protection of the Academy of Sciences of the Republic of Tatarstan) 124 p

[12] *Methodical instructions. A method of integrated assessment of surface water pollution degree by hydrochemical indicators* RD 52.24.643-2002 (approved and introduced by the Federal Service for Hydrometeorology and Environmental Monitoring on December 3, 2002)

IOP Conf. Series: Materials Science and Engineering **663** (2019) 012060 doi:10.1088/1757-899X/663/1/012060

A new efficient method of adaptive filter using the Galois field arithmetic

H H Alaeddine[1], **K Charafeddine**[2] **and M E Vikhrov**[2]

[1]Electronics and Physics Department, Faculty of Sciences I - Lebanese University, HKS Laboratory, Beirut, Lebanon
[2]Department of electric supply of industrial enterprises, National Research University MPEI, Moscow, Russia

E-mail: vikhrovme@mail.ru

Abstract. This paper proposes an efficiently implementation of Multi-Delay Filter Block Recursive Least Squares (MDF-BRLS) algorithm. This implementation uses a particular transform that is defined on a finite ring of integers with arithmetic carried out modulo Fermat numbers. In term of performances, this Fermat Number Transforms (FNT) is ideally suited to digital computation, requiring approximately $N\log_2 N$ additions, subtractions and bit shifts, but no multiplications unlike the Fourier transform (FFT). FNT implementation results confirm that the MDF-BRLS adaptive filter can implement with smallest computational complexity compared with an implementation using the Fourier transform.

1. Introduction

Adaptive filters are widely used in the area of signal processing such as echo cancellation, noise cancellation, channel equalization for communications and networking systems. Necessity of adaptive filters implementations is growing up in many fields. An adaptive filter is a filter that self-adjusts its transfer function according to an optimizing algorithm [1, 2]. Because of the complexity of the optimizing algorithms, most adaptive filters are digital filters that perform digital signal processing and adapt their performance based on the input signal. Adaptive filters are time-varying since their parameters are continually changing in order to meet a performance requirement [3, 4]. The implementation process requires various of performances such as good convergence characteristics, lower execution time and reduce computational complexity. However, it is difficult to satisfy these characteristics simultaneously, so efficient algorithms and efficient architectures are desired.

The design of adaptive filter algorithm is an important part within the design of adaptive filter. The fast Multi-Delay Filter Block Recursive Least Squares (MDF-BRLS) algorithm [5] is one of the efficient adaptive algorithm since here the process of filtering and adaption is done in frequency domain without overlap and occupied the smallest block size using FFT method [6–8].

The linear convolution, used by the MDF-BRLS algorithm, is one of the most important digital signal processing problems. It can be implemented more efficiently based on the Fermat number transform (FNT) than based on the fast Fourier transform (FFT) since the FNT possesses the property of cyclic convolution and low computational complexity. The expensive multiplication in the FFT and inverse (IFFT) can be replaced by bit shifts in the FNT and its inverse (IFNT) with the transform kernel 2 or its integer power [9, 10].

The principal objective of our study is to reduce the computational complexity of the MDF-BRLS algorithm. In order to ensure this, we looked further into the mathematical bases of the Number Theoretic Transform (NTT) and in particular the Fermat Number Transform (FNT), compared to the Fast Fourier Transform (FFT), which allows reduction of several multiplications which are necessary to achieve certain functions such as convolution products.

The rest of the paper is organized as follows. In second section, we will present the MDF-BRLS algorithm. Third Section presents the Number theoretic transform and the implementation technique of the MDF-BLMS algorithm, and in particular the calculation of the circular convolution, via a particular transform, called Extended-FNT. Then, simulation results and computational complexity estimation of the MDF-BRLS algorithm using the Extended-FNT are presented in section four and compared to those obtained using the FFT.

2. A multi-delay BRLS algorithm

2.1 Block recursive least squares (BRLS)

The weight vector $\hat{w}_k$ of BRLS algorithm is given by [5]:

$$\hat{w}_{k+1} = \hat{w}_k + \frac{\lambda^{-1} P_{k-1}}{1+\lambda^{-1}\tilde{x}_k^T P_{k-1}\tilde{x}_k}.\left(\tilde{x}_{-k} * \varepsilon_k\right) = \hat{w}_k + \frac{\lambda^{-1} P_{k-1}}{1+q_k}.\left(\tilde{x}_{-k} * \varepsilon_k\right) = \hat{w}_k + \alpha_k\left(\tilde{x}_{-k} * \varepsilon_k\right) \tag{1}$$

The different parameters and vectors are exposed below:

- $\hat{w}_k = \left[\hat{w}_k(0) \quad \hat{w}_k(1)\ldots\hat{w}_k(L-1)\right]^T$ where L represents the length of the impulse response of the filter.
- $\tilde{x}_{-k}$ is the inverse-time of the input sequence $\tilde{x}_k = \left[\tilde{x}_{kN-L+1} \quad \tilde{x}_{kN-L+2}\ldots\tilde{x}_{(k+1)N-1}\right]^T$ of length $N+L-1$.
- $\varepsilon_k = \tilde{d}_k - \tilde{y}_k = \left[e_k(0) \quad e_k(1)\ldots e_k(N-1)\right]^T$ is the error vector.
- λ represents the forgetting factor.
- $q_k = \lambda^{-1}\tilde{x}_k^T P_{k-1}\tilde{x}_k$, where P_k is a matrix of ($M \times M$) dimensions, and characterizes the inverse correlation matrix by the equation seen below:

$$P_k = \lambda^{-1}\left(P_{k-1} - G_k\left(\tilde{x}_k^T P_{k-1}\right)\right) \tag{2}$$

M represents the lowest power of 2 larger than or equal to $N+L-1$.

- $G_k = \alpha_k.\tilde{x}_k$ represents the Kalman gain.
- * denotes the linear convolution.

2.2 Multi-Delay Filer BRLS (MDF-BRLS)

The decomposition of the two vectors ε_k and $\tilde{x}_{-k}$ without been overlapped was beyond the key idea of the MDF-BRLS algorithm [8], and that is on the contrary of the preceding MDF [6], [7]. That is done mainly according to the following vectors:

$$\tilde{x}_{-k} = \left[\bar{x}_{-k}^{0\,T} \quad \bar{x}_{-k}^{1\,T} \quad \ldots \quad \bar{x}_{-k}^{K'-1\,T}\right]^T$$

$$\varepsilon_k = \left[\bar{\varepsilon}_k^{0\,T} \quad \bar{\varepsilon}_k^{1\,T} \quad \bar{\varepsilon}_k^{K'-1\,T}\right]^T$$

Wherein they turns, $\overline{x}_{-k}^{k'}$ and $\overline{\varepsilon}_{k}^{k'}$ are in fact vectors of length $(M / K') < M'$ defined by:

$$\overline{x}_{-k}^{k'} = \left[\mathbf{x}_{\left((k'+1)\frac{M}{K'}\right)-1} \; \mathbf{x}_{\left((k'+1)\frac{M}{K'}\right)-2} \cdots \mathbf{x}_{\left(k'\frac{M}{K'}\right)} \right]^T$$

$$\overline{\varepsilon}_{k}^{k'} = \left[e_{\left(k'\frac{M}{K'}\right)} \; e_{\left(k'\frac{M}{K'}\right)+1} \cdots e_{\left((k'+1)\frac{M}{K'}\right)-1} \right]^T$$

The exhibited problematic of the matrix multiplication among the two coefficients $\alpha_k^{k'}$ and $\psi_k^{k'} = \left(\overline{x}_{-k}^{k'} * \overline{\varepsilon}_k^{k'}\right)$, and how it can be solved presents in fact the novelty of our choice. Actually, a problem of calculation is presented by the matrix α_k which is of ($M \times M$) dimensions, once an updating of the filter coefficients through the MDF-BRLS algorithm is taken place. There are essentially *N-1* samples difference among the FFT of the two sequences $\tilde{x}_{-k}^{k'}$ and $\tilde{x}_{-k}$. Accordingly, the multiplication between $\alpha_k^{k'}$ and $\psi_k^{k'}$ gives a result differs from that attained by using BRLS algorithm.

Therefore, the MDF-BRLS proposes to decompose the matrix α_k, into K' temporarily matrices $\alpha_k^{k'} (k' = 0,1,\ldots,K'-1)$, of size $\left(\frac{M}{K'} \times M\right)$:

$$\alpha_k = \begin{bmatrix} \alpha_k^0 \\ \alpha_k^1 \\ \vdots \\ \alpha_k^{K'-1} \end{bmatrix} \tag{3}$$

Then, each matrix $\alpha_k^{k'}$ is decomposed to K' matrices $\alpha_k^{k',i} (i = 0,1,\ldots,K'-1)$, of $\left(\frac{M}{K'} \times \frac{M}{K'}\right)$ size according to the following vector:

$$\alpha_k^{k'} = \left[\alpha_k^{k',0} \quad \alpha_k^{k',1} \ldots \alpha_k^{k',K'-1}\right] \tag{4}$$

Also, the inverse correlation matrix P_{k-1} should also be minimized into K' temporarily matrices $P_{k-1}^{k'}$ of dimension $\left(\frac{M}{K'} \times M\right)$:

$$P_{k-1}^{k'} = \left[P_{k-1}^{k',0} \quad P_{k-1}^{k',1} \ldots P_{k-1}^{k',K'-1}\right] \tag{5}$$

Each matrix $P_{k-1}^{k'}$ is composed of K' matrices $p_{k-1}^{k',i}$ of size $\left(\frac{M}{K'} \times \frac{M}{K'}\right)$:

$$p_{k-1}^{k',i} = P_{k-1}\left(k'\frac{M}{K'} \ldots \left((k'+1)\frac{M}{K'}\right) - 1 \;,\; i.\frac{M}{K'} \ldots \left((i+1)\frac{M}{K'}\right) - 1\right) \tag{6}$$

From the above matrix decompositions, we define the calculation of the scalar q_k according to this relation:

$$q_k = \lambda^{-1} \tilde{x}_k^T . (P_{k-1} \tilde{x}_k) = \lambda^{-1} \tilde{x}_k^T . \beta_k \tag{7}$$

As the vector P_{k-1}, we suggest the division of the vector β_k into K' sub-vectors $\beta_k^{k'}$. Each sub-vector $\beta_k^{k'}$ is defined by:

$$\beta_k^{k'} = p_{k-1}^{k'}.\bar{x}_k = \left[P_{k-1}^{k',0} \quad P_{k-1}^{k',1} \ldots P_{k-1}^{k',K'-1}\right].\begin{bmatrix} \bar{x}_k^0 \\ \bar{x}_k^1 \\ \vdots \\ \bar{x}_k^{K'-1} \end{bmatrix} = \sum_{i=0}^{K'-1} P_{k-1}^{k',i}.\bar{x}_k^i \tag{8}$$

where $\bar{x}_k^{k'} = \left[\mathbf{x}_{\left(k'\frac{M}{K'}\right)} \ \mathbf{x}_{\left(k'\frac{M}{K'}+1\right)} \ldots \mathbf{x}_{\left((k'+1)\frac{M}{K'}\right)-1} \right]^T$

Thus, q_k and $\alpha_k^{k'}$ which are consecutively scalar and matrix could be written as related by the two equations below:

$$q_k = \lambda^{-1}.\bar{x}_k^T.\beta_k = \lambda^{-1}.\left[\bar{x}_k^0 \quad \bar{x}_k^1 \ldots \bar{x}_k^{K'-1}\right].\begin{bmatrix} \beta_k^0 \\ \beta_k^1 \\ \vdots \\ \beta_k^{K'-1} \end{bmatrix} = \lambda^{-1}.\sum_{i=0}^{K'-1} \bar{x}_k^i.\beta_k^i \tag{9}$$

$$\alpha_k^{k'} = \frac{\lambda^{-1}.P_{k-1}^{k'}}{1+q_k} = \frac{\lambda^{-1}.\left[P_{k-1}^{k',0} \quad P_{k-1}^{k',1} \ldots P_{k-1}^{k',K'-1}\right]}{1+.\lambda^{-1}.\sum_{i=0}^{K'-1} \bar{x}_k^i.\beta_k^i} \tag{10}$$

Count on the earlier equations, the updated weight equation of L'-points MDF-BRLS algorithm is presented now by:

$$\hat{w}_{k+1}^{k'} = \hat{w}_k^{k'} + \alpha_k^{k'}.\psi_k^{k'} = \hat{w}_k^{k'} + \sum_{i=0}^{K'-1} \alpha_k^{k',i}.\left(\psi_k^i\right) \tag{11}$$

The $L = K'.L'$ samples of weight $\hat{w}_{k+1}$ occurs in the last $n = \frac{K'L}{M}$ products $\alpha_k^j.\psi_k$, $j = n,\ldots,K'-1$.

For this reason and to get K' vectors of length L', the result of each product $\alpha_k^j.\psi_k$ of $\frac{M}{K'}$ size should be divided to $\frac{M}{K'L'}$,. This technique reduces the computational complexity of the MDF-BRLS algorithm because the first $\frac{K'(M-L)}{M}$ products of $\alpha_k^{k'}.\psi_k, k' = 0,1,\ldots\frac{K'(M-L)}{M}-1$, must be excluded, therefore the weight vector $\hat{w}_k$ of MDF-BRLS algorithm takes the following concluding form:

$$\hat{w}_{k+1}^{k'} = \hat{w}_k^{k'} + \alpha_k^{k'}.\psi_k^{k'} = \hat{w}_k^{k'} + u_k^{k'} \tag{12}$$

with $u_k^{k'} = H_k^j\left(k'L'\ldots\left((k'+1)L'\right)-1\right)$, $j = n,\ldots,K'-1$

$$H_k^j = \sum_{i=0}^{K'-1} \alpha_k^{j,i}.\left(\ \psi_k^i\right) \tag{13}$$

On the other hand, to get the inverse correlation matrix P_k (2) updated, a similar decomposition method of that presented in the previous equations must be respected:

$$P_k = \lambda^{-1}\left(P_{k-1} - \frac{\lambda^{-1}(P_{k-1}\tilde{x}_k).(\tilde{x}_k^T P_{k-1})}{1+\lambda^{-1}\tilde{x}_k^T P_{k-1}\tilde{x}_k}\right) = \lambda^{-1}.\left(\left(P_{k-1}^{k'}\right) - \frac{\lambda^{-1}.\left(\beta_k^{k'}.\overline{\beta}_k^{k'}\right)}{1+\lambda^{-1}.\sum_{i=0}^{K'-1}\overline{x}_k^i.\beta_k^i}\right) \quad (14)$$

Each matrix $P_{k-1}^{i,k'}$ $(i = 0,\ 1,\ldots,\ K'-1)$ of size $\left(\frac{M}{K'} \times \frac{M}{K'}\right)$ is defined as follow:

$$P_{k-1}^{k'} = \begin{bmatrix} P_{k-1}^{0,k'} \\ P_{k-1}^{1,k'} \\ \vdots \\ P_{k-1}^{K'-1,k'} \end{bmatrix}$$

$$p_{k-1}^{i,k'} = P_{k-1}\left(i.\frac{M}{K'}\ldots\left((i+1)\frac{M}{K'}\right)-1,\ k'.\frac{M}{K'}\ldots\left((k'+1)\frac{M}{K'}\right)-1\right) \quad (15)$$

Taking into account the matrix P_k, we define the vector $\overline{\beta}_k$ as follow:

$$\overline{\beta}_k = \left[\overline{\beta}_k^0 \quad \overline{\beta}_k^1 \ldots \overline{\beta}_k^{K'-1}\right] \quad (16)$$

$$\overline{\beta}_k^{k'} = \tilde{x}_k^T.P_{k-1}^{k'} = \left[\overline{x}_k^{0T} \quad \overline{x}_k^{1T} \ldots \overline{x}_k^{K'-1T}\right].\begin{bmatrix} P_{k-1}^{0,k'} \\ P_{k-1}^{1,k'} \\ \vdots \\ P_{k-1}^{K'-1,k'} \end{bmatrix} = \sum_{i=0}^{K'-1}\overline{x}_k^{iT}.P_{k-1}^{i,k'} \quad (17)$$

The MDF-BRLS algorithm processes the adaptive filter using the smallest block size. Consequently, it allows reducing, as much as possible, the computational complexity and the execution time of this algorithm.

3. Extended number theoretic transform (ENTT)

3.1 The Fermat number transform (FNT)

The Fermat Number Transform (FNT) belongs to class of transform known as Number Theoretic Transform (NTTs), defined over Galois Field (GF($\hat{M}$)). The onward and opposite FNTs are well-defined as [9]:

$$\mathbf{X}_k(m) = \left\langle \sum_{n=0}^{M-1} x(n).\partial^{n.m} \right\rangle_{F_t} \quad (18)$$

$$x_k(n) = \left\langle M^{-1}\sum_{m=0}^{M-1}\mathbf{X}(m).\partial^{-n.m} \right\rangle_{F_t} \quad (19)$$

$$m,n = 0,\ 1,\ldots,M-1.$$

Respectively, where F_t, is the characteristic of the Galois Field (GF(F_t)), M is the transform length,

∂ is a root of unity of order M in GF(F_t) and $\langle\cdot\rangle_{F_t}$ stands for residue reduction modulo F_t. The characteristic, F_t, is the tth Fermat number, defined as [9], [10]:

$$F_t = 2^b + 1 = 2^{2^t} + 1 \tag{20}$$

The choice of a modulo equal to a Fermat number F_t offers numerous possibilities for the length M of the transform (Table 1). The values of M and ∂ associated with a FNT are given by $M = 2^{t+1-i}$ and $\left\langle \partial = 2^{2^i} \right\rangle_{F_t}$ with $0 \le i < t$.

Table 1. Constraints between F_t ∂ and M.

Modulus	Dynamic range	M		
F_t	Bits	$\partial=2$	$\partial=3$	$\partial=\sqrt{2}$
$F_3 = 2^8+1$	2^7	16	256	32
$F_4 = 2^{16}+1$	2^{15}	32	65536	64
$F_5 = 2^{32}+1$	2^{31}	64	-	128

In an FNT over F_t, the quantity $\partial = \sqrt{2}$ represents the integer $2^{2^{t-2}}\left(2^{2^{t-1}} - 1\right)$.

It can be shown that the FNT possess the Linear Convolution Property (LCP). The FNT can therefore be employed to perform convolution or correlation via point-by-point multiplication in the transform-domain in the same manner as the FFT, thus:

$$f_k = z_k * h_k = FNT^{-1}\left(FNT(z_k) \bullet FNT(h_k)\right) \tag{21}$$

Since the FNT is defined over a fine ring, convolution (or correlation) via (21) has the highly desirable property that no roundoff errors are introduced. Note, however, that the filter output, f_k, must lie in the range $\left[-(F_t/2),\ (F_t/2)\right]$, , which requires that, the relation:

$$|k_n|_{\max},\ |h_k|_{\max} \le \sqrt{2^{2^t}/2^{2^{t-2}}} \tag{22}$$

Be satisfied. Otherwise, severe amplitude aliasing results. A similar restriction applies to the output when the NTT is used for correlation.

3.2 Extended Fermat number transform (EFNT)

To achieve the goal of implementing the MDF-BRLS algorithm without multiplications, the solution is to extend the idea of the FNT. Indeed, in MDF-BRLS algorithm, the linear convolution between two sequences, each of length M, must be calculated by finding $2K'$ FNT and K' IFNT each transform of length M/K'.

$$f_k^{k'} = z_k^{k'} * h_k^{k'} = EFNT^{-1}\left(EFNT\left(z_k^{k'}\right) \bullet EFNT\left(h_k^{k'}\right)\right) \tag{23}$$

$$k' = 0,\ 1, \ldots, K'-1.$$

The extended FNT (EFNT) and its inverse are proposed by the following equations (24 and 25).

$$\mathbf{F}_k^{k'}(m) = EFNT\left(f_k^{k'}(m)\right) = \left(EFNT\left(z_k^{k'}(m)\right) \bullet EFNT\left(h_k^{k'}(m)\right)\right) = \mathbf{Z}_k^{k'}(m) \bullet \mathbf{H}_k^{k'}(m)$$

$$=\left\langle \sum_{n=0}^{2^{t+1-i}-1} f_k(n).\left(2^{2^{t-2}}\left(2^{2^{t-1}}-1\right)\right)^{n.m}\right\rangle_{F_t} \quad (24)$$

$$f_k^{k'}(m)=\left\langle -2^{2^{t-i}-1-(t-i)} \sum_{n=0}^{2^{t+1-i}-1} \mathbf{F}_k(n).\left(2^{2^{t-2}}\left(2^{2^{t-1}}-1\right)\right)^{-n.m}\right\rangle_{F_t} \quad (25)$$

$$m=k'.\frac{M}{K'},\ \left(k'.\frac{M}{K'}\right)+1,\ldots,\left((k'+1).\frac{M}{K'}\right)-1$$

Where $\mathbf{F}_k=\left[\mathbf{F}_k^0 \quad \mathbf{F}_k^1 \quad \ldots \quad \mathbf{F}_k^{K'-1}\right]$.

To highlight, schematically, the efficiency of the Linear Convolution (LC) implementing by the Extended FNT, we present a block diagram depicted in Figure 1 for $L=K'=2$ and $M=4$. The parallel computing of the K' LC, without multiplications, decreases the execution time and the computational complexity of the MDF-BRLS algorithm.

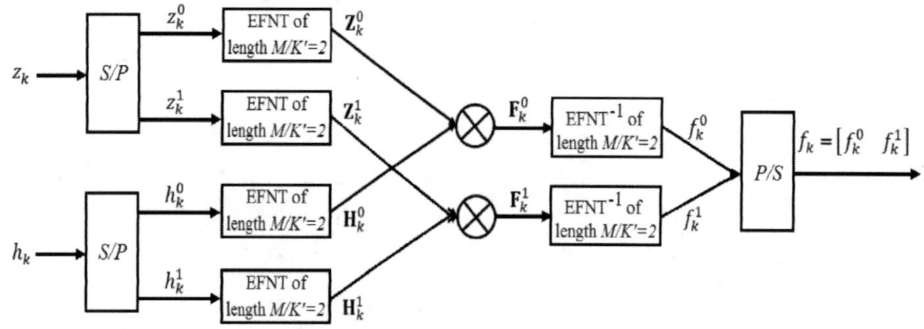

Figure 1. Block diagram of Linear Convolution using EFNT for $L=K'=2$ and $M=4$

4. Simulation results

The performance of the MDF-BRLS algorithm implemented with FFT and Extended-FNT transforms is evaluated by computer simulation using Matlab. In these simulations, the MDF-BRLS algorithm is considered, in single-talk situation, with the parameters listed in Table 2.

Table 2. Simulation parameters.

Parameter	Value
Number of new input samples	$N=129$
Length of filter's impulse response	$L=128$
The forgetting factor	$\lambda=1$
Length of input sequence	$M=256$
Number of sub-vectors	$K'=32$
Length of sub-impulse response	$L'=4$
FFT/EFNT length	$M/K'=M'=8$
Value of Fermat number	$F_3=2^8+1$

The size of the impulse response is $L=128$, that matches a delay of 16 ms used for a sampling rate of 8 kHz. In single-talk state, the acoustic echo cancellation system should reduce the echo to around 24 dB for delays lesser than 25 ms and to around 40 dB for delay beyond 25 ms.

The criteria used to evaluate the performance of MDF-BRLS algorithm implemented with FFT and EFNT are compared as follows:

- The filter weight error convergence N_m :

$$N_m = 10\log_{10}\left(\frac{\|w-\hat{w}\|^2}{\|w\|^2}\right)$$

- The response of the echo track impulse
- The computational complexity

Based on the equation of the weight error convergence, the MDF-BRLS performance is simulated and presented in Figure 2.

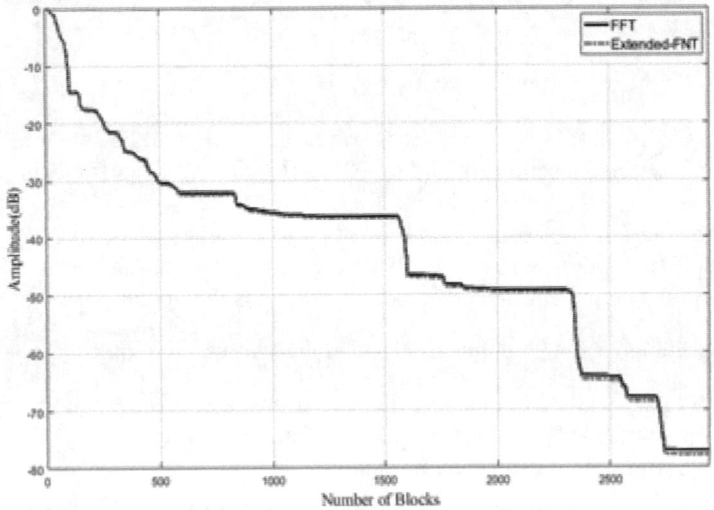

Figure 2. Adaptation normalized misalignement for MDF-BRLS algorithm implemented with FFT (solid line) and Extended-FNT (dashed line)

The result obtained in figure just before presents a perfect construction of echo's path of the MDF-BRLS algorithm (convergence is around –78 dB), using FFT and Extended-FNT transforms. This figure shows also that both transforms, FFT and Extended-FNT, do not reveal any significant difference in terms of convergence.

The impulse response of the echo path calculated by both transforms FFT and Extended-FNT is seen in Figure 3.

This result obtained in figure above, by using the MDF-BRLS algorithm were analyzed with both transforms, FNT and Extended FNT revealed that there was no difference in terms of construction of echo's path (error difference is around 10^{-16} dB). Furthermore, the result obtained presents a perfect construction of echo's path, which means that the residual echo is unheard at the output of the echo cancellation system using both methods of implementation.

Based on the obtained values, seen in Table III, the differences between both methods of implementation reside mainly in the computational complexity of the filter.

As well known, that the FFT and Extended-FNT are fundamentally having the same implementing, our EFNT-based block adaptive filtering has been applied with convolution steps based on the FFT procedure. One difference that distinguishes the Extended-FNT computation from other techniques, arises from the use of finite arithmetic modulo, the Fermat number $F_3 = 2^8 + 1$, that can be realized based on the conventional binary arithmetic.

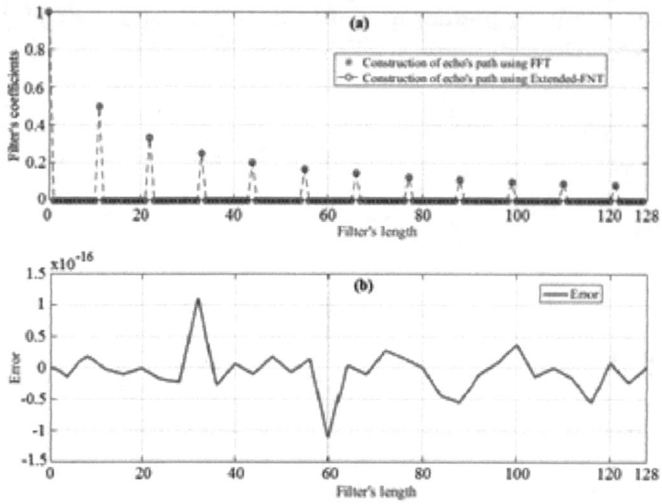

Figure 3. (a) Impulse response of the echo path. (b) Difference between the echo path and the MDF-BRLS's filter coefficients

The global computational effectiveness of the implementation is directly related with the computational complexity of the algorithm MDF-BRLS based on both transforms, FFT and EFNT. The size of FFT and the EFNT used and their numbers, determines the operations number as additions, bit shifts and multiplications. A M'-point EFNT can be shown to require $\frac{3.M'}{2}\log_2(M')$ basic operations such as bit shifts and additions but without multiplication. Based on this hypothesis, Table 3 gives the total operation number per block for each linear convolution used by MDF-BRLS algorithm and implemented with the both transforms, FFT and EFNT.

Table 3. Basic operations and multiplications required for each linear convolution implemented with FFT and extended - FNT

Transformation	Basic operations	Multiplications
FFT	$3.K'.M'\log_2(M') = 2304$	$K'M'\left(1+\frac{3}{2}\log_2(M')\right) = 1408$
Extended-FNT	$\frac{9.K'}{2}M'\log_2(M') = 3456$	$K'.M' = 256$

Besides, this previous realization of the MDF-BRLS filter is computationally further competent because of the computational proficiency of the Extended Fermat number transform.

5. Conclusion

Throughout this paper, an Extended Fermat Number Transform (EFNT) to reduce enormously the computational complexity and in particular the multiplications have been presented. Moreover, an implementation of the MDF-BRLS algorithm using EFNT reduced enormously the number of multiplications of the implantation compared with an implementation of the same algorithm using FFT. The algorithm using EFNT finds a numerous application in several domains include, power electronics control [11], [12], radar system, echo cancelation in teleconference room and in headset. Our proposition can be more advanced to tackle the double talk situation, and other matters.

References

[1] Farhang-Boroujeny B 1998 *Adaptive Filters. Theory and Applications* (New York: Wiley & Sons)

[2] Bashirov M G, Khusnutdinova I G, Khusnutdinova L G and Usmanov D R 2016 Electromagnetic-Acoustic Method for Assessing the Technical Condition of Energy Equipment *Industrial energy* **12** 8–13

[3] Mohammed M S and Kim K S 2016 Preliminary Treatment of a Signal Based on The Compression of Noise for a Filter Using a Recursive Algorithm of The Least Squares in Ultrasonic Material Control *Defectoscopya* **6** 51–9

[4] Benesty J, Gansler T, Morgan D, Sondhi M and Gay S L 2001 *Advances in Network and Acoustic Echo Cancellation* (Berlin: Springer-Verlag)

[5] Alaeddine H H, Ghaith A, Al-Sahili M and Fadlallah M 2017 New Approach of Performance Analysis of RLS adaptive Filter Using a Block-Wise Processing *European Academic Research* **V** 1030–53

[6] Soo J S and Pang K K 1990 Multidelay block frequency domain adaptive filter *IEEE Transactions on Acoustics, Speech, and Signal Processing* **38** 373–76

[7] Buchner H, Benesty J, Gansler T and Kellermann W 2006 Robust Extended Multidelay Filter and Double-Talk Detector for Acoustic Echo Cancellation *IEEE Transactions on Audio, Speech and Language Processing* **14** 1633–44

[8] Alaeddine H H, Houssini M, Baghious E H and Burel G 2018 *IEEE Int. Conf. on Selected Topics in Mobile and Wireless Networking* (Morocco, Tangier) pp 1-6

[9] Julien G A 1991 Number theoretic techniques in digital signal processing *Advances in Electronics and Electron Physics* **80** 69–163

[10] Alaeddine H H, Bazzi O, Alaeddine A and Burel G 2012 Fast convolution using generalized sliding Fermat number transform with application to digital filtering *IEICE Transactions on Fundamentals of Electronics, Communications and Computer Sciences* E95.A **6** 1007–17

[11] Charafeddine K, Sangov K and Tsyruk S 2017 *IEEE-IT-DREPS Conf.* (Jordan, Amman) pp 153–8

[12] Falshina V A and Kulikov A L 2011 Algorithms for Simplified Digital Filtration of Electrical Signals of Industrial Frequency *Industrial Energy* **4** 39–46

IOP Conf. Series: Materials Science and Engineering **663** (2019) 012061 doi:10.1088/1757-899X/663/1/012061

Prevention of sticking during surface casing drilling in Samotlorskoye field

N Aksyonova

Tyumen Industrial University, 38, Volodarsky St., Tyumen, 625000, Russia

E-mail: na-acs@yandex.ru

Abstract. The paper considers the current problem of sidewall sticking during surface casing in Samotlorskoye field. The analysis of geological and field material was carried out and the reasons for sticking during drilling of upper unstable argillaceous deposits of Lulinvorskaya, Talitskaya and Gankovskaya suites were determined. It is found that in most well construction projects, the inhibitory agents preventing clay hydration are used in drilling muds. The authors studied the mechanisms of shale inhibition. The results of experimental studies on the impact of various inhibitor additives on clay samples demonstrated the efficiency of Polysil and Osnova GS (hydrophobizated fluid) additives with sulphoacid. The results of the analysis made it possible to make proposals and recommendations to prevent sticking.

1. Relevance of the problem

Well drilling is a complex and critical technological process, which is sometimes accompanied by various complications and accidents. According to statistics, in Samotlorskoye field the largest number of accidents accounts for sticking (Figure 1). Thus, sometimes there is a situation in the interval of surface casing drilling when the drill string cannot be lifted, ran or rotated, thus causing sticking. The stuck freeing is expensive and causes complications and in many cases may end up with torpedoing of the tool and re-drilling of part of a well bore [4]. Special tools, professional technical services and special mud conditioning to release a drill pipe cause expenses, not to mention the lost time of a drilling rig. The stability of open borehole may deteriorate during time required to release the pipe and restart drilling. For this reason, the prevention of pipe sticking and accelerating of their release if such occurs is critical for well drilling.

2. Theoretical aspects of the problem

Drill pipe sticking may be caused by the following:

1. Keyseating on well walls;
2. Particles in a well;
3. Borehole with a diameter less than nominal;
4. Pressure drop in a wellbore;
5. Pipe sticking to well walls when it is left without movement;
6. Violation of well washing technology;
7. Others

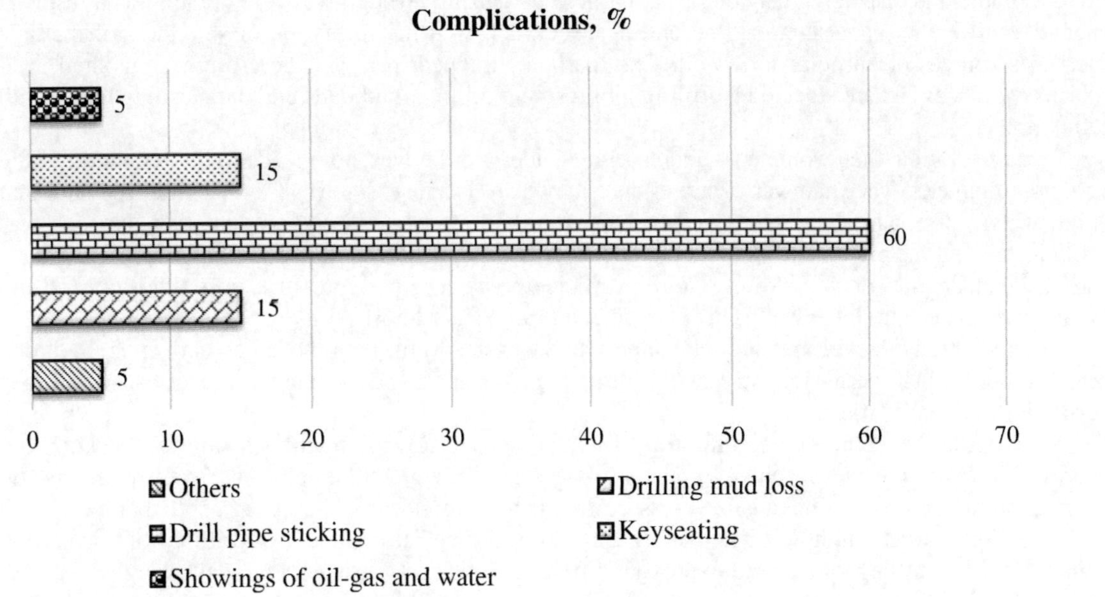

Figure 1. Diagram of distribution of complications and accidents during well drilling

Figure 2 shows the diagram of reasons causing sticking of a drill bit in a well. The largest percentage of sticking cases results from the pressure drop, when a tool is left without movement and the reduction of the nominal well diameter.

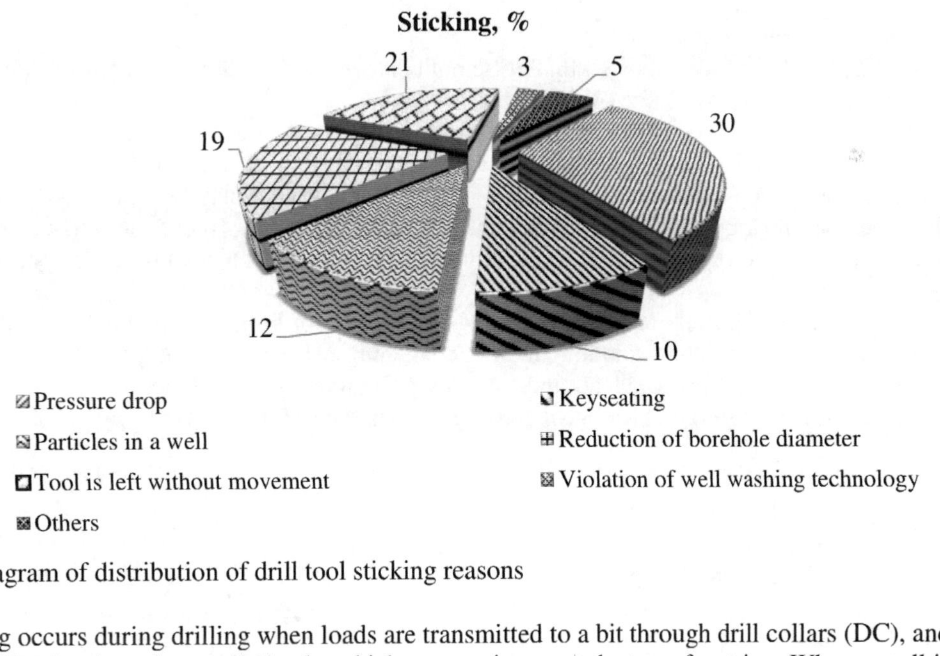

Figure 2. Diagram of distribution of drill tool sticking reasons

Keyseating occurs during drilling when loads are transmitted to a bit through drill collars (DC), and as a rule a drill string is under tensile loads, which are maximum at the top of a string. When a well is vertically deviated and drilling continues this part of a bore is opposite a stretched drill pipe and tends to dig into a well wall, thereby forming a key seat where the drill string is stuck. If a well near the vertical one is drilled, the problems with keyseating on well walls are minimized.

Swellable and dispersing clay deposits, particles occurring through wellbore collapse as a result of negative differential pressure, drilling cuttings, rock debris at the formation of shoulders of the hole serve as sources of particles in a well. Clay particles trapped in a well bore are accumulated in at shrinking places and are stuck on a drilling tool forming drilling mud balls and bars leading to sidewall sticking [1–3].

Shale hydration, their volumetric increase and slide into the well bore leads to the reduction of the nominal diameter. The diameter of a wellbore may also decrease due to abnormal pore pressure and high water content in shale under geostatic pressure. In the process of sedimentation the rock was compacted and the pore fluid was compressed in closed pores of the manifold. When such clay rocks are drilled using low density drilling fluid insufficient to balance pore pressure, pore fluids expand and shale is squeezed out to the well. This phenomenon can occur almost at any depth.

The reduction of a well nominal diameter may occur due to formation of thick clay crust on a rock surface due to high permeability, pressure drop, high filtration and prolonged impact of water-based drilling fluid on clay rocks.

Sticking due to wellbore pressure drop (differential sticking) is defined as a sidewall sticking to a permeable formation as a result of excess pressure of a drilling fluid column in a well bore over the formation fluid pressure. Sticking due pressure drop in a wellbore can occur at any depth, but most often at large depths where high-density washing fluid is used. Special treatment of drilling fluid is carried out and special drilling tool is used to prevent differential sticking. As a preventive measure, strict filter control and addition of surfactants to water-based drilling fluids are carried out [5]. In some drilling areas drilling fluids based on hydrocarbon oil are used. Frequent movement of a drill pipe is a standard operating method to prevent sticking and maintain the nominal diameter of a well bore during deep drilling using heavy washing fluids. It is also effective to change square and spiral DC, which reduce the contact area between DC and a clay crust. The probability of differential sticking increases when a drill string is left for a long time without movement and rotation and in the absence of circulation of a drilling fluid.

3. Field geologic analysis

The analysis of field data showed that at Samotlorskoye field, sticking cases most often occur in the interval of Lulinvorskaya Suite (390–560 m).

In our opinion, clays, which alternate with sand, when a clay content makes 95 % primarily cause sticking during surface casing drilling in Samotlorskoye field (Table 1). It is known that clays are characterized by high dispersion and hydrophilicity. Hydration of clay minerals leads to clay swelling, its dispersion and transition into solution, which leads to softening of rocks forming well walls, increase of clay concentration in a solution, and accordingly to the change of technological properties of solutions [1].

The analysis of drilling fluids used during surface casing drilling in Samotlorskoye field showed that in most cases polymer clay solutions mainly containing pH, viscosity and water recovery regulators, weighting agents, lubricating agents and antisealing additives are used. Only in some projects inhibitory additives "Osnova GS" and "Kolmasil" are included in the drilling mud [6].

4. Experimental analysis

Inhibition of drilling fluids is ensured by the following:

- ion exchange (chlorocalium, calcium or chlorosodium, spersene, formate or potassium solutions, Kla-Gard, Kla-Cure, Kla-Stop);
- encapsulation of particles of drilled clays and well walls by creating polymer films (Poly Plus RD, Polypac R, SP-101);
- encapsulation of drill clay particles and well walls by creating water-resistant silicate and glycol films;
- creation of waterproof barrier in pores and channels of drilled rocks with hydrophilic additives (Asphasol or Glycol);

- encapsulation of drill rock particles and well walls due to multipolymer solutions;
- hydrocarbon-based solutions both having no water in the formulation and hence being neutral to drilled clays.

Table 1. Lithological well profile during surface casing drilling

Stratigraphic unit index	Stratigraphic unit	Vertical interval from (top)	Vertical interval to (bottom)	Short rock name	Clay content, %	Type of complication leading to sticking
Q	Quaternary deposits	0	125	sand clay sand clay loam	95.0	
$P_{2/3}$	Zhuravskaya Suite	125	140	clay sand clay sand	95.0	
$P_{2/3}$	Novomikhailovskaya Suite	140	240	clay sand clay sand	95.0	Packing, differential sticking. Loss of bore hole wall stability
$P_{1/3}$	Atlymskaya Suite	240	350	gaize clay	95.0	
$P_{1/3}$	Cheganskaya Suite	350	390	clay sand	95.0	
$P_{2/2}$	Lyulinvorskaya Suite	390	560	sandstone gaize clay	95.0	
P_1	Talitskaya Suite	560	655	siltstone clay sandstone	95.0-	
K_2	Gankinskaya Suite	655	790	clay siltstone	95.0	

To determine the best inhibitory capacity, clay samples were tested in 20 % aqueous solutions of various reagents (Figure 3). The experiments showed that the increase of clay samples in non-dispersing drilling fluids was different and ranged from 13.6 to 56.4 %.

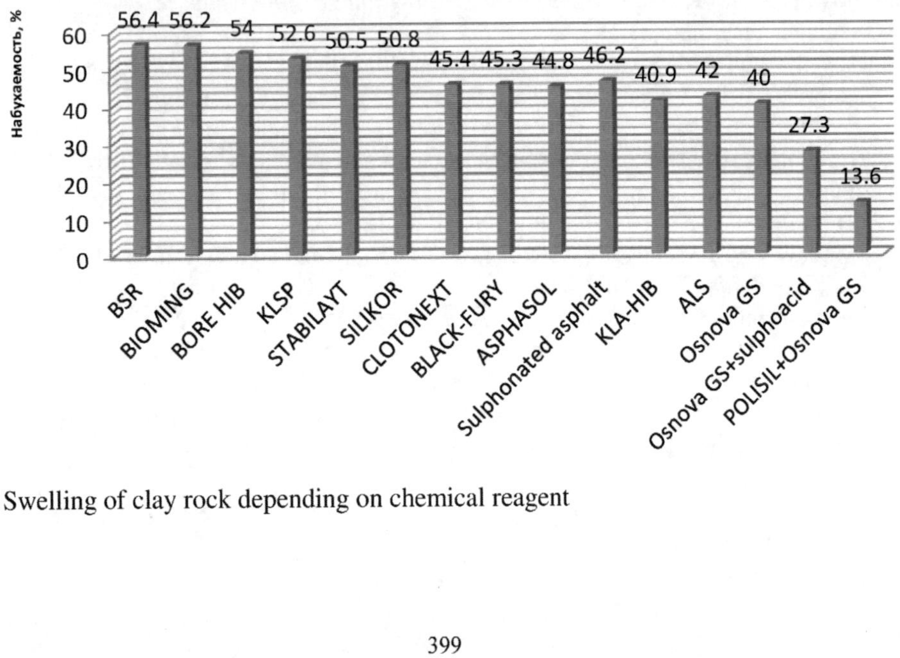

Figure 3. Swelling of clay rock depending on chemical reagent

The obtained results confirm the advantages of solutions based on complex reagent Polysil and Osnova GS with sulfonic acid. The final swelling over 12 hours of this solution makes 13.6 %. A good indicator is that the clay rock samples remained solid after being exposed to the solution.

Osnova GS additive hydrophibizes clay surface of drilled rock and well walls. It prevents the dispersion of slurry, stabilizes unstable rocks prone to collapse and landslide, prevents tool packing. Besides, Osnova GS stabilizes rheological and filtration parameters of the drilling fluid, including under conditions of salt aggression.

According to TU U 26.8-34656408-002:2010, POLISIL is a complex reagent containing sodium or potassium silicate; modified bitumen; nitrogen-containing polymer. The synergetic effect of three reagents ensures high inhibitory capacity of drilling fluid with POLISIL.

Modified bitumen blocks cracks of a well bore, prevents collapse and landslide, as well as hydration of clay shale.

The main components of bitumen: asphaltenes – crystalline solids, resins – amorphous solids and oils, which may contain dissolved waxes. Using the above components the chemical agent blocks microcracks in sloughing argillites. Sodium silicate is used to suppress hydration and swelling of clay materials. Silicate molecules are adsorbed in active areas of a clay surface. By shielding these areas, the hydration and swelling processes of clay materials are suppressed.

Nitrogen-containing polymer increases viscosity of washing liquid filtrate thus affecting the reduction of water yield index. As a result, the penetration of a filtrate into fractured clay rocks is reduced. Besides, nitrogen-containing polymers have low inhibitory capacity.

5. Conclusions and recommendations

Thus, geological-field material and analysis of design documents for well construction showed that despite high content of clay minerals in rocks during surface casing drilling the inhibitory additives are not designed in drilling mud solutions in most well construction projects.

This analysis allows recommending the following:

- To prevent washing of unstable sands, maintain the relative viscosity to a depth of 350-400 meters (up to Lyulinvorskaya Suite) shall be maintained not less than 70 sec.
- During drilling of clay deposits POLISIL and Osnova GS inhibitory reagents shall be introduced.
- Talitskaya Suite shall be uncovered using bulk density drilling mud.
- In order to limit the drilling mud impact on clay deposits and reduce the probability of well wall washing, the interval of surface casing drilling shall be drilled at the maximum possible speed.

References

[1] Aksenova N A and Rozhkova O V 2016 *Drilling and drilling wells: textbook for universities* In 3 volumes (Tyumen: TIU)

[2] Konesev G V and Aksenova N A 2017 Drilling fluids In the book: *Drilling technology of oil and gas wells: textbook for university students* In 5 volumes (Tyumen) pp 7–301

[3] Lipatov E Y and Aksenova N A 2017 Experience of application of biopoly-emulsion field while drilling in the Koshilskoye field *Scientific works of NIPI Naftogaz SOCAR* **4** 36–41

[4] Ovchinnikov V P, Ismakov R A, Oganov A V et al 2014 *Oil and gas well drilling technology* ed V P Ovchinnikova vol 4 (Tyumen: TSOGU) 418 p

[5] Ovchinnikov V P, Aksenova N A, Rozhkova O V and Grosheva T A 2013 *Modern compositions of drilling fluids* (Tyumen: Express)

[6] Khaibdrakhmanov D T, Aksyonova N A and Enginoev K B 2018 Analysis of sticking causes during surface casing drilling in Samotlorskoye field. In the collection: Experience, actual problems and prospects for the development of the oil and gas complex *Mat. of the VIII Int. Scientific and Practical Con. of Students, Postgraduates and Scientists* In 2 volumes, ed Yu B Chebykina pp 116–21

Efficiency analysis of hydrodynamic enhanced oil recovery methods on the example of Las-Eganskoye field

V Dyagilev, N Lazutin, T Dyagileva and V Baksheev

Tyumen Industrial University, 38, Volodarsky St., Tyumen, 625000, Russia

E-mail: glibazval.dyagilev@yandex.ru, kpw72@yandex.ru, LazutinaTV@yandex.ru, baksheev46@yandex.ru

Abstract. The majority of oil fields in Western Siberia are at the late stage of their development. Therefore, the minimization of economic and technological costs aimed at enhanced oil recovery is becoming a primary task. Hydrodynamic methods of enhanced oil recovery (HM EOR) fully satisfy these requirements, which makes them the most promising in further development of oil fields in Western Siberia. The methodology includes measurement and description, which use allows analyzing and assessing the study of HM EOR application. The purpose of this study is to demonstrate the efficiency of HM EOR on the example of Las-Eganskoye field. The study of hydrodynamic methods of enhanced oil recovery included the analysis of results, which allowed developing recommendations for the implementation of HM EOR in the oil fields of Western Siberia. It is shown that the technological opportunity of HM EOR requires clarification based on the current state of field development taking into account the peculiarities of reservoir pressure maintenance system, technical condition of pipelines and the stock of active injection wells proposed for cyclic injection.

1. Introduction

At present, the search for effective methods and technologies that provide for qualitative and quantitative increase of the volume of extracted natural reserves is becoming ever more critical in the field of oil production and refining technologies. Therefore, the core object of this study is Las-Eganskoye field located within the territory of Nizhnevartovsk district of Khanty-Mansi Autonomous District – Ugra of Tyumen Region, 120 km north-west of the district center Nizhnevartovsk and 30 km north of Langepas.

As of 01.01.2015 the efficiency analysis of hydrodynamic methods (HM) for enhanced oil recovery (EOR) in Las-Egan field was carried out from January 2010 to December 2014.

2. Literature review

The scientific community was centered around the problem of the rational use of hydrocarbon resources [10] that was theoretically solved in the works of domestic [1, 5] and foreign studies [6, 7], which are focused on the search for methods to forecast the technological efficiency of oil production and evaluation. The analysis of modern methodology of resource development within oil and gas complexes [3, 4] showed the timeliness of the study of existing techniques and principles of assessing the current state of oil reserves in northern territories of the Russian Federation. Modern researchers overestimate the existing methods to forecast oil production [1, 2, 8] thus suggesting the latest techniques aimed at predicting the technological efficiency of oil production and evaluation.

IOP Conf. Series: Materials Science and Engineering **663** (2019) 012062 doi:10.1088/1757-899X/663/1/012062

3. Methods

Hydrodynamic methods of enhanced oil recovery (HM EOR) included the analysis of shutdowns of injection wells, forced fluid production (FFP) and development of production wells for injection (DWI).

4. Results and discussion

The non-stationary (cyclic) water flooding (NWF) at the facilities of Las-Eganskoye field TPP Langepasneftegaz under special programs based on the study of geological features of the field and technological possibilities of their implementation was carried out in 2013 and 2014.

Since the number of planned measures at field facilities is not sufficient for statistical assessment of NWF efficiency and subsequent calculation of predicted oil production, the analysis of this measure regarding the shutdown of injection wells was carried out. Under certain conditions the shutdown of injection wells can be considered as NWF. Such conditions include the following: 1) shutdown of the injection well for more than 7 days (in this case the shutdown time of the injection well is considered as the half-cycle of NWF); 2) injectivity at the moment of shutdown more than 50 m^3/day. Such shutdowns are carried out to ensure various types of well interventions (WI), which are not intended to obtain additional oil production from NWF, hence no preliminary study of geological and technological features favorable to NWF were carried out. In this case, this action is mainly characterized by the side effect of the main purpose of the injection well shutdown.

A 50 % increase in the liquid rate in production wells was taken as FFP. Besides, the fluid flow rate after FFP shall on average remain at the same level within at least 6 months after WI. The fluid withdrawal is usually increased due to replacement of a pump with a pump that is more efficient, less often due to the change of the pump speed (ESP) or reciprocating speed (SRP). In documents such activities are usually referred to as Operation Mode Optimization (OMO), sometimes as Pump Revision and Change (PRC).

Based on these conditions during the period from January 2010 to December 2014, as of 01.01.2015, 434 activities were performed in developed objects of the field, $AV_{1\text{-}2}$, BV_6, BV_8, AC and JV_1, which resulted in the effect from HM EOR. Among them 29 well operations were performed according to a specially developed and justified program.

Since 2013, all NWF works at Las-Eganskoye field were planned, so the maximum number of activities fell within 2013 and 2014, which is 104 and 85 WI respectively. The minimum number of activities was carried out in 2010 (39). In 2011 and 2012, approximately equal number of activities were carried out (50 and 49 respectively).

Figure 1 shows the distribution of HM EOR by year of the analyzed period of development and objects of Las-Yeganskoue field.

We see that the maximum number of NWF was performed in 2013 (104) and in 2014 (85), the minimum – in 2010 (39). In 2011 and 2012, approximately the same number of activities were carried out (50 and 49 respectively), of which 226 WI in $AV_{1\text{-}2}$, 76 in JV_1 and 23 in BV_6.

FFP was most often performed in 2011 and 2014 (15 and 13 activities respectively). In 2010, 2012 4 activities were carried out, in 2013 – 7. At the same time, the maximum number of FFP was carried out in JV_1 (20) and $AV_{1\text{-}2}$ (16), the minimum – in BV_6.

Similar to FFP, the maximum number of DWI was carried out in 2013 – 23 WI. Since 2011, the number of DWI was reduced from 23 to 5 in 2014. The maximum number of DWI for the entire period under review was performed in JV_1 (34) and $AV_{1\text{-}2}$ (26). 2 FFP actions were carried out in BV_6. In AC there were 2 NWF and 1 DWI.

The effect from each performed action was calculated. The effect of HM EOR refers to additional actions in a formation site, which is influenced by the change of the operation modes of injection wells (NWF, DWI) or production wells (FFP).

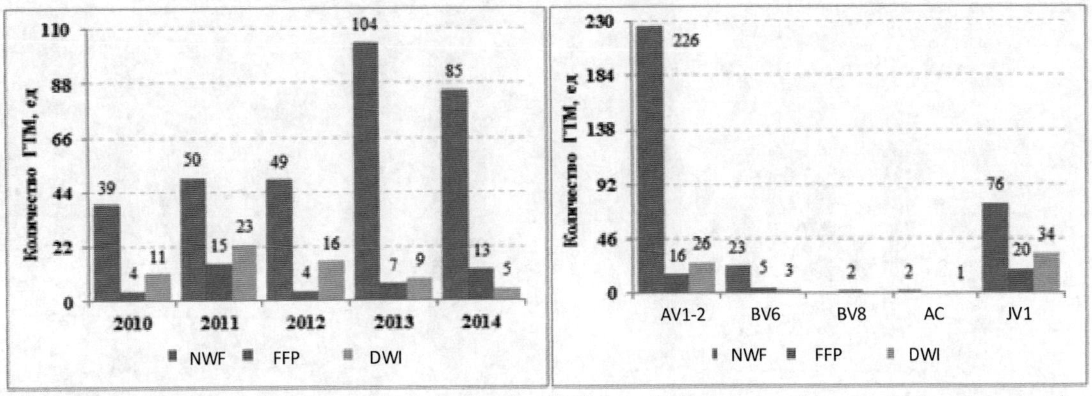

Figure 1. Distribution of hydrodynamic methods of enhanced oil recovery by year of development in Las-Eganskoye field [5]

The number of observation wells on the site was selected from the nearest wells where the operations were carried out based on peculiarities of object development in different periods. If the observation wells were in the zone of influence of several wells, where HM EOR were performed simultaneously (or sequentially, in 1–5 months), then such sites were combined into one and additional oil production was calculated based on their joint impact.

223 sites were selected for efficiency calculations.

The effect was calculated for all observation wells of the site for the next six months after the operations were completed. If well operations were carried out with some periodicity (from 15 days to 4 months) or the site was influenced by two or more shutdown wells, the effect was usually calculated after six months from the last shutdown.

The overall effect was calculated according to Medvedsky's characteristic of displacement in Delta software, which allows dividing the overall effect into two components: oil recovery and development stimulation. The positive effect of oil recovery is understood as additionally extracted oil due to WI, which under the basic development would not be extracted (WI resulted in the improvement of oil displacement characteristics). The positive effect of development stimulation is understood as additionally recovered oil in a given period of time (at calculations for a period of 6 months), which under the basic development would be extracted from the formation for a longer period of time.

Alongside with NWF, other WI were carried out on some selected areas, for example, perforation (isolation of separate intervals), start-up of a production well, transfer to another object with subsequent perforation, well fracturing, bottom hole treatment, etc. If their obvious effect on additional oil production from HM EOR was observed, then such wells were excluded from the calculation of additional oil produced in the site.

Calculations show that HM EOR for the last 5 years of development allowed producing additional 101.34 thousand tons of oil. Figure 2 shows the distribution of the total quantity of additional oil production and the quantity of HM EOR in Las-Eganskoye field by years (left) and by objects (right).

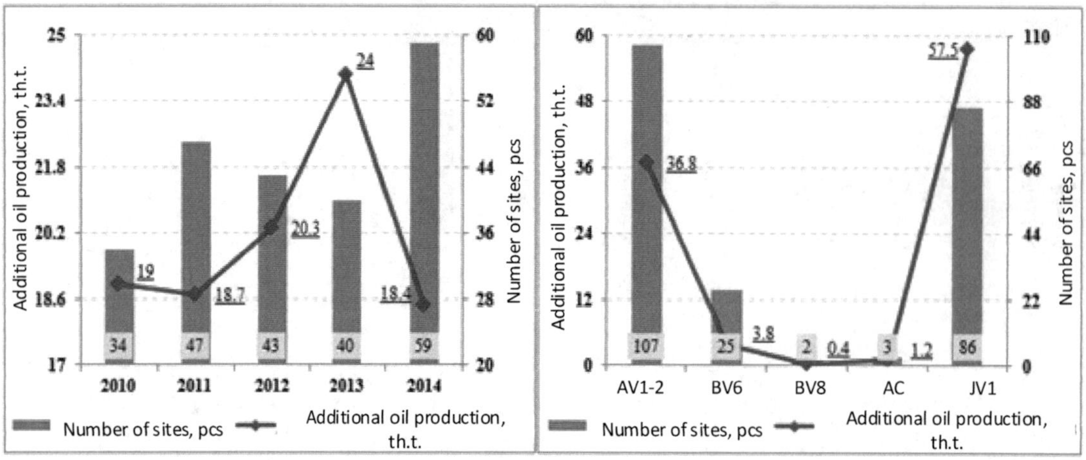

Figure 2. Distribution of the total number of additional oil production and number of hydrodynamic methods in Las-Eganskoye field for the analyzed period of development (left) and by objects (right) [5]

Figure 3 shows the distribution of specific additional oil production by types of HM EOR.

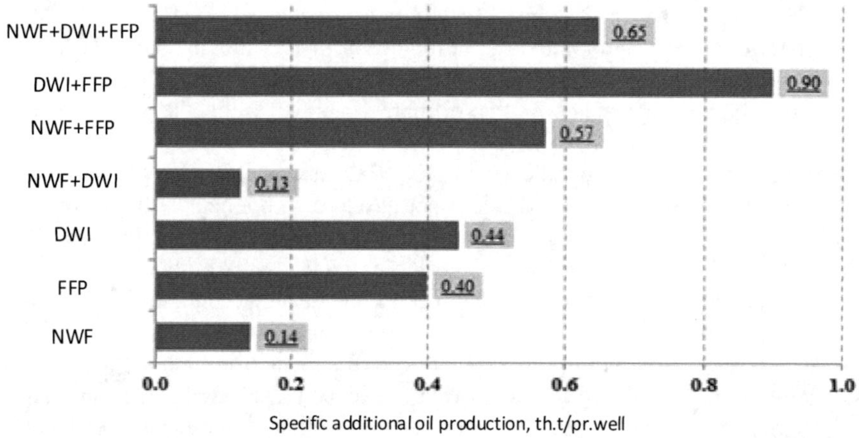

Figure 3. Distribution of specific additional oil production by types of hydrodynamic methods of enhanced oil recovery [5]

Table 1 shows the distribution of additional production by objects, years and number of operations. The last column of the table shows specific additional oil production for each object and type of WI.

Table 1. Distribution of activities and additional oil production by types of hydrodynamic methods, objects and years for the analyzed period of development [5]

Object	2010		2011		2012		2013		2014		Total		
	Additional oil production, th.t.	Number of operations	Additional oil production, th.t.	Number of operations	Additional oil production, th.t.	Number of operations	Additional oil production, th.t.	Number of operations	Additional oil production, th.t.	Number of operations	Additional oil production, th.t.	Number of operations	Specific additional oil production, th.t./pr.well
Forced fluid production													
AV_{1-2}	-	-	1.14	4	0.02	1	0.53	3	0.29	4	1.98	12	0.17
BV_6	-	-	2.15	3	-	-	-0.53	1	-0.01	1	1.63	5	0.32
BV_8	-	-	-	-	-	-	-	-	0.41	2	0.41	2	0.21
JV_1	4.18	3	-	-	2.61	2	0.15	2	0.62	3	7.56	10	0.76
Total:	4.18	3	3.29	7	2.63	3	0.15	6	1.31	10	11.56	29	0.4
Well development for injection													
1	2	3	4	5	6	7	8	9	10	11	12	13	14
AV_{1-2}	0.11	1	1.1	3	2.11	6	-	-	-	2	3.3	12	0.28
BV_6	-	-	-	1	-	-	-	-	-	-	-0.03	2	-0.02
BV_8	0.28	1	-	-	-	-	-	-	-	-	0.28	1	0.28
JV_1	3.93	6	-	3	3.35	8	6.8	6	-	-	13.3	23	0.58
Total:	4.32	8	0.0	7	5.46	14	7.0	7	-	2	16.8	38	0.44
Non-stationary water flooding													
AV_{1-2}	2.92	29	2.3	24	7.52	30	7.1	76	6.7	47	24.4	206	0.12
BV_6	0.32	4	1.2	3	0.16	5	0.1	7	0.15	3	2.03	22	0.09
BV_8	-	-	0.9	1	-	-	0.0	1	-	-	0.95	2	0.48
JV_1	-0.56	2	0.1	5	1.5	6	6.0	17	3.99	28	11.1	58	0.19
Total:	2.68	39	4.6	33	9.18	50	13.	101	10.8	78	40.8	288	0.13
Non-stationary water flooding and well development for injection													
AV_{1-2}	0.25	2	-	18	0.94	7	-	-	-	-	0.64	27	0.02
BV_6	-	-	0.2	2	-	-	-	-	-	-	0.22	2	0.11
JV_1	0.73	2	1.5	3	0.55	2	0.7	4	1.13	3	4.64	14	0.33
Total:	0.98	4	1.1	23	1.49	9	0.7	4	1.13	3	5.5	43	0.13
Non-stationary water flooding and forced fluid production													
JV_1			2.7	4	1.56	2	1.9	2	1.82	6	8.01	14	0.57
Well development for injection and forced fluid production													
AV_{1-2}	-	-	2.2	2	-	-	-	-	1.65	2	3.85	4	0.97
JV_1	-	-	4.2	5	-	-	-	-	1.75	2	6.02	7	0.86
Total:	-	-	6.4	7	-	-	-	-	3.4	4	9.87	11	0.9
Non-stationary water flooding, well development for injection and forced fluid production													
AV_{1-2}	-	-	0.3	7	-	-	-	-	-	-	0.3	7	0.04
JV_1	6.81	4	-	-	-	-	-	-	-	-	6.81	4	1.7
Total:	6.81	4	0.3	7	-	-	-	-	-	-	7.11	11	0.65

In general, it can be said that additional oil production for each year of the analyzed period is approximately the same and varies from the minimum of 18.4 thousand tons (in 2011) to 24 thousand tons (2013). At the same time, the maximum additional oil production accounts for JV_1 (57.5 thousand tons) and AV_{1-2} (36.8 thousand tons). The maximum additional oil production was received from NWF

(40.8 thousand tons in 141 sites). However, the specific additional production from FFP and DWI made 0.14 thousand tons/pr.well, which is 11.6 and 16.9 thousand tons of oil respectively, with higher specific additional oil production (0.40 and 0.44 thousand tons/pr.well respectively).

In sites where the operations were carried out jointly, the maximum amount of additional oil production was obtained where FFP and DWI were performed simultaneously (9.9 thousand tons in 5 sites).

The lowest production was obtained from the joint influence of NWF and DWI (5.5 thousand tons in 14 sites). In areas where specific additional production was carried out simultaneously with NWF and DWI the specific additional production was high (up to 0.9 thousand tons/pr.well). However, it is not possible to draw conclusions and make recommendations on the joint application of FFP with other HM EOR due to insufficient statistical data.

The impact of hydrodynamic methods of enhanced oil recovery on development indicators of Las-Eganskoye field is shown below.

In order to answer the question, how HM EOR affected field development indicators, the frequency of distribution of positive and negative effects of oil recovery and stimulation was carried out. The oil recovery effect is understood as additional oil produced due to WI, which under basic development would not be covered by development. The stimulation effect means additionally produced oil in a given period of time (calculated for a period of 6 months), which under basic development would be extracted from the formation for a longer period of time [2, 5].

For this assessment, the number of positive, negative and unclear HM effects at Las-Eganskoye field were calculated.

Diagrams of distribution of oil recovery and stimulation effects were designed. In calculating the number of positive and negative effects, sites with additional oil production ranging from –200 to +200 tons were removed to improve the quality of sampling. These effects are classified as unclear. The figures apply the following symbols: "+" – positive, "–" – negative and "UC" – unclear effect [4].

In the majority of cases NWF has a positive effect in terms of oil recovery – 72 sites (or 51 % of sites) (Figure 4). The number of sites with positive stimulation effect is small – 14 (10 % of sites), at the same time in 98 sites this effect at NWF is unclear.

FFP can both increase oil recovery and stimulate oil production, but the number of sites with a positive stimulation effect is greater than the number of sites with a positive recovery effect – 15 (55 %) and 11 (40 %) sites respectively. In most cases, FFP results in stimulation of oil production.

Similar to NWF, DWI in most cases has a positive effect in terms of oil recovery – 18 sites (62 %). The number of sites with positive stimulation effect is 7 (24 %), with negative – 6 (20 %), at the same time in 16 sites out of 29 the effect is unclear.

In sites where NWF and DWI were performed simultaneously, the number of sites with the positive effect of oil recovery is also greater than the number of sites with the positive effect of stimulation, 9 and 4 sites respectively. The number of sites where NWF and FFP were performed simultaneously (5 sites), DWI and FFP (5 sites) or NWF, OSN and FFP (3 sites) is not sufficient to draw a statistically reliable conclusion.

Based on statistically reliable data, it can be concluded that NWF and DWI operations in Las-Eganskoye field in most cases allow obtaining additional oil production due to enhanced oil recovery. In most cases FFP (OMO, PRC) operations intensify oil production.

Figure 5 shows the trend of the main development indicators during NWF operations in Las-Eganskoye field.

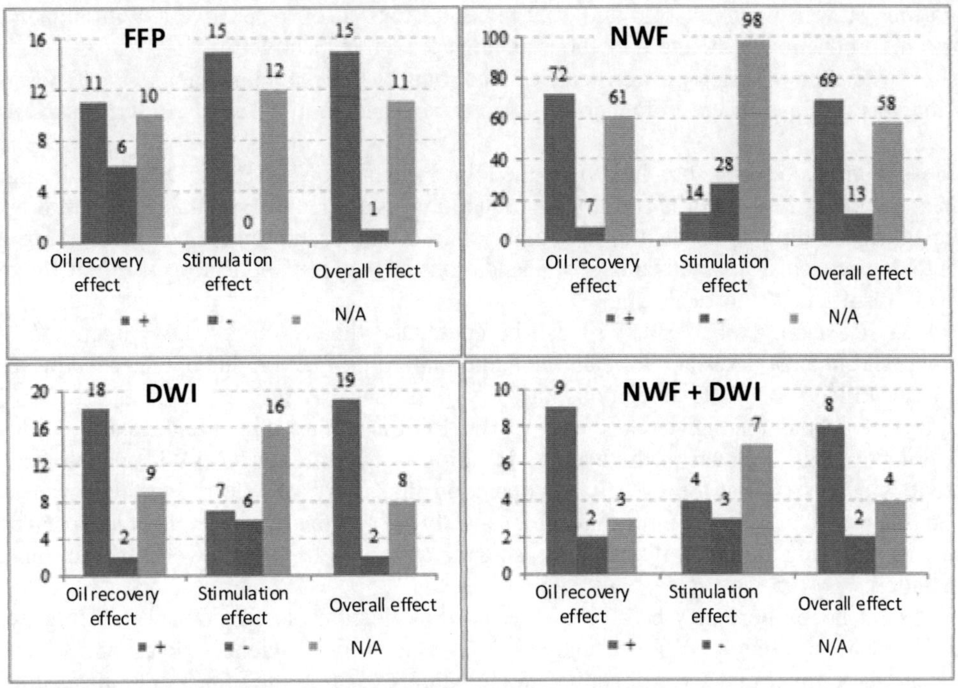

Figure 4. Distribution of the effect of HM EOR by category [5]

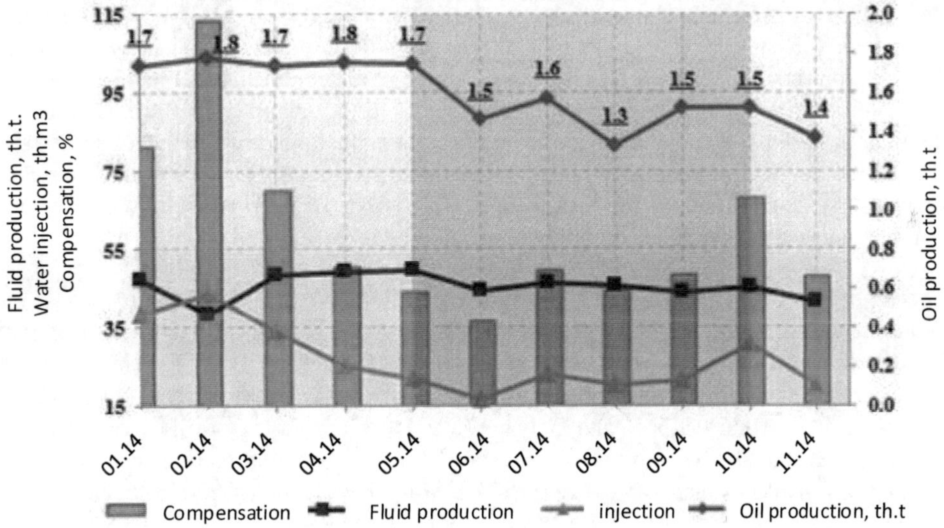

Figure 5. Dynamics of major indicators of development during NWF period in Las-Eganskoye field [5]

5. Conclusion

Additional oil production as a result of HM EOR for each year of the analyzed period is approximately the same and varies from a minimum of 18.4 thousand tons (in 2011) to 24 thousand tons (2013). At the same time, the maximum additional oil production accounts for objects: JV_1 (57.5 thousand tons) and $AV_{1\text{-}2}$ (36.8 thousand tons). The maximum additional oil production was received from NWF (40.8 thousand tons in 141 sites). However, the specific additional production from FFP and DWI made 0.14

thousand tons/pr.well, which is 11.6 and 16.9 thousand tons of oil respectively, with higher specific additional oil production (0.40 and 0.44 thousand tons/pr.well respectively).

In sites where the operations were carried out jointly, the maximum amount of additional oil production was obtained where FFP and DWI were performed simultaneously (9.9 thousand tons in 5 sites).

The lowest production was obtained from the joint influence of NWF and DWI (5.5 thousand tons in 14 sites). In areas where specific additional production was carried out simultaneously with NWF and DWI the specific additional production was high (up to 0.9 thousand tons/pr.well). However, it is not possible to draw conclusions and make recommendations on the joint application of FFP with other HM EOR due to insufficient statistical data.

Based on statistically reliable data, it can be concluded that NWF and DWI operations in Las-Eganskoye field in most cases allow obtaining additional oil production due to enhanced oil recovery. In most cases FFP (OMO, PRC) operations intensify oil production.

It shall be noted that the highest efficiency of HM EOR is achieved in case the average injection in the site prior to NWF is equal to the average injection in the site during NWF operations. For this purpose, it is necessary that the capacity of operating injection wells during the half-cycle of their operation is approximately twice more than their capacity prior to the start of cyclic injection. Therefore, it is necessary to study the possibility of increasing the capacity of injection wells in the semi-cycle of their operation.

If this is not done, there may be a problem with the injection of formation water. In general, the amount of formation water at NWF is usually reduced. One of the efficiency criteria of cyclic injection of water into the formation is the reduction of water content of observation production wells of the site and the reduction of the quantity of formation water. Besides, the quantity of formation water may decrease in production wells located in the homogeneous part of the formation and having direct hydrodynamic connection with shut-off injection wells, which did not react to non-stationary flooding. As a result, such wells may be characterized by the decrease in fluid and oil flow rates. As the experience of cyclic injection shows, in Las-Yeganskoye field and in other fields, this problem is usually solved by the redistribution of water through other injection wells not involved in cyclic injection. If it is not possible to do so for any reason, then inactive or temporary abandoned injection wells (especially those located in the border zone) may be used for the injection of formation water. If water from the wells of PC formation (the Cenomanian) is used for injection, it is recommended to shutdown such wells for the period of non-stationary flooding.

References

[1] Vakhrushev V V, Melnikov V N and Moskvitin S A 2016 Generalization of experience of development of the target area YUV1 of fields of LLC "LUKOIL – Western Siberia" to increase the validity of the forecast of technological parameters *Oil industry* **8** 44–7

[2] Dyagilev V F, Abramov N V and Belyaev O V 2018 Technology of development of water-oil zones in the North Orekhov field *Opyt, aktual'nye problemy i perspektivy razvitiya neftegazovogo kompleksa (Materialy Mezhdunarodnoj nauchno-prakticheskoj konferencii obuchayushchihsya, aspirantov i uchenyh* no 1*)* pp 96–100

[3] Seigman Yu V, Gumerov O A, Karimov R M and Shamaev G A 2010 Hydrodynamic methods of regulation of oil field development *Aktual'nye voprosy razrabotki neftegazovyh mestorozhdenij na pozdnih stadiyah. Tekhnologii. Oborudovanie. Bezopasnost'. Ekologiya (Materialy nauchno-prakticheskoj konferencii)* 258 p

[4] *Report on the main technical and economic indicators of exploitation of deposits of CCI "Langepasneftegaz" for the period 1999–2014* 2015 245 p

[5] Surguchev M L and Sharbatova I N 1988 *Cyclical effects on heterogeneous oil reservoirs* (Moscow: Subsoil) 121 p

[6] Wang L et al 2017 Advances in improved/enhanced oil recovery technologies for tight and shale reservoirs *Fuel* **210** 425–45

IOP Conf. Series: Materials Science and Engineering **663** (2019) 012062 doi:10.1088/1757-899X/663/1/012062

[7] Xian'e L, Zhi X, Hongyao Y, Yujun F and Hongping Q 2017 Comparative Studies on Enhanced Oil Recovery: Thermoviscosifying Polymer Versus Polyacrylamide *Energy fuels* **31(3)** 2107–3328
[8] Xian'e L, Zhi X, Hongyao Y, Yujun F and Hongping Q 2017 Comparative Studies on Enhanced Oil Recovery: Thermoviscosifying Polymer Versus Polyacrylamide *Energy fuels* **31(3)** 2479–87
[9] Wang X and Sheng J J 2017 Effect of low-velocity non-Darcy flow on well production performance in shale and tight oil reservoirs *Fuel* **190** 41–6
[10] Etminan S R, Maini B B and Chen Z 2017 Determination of mass transfer parameters in solvent-based oil recovery techniques using a non-equilibrium boundary condition at the interface *Fuel* **120** 218–32

IOP Conf. Series: Materials Science and Engineering **663** (2019) 012063 doi:10.1088/1757-899X/663/1/012063

Reserves of energy saving and reduction of oil recovery costs

S V Kolesnik and N R Krivova

Tyumen Industrial University, 38, Volodarskiy str., Tyumen, 625000, Russia

E-mail: kolesniksv@bk.ru, nrkrivova@rosneft.ru

Abstract. The paper considers the use of controlled drive for oil recovery and pressure maintenance, which represents the oscillating system (sucker-rod pump drive) and adjustable drive of a pump unit (for pressure maintenance system) thus making it possible to reduce power consumption through oil recovery by 25...30 %. The development of electric drives, increase of their controllability, reduction of energy losses is considered alongside with the method of rotation speed control by redistribution of angular speed and power flows through a differential. In order to provide modern technological systems with cost efficient electric drives controlled by speed in a wide range with constant power on the output shaft, the new control principle (birotating) can be considered promising due to economy, simple design, low cost caused by lack of power electronics and high reliability.

1. Introduction

The analysis of methods and technological means of oil recovery first of all shows uneven development of technical control means of different levels, for example, advanced system of information collection and processing compared to the systems of control of lower level, which remained unchanged. Second, the share of energy costs in the total cost of oil production is increasing due to increasing costs of pressure maintenance and, as a result, water content of borehole fluid. Besides, legacy pump drives of a walking beam type do not allow setting pumping modes in accordance with the actual well flow rate. Therefore, in order to reduce energy consumption, there is a need for technical solutions to create an electric drive that meets the requirements of energy saving and controllability.

In this regard, the introduction of automation and DCS in oil fields becomes particularly important since it ensures efficient operation of oil fields in the specified modes with limited human resources. Mass introduction of modern control systems with computer, measurement and communication facilities using the latest achievements of electronic engineering requires systematic analysis of control tools and systems.

In oil production pump rate shall be adjusted to actual production rate of a well. This, in turn, requires the use of a speed-adjustable electric drive that is able to continuously change the operating mode of a pump and thereby reduce energy consumption. When water is extracted from well fluid, the discounted income increases by 8–10 % for each oil recovery method.

A drive, which regulates mechanical conductivity of a kinematic chain based on setting a gearing angle of a pair of gears equal to a friction angle is used to control pump operating modes, will help to solve this task. This makes it possible to extend the control the drive rotation speed at minimum power impact on transmission, i.e. to ensure the principle of "mechanical amplifier".

2. Birotating drive

IOP Conf. Series: Materials Science and Engineering **663** (2019) 012063 doi:10.1088/1757-899X/663/1/012063

A drive of fundamentally new type – a double rotation (birotating) drive has been developed [1–4]. The task was to develop the drive control principle in the widest speed range at constant rotational speed of the asynchronous motor. Such drive is based on the principle of invariance of relative angular speed of rotor and stator at the change of the output shaft speed in a wide range.

The asynchronous motor can rotate both the rotor and stator (Fig. 1).

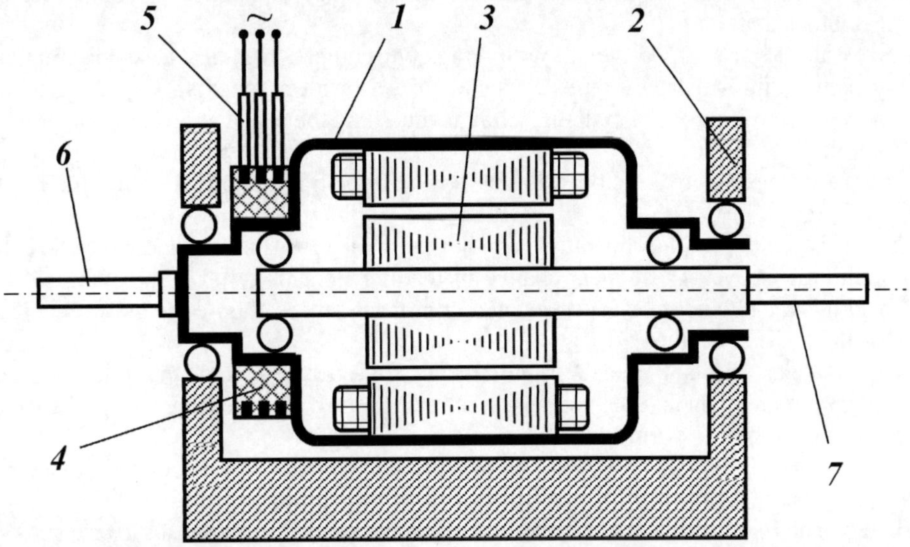

Figure 1. Scheme of birotating motor

The engine has a stator 1 installed on bearings on a frame 2. A rotor 3 is installed on bearings inside the stator 1. A collector 4 is fixed on the stator 1. Electric power is supplied to contact rings of the collector through sliding brushes 5. An output shaft 6 connected to the stator 1 and a shaft 7 of the rotor 3 can be connected to loading. The properties of this engine are different from traditional engines (with fixed stator):

1. The property of the engine is its algebraic addition of angular speed of its rotating parts – a rotor and a stator:

$$\omega_{\text{ст}} + \omega_{\text{р}} = \omega_{\text{ном}}, \tag{1}$$

where $\omega_{\text{ст}}$ – angular speed of a stator; $\omega_{\text{р}}$ – angular speed of a rotor; $\omega_{\text{ном}}$ – nominal speed of an engine (relative speed of a rotor and a stator).

Directions of rotor and stator rotation in opposite directions are taken as positive directions. From expression (1) we see that the equality is maintained if one of the rotating parts (rotor or stator) changes the direction of rotation, i.e. both parts of the motor will rotate in the same direction at angular speeds different in magnitude by $\omega_{\text{ном}}$. This property of the birotating motor allows the speed of the output shaft to be controlled within wide limits (from 0 to speeds several times exceeding the nominal). Thus, the equality (1) is maintained in the whole range of control.

2. In free state, when neither rotor nor stator are connected to the load (friction forces in bearings and on contact rings of the manifold are not considered as insignificant), the angular speeds of rotation of rotor and stator are determined by their inertia moments while maintaining equality (1).

3. There is no reactive torque on structural elements of the drive external to birotating engine (for example, on frame 2, Fig. 1). The reactive torque generated by each of the rotating parts (the rotor creates a reactive torque on the stator and the stator on the rotor) is used to create the required torque on the output shaft at given angular speeds.

4. When rotor and stator torques rotating in opposite directions are equal, Coriolis forces influencing each rotating part are mutually balanced. This property can have a significant impact on the accuracy of gyroscopic systems by reducing the area of precession and thus increasing the gyroscope stability.

5. Selection of the type and characteristics of synchronization system in the birotating drive can ensure the necessary mechanical characteristic of loading regardless of the type of engine used. This is guaranteed by the fact that the engine maintains a nominal value of its energy at all values of angular speeds of the output shaft.

These properties of birotating drives allow solving quite complex technical tasks via simpler methods than those used in traditional drives, for example, to ensure a wide range of speed change without using expensive power electronic control systems, while maintaining the constant level of power in the whole range of control.

A number of methods were developed to control the birotating drive that ensure the required load characteristic [4, 5].

Taking into account complex operating conditions of drives in oil-producing wells (changes of weather conditions), as well as frequent change of loading (in sucker-rod pumps the load on a drive varies from zero to the maximum every 5 seconds), it is necessary to choose the most reliable and easy-to-operate method of drive control [6, 7].

Such drive, which meets the main requirements for energy saving and reliability, can be a "mechanical amplifier", which uses the effect of controlling the mechanical conductivity of the kinematic pair "worm gear – worm shaft".

The scheme in Fig. 2 explains the principle of mechanical conductivity control implemented in the form of a differential mechanism with continuously variable gear ratio.

Electric motor 1 (Fig. 2) rotates a differential mechanism 7, and the output shaft 9 remains stationary at fixed housing 4. To transmit rotation to the output shaft 9 it is necessary to set the rotation of housing 4 in trigger mode by servo motor 10 and worm gear – worm shaft 3 and gear wheel 2. Then the output shaft 9 will start rotation with a speed, precisely corresponding to the rotation speed of the housing 4. Such mode is guaranteed by the fact that the gear wheel 6 transfers rotation to a wheel 11 rotating a differential carrier 7 with twice smaller speed in direction opposite to a gear wheel 6. In this case, the sum of angular speeds in the summing differential equals zero. The transmission coefficient by the servo motor 10 is controlled by selecting a gearing angle between a worm 3 and a worm wheel 2 equal to the friction angle. This reduces the radial force that rotates worm 3 to zero. By changing the rotational speed of the servo motor 10, the rotational speed of the output shaft 9 can be smoothly changed.

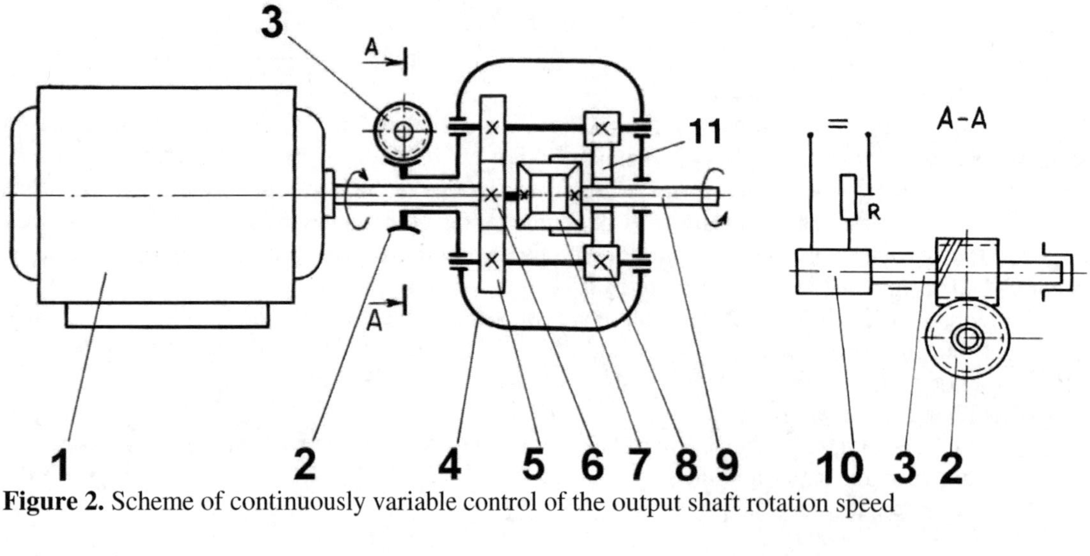

Figure 2. Scheme of continuously variable control of the output shaft rotation speed

IOP Conf. Series: Materials Science and Engineering **663** (2019) 012063 doi:10.1088/1757-899X/663/1/012063

The controlled mode of sucker-rod pumps is connected with the choice of the rational scheme of a ground drive. Below is a scheme of creating periodic vibrations of a balance-free drive based on the use of a roller-screw pair (Fig. 3).

The balance-free drive is a container accommodating telescopic post 1 and two vertical screws 2 and 3 (Fig. 3). The screws are connected to each other by gear wheels 4 and 5, and with electric motor 6. Nuts 7 of vertical screws 2 and 3 are connected with rollers 8 for chain 9 resting on support sprockets 10 and 11.

The drive operates as follows.

After the electric motor 6 (Fig. 3) is activated, the rotation is transmitted through gear wheels 4 and 5 to vertical screws 2 and 3. At the same time, they start to rotate in opposite directions, and their nuts 7 will move in different directions – one up, the other down. Rollers 8 with chain 9 move with them. As a result of these movements chain 9 shortens and lifts load P (i.e. a polished rod of a well rod pump). When the nuts 7 reach their extreme positions, the position sensor (not shown) transmits signal to the motor 6, which turns off and the weight P starts to descend under its own weight (roller-screw pairs are selected as non-self-braking). When chain 9 takes position parallel to straight N-N, the engine 6 turns on and, continuing to lift load P expands nuts to opposite positions until stop, after which the engine 6 turns off. In one full stroke of nuts 7 from the extreme upper to the extreme lower position the load (P) rises twice.

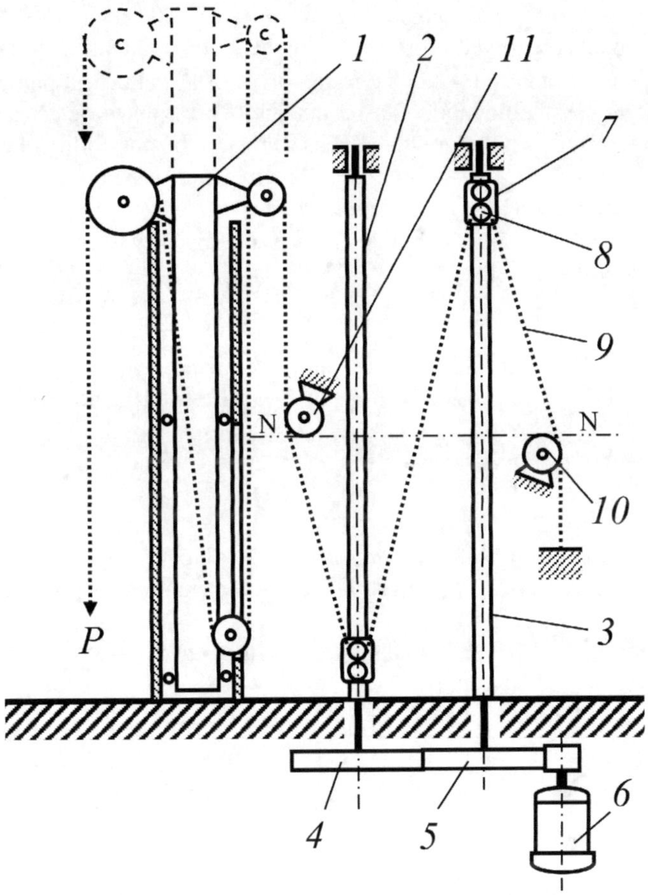

Figure 3. Kinematic scheme of a screw drive mechanism of a downhole rod pump: 1 – telescopic rack; 2, 3 – vertical screws; 4, 5 – gear wheels; 6 – electric motor; 7 – nut; 8 – roller; 9 – chain; 10, 11 – support sprocket

At the maximum load on suspension 120 kN the drive has electric motor power – 7.5 kW, height of pump rod lifting – 1.5–3 m, adjustable number of double strokes from 0.5 to 8 min^{-1}. Without increasing drive dimensions, the height of the rod lift can be increased to 6 meters by reworking the telescopic rack 1. The speed of pump descent and ascent is controlled by the rotation speed of the electric motor 6, which has frequency control and can vary the speed within wide limits. The height of the pump lift is adjusted by changing the operating motion sensors of the controller. End switches of explosion-proof type are used as track sensors.

Compared to similar devices, for example, sucker-rod pumps, the proposed technical solution has the following advantages:

- continuous oil pumping from low-yield wells;
- reduction of metal consumption by 3.5 times;
- 2x less power consumption;
- possibility of use in weak boggy ground;
- simple design ensures technological efficiency of production and reduces the prime cost of drive manufacturing.

3. Conclusion

The use of controlled drive for oil recovery and pressure maintenance, which represents the oscillating system (sucker-rod pump drive) and adjustable drive of a pump unit (for pressure maintenance system) makes it possible to reduce power consumption through oil recovery by 25...30 %.

Thus, the following methods can be considered promising in reducing oil production costs:

- commissioning of inactive wells using a downhole ground drive of a sucker-rod pump (inactive wells due to swampy soil or due to low-debit mode can be put into operation via electric drives of container type, which pressure on soil makes not more than 0.2 kg/cm^2, and the possibility of continuous control of its operating modes within wide limits – from 0.5 to 8 swings per minute – allows selecting the mode corresponding to individual characteristics of wells);
- equipping the operating sucker-rod pumps with adjustable birotating drives will prolong active oil recovery from highly watered low-debit wells.

References

[1] Kasimov L N and Shangin E S 1997 Birotating electric drive *Electrical engineering* **9** 9–13
[2] Kasimov L N and Shangin E S 2000 *Subsurface pump drive* Patent RU no 2163311. MPC F04B47/02. BI no 5
[3] Shangin E S 2002 *Recovery mechanism and its implementation* Patent RU no 2184838, cl. E2143/00. BI no 19
[4] Shangin E S 1998 *Theory of birotating electric drive* (Ufa: UTIS) 264 p
[5] Adonin A N 1979 *Oil production via sucker-rod pumps* (Moscow: Nedra) 386 p
[6] Ilyasov B G and Shangin E S 2012 Automated drive of pendulum sucker-rod pumps *Mechatronics, Automation, Control* **1** 18–24
[7] Valeev M D and Khasanov M M 1992 *Deeppumping production of viscous oil* (Ufa: Bashknigoizdat) 122 p

IOP Conf. Series: Materials Science and Engineering **663** (2019) 012064 doi:10.1088/1757-899X/663/1/012064

Improving the cost-performance ratio of well construction using a multi-stage hydrostatic pressure tester

M I Korabelnikov[1], N A Aksenova[1], E Yu Lipatov[1] and A M Korabelnikov[2]

[1]Tyumen Industrial University, 38, Volodarsky St., Tyumen, 625000, Russia

[2]RN-Vankor Oil and Gas Production Directorate

E-mail: na-acs@yandex.ru

Abstract. The article considers the issue of improving the cost-performance ratio of well drilling through the use of a device for drill and production pipes, which allows for multi-stage hydrostatic pressure testing during preparatory work in the well before liner completion, well lining and other technological operations. The authors present the results of a patent analysis on the problem under study and prove the need to create a device with a simpler design with a reduced probability of breaking and allowing multiple pressure testing of pipes in the well without lifting them to the surface. The article presents a scheme of the device and the technology of its operation in the well. The results of factory tests of a prototype device with a diameter of 105 mm showed its efficiency and practicality for technological processes during the drilling and exploitation of deep oil and gas wells.

1. Work timeliness

Well construction is a capital-intensive and time-consuming process requiring large investments, including expenses for unforeseen operations to eliminate the consequences of accidents and complications arising during the drilling and completion of wells. Therefore, it is very important to use new technologies and technical means to improve the cost-performance ratio of well construction.

In technological operations using a drill string and production string: lowering the liner, drilling sidetracks, repair and insulation work, treating the bottom-hole zone (BHZ), developing wells, it is necessary to test the strings with internal pressure for leaks. Existing testing technologies and applied devices (valves) do not allow repeatedly conducting leak tests, and when the valve is actuated, it is necessary to lift it from the well and replace to resume circulation of the working fluid and conduct subsequent pressure testing. All this leads to additional costs of time, money and to well appreciation.

2. Literature and patent analysis on the studied issues

Analysis of scientific and technical literature and patents for inventions showed that a large number of technical solutions related to operations for pressure testing of drill and production pipes in wells is based on the use of valves of various designs. Table 1 in chronological sequence shows the evolution of technical solutions to improve the design of valves for pressure testing of pipes inside the well with overpressure.

Table 1. Brief description of devices for pressure testing of strings in well

Name	Characteristics	Drawbacks
Controlled valve for blocking filter well [1]	Valve for blocking the filter part of the well with a packing element, made in the form of a cylindrical body with a valve seat in the upper part and a movable sleeve with a valve head and a closed zigzag groove on its outer surface, which includes fixed body pins; the groove is made of alternating in size sides of isosceles V-shaped sections equipped with leaf springs that allow the sleeve to rotate in only one direction with axial displacement	Complex design and low reliability
Device for pressure testing of pipe string in well [2]	The device consists of a housing, a shutter and a plate. The housing has a central channel, a membrane, side channels and return channels for lowering the string into the well. When the pressure in the string increases to the calculated one, the membrane calibrated for a certain pressure is destroyed, passing liquid through the side channels	A large spread of destructive pressure, which affects the accuracy of the tests. The complexity of housing manufacturing
Device for pressure testing of drilling string pipes [3]	Device for pressure testing of drill pipes, comprising a housing with connecting threads and a stationary sleeve with axial and peripheral holes and a movable sleeve with a saddle for a wireline ball both located in the housing's cavity, the movable sleeve being fixed in the initial position relative to the housing by shear elements	Inability to use in case of contamination of the solution with sludge. Disposable
Device for pressure testing of drilling pipes [4]	Device includes a housing, a saddle, a cap with a rod, a spring and a locking device made in the form of interacting rod, sleeve with a labyrinth track and a ball, where the rod and sleeve of the locking device have non-circular cross-section and are connected together by a pin in a movable manner	Low reliability due to the spring that during deformation loses its elasticity, and the valve fails
Device for pressure testing of drilling pipes [5]	Device of a more simplified design consists of a housing with a bore, a sleeve with an axial channel, and a restriction plate with a saddle under the sleeve, which is freely installed in the bore of the housing and has peripheral axial channels covered by a membrane, and the saddle of the restriction plate has a seal	Disposable
Device for pressure testing of pipes in well [6]	Device includes a housing, a spring-loaded sleeve with radial holes and a locking element mounted in the cavity of the housing concentrically to its inner surface, and a saddle for the locking element; the distinctive feature is that to increase the reliability of the device, the spring-loaded sleeve has a fitting in the upper part, and a saddle for the locking element is located in the lower part of the housing	Contributes to a sharp change in pressure in the pipe string, which can lead to their destruction
Device for pressure testing of pipes in well [7]	Device consists of a housing, a spring-loaded sleeve with radial holes and a locking element, the sleeve being installed in the cavity of the housing concentrically to its inner surface, and a saddle for the locking element, made in the form of a transition pipe with radial holes	The need for additional operations for the delivery of the locking element for pressure testing of pipes
Device for pressure testing of drilling pipes in well [8]	Device consists of a hollow body with connecting elements, with a sleeve with a saddle for a wireline ball mounted on shear elements and a restrictive sleeve located under it, rigidly connected to the body and having central and peripheral axial channels. The device is equipped with a spring, one end of which is fixed in the axial channel of the restrictive sleeve, and the other is freely placed in the axial channel of the sleeve with a saddle	Complexity of the design due to the large number of components and parts, and complex production
Device for pressure testing of drilling pipes [9]	Device comprises a housing and a stationary sleeve with axial peripheral holes and a movable sleeve with a saddle for a wireline ball both located in the housing's cavity, the movable sleeve being fixed in the initial position by shear elements. The device is equipped with valves in the form of a ring to overlap the holes of the stationary sleeve, and the housing has radial channels	Complexity of housing manufacture, due to the presence of radial channels which requires milling and drilling operations

Table 1. Ending

Name	Characteristics	Drawbacks
Device for pressure testing of drilling string in well [10]	Device consists of a hollow body with an annular groove on the inner surface with connecting threads mounted in the cavity of the body with the possibility of axial movement of the sleeve with slots and with a saddle in the lower part under the wireline shut-off element. The slots of the sleeve go to its upper end as windows, with rotary latches located in them in the sleeve's wall above the end of its slots. The sleeve location enables interaction of the latch with the housing wall above its groove in the initial position, and in the working position it allows the latch to be placed in the groove of the housing	Complex design, instable operation
Device for blocking the cavity of the drilling string [11]	Device consists of a housing, a saddle and a wireline shut-off element containing movably mounted sleeves, one of sleeves being equipped with a poppet spring-loaded valve; in the shutter housing there are side windows in one of which a spring-loaded rotary lock is placed, and in the other a spring-loaded axis of the poppet valve is located	Unsatisfactory reliability. Insufficient impermeability of evacuated cavity. Disposable
Device for pressure testing of drilling pipes [12]	Two sleeves are movably mounted in the housing cavity. In the cavity of the inner sleeve, there is a valve made in the form of a bushing spring-loaded relative to the outer sleeve with windows in its wall. The bushing and the outer sleeve have spring-loaded clamps, and the inner sleeve for their placement on the inner and outer surfaces is made with annular grooves	Low reliability during repeated pressure tests of the pipe string in small-diameter wells
Pressure test packer [13]	Device includes a housing, a sealing assembly with a sealing element, a lower sleeve and a sealing chamber, a bypass device, a check valve and a valve assembly. The valve assembly has shear elements, a ball valve and a seat under it in the form of a cylindrical sleeve. The housing has an annular collar restricting the movement of the lower sleeve, an annular groove, a stepped annular undercut and radial holes for installing shear screws. The bypass device is made in the form of a spool with an internal step recess, a spring and an adjusting nut. It sets the spring force and stroke length of the spool. The seal chamber is formed by a sealing element and an annular groove of the housing	Design complex to fabricate. Pressure test requires filling the annulus with the working medium
Device for pressure testing of pipe string in well [14]	Device includes a housing and a shutter. The housing is provided with an internal cylindrical bore, and the shutter is made in the form of a central rod with a channel made in the form of a lifter saddle, a cylindrical part, a narrow cylindrical part and a stop placed from top to bottom, while a hollow sleeve is installed on the cylindrical part of the central rod with the possibility of limited downward movement with radial windows at the bottom, which on top is equipped with a self-sealing cuff that allows fluid to pass from the bottom up, and is fixed in the transport position relative to the center ceiling elements rod shear element sealingly closing the channel, connecting spaces above and below the self-sealing cuff when the gate is extracted. Into the radial windows of the hollow sleeve dies are installed with the possibility of radial movement inward and interaction in the working position with the inner cylindrical bore of the housing and the cylindrical part of the central rod	Design complexity. Operation of the design requires upward and downward movements. The device passes through only from bottom up

The analysis showed that the designs were improved by complicating design with the addition of new structural elements and complicating production with the need for high-precision milling and drilling operations. The implementation of a large number of structural parts and interconnected elements is known to increase the likelihood of their breaking. In addition, most of the technical developments presented have a considerable drawback: after pressure testing of the pipes and "actuation" of the valve to resume circulation of the working fluid, subsequent pressure testing of these

IOP Conf. Series: Materials Science and Engineering **663** (2019) 012064 doi:10.1088/1757-899X/663/1/012064

pipes in the well can be achieved only after raising their wells and replacing the necessary elements (saddle, plug, pins, etc.) of known valves. This drawback leads to the need for additional technological operations, which increases the duration of the main works and increases their cost.

3. Technology of multi-stage pressure testing of pipes in well

The authors developed a device [15], which is equipped with a die sleeve equipped with an additional female ring. When pressure testing the string, a wireline ball is placed in the inner cavity of the pipes, which is designed to overlap the central channel of the die sleeve. When testing a pipe string for leaks, with a central channel blocked by a wireline ball, an internal test pressure (Pt) is created. The time and pressure value is determined by the technological regulations for pressure testing. After the leak test of the string, by excess pressure (Rs) in the pipes of 1.5 times, the wireline ball is "extruded" from the central channel of the die sleeve.

After extruding the wireline ball from the die sleeve of the proposed device, the circulation of the flushing fluid in the well is restored according to the accepted technology.

Then, in case of necessary additional pressure testing of pipes in a well, caused, for example, by their partial replacement, the next wireline ball is dropped into the pipe cavity, which is pumped to the seating face of the die sleeve, and then the calculated internal test pressure (Rt) is created. After the pressure test, according to the technology above, the circulation in the well is restored by creating an excess pressure that is 1.5 times higher than the calculated pressure of the pressure test (Rt) and extruding another wireline ball from the die sleeve of the device. This technological technique is new and can be assessed as essential, since it has positive effect on the solution of the formulated problem. Wireline balls after pressure testing of the pipes and forcing them through the die sleeve are accumulated in a container, which is a mechanical extension of the device body.

4. Multi-stage hydrostatic pressure tester

Multistage pressure testing of pipes is achieved by using a cylindrical die sleeve coaxially located with an annular gap in the device body and having strictly maintained geometric dimensions, in particular, in terms of the inner (d_{is}) and outer diameters (d_{os}) of the die sleeve, which are assigned according to the calculation results, taking into account the diameters of the wireline balls (d_b), the yield strength (σ_y) of the die sleeve metal, the testing overpressure (P_i) and the overpressure required to “extrude” the wireline ball (P_e).

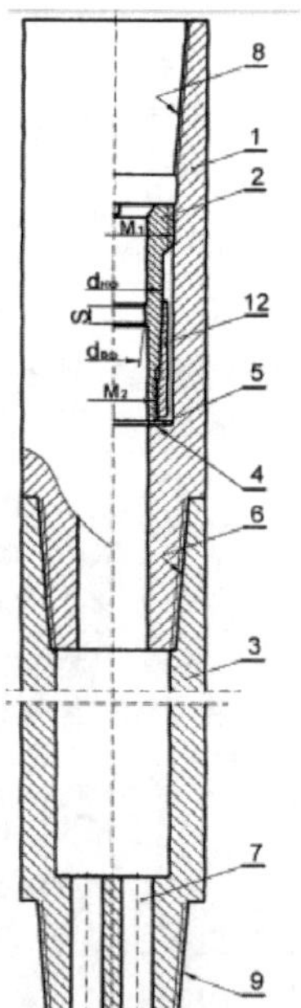

Figure 1. Multi-stage hydrostatic pressure tester

The device for multi-stage pressure testing of pipes in a well (Figure 1) includes a housing 1, with die sleeve 2 placed therein, and a container body 3 for accommodating spent wireline balls. In the inner part of the housing 1, a thread M_1 was made for screwing up with the die sleeve 2. The screwing of the die sleeve 2 in the housing 1 is made until it stops with its end ledge 4, using the o-ring 5. The housing 1 and the container body 3 are interconnected by threaded connection 6. In the lower part of the container body 3, through longitudinal holes 7 are made, providing unhindered passage of the process fluid. The device housing 1 and the container body 3 are provided with threads 8 and 9 for incorporating the device into the pipe arrangement lowered into the well.

The figure shows a device for multi-stage crimping of pipes in a well in working condition. Three spent wireline balls 10 are shown in the container body 3. One wireline ball 11 is located in the inner cavity of the die sleeve and adjusting ring 12 is shown. An excess pressure P_t is created in the inner cavity of the pipes; in the annular space between the die sleeve and the cylinder wall, the excess pressure in the pipes is leveled up along the threaded channel to pressure P_t.

IOP Conf. Series: Materials Science and Engineering **663** (2019) 012064 doi:10.1088/1757-899X/663/1/012064

5. Testing of multi-stage hydrostatic pressure tester

A prototype device with a diameter of 105 mm was manufactured at an industrial enterprise in Tyumen. In order to verify the operability of the device and adjust it to the calculated operating parameters, the authors conducted preliminary factory acceptance tests, which showed the promise of using the developed device for multi-stage pipe testing in field conditions. At the same time, the need for more accurate manufacturing of mating (moving) parts of the device and the use of wear-resistant components (balls) was noted.

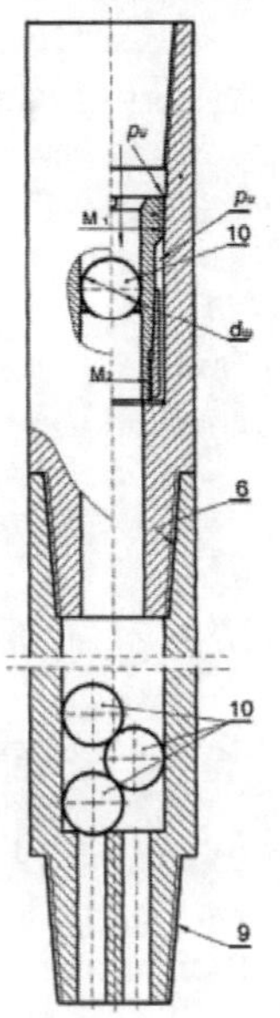

Figure 2. Pressure tester in operating state

Figure 3. Photo of the pressure tester with wireline balls

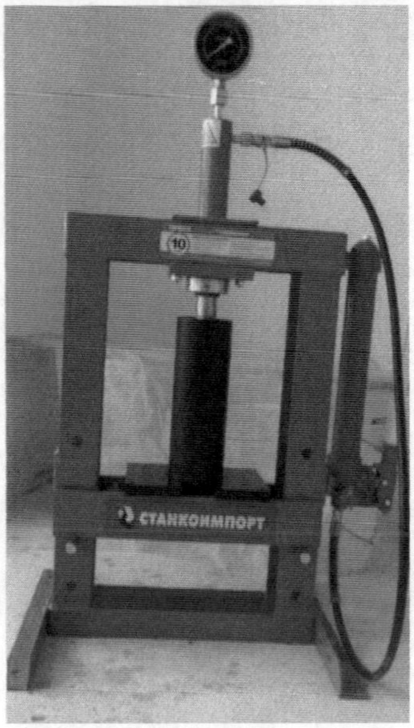

Figure 4. Press for pressure adjustment

6. Advantages of multi-stage hydrostatic pressure tester

- Possibility of multiple pressure testing of drill pipes without lifting;
- The ability to conduct direct and reverse flushing while drilling;
- Simplicity and reliability of the design;
- To determine the place of leakage of drill pipes, it is not necessary to activate the device and completely raise the drilling tool;
- Saving time by reducing the number of trips;
- Cost saving in terms of the installation of new devices.

The expected economic effect from the introduction of the device is conditioned by a reduction in unproductive time by an average of 8 hours when drilling one well.

With the cost of one hour of drilling rig operation of 30,000 rubles, the economic effect of the use of the proposed device is 240,000 rubles per well.

7. Conclusions

The device [15] developed by the authors allows for multiple pressure testing of pipes in a well without lifting them to the surface by simplifying the design of the device, pre-adjusting the device for pressure testing, increasing the reliability of the work by the absence of complex movable elements, as well as reducing the time of pressure testing.

The number of stages of pipe pressure testing, without lifting them to the surface, according to the proposed technology, can be limited only by the number of wireline balls and the volume of the container for collecting them, which is included as a component of the present device, while the absence of interconnected moving parts in the device ensures its high reliability.

References

[1] Baboshkin N N, Pinkenson D B and Potekhin L F 1965 *Controlled valve for blocking filter well* Authorship certificate no 287864 USSR, IPC E21B43/00
[2] Efremov Yu V, Starikov A A and Halyk-zade E A 1968 *Device for pressure testing of pipe string in well* Authorship certificate no 369242 USSR, IPC E21B17/00
[3] Ivanitsky E A., Strue Ya M and Tychinsky R D 1971 *Device for pressure testing of drilling string pipes* Authorship certificate no 370334 USSR, IPC E21B17/00
[4] Yadullaev N N and Vinitsky Yu N 1978 *Device for pressure testing of drilling pipes* Authorship certificate no 610973, USSR, IPC E21B17/00
[5] Karash E B, Abdulzade A N G, Tarnovsky V P and Nazarov S B 1980 *Device for pressure testing of drilling pipes* Authorship certificate no 973796 of the USSR, IPC E21B34/06
[6] Kuskov A I 1981 *Device for pressure testing of pipes in well* Authorship certificate no 794167 USSR, IPC E21B17/00
[7] Kurtov V D 1981 *Device for pressure testing of pipes in well* Authorship certificate no 794166 USSR, IPC E21B17/00
[8] Karmalsky V B and Spiridonov O N 1985 *Device for pressure testing of drill pipes in well* Authorship certificate no 1344890 USSR, IPC A1E21B17/00, 34/06
[9] Zelenin V M andMalikov I V 1985 *Device for pressure testing of drill pipes* Authorship certificate no 1314004 USSR, IPC E21B17/00
[10] Krykh B V 1986 *Device for pressure testing of drill string in well* Authorship certificate no 1411424 USSR, IPC E21B17/00, 34/06
[11] Krykh B V 1988 *Device for blocking the cavity of the drilling string* Authorship certificate no 1530739 USSR, MKI E 21B17/00
[12] Ishmukhametov A T and Kushnarev V I 1990 *Device for pressure testing of drill pipes* Authorship certificate no 2018633 RF, IPC E21B34/10
[13] Verzhbitsky A S, Sgibnev A D and Kuznetsov V G 2000 *Pressure test packer* Authorship certificate no 2153570 RF, IPC E21B33/12
[14] Makhmutov I Kh, Strakhov D V, Osnos V B et al 2006 *Device for pressure testing of pipe string in well* Authorship certificate no 2278944 RF, IPC E21B17/00
[15] Korabelnikov M I, Aksenova N A, Lipatov E Yu and Korabelnikov A M 2019 *The method of multi-stage pressure testing of pipes in the well and a device for its implementation* Authorship certificate no 2691037 RF, IPC E21B33/10

IOP Conf. Series: Materials Science and Engineering **663** (2019) 012065 doi:10.1088/1757-899X/663/1/012065

The simulation of high-pressure water vapor injection using core material

N R Krivova, S V Kolesnik and Yu B Chebykina

[1]Tyumen Industrial University, 38, Volodarsky St., Tyumen, 625000, Russia

E-mail: nrk4@yandex.ru

Abstract. Effective reservoir development and economically viable development of high-viscosity oil require experience and comprehensive experimental studies on the physical modeling of oil recovery processes using various technologies and displacement agents. The article discusses the filtration processes occurring in the reservoir during the injection of high-pressure water vapor. The results of the filtration experiments show that increasing the temperature to 90 °C makes it possible to effectively displace viscous oil from the reservoir models, while the displacement coefficient reaches 76 % after pumping 1.59 pore volumes of distilled water. It should be noted that the injection of low or demineralized aqueous fluids at reservoir temperature and pressure leads to rapid swelling of clay minerals of the formation rock, attenuation of filtration and loss of reservoir properties of the rock.

1. Introduction

In recent years, given the decline in oil production, much attention has been paid to the development of deposits with hard-to-recover reserves. Effective reservoir development and economically viable development of high-viscosity oil demand experience and comprehensive experimental studies on the physical modeling of oil recovery processes using various technologies and displacement agents. To date, thermal methods for increasing oil recovery are considered as the only industrial alternative.

2. Literature review

Thermal methods are oil recovery enhancement methods for productive formations based on additional heating of oil-saturated reservoirs. The thermal effect on the reservoir environment covers all its components (solid, liquid, gaseous); as a result, the reservoir filtration characteristics change due to the weakening of structural and mechanical properties. Improving the conditions for capillary impregnation, reducing the viscosity of oil, increasing its mobility, the transition of oil components to a gaseous state, improving the wettability of the displacing agent lead to an increase in the coefficient of displacement and final oil recovery [5, 6, 7, 9, 14, 18]. Thermal methods involving exposure to the reservoir are accompanied by phase transitions, as well as changes in the characteristics of the reservoir and its saturating fluids, the management of which in specific geological conditions is the most important task facing development specialists [1, 2, 10–13].

Currently, thermal methods are most effective in the development of high-viscosity oil fields and are priority among other methods; however, the development of such fields is currently complicated due to insufficient development of cost-effective technologies for the operation of such fields [8].

For cost-effective steam development on an industrial scale, two criteria systems were developed and published in 1984: one created at the University of Oklahoma under the name PDS (Petroleum Data System) [3]; the second was proposed at the Institute for Subsoil Development and Technology in Socorro, New York State [4].

Indeed, in the United States, commercial development using steam on an industrial scale is carried out in fields with oil viscosity of <7000 mPa·s, formation thickness of >15 m, and depth of >100 m.

But the quantitative values of the criteria for different authors do not always coincide, and sometimes significantly differ. V.D. Morozov and R.Kh. Safiullin (1988) analyzed data on 265 thermal projects in Venezuela, the USA, Canada and other countries [17]. Based on the results, it was found that the industrial development of high-viscosity oil and natural bitumen deposits is carried out in a rather narrow range of geological and physical parameters.

Thermal methods are based on a decrease in the viscosity of formation fluids due to their heating by heat transfer agents. Thermal methods of exposure based on the continuous injection of coolant into injection wells have a significant drawback. The heat supplied to the formation with a coolant is distributed in the formation between the rock forming the formation and the saturating fluids approximately in the ratio of 7–8 to 1. This heat distribution leads to low efficiency of such methods of exposure, since 70–80 % of the heat introduced into the formation are heat losses to the skeleton of the saturated porous medium, as well as to the top and bottom of the formation [6, 9, 11].

The mechanism of the processes that occur during the injection of steam into the bottomhole formation zone is rather complicated. The effectiveness of thermal treatment depends on several factors. Increasing the temperature of the reservoir leads to a decrease in the viscosity of oil, a decrease in interphase tension and adsorption of the active components of oil. As a result of the dissolution of paraffin-tar deposits, the bottom-hole zone is cleaned and its initial filtration properties are restored.

During steam and heat treatments under conditions of a high degree of heterogeneity of the collectors, the activation of the processes of capillary absorption of condensate in the pores of oil-saturated low-permeability blocks also acquires special significance.

Steam is an ideal heat agent because it has a high heat content. For example, water at a temperature of 148.9 °C contains 630 kJ/kg, and saturated steam at the same temperature contains 2751 kJ/kg. Therefore, the heat transferred to the formation by steam is more than four times the heat transferred to the formation by hot water. If the reservoir temperature is 650 °C, then 1 kg of water heated to 148.9 °C will transfer 357 kJ to the reservoir, while 1 kg of steam under the same conditions will transfer 2478 kJ, i.e. almost 7 times more [6].

Despite the fact that the injection of hot water is the most technologically simple, it is less effective than the injection of steam. Due to the low specific heat content of hot water, heating up a typical formation to a temperature close to the temperature of the injected water may require 2.5 to 3 pore volumes of hot water.

Steam as a thermal agent has the following advantages:

- high specific heat amount due to latent heat of vaporization; with a degree of dryness of steam of 0.8 (80 % of steam and 20 % of water, by weight), significantly more heat can be introduced into the formation (calculated per unit mass of the injected agent) than when hot water is injected;
- can occupy a volume in the reservoir 25–40 times greater than water;
- can displace up to 98 % of oil from a porous medium.

To quickly continuously increase the vapor zone, it is necessary to minimize heat loss in the wellbore. These losses during the steam injection process depend on the temperature of the injected steam and the equipment used [6, 11].

There are two ways to produce steam using steam generators and steam and gas generators. Steam and gas generators are characterized by good mobility, high efficiency, the possibility of layout for downhole applications, but relatively low productivity. Steam generators are usually massive plants that require complex water treatment systems with relatively low efficiency (to produce one ton of steam, you need to burn a lot of fuel), but with a sufficiently high productivity.

The principal physical mechanism for producing a thermal agent in a steam and gas generator is the injection of water into the fuel combustion products, followed by injection of the resulting mixture into the formation. The combustion process occurs at a pressure exceeding the reservoir, i.e. no additional devices are used to compress the resulting mixture to the required injection pressures. At the same time, unproductive heat losses are minimal, and useful heat can exceed 90 %. The performance of modern steam generating units with respect to the thermal agent is small and amounts to 2 t/h.

With increasing formation depths and increasing injection pressure, it is technologically and economically feasible to inject high-temperature water into the formation without bringing it to boiling point, since at high pressures the enthalpy of steam, hot water or steam-water mixture are close in value. As shown in the work of Bourget et al, if the oil reservoir lies at sufficiently large depths, then with increasing injection pressure in a certain temperature range, the enthalpy of the steam-water mixture decreases. It is noted that when using high-temperature water as a heat agent, the operating characteristics of both ground and underground equipment are much higher than when using steam [11].

Of course, the efficiency of the hot water flooding process is lower than the steam-thermal effect, since the required volume of water must be twice as large as the pore volume of the collector to wash out a unit volume of the oil-bearing formation. However, the efficiency of hot water injection is much higher than normal water flooding and is used in cases where steam injection is unacceptable. This includes great depths, and the presence in the formation of clays that swell when in contact with fresh water, as well as the need to maintain temperature in the formation to prevent solid hydrocarbons from precipitating in oil, etc. [6].

3. Research methods

The considered stratum $PK_{1\text{-}2}$ of the West Siberian deposit is the largest in area, represented by the alternation of sand, sandstone, siltstone with clays and clay siltstones, in the section of the stratum there are interlayers (up to 1.5 m) of rocks with calcareous cement. Oil and gas collectors are sands, sandstones and silts, gray loose or slightly compacted, fine, small- and medium-grained, clay interlayers. The formation is characterized by an average oil-saturated thickness of 8.6 m, gas-saturated thickness of 20.5 m, the depth of the formation is an average of 890 m. The viscosity of the oil in the reservoir is 377 mPa·s. The oil reserves of the facility are fully difficult to recover: the highly viscous oil rim is completely lined with water, the gas cap covers 82 % of the oil area. The deposit has a block structure and is represented by a weakly cemented reservoir [15].

In order to determine further plans for the development of the PK1-2 formation, the task was set to simulate the filtration processes occurring in the formation as a result of injection of high-pressure water vapor.

The methodological basis for preparing samples and fluids for testing and for conducting experimental work on oil displacement is OST 39-195-86 [16].

In the studies described below, core material of the PK1-2 formation was used, which is characterized by the following reservoir properties: porosity – 0.34 units, permeability – $125 \cdot 10^{-3}$ μm^2, oil saturation – 0.63 units, sandiness coefficient – 0.57 units.

As a model of water vapor condensate, distilled (demineralized) water was used.

To prepare an isoviscose model of oil in the PK_1 formation, dehydrated and filtered oil was used. For this purpose, purified kerosene in an amount of 8.5–9.0 % was added to the oil, selecting the kerosene content so that the oil viscosity at 33 °C was equal to 204 mPa·s. The characteristics of the oil model are given in Table 1.

It was planned to conduct the entire study in the course of one experiment, and to use a stepwise increase in the temperature of the displacing agent and the reservoir model. However, the rock properties of the PK1-2 formation did not allow this study to be conducted as part of a single experiment. During the research on core material, clay particles of the rock swelled and dispersed, which led to attenuation of the filtration and made the ultrasonic separator fail to measure the amount of oil displaced, and the clay suspension blocked flow lines and disabled the pressure regulator valve.

IOP Conf. Series: Materials Science and Engineering **663** (2019) 012065 doi:10.1088/1757-899X/663/1/012065

Table 1. Oil model characteristics

Parameter	Measurement temperature [°C]	
	21	33
Viscosity [mPa·s]	560	204
Density [kg/m^3]	938-939	930

The purpose of filtration experiments No. 1 and No. 2 was to develop a methodology for preparing reservoir models and saturating them with oil. The characteristics of the reservoir models are given in Table 2.

Table 2. Characteristics of models of reservoirs No. 1 and No. 2

Experiment number (reservoir models)	Pore volume [cm^3]	Permeability [μm^2]	
		For gas	For mineralized water
1	155	1.50	0.408
2	153	1.20	0.371

In experiment No. 1, the following technique was used to saturate the reservoir model with oil. The installation was assembled, while the oil in the pressure vessel was heated to 55 °C, and the reservoir model, to reservoir temperature. After thermostating for 2 hours, we started pumping the isoviscous oil model into the reservoir model with a volumetric feed rate of 12 cm^3/hour.

Initially, water was extracting from the reservoir model, which is typical for this type of experiment. However, after 0.2 porous volume of oil was filtered, at the exit of the reservoir model oil broke through. Such an early oil breakthrough is usually observed only in the case of a fractured core.

Subsequent oil filtration was accompanied by a gradual displacement of water (oil and water were simultaneously observed at the outlet). After pumping 1.38 of pore volume of oil, saturation reached 33 %. Then, the oil filtration rate was reduced to 6 cm^3/h in order to equalize the displacement front. Subsequent filtration of 0.61 of pore volume of oil did not increase displaced water volume. The model oil permeability estimation yielded a value of 4.2 μm^2, which is significantly higher than gas permeability.

Subsequently, model No. 2 was saturated with oil. The saturation technique was changed. Initially, oil was pumped at a temperature of 23 °C (room temperature), which increased its viscosity from 204 to 560 mPa·S. The feed rate was 6 cm^3/h. Oil breakthrough through the reservoir model occurred after pumping 0.28 of pore volume of oil, i.e. later than in experiment No. 1. After the oil breakthrough, the feed rate was increased to 12 cm^3/h, which was accompanied by an increase in the rate of water displacement. After pumping 1.87 of pore volume of oil initial oil saturation reached 64 %.

After the experiments, when removing sandstone from the reservoir of the reservoir model, it was found that the oil is evenly distributed over the volume of core material, i.e. the porous medium is uniformly saturated.

The experiments performed made it possible to make the following assumption that the clayed sandstone of the PK1-2 formation has double porosity:

- intergranular porosity, i.e. the void space between the particles of the rock,
- intragrain porosity, i.e. clay particles contain capillary pores filled with water.

Initially, oil displaces water from the void space between the rock particles, which leads to a rapid breakthrough of oil, and then, at a higher filtration rate, oil is displaced from the capillary pores inside the particles of clay minerals. Perhaps this is partially due to a change in the wettability of some clay minerals under the influence of viscous oil.

At the final stage of the research complex of filtration experiment No. 1, the effect of distilled water injection (a model of water condensate) on the permeability of the porous medium from the core of the $PK_{1\text{-}2}$ formation was tested. The average filtration rate was 30 cm^3/h (1.8 m/day). The results of the last stage of experiment No. 1 are presented in Figs. 1 and 2.

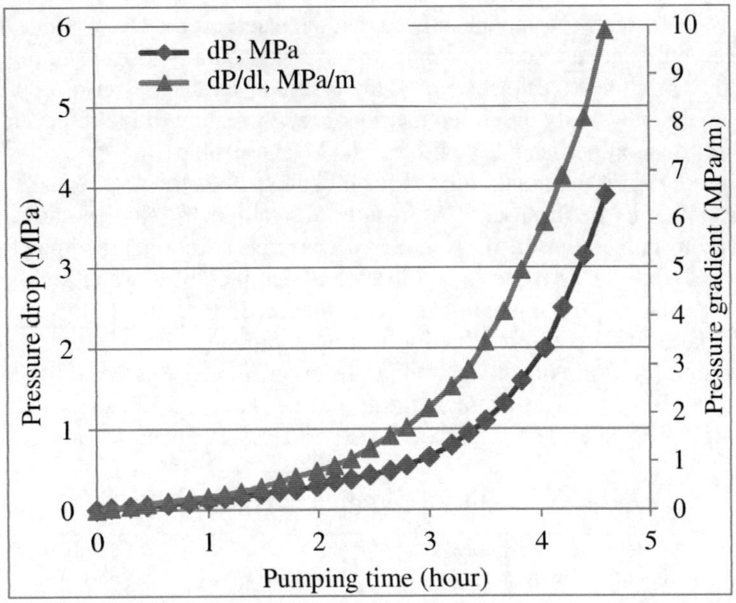

Figure 1. Dependence of the differential pressure and pressure gradient on the time of injection of distilled water in experiment No. 1 (33 °C, back pressure is 9.8 MPa)

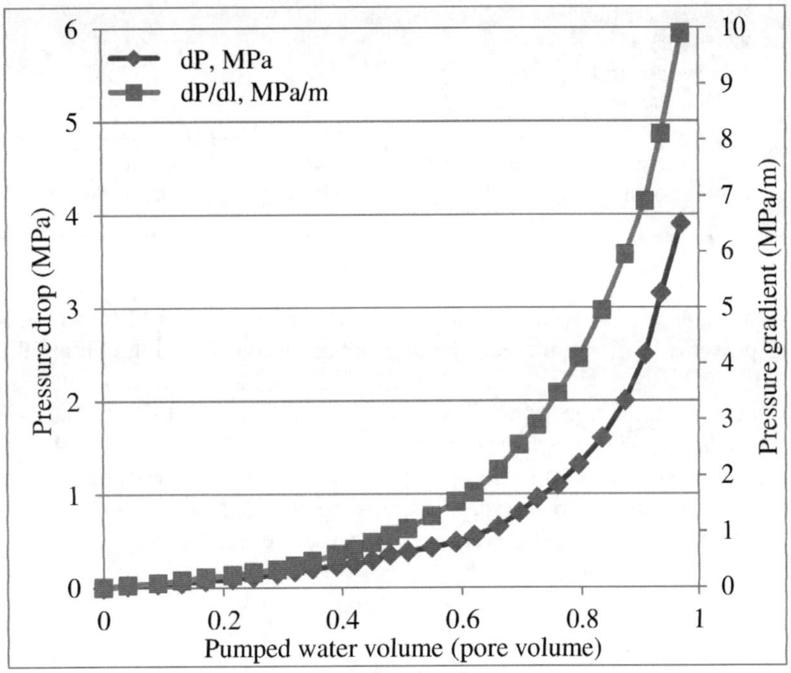

Figure 2. Dependence of the differential pressure and pressure gradient on the volume of injection of distilled water in experiment No. 1 (33 °C, back pressure is 9.8 MPa)

Figs. 1 and 2 show that in a very short time (3–4.5 hours), the clay components of the rock swell, which leads to a decrease in filtration. During the experiment, a pressure gradient of about 10 MPa/m was achieved, i.e. in the near-wellbore zone of the formation, as a result of the injection of demineralized fluid (water vapor condensate), an impermeable screen of high strength will form.

IOP Conf. Series: Materials Science and Engineering **663** (2019) 012065 doi:10.1088/1757-899X/663/1/012065

The disassembly of the reservoir model showed that the reservoir rock, as a result of interaction with demineralized water, turns into a viscous mushed mixture, i.e. actually loses reservoir properties [19].

When conducting filtration experiment No. 3, the reservoir model was also prepared and saturated with oil according to the previously described methodology. A feature of the reservoir model is that gas permeability is lower than oil permeability with residual water (Table 3).

The data obtained confirm the assumption that oil changes the structure of the void space and the wettability of clay minerals of the rock. The results of preliminary studies showed that prolonged injection at a reservoir temperature of demineralized water into a porous medium from a core of the PK_{1-2} formation is not possible. Therefore, the filtration of demineralized water at reservoir temperature was not carried out. The experiment was carried out according to the following procedure. After assembly, the installation was pressure-tested at reservoir pressure. Then, at the same time, distilled water was pumped into the reservoir model and the reservoir model was heated to 90 °C. The rate of temperature rise was 60 °C/h. The average filtration rate was 14.5 cm³/h, which ensured a linear displacement rate of 0.89 m/day. The results of the experiment are given in Table 3 and in Fig. 3.

Table 3. Results of filtration experiment No. 3

Reservoir model characteristics				Indicator	
Initial oil content, decimal units				0.737	
Permeability [μm²]:					
for gas				1.38	
for water				0.426	
for oil (with residual water)				1.53	
Water density [kg/m³]				1012	
Pore volume [cm³]				155	
Length [cm]				39.5	
Diameter [cm]				3.2	
Study of distilled water injection effect					
Fluid	Pumped volume [pore volume]	Temperature, °C	Pressure drop in the end of experiment [MPa]	Oil displacement coefficient [%]	Average filtration rate [m/day]
Distilled water	1.59	90	0.212	76	0.89

Water filtration began with a rapid increase in pressure drop to 0.154 MPa, which is explained by the fact that high pressure gradients are required to displace viscous oil. In the future, the pressure drop rapidly decreased, which is associated with the heating of oil and a decrease in its viscosity. A water breakthrough occurred after the injection of 0.425 of pore volume of distilled water. At the same time, an oil displacement rate of 57.6 % was achieved. Subsequently, the increase in the oil displacement coefficient continued at a slower rate and after pumping 1.59 of pore volume of distilled water, it reached 76 %. In general, the shape of the dependence of the oil displacement coefficient on the injection volume shows that the conditions for oil displacement at 90 °C are close to piston ones.

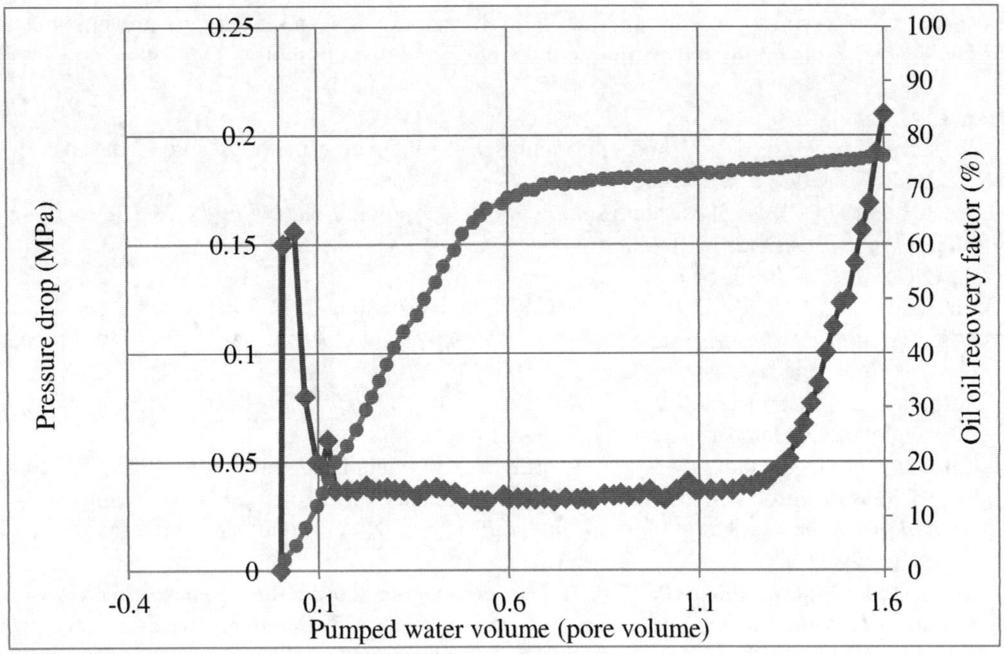

Figure 3. Filtration dynamics in experiment No. 3 at 90 °C

4. Results

Thus, increasing the temperature to 90 °C allows efficiently displacing viscous oil from the reservoir model. The dependence of the pressure drop on the injection volume is complex.

After the initial burst, the pressure drop decreases and stabilizes at sufficiently low values (about 0.023–0.030 MPa). However, after pumping 0.8 of pore volume of water the pressure drop begins to increase especially accelerating by the end of the experiment.

The obtained dependence indicates that clay minerals of the rock swell in distilled water and at 90 °C.

After pumping, approximately 1.1 of pore volume of water, the performance of the ultrasonic separator meter deteriorated, which determined the volume of the oil displaced from the model. Therefore, the experiment was temporarily stopped. At the same time, the inlet and outlet of the reservoir model were closed, and thermostating was continued. After cleaning the separator-meter and washing the discharge tank, the installation was pressure-tested at 9.8 MPa and prepared to continue the experiment. Initially, distilled water was started to be pumped through the reservoir model at 90 °C, but almost immediately the separator-meter again failed, the valve-pressure regulator at the outlet of the installation deteriorated (the back pressure began to fall). All this did not allow the experiment to continue in the region of higher temperatures.

Dismantling the reservoir model showed that the core material turned into a viscous mushed mixture ("quicksand"), as in experiment No. 1.

5. Conclusions

Thus, an increase in temperature to 90 °C makes it possible to effectively displace viscous oil from the reservoir models PK_{1-2} of the field. The displacement coefficient reached a value of 76 % after pumping 1.59 of porous volume of distilled water. When the formation is heated to a temperature of 90 °C, the clay minerals of the formation rock swell, but the swelling rate is lower than at the reservoir temperature [19].

Summing up the results of the experiments, it should be noted that the injection of low or demineralized aqueous fluids at reservoir temperature and pressure leads to rapid swelling of clay minerals of the PK_{1-2} formation rock, attenuation of filtration and loss of reservoir properties of the rock.

To preserve the reservoir properties of the $PK_{1\text{-}2}$ formation, it is necessary to prevent the use of desalinated water agents during the drilling, construction and operation of wells.

References

[1] Bokserman A, Mamedov Y and Antoniady D 1991 Diverse methods spread thermal EOR in USSR *Oil & Gas J.* **89(40)** 82–4
[2] Giusti L E 1974 CSV makes steam soak work in Venezuela field *Oil and Gas J.* **4**
[3] *Oil & Gas J.* 1984 **88(23)** 138–40
[4] *Oil & Gas J.* 1984 **88(26)** 49–53
[5] Stashok Y I, Antoniady D G, Drampov R T and Garushev A R 1991 Cyclic gas-steam well stimulations *Sixth Eurupean Symposium on Improved Oil Recovery* 21–23 May (Proceedings vol 1 Book I) (Stavanger, Norway) pp 105–14
[6] Antoniadi D G, Garushev A R and Ishkhanov V G 2000 *Handbook on thermal methods of oil production* (Krasnodar: Soviet Kuban) 464 p
[7] Antoniadi D G, Bichkevsky A D, Zavertaylo M M and Ishkhanov V G 1985 The state and problems of introducing thermal methods to enhance oil recovery *Theory and practice of oil field development using thermal methods* (Proc. NGO Soyuztermneft) (Moscow: Publ. VNIIOENG) pp 3–7
[8] Arzhanov F G and Ishkhanov V G 1990 *The role and place of thermal methods in the problem of increasing oil recovery. State and ways of improving technological processes and technical means of thermal impact on the reservoir* (Moscow: Publ. VNIIOENG) pp 3–10
[9] Arzhanov F G, Antoniadi D G, Garushev A R, Ishkhanov V G and Bekukh I I 1995 *Thermal methods of influence on oil reservoirs* (Nedra) 192 p
[10] Boxerman A A, Zhdanov S V and Kopanev D G *Antoniadi A method of developing an oil field* Authorship certificate. 16331166 (USSR)
[11] Bourget J, Bourget J, Surio P and Combarnu M 1988 *Thermal methods of increasing oil recovery* (Moscow: Nedra) 422 p
[12] Docher T M and Hassemi F 1984 The effect of oil viscosity and the thickness of the reservoir on the effectiveness of steam and heat exposure *Express-inform. Ser. Oil field business* Iss 1 (Moscow: Publ. VNIIOENG) 10 p
[13] Zheltov Yu P 1990 Mathematical modeling of thermal processes of stimulation *Thermal oil recovery enhancement methods* (Moscow: Nedra) pp 46–65
[14] Ivanov V A, Boxerman A A and Ishkhanov V G 1986 Development of high-viscosity oil fields by thermal methods *Review. inform. Ser. Oil field business* **3** 55
[15] Krivova N R and Dzhamalutdinov A B 2018 A generalization of the experience of pilot works and the formation of a strategy for developing the rim of high-viscosity oil *Experience, Urgent Problems and Prospects for the Development of the Oil and Gas Complex: (Materials of the VIII Int. Scientific and Practical Conf. of Students, Postgraduates and Scientists* vol 1) (Tyumen: TIU) pp 134–9
[16] *OST 39-195-86 Oil. A method for determining the coefficient of oil displacement by water in laboratory conditions* (Moscow: Minnefteprom) 1986
[17] Development of heavy and highly viscous oil fields abroad 1989 *Proc. of VNIIOENG* Iss 13(33) (Moscow)
[18] Sidorov I V 2014 *Justification for the development of high-viscosity oil deposits by horizontal wells* (Tyumen) 132 p
[19] Toporkov V G, Rudakovskaya S Yu, Khlebnikov V N et al 2008 Progress report on the topic *"Comprehensive core surveys from wells of fields 2050, 1002, 2031 and 2025"* (Tver) 51 p

IOP Conf. Series: Materials Science and Engineering **663** (2019) 012066 doi:10.1088/1757-899X/663/1/012066

Influence of drilling fluids on the filtration properties of reservoirs

E Yu Lipatov and N R Krivova

Tyumen Industrial University, 38, Volodarsky St., Tyumen, 625000, Russia

E-mail: nrk4@yandex.ru

Abstract. Currently, during the opening and testing of productive formations, as well as during repair work in wells, specialized water- or oil-based technical fluids are used. All of them are characterized by compatibility to varying degrees with the lithological and physical properties of reservoirs in Western Siberia and have different effects on the filtration properties of the porous reservoir of productive formations. It is known that when using fresh water-based drilling fluids, due to the interaction of the fresh water filtrate with rock cement and some rock-forming minerals, the permeability of the bottom-hole zone changes. The paper presents the results of an experimental study of the effect of drilling fluids on terrigenous rocks of one of the fields. The impact of drilling fluids on the core, with the exception of FLO-PRO and NaCl, led to a decrease in the oil-permeability of core samples. In the core of the terrigenous reservoir, filtering FLO-PRO does not cause a negative result. The curves of permeability changes with increasing filtration pressure before and after FLO-PRO flow are identical in dynamics. According to the results of the study, it can be concluded that technical fluids FLO-PRO and IES (invert emulsion solutions) can be recommended for use as they do not violate the structure of the pore space of the reservoir and do not lead to a significant decrease in oil-permeability of all core layers.

1. Introduction

Currently, during the opening and testing of productive formations, as well as during repair work in wells, specialized water- or oil-based technical fluids are used. The chemical composition of fluids used for working in wells of multilayer fields is diverse: from simple salt solutions (NaCl and $CaCl_2$) to multicomponent polymer-containing compositions, such as FLO-PRO [1, 13]. Common for these liquids is their low viscosity and relatively high specific gravity with almost complete absence of weighting agents in the form of a solid phase (bentonite, barite, etc.). All of them are characterized by compatibility to varying degrees with the lithological and physical properties of reservoirs in Western Siberia and have different effects on the filtration properties of the porous reservoir of productive formations [2]. Especially during the opening of potentially unstable rocks of the Bazhenov and Georgievskaya Formations together with the productive formation JV1, which has a low reservoir pressure anomaly coefficient (0.85–0.9), inhomogeneous permeability across the section (from 1–2 to 15–20 md) and high fractured porosity imposes a number of serious restrictions on the choice of the type of drilling flushing fluid [3].

IOP Conf. Series: Materials Science and Engineering **663** (2019) 012066 doi:10.1088/1757-899X/663/1/012066

2. Literature review

In general terms, the reasons affecting the reservoir filtration and capacitive properties of the bottom-hole zone (BHZ) of porous reservoirs when exposed to industrial fluids can be divided into two groups: technological and geological [4]. Technological reasons, caused mainly by the mudding of the bottom-hole zone by solid particles of drilling fluids, are not considered in this work, since there is no solid phase in all tested fluids.

The geological reasons for the change in reservoir properties of reservoir rocks during their interaction with drilling drilling fluids include [9, 10]:

- swelling of the clay phase of the reservoir due to the water-based drilling fluid filtrate penetrating into it;
- insoluble precipitation in the pores of the formation during the chemical interaction of the formation water with the filtrate of the technical/drilling fluid;
- a change in phase oil-permeability after the drilling fluid penetration into the reservoir;
- the formation of oil-water emulsions that reduce the relative permeability of the BHZ.

Being inherently a heterogeneous mineral system, reservoir rocks are differently resistant to various drilling fluids. The chemical reagents in their composition cause physicochemical reactions in the reservoir rocks, the final result of which is a change in the properties of the pore-fissure space [7, 8]. To prevent screes and collapses of the borehole walls, taking into account the possibility of other types of complications and ensuring high rates of penetration of the hole, it is necessary to comprehensively approach to the choice of the optimal density [14].

It was established that when using fresh water-based drilling fluids, due to the interaction of the fresh water filtrate with rock cement and some rock-forming minerals, the permeability of the bottom-hole zone changes. Using water-based salt drilling drilling fluids, such as NaCl and $CaCl_2$, causes crystallization and formation of salt crystals in a porous medium due to the high degree of filtration of water-based drilling fluids into the formation. The effect of multicomponent drilling fluids, including oil-based invert-emulsion solutions (hydrocarbon-based solutions), for instance BARADRIL-N/XP-07 solution, when drilling a horizontal section of a well in a producing reservoir JV1, enables minimal repression on the formation and minimal risk of absorption which prevents pollution of the reservoir [5].

The use of hydrocarbon-based solutions to a lesser extent affects the change in reservoir properties of reservoir rocks, but can lead to hydrophobization of part of their pore space.

The oil-permeability of the rocks after oil interaction with the drilling fluid used to work in the BHZ of the well will determines their applicability.

Methods for monitoring and controlling the filtration properties of drilling fluids, especially for deep wells, also require clarification. In this case, it is necessary to distinguish between the problems that arise when developing a drilling fluid as a coagulation-thixotropic system of a certain type (which is associated with the stability of the fluid and the nature of the interaction with drilled rocks) and the problems of choosing its formulation (fractional composition) for controlling the technological properties [15].

Salt solutions (brines) are technological fluids prepared on the basis of filtered fresh water and NaCl and $CaCl_2$ salts used to control the density of the solution. The specific gravity of solutions in the range of 1.1–1.25 g/cm^3 is selected depending on the task. Mineralization of brines significantly exceeds the salinity of formation water.

Solutions (brines) are used as a working medium during repair and restoration work in wells and during the secondary opening of formations by perforation. The use of solid-phase perforation liquids based on NaCl and $CaCl_2$ salts significantly prevents an intensive decrease in the natural permeability of the formation during perforation, but does not exclude it, especially in low-permeability terrigenous reservoirs with small pore sizes and high clay content.

The main properties of highly saline solutions of salts are ease of preparation and low cost. Solutions do not cause swelling of clay minerals of the rock and penetrate no deeper than 5-10 meters into the reservoir.

An indifferent solution is a technical fluid used as a working medium during the secondary opening of formations by perforation and for killing wells during repair and insulation works. In its composition, the solution contains water; calcium chloride (1–40 wt. %); organic solvent (lower aliphatic alcohols, glycols, acetone; 10–50 wt. %), inhibitory salt (potassium or ammonium chloride; 1–10 wt. %).

The solution possesses improved drainage and inhibitory properties, increases the efficiency of opening productive formations by perforation, and facilitates the initiation of fluid flow after perforation. A positive effect is achieved by the fact that the composition of the kill fluid includes, in addition to the traditional calcium chloride ($CaCl_2$) used to regulate the solution, an organic solvent that promotes the dehydration of swelling clays, reduces interphase tension at the border with hydrocarbons, and destroys persistent emulsions in the bottom-hole zone. The presence of potassium chloride (or ammonium) in the solution prevents swelling and stabilizes clay minerals in the reservoirs. The density of the solution is controlled by the concentration of chemicals and can range from 1.01 to 1.28 g/cm^3.

FLO-PRO solution is a water-free clay-free system with a minimum content of solid phase. The main components of the FLO-PRO solution are: fresh (or sea) water, calcium carbonate, polymer preparations FLO-VIZ and FLO-TROL, sodium or potassium salts, LUB lubricating additives. FLO-PRO is used to provide high quality opening of productive horizons with deviated and horizontal wells. The main properties that underlie the use of the FLO-PRO system are:

- high viscosity of the filtrate of the FLO-PRO solution at low stress and shear rate. FLO-PRO simultaneously has the properties of both a liquid (during flow) and a solid (at rest). The effective viscosity of the solution is minimal at high flow rates, which reduces pressure losses in bottom-hole hydraulic devices (downhole motors, bit nozzles, etc.). At the same time, the viscosity of the solution increases to anomalously high values at low shear rates (near the walls of the well, in pores and fractures of the rock), which significantly increases the holding and carrying capacity of this technical fluid;
- the absence of a finely divided solid phase, which prevents clogging of the pore channels of the reservoir rock;
- low fluid loss of the solution, which can significantly reduce the radius of the zone of penetration of the filtrate into the BHZ;
- high thermal stability of the solution.

The IES (invert emulsion solution) is a technical fluid that is used to limit water inflow (killing) of the filtrates of working solutions in the BHZ of a well during the secondary opening of reservoir formations.

The invert emulsion solution is based on a specially prepared drip-liquid mixture of diesel fuel and industrial water in the presence of chemical stabilizers, and sometimes surfactants.

The main properties that are indicators for the operational use of the solution of IES are:

- the minimum fluid loss of the solution, which can significantly reduce the radius of the zone of penetration of the filtrate into the BHZ;
- a relatively easy removal of the solution from the BHZ during depression;
- high thermal stability of the solution;
- the absence of a finely dispersed solid phase, which prevents the clogging of the pore channels of the reservoir rock.

The results of field studies indicate that when wells are perforated using IES solutions, which prevent the development of negative colmatation and physicochemical processes in the BHZ, the permeability of productive formations remains up to 80% of the initial one.

To select the optimal properties of drilling drilling fluid, it is necessary to conduct standard petrophysical studies to study their filtration-capacitive properties and chemical composition.

This paper presents the results of an experimental study of the effect of drilling drilling fluids on terrigenous rocks of one of the fields. From the total amount of core samples studied in the laboratory, the samples were selected that are characterized by average (or close average) values of porosity and permeability (Table 1).

Table 1. Brief information on the physical properties of core samples

Indicator		JV1
1		5
Porosity	Min	3.11
	Max	20.76
	Mean	13.19
Klinkenberg-corrected permeability	Min	0.007
	Max	> 200
	Mean	3.44
Sample number		37

3. Methods for study

The reservoir properties of the Jurassic horizon are low. The collectors are gray sandstones, dark gray with a brownish tint, fine-grained, silty. The texture is homogeneous, oblique, intermittently wavy, due to the distribution of clay and carbonaceous material. Clastic material makes up 85–90 % of the rock, medium and poor sorting.

Cement (5–10 %) of sandstones is film-pore in type, heterogeneous in composition. Open pores are frequent (up to 1 %). A distinctive feature of the rocks of the JV1 horizon as a whole is finely dispersed crystalline pyrite (up to 15 %), which often impregnates the rock mass, sometimes forms condensations, and also replaces the central parts of the radiolarians in the form of continuous masses.

Impermeable interlayers are represented by fine-grained sandstones and siltstones, clayey with basal cement, which includes kaolinite, hydromica, chlorite, and the formation of the montmorillonite series. Interlayers with carbonate cement are often encountered, the main component of which is calcite, often dolomitized. The mudstones from both parts of the reservoir and the lintels separating them are dark gray, clay material of a chlorite-hydromica composition.

The cover of the JV1 horizon is mudstones of the Bazhenov and Georgiev suites up to 30-meters thick with high shielding properties. The mudstones are black and dark gray, bituminous, tiled and finely elutriated with a large number of radiolarians. The texture is homogeneous, layered, sometimes micro-spotted, due to the uneven distribution of organic matter.

The considered reservoir JV1 (1) is represented by siltstone >70 %, sandstone ~15 %, clay minerals (10–15 %), and is complicated by carbonates (Fig. 1). Clay cement (according to the XRD analysis) is mainly represented by kaolinite and the illite group; it is complicated by chlorite in terms of spar grains. Swelling minerals (smectite and montmorillonite) are practically absent.

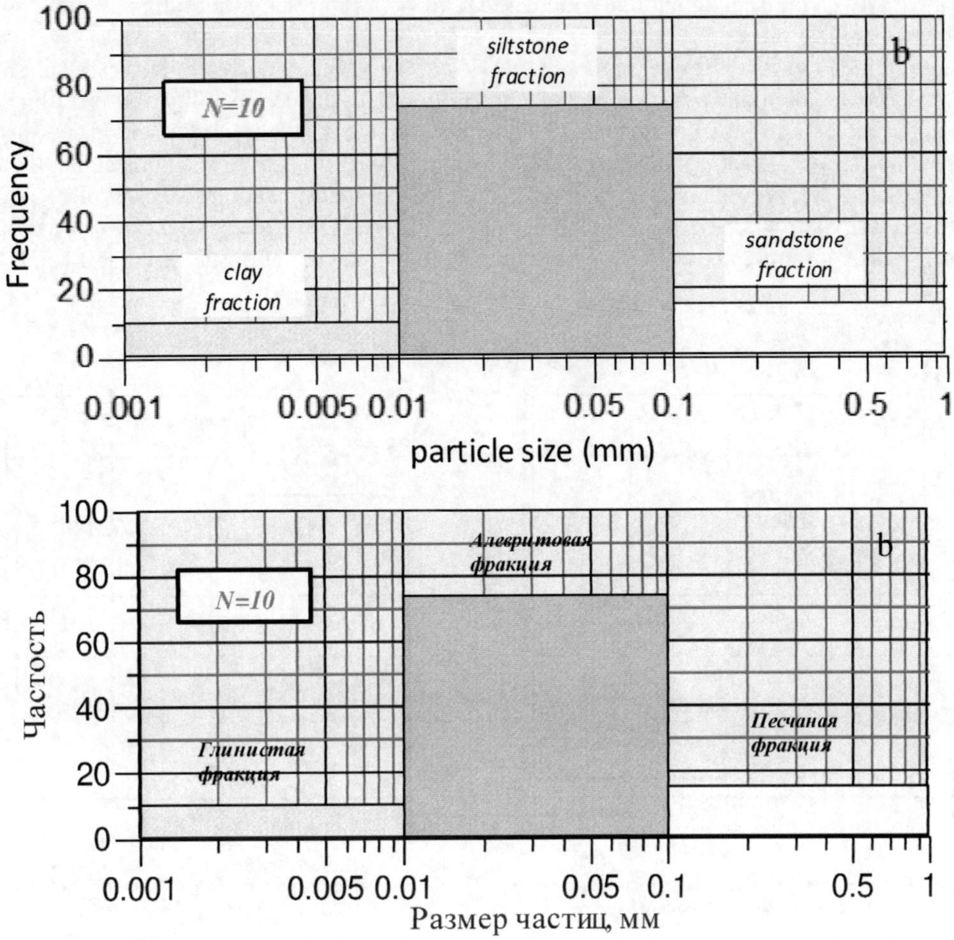

Figure 1. Reservoir Particle Size Histogram

The porosity of the rocks varies in a wide range from 2 to 22 %; low porosity is due to the presence of carbonate cement. The average value of porosity is 13.2 %. The gas-permeability of the formation varies over a very wide range, from hundredths of md to tens of md with an average permeability of 0.64 md. Oil saturation in the reservoir is episodic, with rare interlayers. In practice, the formation, if it is a reservoir, has a very low potential in terms of filtration and capacitive properties. Obviously, in this section of the field there were the most unfavorable conditions for the formation of the reservoir.

In terms of well 162, the JV1 (1) layer, according to well logging data, had more favorable conditions for the formation of the reservoir, but it was not possible to take out the core sample from the JV1 group. Only one sample with a porosity of 29 % and good permeability of 300 md was selected from the productive part of the formation. This sample was used to study the effect of FLO-PRO drilling fluid on the reservoir properties.

The rocks of the JV1 formation (1) react to the drilling fluid differently than the rocks of the overlying layers. The permeability of core samples, through which NaCl drilling fluids (sample k-34712), $CaCl_2$ (sample k-34715), an indifferent solution (sample k-34742, Fig. 2), and the IES filtrate (sample k-34743, Fig. 3) were filtered), in the pressure drop ranging from 0.2 to 0.8 atm. decreased, and with a pressure drop of more than 1.0 atm. stabilized for almost all samples.

Otherwise, there was a change in the permeability of the core sample through which the process water was filtered (sample k-34744, Fig. 4). Since the start of filtration (at a pressure drop of 1.8 atm.),

IOP Conf. Series: Materials Science and Engineering **663** (2019) 012066 doi:10.1088/1757-899X/663/1/012066

the permeability of the sample naturally increased, and with repression more than 3.0 atm. showed a tendency to stabilization.

Studies of the FLO-PRO filtrate. In the sample (k-21796), characterized by high reservoir properties (Kp for gas is over 200 md, Kp for solution of NaCl is 79.6 md), the permeability of this drilling drilling fluid with an increase in pressure drop from 0.2 to 1.4 atm. increased by more than 5 times. At higher pressure drops (above 1.6 atm.), the permeability of the FLO-PRO filtrate is stabilized at 40 md. For the sample (k-34738), whose reservoir properties are close to average and in the reservoir (Kp for gas = 5.16 md, Kp for NaCl = 2.82), the permeability of the FLO-PRO filtrate at slight pressure drop (up to 1.0 atm.) slightly lowered, and with a pressure drop of more than 1.5 atm. stabilized at 0.1 md (Fig. 5).

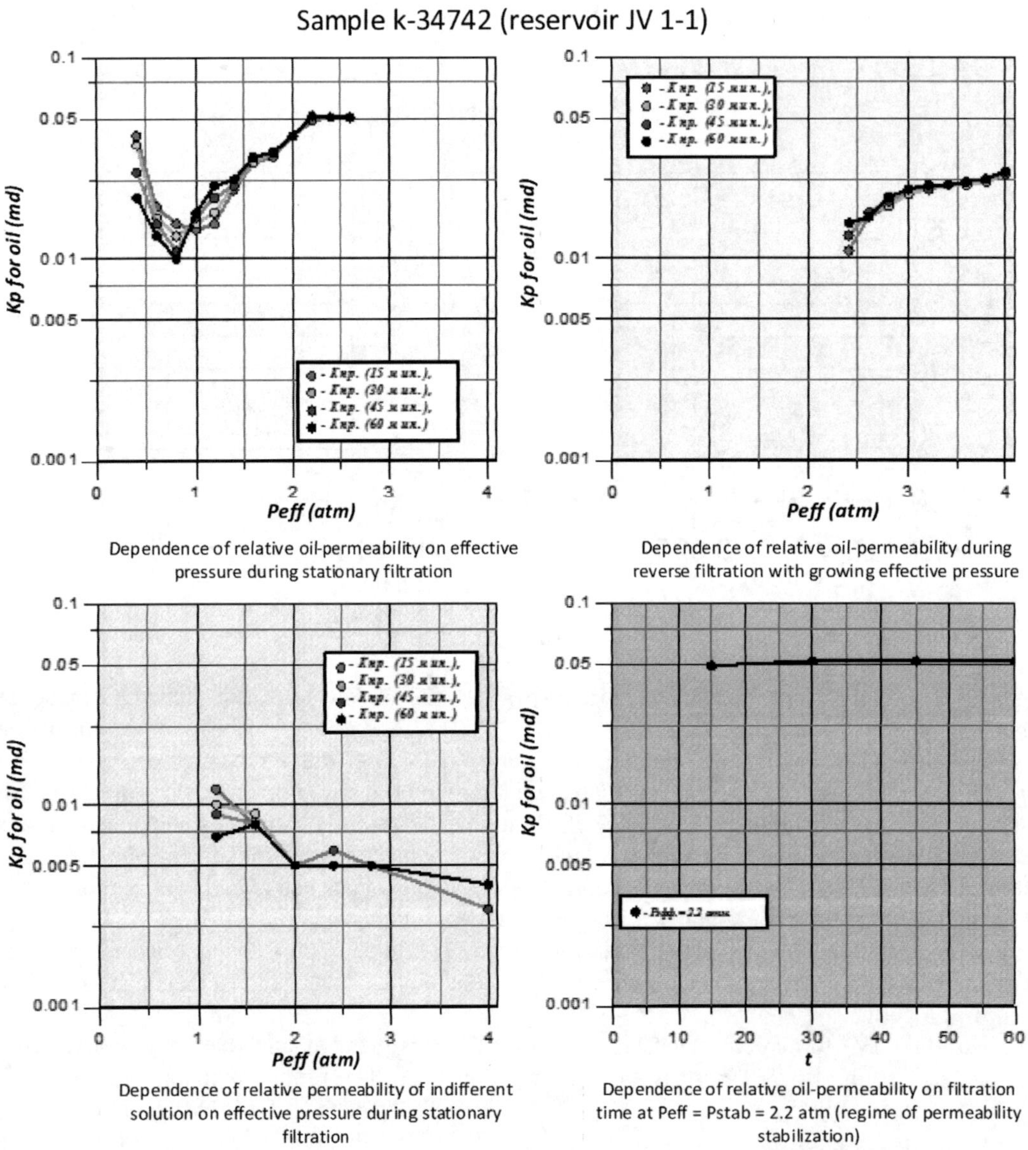

Figure 2. Change in permeability during injection of an indifferent solution

IOP Conf. Series: Materials Science and Engineering **663** (2019) 012066 doi:10.1088/1757-899X/663/1/012066

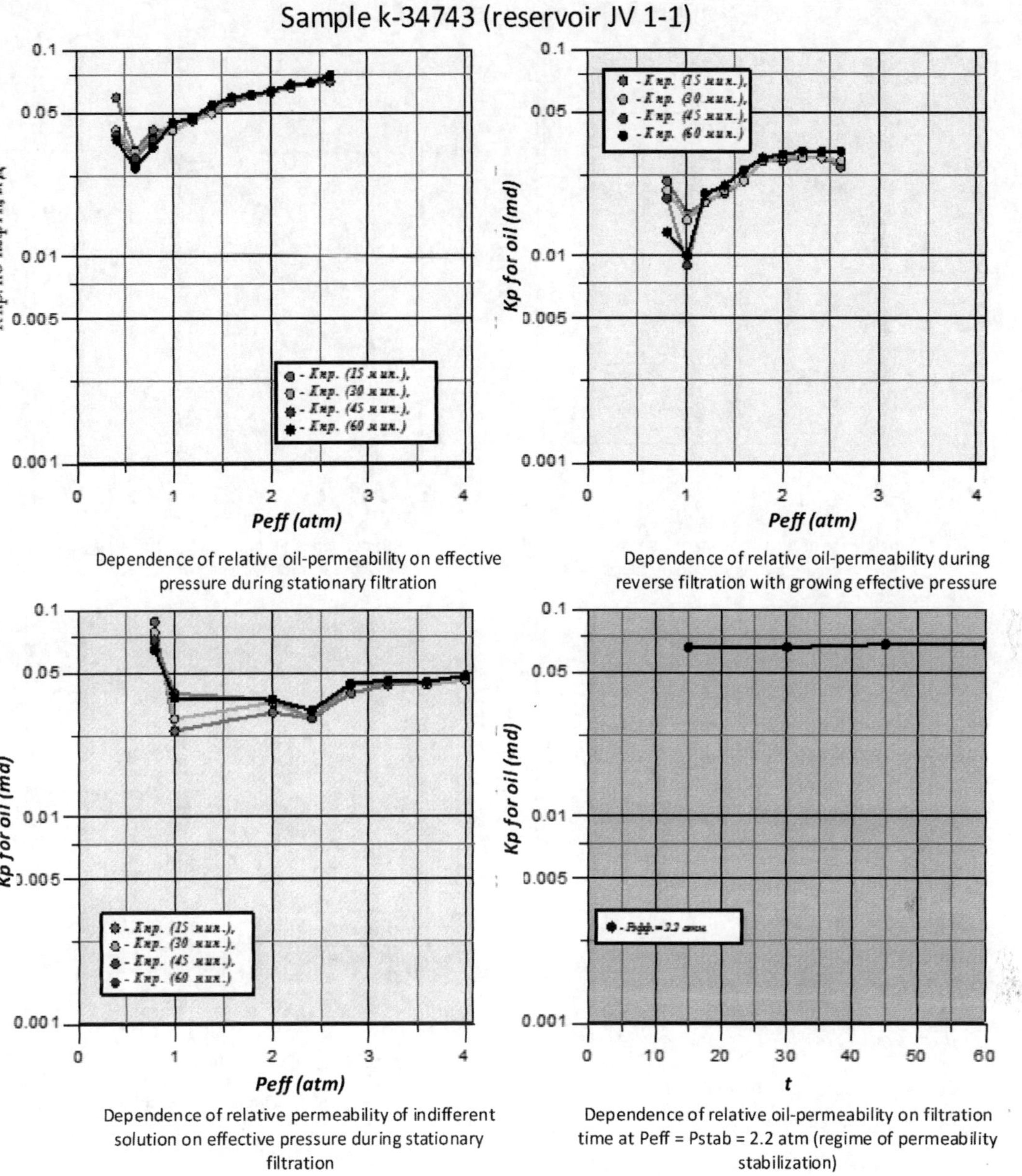

Figure 3. Changes in permeability during injection of the IES filtrate

IOP Conf. Series: Materials Science and Engineering **663** (2019) 012066 doi:10.1088/1757-899X/663/1/012066

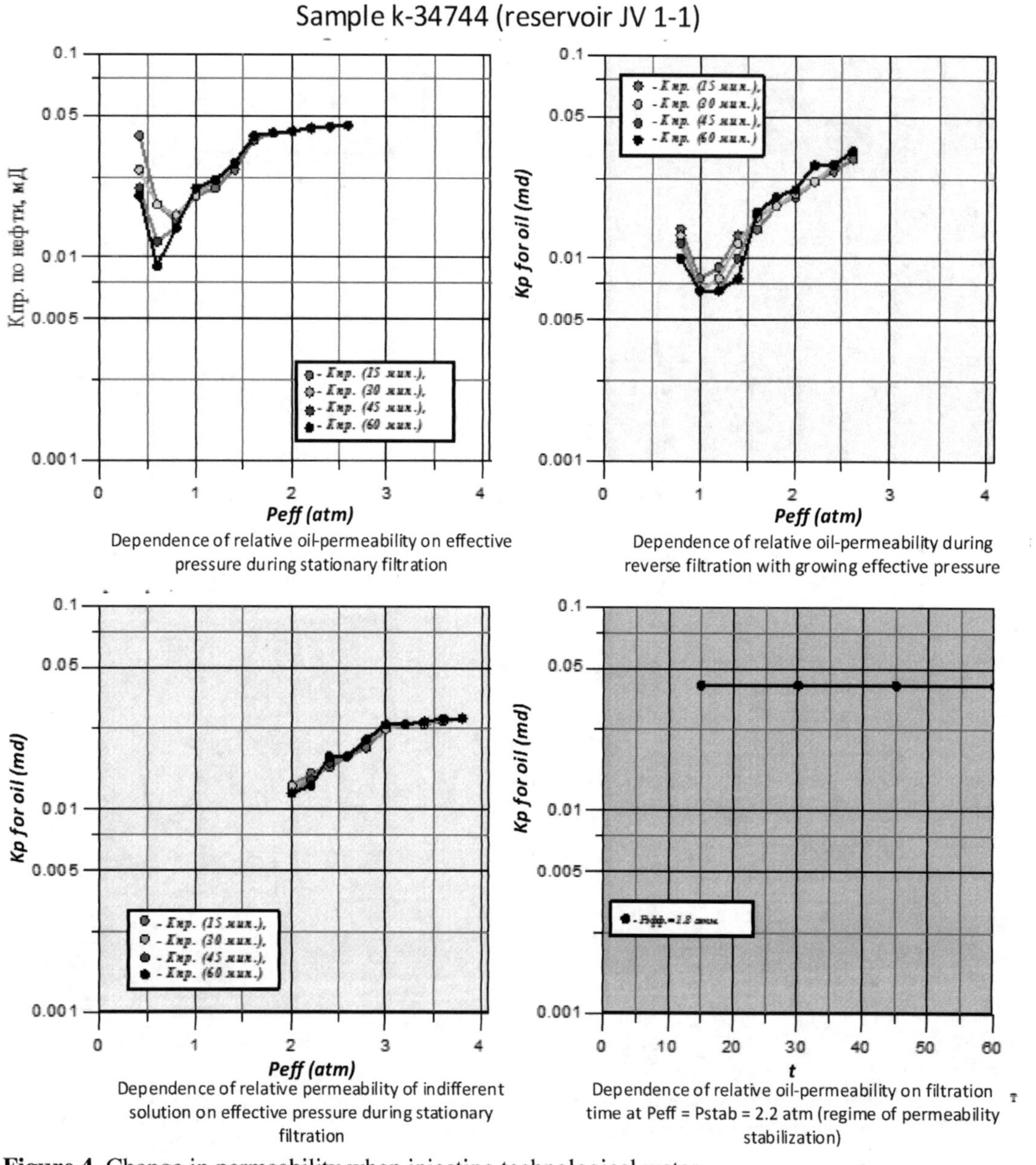

Figure 4. Change in permeability when injecting technological water

The impact of enumerated drilling fluids on the core, with the exception of FLO-PRO and NaCl, led to a decrease in the oil-permeability of core samples.

Indeed, for the core samples through which an indifferent solution, IES, $CaCl_2$ and technological water were filtered, the core permeability during reverse oil filtration decreased by 1.5–3.0 times. In this case, the smallest decrease is observed for thye samples through which $CaCl_2$ solutions and technological water were filtered.

IOP Conf. Series: Materials Science and Engineering **663** (2019) 012066 doi:10.1088/1757-899X/663/1/012066

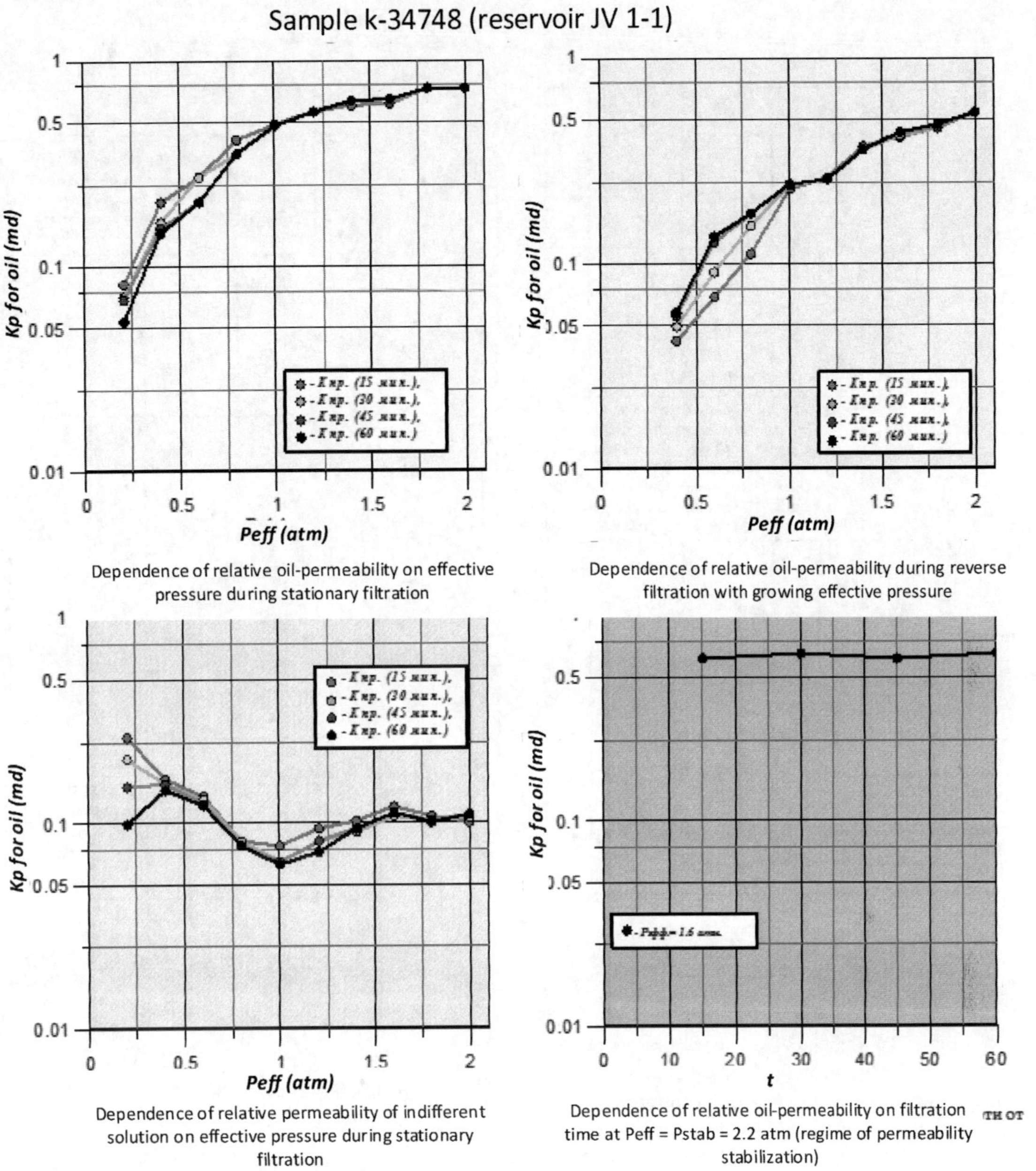

Figure 5. Changes in permeability during injection of the FLO-PRO filtrate

FLO-PRO drilling fluid (samples k-21796, k-34738) and highly mineralized NaCl solution (k-34712) practically did not cause changes in the oil-permeability of core samples. The oil-permeability of these samples was intensively and almost completely restored after filtration through the sample of the corresponding technical fluid.

The generalized results of the effect of drilling drilling fluids on the JV1 formation (1) are shown in Fig. 6.

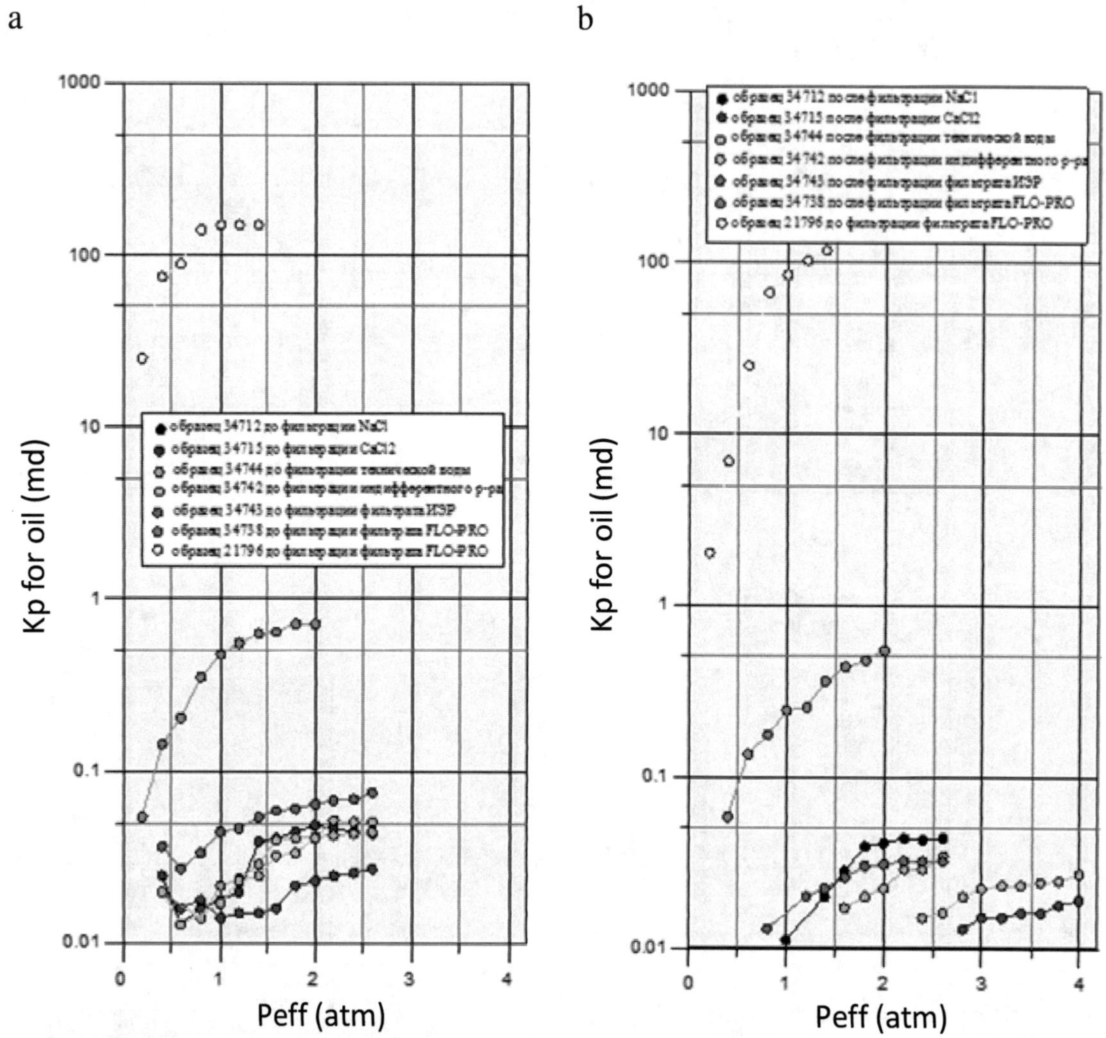

- sample 34712 after filtration of NaCl
- sample 34715 after filtration of $CaCl_2$
- sample 34744 after filtration of technological water
- sample 34742 after filtration of indifferent solution
- sample 34743 after filtration of IES
- sample 34738 after filtration of FLO-PRO
- sample 21796 after filtration of FLO-PRO

Figure 6. The dependence of the relative oil-permeability of the samples on the effective pressure: a) before filtering technological fluids; b) after technological fluids. Filtration time is 60 min

4. Results

Photographs obtained using a scanning electron microscope do not give an unambiguous picture of the reasons for the decrease in the permeability of samples in oil after filtering technological water, $CaCl_2$ solution, and IES through them. On the one hand, it is clearly seen that the filtration of these liquids did not cause swelling of the soaking clays, and on the other hand, the surface of the mineral grains was even somewhat cleansed of small crystals of the second and third generation.

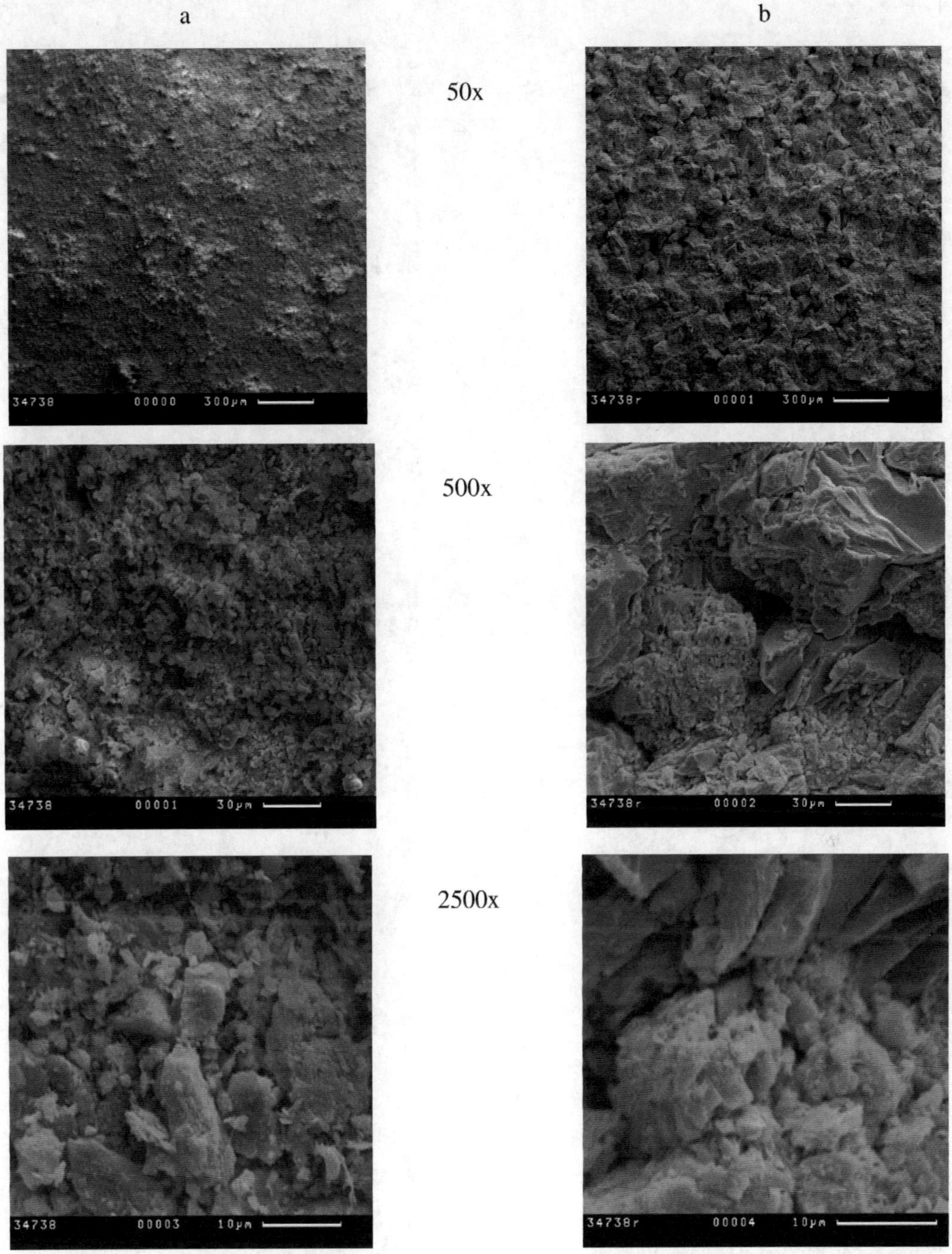

Figure 7. Photos of cleaved rocks (a) before and (b) after FLO-PRO filtering: sample k-34738, layer JV1 (1) (50x, 500x, 2500x magnification)

The pore space is relatively clean. Pore channels are clearly visible. However, this did not lead to a complete restoration of the permeability of core samples during reverse oil filtration. At the same time, in the photographs of the samples (k-34738 and k-34712) (Figs. 7 and 8), through which the NaCl and

FLO-PRO solution were filtered, it can be seen that the surface of the large grains remained complicated with small crystals and scales of clay minerals. Changes in the pore space structure of these samples are not observed [6].

Figure 8. Photos of cleaved rocks (a) before and (b) after NaCl filtering: sample k-34712, layer JV1 (1) (50x, 500x, 1000x, 2500x magnification)

IOP Conf. Series: Materials Science and Engineering **663** (2019) 012066 doi:10.1088/1757-899X/663/1/012066

5. Conclusions

In the core of the JV1 (1) formation, the FLO-PRO filtering does not cause a negative result. The curves of permeability changes with increasing filtration pressure before and after FLO-PRO flow are identical in dynamics.

Technical fluids FLO-PRO and IES can be recommended for use, as they do not violate the structure of the pore space of the reservoir and do not lead to a significant decrease in the permeability of core of all layers in oil.

References

[1] Aksenova N A and Rozhkova O V 2016 *Drilling flushing fluids and well flushing: a textbook for universities* In 3 volume vol 1 (Tyumen: TIU) 168 p

[2] Bulach M Kh, Belonovskaya L G and Gmid L P 2007 Low-pore reservoirs of oil and gas provinces of Russia and the CIS *Oil and gas geology. Theory and practice* **2** Retrieved from: http://www.ngtp.ru/rub/10/029.pdf

[3] Lipatov E Y and Aksenova N A 2017 Experience of Application of Biopolymer Emulsion Drilling Mud While Drilling Horizontal Wells in the Koshilskoye Field *SOCAR Proc.* **4** 36–41

[4] Gray J and Darley G 1985 *Composition and properties of drilling agents (flushing fluids)* (Moscow: Nedra) 509 p

[5] Lipatov E Yu and Aksenova N A 2017 Analysis of the effectiveness of the application of the BARADRIL-N/XP-07 system at the Koshilskoye field *Oil and Gas of Western Siberia (Proc. of the Int. Scientific and Technical Conf.* vol 1*)* (Tyumen: TIU) pp 55–7

[6] Toporkov V G, Astafiev O V, Orlov V N et al 2004 *The influence of technical fluids and drilling fluids on the operational characteristics of reservoirs*: report on R&D work (Tver) 206 p

[7] Classifier of drilling fluids, 2009 *Oil and gas technology* **10** 23–79 Retrieved from: https://rucont.ru/efd/261871

[8] Baroid N L 1979 *Industries Inr. Manual of Drilling Fluids Technology* (Industries Publishers)

[9] Gray G and Darley H 1980 Composition and Properties of Oil Well *Drilling Fluids* (Gulf Publishing Company)

[10] Howard G and Scott P 1951 An analysis and the control of lost circulation *Annual meeting of AIME* (St. Louis, Mo)

[11] Wilson G 1976 How to drill a usable hole *World Oil*

[12] Chencvert M E 1973 Shale Control with Balanced Activity *Oil-continuous muds Society of Petroleum Engineers Reprint Series* **N6a** (SPE Houston TX)

[13] MI–SWACO 2009 *School on Drilling Fluids and Sludge Treatment Technology* (Volgograd: IKF-Servis) 340 p

[14] Martin M, Vaussard A and Patroni J M 1984 Filtration des fluides de forade a travers les parois du puits *Revue de L'institut Francais du Petrofe* **3(6)** 735–8

[15] West G and Pharis B Dewatering cuts drilling mud and disposal costs *Oil and Gas J.* 1991 **89(39)** 84–8

IOP Conf. Series: Materials Science and Engineering **663** (2019) 012067 doi:10.1088/1757-899X/663/1/012067

Fisher–Tropsh synthesis technology evolution

N I Loseva, I I Niyazbakiev and A V Silman

Tyumen Industrial University, 38, Volodarsky St., Tyumen, 625000, Russia

E-mail: lni99@yandex.ru

Abstract. the article considers the development stages of technologies for producing hydrocarbons based on syngas. XTL technology is based on the Fischer–Tropsch process. Carbon-containing substances in any aggregate state can be used as a raw material of XTL technology, which will affect the chemistry and process parameters. Currently, interest in these technologies has increased in connection with the advent of scientific and technological innovations. The latest generation of XTL technology has already entered the industrial production phase. The article presents the results of a comparative analysis of XTL technology of different generations, the main indicators of which were the state of raw materials, the type of reactor devices, the chemical composition of the catalytic systems and the resulting product. The advantages of the technology proposed by INFRA Technology Company involving an innovative reactor design for a high-performance multifunctional catalyst, are shown. At the same time, the necessity of combining gas and liquid flows in an innovative microchannel reactor, increasing the conversion of synthesis gas, as well as the high cost of the fourth-generation XTL technology are urgent problems that need to be addressed at the current stage of the evolution of Fischer–Tropsch processes.

1. Introduction

Fischer–Tropsch process technology (XTL) has been used to synthesize fuels and synthetic oil for nearly a century. At the stage of modern development, much attention is focused on this synthesis and the search for new developments to improve the process continues. According to recent literature, there are four generations of Fischer-Tropsch synthesis technologies, each of which has characteristic features.

2. Comparative analysis of Fischer–Tropsch synthesis technology of different generations

In 1919, German scientists Franz Fischer and Hans Tropsch discovered that from synthesis gas (a mixture of CO and H_2) obtained during the processing of carbon-containing raw materials, it is possible to synthesize liquid synthetic hydrocarbons, which is usually termed synthetic oil. The main idea of this process was to obtain a liquid from a gas.

$$nCO + (2n+1)\,H_2 \xrightarrow{Co} C_nH_{2n+2} + nH_2O$$

$$2nCO + (n+1)\,H_2 \xrightarrow{Fe} C_nH_{2n+2} + nCO_2$$

In the years 1920–1930 the era of coal gave way to the oil age. The production of cheap oil began, rail and shipping routes for its delivery developed resulting in rapid transition from one fuel to another.

And the opportunity to make liquid fuel out of the aging coal, which Germany was rich in, was seen as a chance for the coal industry.

The Second World War led to the appearance of the **first generation** of the Fischer-Tropsch synthesis due to the emergence of German ersatz fuel. The equipment for the synthesis plant, with annual productivity of about 50 thousand tons, was exported from Germany in the post-war period and found its application in Novocherkassk since 1954. Coal was used as the raw material for the plant, and the plant's products included not only gasoline and diesel fuel, but also other products: special reagents, solvents, soap, etc. The process took place in a tubular reactor with a stationary catalyst bed, which was a German deposited cobalt-thorium (Co-Th) catalyst [1]. But the Fischer–Tropsch process proved to be inappropriate, since the cost of producing fuel from synthetic oil exceeded that of extracted oil. The installation worked until 1992 and was stopped for economic reasons, although in a modernized form [2].

The **second generation** was booming in South Africa. Almost forgotten German technology got a chance to continue its further development in connection with the imposition of sanctions in South Africa, the absence of its own oil fields and the presence of impressive coal reserves. The country rebuilt industry, creating, in particular, the production of synthetic oil from coal in order to live solely on its own resources. The catalyst was refined, the process became more efficient and environmentally friendly. The synthesis is carried out at 2–2.5 MPa and 340 °C in a reactor in a layer of suspended powdery iron catalyst, with a total capacity of 180 thousand tons/year. This technology made it possible to obtain a spectrum of hydrocarbons of the gasoline fraction, olefins (mainly C_2–C_4) and oxygen-containing compounds, etc. [3].

The impetus for the development and implementation of the **third generation** of technology was the oil crisis in the early 1970s. Then the states of the Middle East sharply limited access of the USA and Europe to hydrocarbon reserves. Well-known companies BP and Shell have begun active development of synthetic oil production technology. But the crisis was over, oil has fallen in price again, and industrial production of synthetic oil was never launched. They did not abandon the created technology, and Shell in 1993 in Bintulu (Malaysia) built a plant that included 4 Fisher–Tropsch multi-tube reactors with a capacity of 125 thousand tons/year each. Each reactor contains about 10,000 tubes. The synthesis is carried out in a three-phase reactor with a suspended layer of cobalt catalysts. The process was given the name "Synthesis of Middle Distillates" [4, 5]. This synthesis was distinguished by products of mainly long-chain hydrocarbons – waxes, requiring the presence of additional plants and equipment at the plant for their processing. A truly large, with a capacity of over 1.5 million tons/year, the **third-generation** Oryx plant reached its operational capacity in 2009 in Qatar using Sasol technology. At the end of 2011, Shell completed the construction of the largest in the history of the Pearl plant for the production of synthetic fuel with a capacity of 6 million tons/year [6].

The prospect of a wider introduction of GTL technology ("gas to liquid") led to the improvement of all technological stages of the process, affected the direct production of synthetic oil and conditioned an increase in the productive forces of this process. Performance is one of the key parameters of the Fischer-Tropsch synthesis, because it determines the economic benefits of all stages, the result of which is the necessary synthetic fuel. In modern installations, the Fischer-Tropsch catalyst, which has high performance and selectivity, allows you to change the economic performance of XTL technology in a positive direction, as well as reduce the dimensions of the reactor, which has a height of 30 m and a diameter of 8 m.

The **fourth generation** of Fischer–Tropsch synthesis technology, which is tested in pilot plants, seeks to reduce to zero the disadvantages of the third. Scientists who are developing, resort to high-performance structured catalysts [7] or microchannel reactors with intensified heat transfer [8], but there are other proposals of varying degrees of sophistication.

The main changes introduced into the Fischer–Tropsch synthesis stage entail significant changes in other stages and in general energy-technological solutions. Despite the differences in the technological methods used by different developers, the fourth-generation XTL is characterized by common features:

- a multiple increase in the productivity of catalysts in an industrial environment;

- the possibility of compact modular placement of installations;
- radical simplification of the technological scheme by obtaining a single product directly in the Fischer–Tropsch reactor.

At present, the technologies of the new generation are going through the transition stage: experiments should begin to be implemented on an industrial scale. The developers of the fourth generation of XTL technology see as their goal the creation of a perspective that can open new industrial horizons through simplicity, compactness and a significant reduction in economic performance.

3. Conclusions

A comparative analysis of XTL technology of different generations is given in Table 1 [3, 9].

Table 1. Comparative analysis of Fischer-Tropsch synthesis technologies of different generations

Generation of XTL Technology	Raw Material	Reactor Type	Catalyst	Prevailing Product
I (1930–1950)	coal	with a stationary catalyst bed	Co-Th (Co-Zr)	gasoline+diesel fuel+olefins
II (1950–1980)	coal	fluidized bed catalyst	Fe	gasoline + olefins + oxygenates
III (1980–1990)	syngas	with a suspended catalyst bed	Co	diesel fuel + naphtha + LNG + oxygenates
IV (2000–2019)	HC⟶syngas	microchannel	Co/zeolite	single product (light fractions <3600 °C)

Unlike previous generations, the advantages of the fourth generation representative–INFRA technology–lie not only in obtaining a single product directly in the reactor, high productivity and selectivity of the catalytic system, but also in the intensification of heat transfer and environmental friendliness of the process [9].

References

[1] Krylova A Yu and Kozyukov E A 2007 The state of the processes for producing synthetic liquid fuels based on Fischer-Tropsch synthesis *Chemistry of solid fuels* **6** 16–25

[2] Eliseev O L 2008 *Russian Chemical J. (J. Russian Chemical Society of D.I. Mendeleev)* **LII(6)** 53–62

[3] Mordkovich V Z, Sineva L V, Kulmachakovskaya E V and Asalieva E Yu 2015 Four generations of the technology for producing synthetic liquid fuel based on Fischer–Tropsch synthesis *Catalysis in industry* **5** 23–45

[4] *Popular mechanics. GTL technology* Retrieved from: https://www.popmech.ru/technologies/44841-tekhnologiya-gtl-izbavlyaet-ot-strakha-chto-neft-zakonchitsya/ (accessed on 06.04.19)

[5] *Shell website* Retrieved from: https://www.shell.com/ (accessed on 06.04.19)

[6] *Oil and Gas Journal "Gastech": "Pearl GTL receives first gas"* Retrieved from: https://www.ogj.com/articles/2011/03/gastech--pearl-gtl.html (accessed on 06.04.19).

[7] Sarkar A, Graham U M, Neathery J K, Spicer R L and Davis B H 2006 Fischer-tropsch synthesis with ultrafine iron-based catalyst: nano-scale growth of particles *Aiche annual meeting, conf. proc. Ser. Aiche spring national meeting*

[8] Sun Y, Jia Z, Yang G, Zhang I and Sun Z 2017 Fischer-tropsch synthesis using iron based catalyst in a microchannel reactor: performance evaluation and kinetic modeling *Int. J. of Hydrogen Energy* **42(49)** 29222–35

[9] *INFRA Technologies* Retrieved from: https://ru.infratechnology. com/technology (accessed on 06.04.19)

[10] Kee R J, Karakaya C and Zhu H 2017 Process intensification in the catalytic conversion of natural gas to fuels and chemicals *Proc. of the combustion institute* **36(1)** 51–76

IOP Conf. Series: Materials Science and Engineering **663** (2019) 012068 doi:10.1088/1757-899X/663/1/012068

Automation of industrial processes and everyday life

Ya V Saveliev[1] and N N Savelieva[2]

[1]Tomsk Polytechnic University, 30, Lenina ave., 634050, Tomsk, Russia
[2]Tyumen Industrial University, 38, Volodarskii str., 625000, Tyumen, Russia

E-mail: jakoff.saveljev@yandex.ru, nnsavelieva@yandex.ru

Abstract. One of the main features of our information era is “digitization”. In the oil industry, there is almost no easily produced oil left, so we have to constantly create the latest technologies for unconventional reserves and conduct production in hard-to-reach regions. The "digitization" of all the production components is not a tribute to fashion, but a means of maintaining efficiency and success. The principle of "toothbrush". Toothbrush monitoring is used in the extraction, refining and marketing of petroleum products by monitoring the serviceability of industrial equipment. The principle of mobility and utility. In the field of exploration and production, this means that all information about the field development is processed and, using special applications, is displayed on mobile devices of the relevant people in real time. Portable mobile devices for checking the quality of welds at the junction of pipelines directly at the place of their operation, without stopping the operation process and without damaging the inspected object, can be necessary as well. The developed device for the arc welding of magnetized products — the IST-201 welding current inverter — allows pipes to be welded without their preliminary demagnetization. The device allows one to weld pipes in a matter of hours, and instead of a whole group of personnel, only welders are involved. This development can scan not only finished seams or products, but also details during the manufacturing process, which allows avoiding the risk of defective products already at the production stage.

1. Introduction

According to modern studies, one of the main features of our information era is digitization of data to ensure fast development. Yesterday’s emerging technologies are becoming commonplace today and tomorrow they will be replaced with new, more efficient and productive approaches and ideas [1]. This process of quick global information transfer involves vast numbers of professionals who may improve existing or create a new more advanced technology, thus, the advancement is ever speeding up, forcing companies that are unable to keep pace to the market periphery. The situation is even more acute in the oil industry, as it involves more than competition; there is practically no easily extractable oil left, thus it is necessary to constantly create, then improve, and develop technologies oriented towards extraction of non-conventional reserves as well as to operate in hard-to-get regions. Digitization of all the components of production process in this industry is not jumping a bandwagon, but a means to preserve efficiency and success.

The main trend in IT development during the last decade have been a movement from quantity to quality. Indeed, today 16 Gb of RAM or a HDD of several terabytes are no longer considered wonderful feats of technology. The volumes of obtainable information grow geometrically, but for a person or a company a huge part of these data are useless, thus, it is necessary to learn how to both gather and process necessary information. This is a problem that when solved may bring up significant dividends to all the people.

2. Toothbrush Principle

Today, information gathering is no longer a problem, as anything may be provided with sensors, from every-day things and industrial equipment to a human body. For example, according to market researchers, the last year's sensation was fitness gadgets allowing round-the-clock monitoring of their owner's health. The smart wristband can count your pulse and kilometers walked, while also forming an opinion on your way of life in general and providing you with relevant advice. Experts think that regular toothbrush is going to undergo even a more drastic change. They say that in a couple of years toothbrushes will start collecting data on our teeth and transfer them directly to your dentist's office. Then, you shall only wait for an appointment. Here we may see the main trend in automation of things: it is no longer of merely indicative nature, but gets new functions thanks to a new possibility to gather, process and transmit data without concerns for distance. Little by little, it leads to automation of daily life processes.

In the oil industry, this trend presents in all its sectors, from production to sales. A modern oilfield is inconceivable without a constant data gathering concerning the state of wells, underground equipment, pipelines, surface infrastructure, digitization of these data and their further processing. The data obtained this way allow for real-time monitoring of production indicators and timely reaction to changes, preventing failure events, saving energy and other resources.

Toothbrush-like monitoring is used in processing and sales to monitor equipment operability. As for future development in sensors, while improvement of their technical characteristics and efficiency is important, nevertheless the main problem lies in a more intellectual area. All the big data arrays shall be useful and bring profit to the company. This is the only case where automation costs are fully justifiable. Ideally, there shall be a completely automated system where data from all sensors and systems are collected, classified, analyzed and form backbone of models allowing making important decisions in a timely manner [2]. That is, the value of data shall be multiplied by application of IT, while the data themselves shall be transformed into a commercial resource.

3. Principle of Mobility and Usefulness

Another trend that has already become popular globally and in the industry is a total shift to mobile devices. What do you personally use for email, social networks and news? When answering this question, most people name their phone or tablet, rather than a PC. Independence of fixed communications is not just convenient; it brings real profits and savings to the company. Usefulness of mobile technologies is evident for regular office workers, fast food workers and the like, and it is already hard to imagine a manager without a cellphone as a minimum. Within the mobile office framework, many companies start using teleworking, saving on rent and other services.

Multiple mobile apps are being created, and each of them may find application in any field. It makes apps universal and thus in demand for working with industrial assets. Shell, ExxonMobil and other giants of the industry already use the digital control technology in project management with mobile devices [3]. In prospecting and production, it means that all the information on field development is processed and output by special mobile apps onto devices of interested parties in real-time. Project operator may use their tablet to watch drilling or production processes, while a top manager may browse through business analytics. In this case, mobility provides round-the-clock control and swift decision-making.

Portable mobile devices may be just as useful in the field, to be used in pipeline weld quality control directly at the operation location, without breaks to the process and without inflicting any damage to the tested object.

Among such outstripping developments are induction accelerators (betatrons) developed by researchers from Tomsk Polytechnic University (TPU).

Magnetic testing leaves welded elements magnetized thus making arc welding impossible.

IOP Conf. Series: Materials Science and Engineering **663** (2019) 012068 doi:10.1088/1757-899X/663/1/012068

Oil and gas industry employs special expensive equipment to de-magnetize pipes. It requires a long preparation process and shutting off the pipeline for a day or two. When the magnetic flux passes through the part in defective locations, magnetic force lines become concentrated, which allows finding a damaged area of the pipe. The damaged part is cut away and replaced with a new one. After that, the pipes shall be welded to each other [4, 5].

TPU-based researchers proposed an innovative solution of the problem. They developed an arc welding instrument (Arc current invertor IST-201) which allows welding pipes without prior de-magnetization. The instrument allows completing pipe welding in several hours and its operation involves only welders, rather than a diverse group of personnel usually required currently. Additionally, the enterprise is able to save significantly on purchase and maintenance of special pipe de-magnetizing equipment. Experimental prototypes of IST-201 have been employed at facilities of Surgutneftegaz, Tomskneft, KAO Azot and other companies. Besides, it was tested at operating facilities of Transneft-Central Siberia and Gazprom Transgaz Tomsk, confirming increased efficiency.

For example, a person responsible for welding or casting quality control at a certain facility may easier detect and eliminate defects. This R&D product may scan not only completed welds or products, but parts being manufactured as well, thus allowing eliminating risk of faulty production at the manufacturing stage. Advantages of the device are grounded in its characteristics: Power source of SEA-7 has a mass of about 30 kg, while its predecessor PBX is about 70 kg; emission power of the SEA induction accelerator is 7 MeV, while that of the previous version is two times lower at 5 roentgen per meter; these new characteristics allows examining steel with the thickness of up to 40 cm. Such a mobile device will be very helpful for a person and creates no new issues in operation.

Figure 1. Induction accelerator for arc welding of magnetized articles – Arc current invertor IST-201

Additionally, if we consider sales of production, mobility here results in direct income by means of optimized logistics and reduced losses.

4. Conclusion

Mobile technologies provide possibilities of new communications and change the appearance of any average employee, whether industrial or social worker [6, 7]. The coming changes will affect everyone. Integration of information and industrial technologies leads to a situation where human participation is limited to remote monitoring and control of operation and timely reaction in case of arising issues [8, 9]. Additionally, development of mobile apps allows us to predict upcoming changes in the very foundation of decision-making, where the main role is going to be not with an opinion of one subject matter expert, but that of a competent expert community.

References

[1] Cherkasov M V 2015 Smart technologies in oil and gas industry *Control Engineering* 20–2
[2] Dolotovskii I V, Larin E A and Dolotovskaia N V 2015 Information and analytical support of

IOP Conf. Series: Materials Science and Engineering **663** (2019) 012068 doi:10.1088/1757-899X/663/1/012068

energy management systems in oil and gas enterprises *Int. J. of Experimental Education* **11** 132–5 (Moscow)

[3] *Upstream Oil and Gas* 2014 Retrieved from: http://www.upstreamoilandgas.com/

[4] Moskalev V A and Chakhlov V L 2009 *Induction Accelerators* (Tomsk: Tomsk Polytechnic University)

[5] Moroshin A 2016 Ready to work and operate an induction accelerator *Kommersant-Dengi* **8** 44–6 (Moscow: Kommersant)

[6] Bitner E, Mounts T L and Dutton H J 1980 Monitoring of oil composition by an automated system for esterification GC analysis *J. of the American Oil Chemists' Society* **57(7)** 209–12

[7] Tarabanis K A, Allen P K and Tsai R Y 2009 A survey of sensor planning in computer vision, *IEEE transactions on Robotics and Automation* **11(1)** 86–104

[8] Sheridan T B 1995 Teleoperation telerobotics and telepresence: A progress report *Control Engineering Practice* **3(2)**

[9] Gary J H and Handwerk G E 1984 *Petroleum Refining Technology and Economics* 2nd ed (Marcel: Dekker, Inc.; New York)

IOP Conf. Series: Materials Science and Engineering **663** (2019) 012069 doi:10.1088/1757-899X/663/1/012069

Ways to improve production efficiency problems and ways of their solution

P M Kosianov

Tyumen Industrial University, 38, Volodarsky St., Tyumen, 625000, Russia

E-mail: kospiter2012@yandex.ru

Abstract. The work shows various ways to improve the efficiency of oil production. A comparative analysis was carried out. Considered the most environmentally friendly ways. Considered physical models and the most important averaged parameters for them are defined. Identified ways to influence these parameters. The expressions are shown that allow to calculate the volumes of oil filtered through the sections of the bottomhole zone of the reservoir. The problem of determining the viscosity of oil. The methods of steam-thermal impact on the formation, their advantages and opportunities are considered in detail. Identified problems, shows how to solve them. Solutions were found for turning water into saturated steam in a closed volume of a drilling tool. The quantities that evaluate the effectiveness of the device are determined and calculated quantitatively. The optimal size and shape of the nozzles are determined. Calculated steam flow rate from the nozzles of the device, the reactive forces arising in the device and the torque generated by the forces. The relevance and prospects of these methods are shown. at development of deposits with high viscosity oils.

1. Introduction

The main indicator of the efficiency of oil production is the oil recovery efficiency (ORE):

$$K = \frac{V_{\text{produced}}}{V_{\text{balanced}}}$$

Studying complex multicomponent systems as collectors [1], it is impossible to reduce everything to a single model, the Cauchy multi-problem, which models are described by nonlinear autonomous ordinary differential equations (ODE) of arbitrary dimension [2]. It is necessary to conduct a comprehensive research based on well logging (WL). Nevertheless, for disclosing the mechanisms that enable enhanced oil recovery, it is useful to consider physical models that are reducible to several, the most important parameters that can be measured by physical fields. So Kin can be represented as a kind of operator of the average data:

$$K=F(p_{\text{пл}},\mu,k_{\text{пр}},T,t), \qquad (1)$$

$p_{\text{пл}}$ – reservoir pressure; μ – oil viscosity; $k_{\text{пр}}$ – reservoir permeability; T – temperature; t – time.

Using the laws of Poiseuille and Darcy, we can express the volume of oil dV filtered through the cross section dS per unit time:

$$dV = A\frac{k_{\text{пр}}\,\Delta p_{\text{пл}}}{\mu\Delta L}\,dS, \qquad (2)$$

$\Delta p_{\text{пл}}$ – pressure difference at the ends of the capillary with a length ΔL; A – constant for this collector.

Then the volume of oil filtered per unit of time through the section S of the reservoir near well bore:

$$V_S = \int_0^S A\frac{кпр\,Δрпл}{μΔL}\,dS = A\frac{кпр\,Δрпл}{μΔL}\,S. \quad (3)$$

Assuming the pressure in the volume of the well is atmospheric, the volume of oil filtered per unit of time through the cross-section S:

$$V_S \approx A\frac{кпр\,рпл}{μΔL}\,S. \quad (4)$$

Then $V_{produced}$ for the final time of oil in the first approximation:

$$V_{produced} \approx \int_0^t A\frac{кпр\,Δрпл}{μΔL}\,S\,dt \approx A\frac{кпр\,Δрпл}{μΔL}\,St. \quad (5)$$

One of the main problems is the expression of viscosity of oil in an explicit form, which is a large and urgent scientific task [3, 4]. Nowadays, the viscosity is determined in laboratory conditions for filtered volumes. In this model, an increase in oil production implies an increase in reservoir pressure, a decrease in viscosity and permeability, and the influence of external physical factors. Modern geophysics uses all kinds of physical fields. From the most environmentally friendly ways to develop we indicate:

- Gravitational, $K < 5$ %
- Elastic, $K \leq 5$ %
- Dissolved gas, K from 5–7 to 20–25 %
- Water pressure K from 20–30 to 60–70 %
- Gas pressure K 50–70 %
- Thermal steam on the reservoir, $K < 70$ %
- Electromagnetic, $K \leq 70$ %

It is obvious that the prospect of increasing oil production is the integration and combination of various methods of development.

2. Materials and methods

2.1 Thermal steam

Heating of the reservoir leads to a decrease in the permeability and viscosity of the formation fluids: oil, water and gases (Fig. 1, a–c).

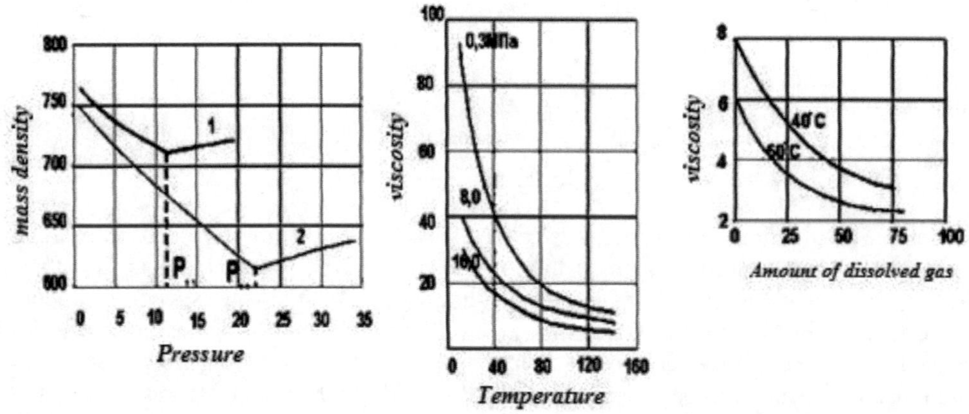

Figure 1. Correlations between the main parameters: a) Dependence of density of reservoir oil on pressure and amount of dissolved gas at t=70 °C; b) Dependence of viscosity on temperature; c) Dependence of viscosity on temperature and amount of dissolved gas

When the collector is heated, a decrease in viscosity and volume expansion of all components of the reservoir, a change in the permeability and mobility of fluid fluids occurs. In [5, 6] considered non-traditional drilling methods, the “drilling tool” integrating the energy of steam.

2.2 Principle of operation

The operation of the drilling tool occurs in the following sequence. When voltage is applied to the heating elements, the latter are heated, increasing linearly. A linear increase in the heating element leads to a shift of the stop. Accordingly, a retainer rigidly connected to the stop is moved along the through groove. Under the pressure of water in the channel valve-piston moves. Water fills the cavity vacated by the valve-piston and enters the channels. Then the water through the outlets from the channels gets on the heating element and begins to evaporate. The pressure inside the body cavity increases. As a result of an increase in pressure, the valve-piston returns to its original position, in this case water is supplied to the channels and intensive water is injected into the heating element. Abundant evaporation of water on the heating element leads to the formation of the working fluid – steam inside the housing. The channels – nozzles, the working fluid expires out, creating a rotational effect, due to the reactive force. Rotation of the body leads to the destruction of the ground with carbide inserts. After evaporation of water on the heating elements, the latter cool down and assume the same linear dimension, i.e. the system returns to its original position and then the process can be repeated.

3. Quantitative estimation of device parameters

To evaluate the effectiveness of the device, it is necessary to identify the values that determine the torque and calculate them quantitatively. Mechanical work is accomplished by generating steam energy inside the body of the drilling tool and using this energy through the body output channels.

We use the Clapeyron–Clausius equation [7] for processes with phase transformations:

$$\frac{dP}{dT}=\frac{q}{T(v_1-v_2)}, \qquad (6)$$

q – specific heat of vaporization, v_1, v_2 – specific volumes of water and steam.

Turning water into saturated steam in a closed vessel, where the temperature T is variable, we get:

$$\frac{dP}{dT}=\frac{\mu q}{RT^2}P, \qquad (7)$$

Separating the variables and integrating the resulting expression:

$$\int_{P_0}^{P}\frac{dP}{P}=\int_{T_0}^{T}\frac{\mu q}{RT^2}dT \quad \ln\frac{P}{P_0}\frac{\mu q}{R}\left(\frac{1}{T_0}-\frac{1}{T}\right)$$

And for the final pressure:

$$P=P_0e^{\frac{\mu q}{R}\left(\frac{1}{T_0}-\frac{1}{T}\right)}, \qquad (8)$$

were P_0 и T_0 – initial pressure and temperature; P и T – final pressure and temperature.

For instance, during the evaporation of 1 kg of water for 1 second, inside the cavity of the device with a volume of V=10^{-3} m^3, superheated steam pressure P=460·10^5 Pa, temperature T=998 K≈$T_{heat.}$.

When a part of its mass is ejected, where the ejected part changes its impulse, a reactive force acts on the system, equal to the change in the impulse of the ejected part per unit of time. The torque of the device occurs under the action of reactive forces, with the expiration of steam from the horizontal output channels – nozzles. In [6] was determined the pressure – p and vapor temperature – T when the specified mass of water evaporates, but to calculate the reactive forces – F_p and flow rate – M_k, it is necessary to first determine the flow rate of steam from the nozzles of the device (Fig. 2).

During the evaporation of 1 kg of water for 1 second, inside the cavity of the device with a volume of V=10^{-3} m^3, the pressure of superheated steam in front of the nozzle is p_1=460·10^5 Pa, temperature T=998 K.

The pressure behind the nozzle p_2 =10^5 Pa. The ratio of pressures in the environment and in front of the nozzle β=p_2/p_1=2,17·10^{-3},

The critical value of the ratio (with a sound outflow of steam) is determined by the expression [8]:

$$\beta_{кр}=\left(\frac{2}{k+1}\right)^{\frac{k}{k-1}}, \qquad (9)$$

k – Expiration adiabatic index, for superheated steam k=1,29 and $\beta_{кр}$=0,547.

As $\beta << \beta_{кр}$ in this case, the flow is supersonic.

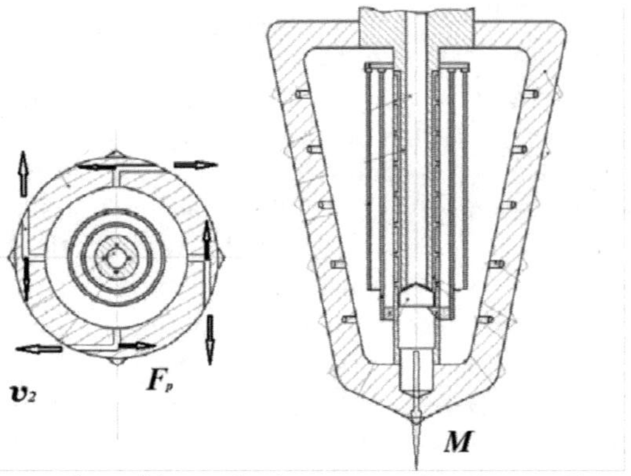

Figure 2. Steam outflow from nozzles with a velocity $\boldsymbol{v}_2$, reactive forces $\boldsymbol{F}_p$ and torque (moment of forces) $\boldsymbol{M}$ acting on the device

The speed at the nozzle exit for real gas is determined by the expression [8]:

$$v_2=\sqrt{2(i_1 - i_2)}, \tag{10}$$

i_1 – Vapor enthalpy at the nozzle entrance; i_2 – vapor enthalpy at the nozzle exit.

Enthalpy is on the is – chart.

For the task: i_1=3785 KJ / kg; i_2=2935 KJ / kg. The speed at the exit of the nozzle, respectively, v_2=1304 m/s. With an expanding nozzle shape with an opening angle $\gamma\approx 10°$ (fig. 3), the nozzle velocity coefficient is maximum φ=0.98 and the actual flow velocity is $v_2^d=\varphi v_2$=1278 m/s.

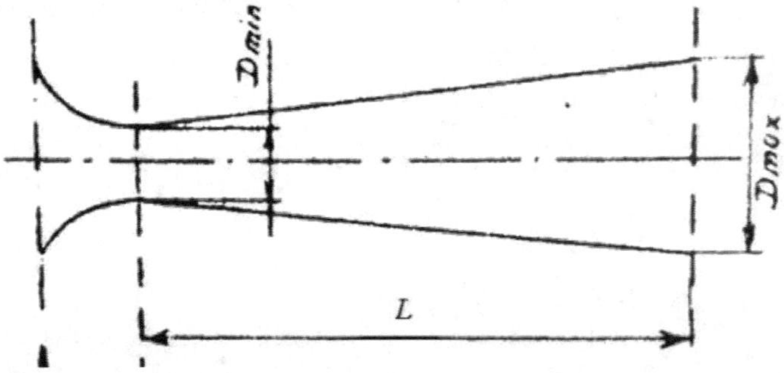

Figure 3. Expanding nozzle, D_{min} – minimum nozzle diameter; D_{max} – maximum nozzle diameter; L – nozzle length

The nozzle parameters are related by the following expression:

$$\tan(\frac{\gamma}{2}) = \frac{Dmax - Dmin}{2L}. \tag{11}$$

Reactive power $F_p=v_2^{д}\ \mu$, where $\mu=\frac{dm}{dt}$ – ejected mass of steam per unit time from a single nozzle. Task μ=0,05 кг/с и F_p=63,9 Н. Assuming the nozzles are identical, the resultant force $F=NF_p$. If N=20 то F=1278 Н. Torque (force moment) $M_к=Fd=NF_pd$, here d=10 cm. Thereafter $M_к$=127,8 Н·m. Reactive forces and torque can be adjusted by changing the amount of water supplied to the device. With an increase in water consumption by an order of magnitude, the torque increases by the same amount.

4. Conclusion

The tasks on the definition of environmentally friendly ways to improve oil recovery, in particular, steam and thermal effects. The parameters of physical models are determined, and the solution is shown in the first approximation. The problems and their solutions are indicated. Details considered the device "Drilling tool". Calculated steam flow rate from the nozzles of the device, the reactive forces emergent

IOP Conf. Series: Materials Science and Engineering **663** (2019) 012069 doi:10.1088/1757-899X/663/1/012069

in the device and the torque generated by the forces. The ways of parameters optimization are shown. The relevance and prospects of the drilling method under consideration is due to the development of deposits with highly viscous oils: Van-Egansky, Vynga-Purovsky, Russky and other fields. Emitted steam (at a temperature of $T \approx 998$ K), in addition to an additional destructive effect on the rock, has a steam-thermal effect on oil, which ultimately allows for enhanced oil recovery.

References

[1] Dobrynin V M, Wendelstein B Yu and Kozhevnikov D A 2004 Petrophysics (rock physics) (Moscow: Oil and Gas) 367 p

[2] Fedotov V Kh, Koltsov N I and Kosianov P M 2018 Nonlinear autonomous quasi-invariants of dynamical systems *Bull. of the Technol. University* **21(12)** 181–5

[3] Beggs H D and Robinson J R 1975 Estimating the Viscosity of Crude Oil Systems *J. of Petroleum Technol.* **27(09)** (SPE-5434-PA)

[4] Vazquez M and Beggs H D 1980 Correlations for Fluid Physical Property Prediction *Society of Petroleum Engineers* **32(06)** (SPE-6719-PA)

[5] Kolomiets S S, Krasnov V G and Kosianov P M 2015 Unconventional drilling methods. *Experience, actual problems and prospects for the development of the oil and gas complex (materials V region. scientific-practical conf.)* (Tyumen) pp 67–77

[6] Krasnov V G, Krever A V and Kosianov P M 2015 *Drilling tools* Patent 155161 Russian Federation, IPC: E21B7/14, E21C37/16 no 2015112010/03; declare 04/02/2015; publ. 09/27/2015, Bull. no 27

[7] Sivukhin D V 1979 *Thermodynamics and Molecular Physics: studies for universities* 2nd ed reclaiming and additional (Moscow: Science) 552 p

[8] Kosianov P M and Krasnov V G 2019 *Optimization of parameters of a drilling tool to increase its efficiency Subsoil use of the XXI century* **1 (77)** 62–6

AUTHOR INDEX